SIGNAL AND POWER INTEGRITY—

SIMPLIFIED

THIRD EDITION

SIGNAL AND POWER INTEGRITY—

SIMPLIFIED

Third Edition

Eric Bogatin

Pearson

Boston • Columbus • Indianapolis • New York • San Francisco • Amsterdam
Cape Town • Dubai • London • Madrid • Milan • Munich • Paris • Montreal
Toronto • Delhi • Mexico City • São Paulo • Sydney • Hong Kong • Seoul
Singapore • Taipei • Tokyo

For information about buying this title in bulk quantities, or for special sales opportunities (which may include electronic versions; custom cover designs; and content particular to your business, training goals, marketing focus, or branding interests), please contact our corporate sales department at corpsales@pearsoned.com or (800) 382-3419.

For government sales inquiries, please contact governmentsales@pearsoned.com.

For questions about sales outside the U.S., please contact intlcs@pearson.com.

Visit us on the Web: informit.com/ph

Library of Congress Control Number: 2017956506

ISBN-13: 978-0-13-451341-6
ISBN-10: 0-13-451341-X

15 2023

Publisher
Mark Taub

Editor-in-Chief
Greg Wiegand

Acquisitions Editor
Kim Spenceley

Managing Editor
Sandra Schroeder

Senior Project Editor
Lori Lyons

Production Manager
Dhayanidhi Karunanidhi

Copyeditor
Catherine Wilson

Indexer
Erika Millen

Proofreader
H S Rupa

Cover Compositor
Chuti Prasertsith

Compositor
codeMantra

*The dedication of a book is called "The Dedication"
because it requires dedication by the author to complete the
work. As every author will tell you, writing is a solitary,
alone process, the opposite of social activity. It's easy to
become engulfed in the writing and researching and exclude
more and more of the real world. Successful authors are
either not married or married to an understanding, supportive
spouse who sees his or her role as providing a nurturing
environment in which creative juices can ferment.*

*Susan, my wife, patiently put up with my solitary writing,
giving me the space to put in the dedication to finish the first
and second and third editions. At the same time, she was also my
anchor to the real world and forced me to keep a healthy
balance between alone work and social life. The third edition
is as much due to her efforts as from mine, and as I get
to write the words, I am dedicating the third edition to her.*

CONTENTS AT A GLANCE

CONTENTS

PREFACE TO THE THIRD EDITION

This third edition of *Signal and Power Integrity—Simplified* has one significant new addition over the previous edition: questions and problems at the end of each chapter to test and reinforce your understanding.

Students and engineers will benefit from this chance to see if they took away the important messages from each chapter and are able to apply the principles to tackle real problems.

Since the publication of the second edition of *Signal and Power Integrity—Simplified*, the principles and applications of signal and power integrity have not changed. As I travel around the world lecturing on these topics, I find that the essential principles are more important than ever.

It often astonishes me how even seemingly complex problems can be tackled and valuable insights gained by applying fundamental engineering principles and simple estimates to real-world problems. That's what this book highlights.

The first and second editions were based on the input from teaching signal and power integrity classes to thousands of engineers around the world. This third edition has been updated with a few more recent examples using high-speed serial links and based on input from more than 130 graduate students who took my course at the University of Colorado, Boulder. The most common feedback from students was they wanted exercises to work through to help them reinforce and test their understanding. When I mentioned to professional engineers that I was adding questions at the end of each chapter, I was met with overwhelming support.

We engineers love puzzles and problems that challenge us. Each time we encounter a difficult problem and solve it, we turn a stumbling block into a stepping stone.

I often hear people say that they learn best by doing. I think this is not stated quite right. I don't think we learn from hands-on activities. I think that hands-on, "doing" exercises can really reinforce the understanding we gain from studying essential principles. When we apply essential principles to real-world effects, whether it is interpreting a measurement in the lab or running a simulation or working through a problem, we exercise the process in our head and see how to apply the principles. Engineers have told me that when working through problems, "it finally clicked."

At the end of each chapter in this book are 20–35 questions that will force you to think about the principles in a way that will help them click for you. My answers to these questions are posted, along with each question, in Appendix D. Additional supplemental information is available on my website, www.beTheSignal.com.

Eric Bogatin, September 2017

PREFACE TO THE SECOND EDITION

Since the publication of the first edition of *Signal Integrity—Simplified*, the principles of signal integrity haven't changed. What has changed, though, is the prolific use of high-speed serial links and the critical role power integrity now plays in the success or failure of new product introductions.

In addition to fleshing out more details and examples in many of the chapters, especially on differential pairs and losses, two new chapters have been added to this second edition to provide a strong foundation to meet the needs of today's engineers and designers.

This first new chapter—Chapter 12—provides a thorough introduction to the use of S-parameters in signal-integrity applications. If you deal with any high-speed serial links, you will encounter S-parameters. Because they are written in the foreign language of the frequency domain, they are intimidating to the high-speed digital designer. Chapter 12, like all the chapters in this book, provides a solid foundation in understanding this formalism and enables all engineers to harness the great power of S-parameters.

Chapter 13, the second new chapter, is on power integrity. These issues increasingly fall in the lap of the design engineer. With higher speed applications, interconnects in the power distribution path affect not just power delivery, but also signals' return paths and passing an EMC certification test.

We start at the beginning and illustrate the role of the power distribution interconnects and how design and technology selection can make or break the performance of the power distribution network. The essential principles of plane impedance, spreading inductance, decoupling capacitors, and the loop inductance of capacitors are introduced. This valuable insight helps feed the intuition of engineers enabling them to apply the power of their creativity to synthesize new designs. Hand in hand with the creation of a design is the analysis of its performance so that cost-performance trade-offs can be explored and the PDN impedance profile can be sculpted to perfection.

If you are new to signal integrity, this second edition of *Signal and Power Integrity—Simplified* provides your starting place to build a strong foundation and empowers you to get your new signal integrity designs right the first time, every time.

PREFACE TO THE FIRST EDITION

"Everything should be made as simple as possible, but not simpler."

Albert Einstein

Printed circuit-board and IC-package design used to be a field that involved expertise in layout, CAD, logic design, heat transfer, mechanical engineering, and reliability analysis. With modern digital electronic systems pushing beyond the 1-GHz barrier, packaging and board designers must now balance signal integrity and electrical performance with these other concerns.

Everyone who touches the physical design of a product has the potential of affecting the performance. All designers should understand how what they do will affect signal integrity or, at the very least, be able to talk with engineers who are responsible for the signal integrity.

The old design methodology of building prototypes, hoping they work, and then testing them to find out is no longer cost effective when time to market is as important as cost and performance. If signal integrity is not taken into account from the beginning, there is little hope a design will work the first time.

In our new "high-speed" world, where the packaging and interconnect are no longer electrically transparent to the signals, a new methodology for designing a product right the first time is needed. This new methodology is based on predictability. The first step is to use established design guidelines based on engineering discipline. The second step is to evaluate the expected performance by "putting in the numbers." This is what distinguishes engineering from guesswork. It takes advantage of four important tools: rules of thumb, analytic approximations, numerical simulation tools, and measurements. With an efficient design and simulation process, many of the trade-offs between the expected performance and the ultimate cost can be evaluated early in the design cycle, where the time, risk, and cost savings will have the biggest impact. The way to solve signal-integrity problems is to first understand their origin and then apply all the tools in our toolbox to find and verify the optimum solution.

The design process is an intuitive one. The source of inspiration for a new way of solving a problem is that mysterious world of imagination and creativity. An idea is generated and the analytical powers of our technical training take over to massage the idea into a practical solution. Though computer simulations are absolutely necessary for final verification of a solution, they only rarely aid in our intuitive understanding. Rather, it is an understanding of the mechanisms, principles and definitions, and exposure to the possibilities, that contribute to the creation of a solution. Arriving at that initial guess and knowing the places to look for solutions require understanding and imagination.

This book emphasizes the intuitive approach. It offers a framework for understanding the electrical properties of interconnects and materials that apply across the entire hierarchy from on-chip, through the packages, circuit boards, connectors, and cables.

Those struggling with the confusing and sometimes contradictory statements made in the trade press will use this book as their starting place. Those experienced in electrical design will use this book as the place to finally understand what the equations mean.

In this book, terms are introduced starting at the ground floor. For example, the impedance of a transmission line is the most fundamental electrical property of an interconnect. It describes what a signal will see electrically and how it will interact with the interconnects. For those new to signal integrity, most of the problems arise from confusion over three terms: the *characteristic* impedance, *the* impedance, and the *instantaneous* impedance a signal sees. This distinction is even important for experienced engineers. This book introduces the reader to each of these terms and their meanings, without complex mathematics.

New topics are introduced at a basic level; most are not covered in other signal integrity books at this level. These include partial inductance (as distinct from loop inductance), the origin of ground bounce and EMI, impedance, transmission line discontinuities, differential impedance, and attenuation in lossy lines affecting the collapse of the eye diagram. These topics have become critically important for the new high-speed serial links.

In addition to understanding the basic principles, leveraging commercially available tools is critical for the practicing engineer who wants to find the best answer in the shortest time. Tools for solving signal integrity problems fall in two categories: analysis and characterization. Analysis is what we usually refer to as a calculation. Characterization is what we usually refer to as a measurement. The various tools, guidelines on when they should be used, and examples of their value are presented throughout the book.

There are three types of analysis tools: rules of thumb, analytic approximations, and numerical simulation. Each has a different balance between accuracy and effort to use. Each has a right and a wrong place for its appropriate use. And each tool is important and should be in the toolbox of every engineer.

Rules of thumb, such as "the self inductance of a wire is about 25 nH/inch," are important when having a quick answer NOW! is more important than having an accurate answer later. With very few exceptions, every equation used in signal integrity is either a definition or an approximation. Approximations are great for exploring design space and balancing design and performance trade-offs. However, without knowing how accurate a particular approximation really is, would you want to risk a $10,000 board-fabrication run and four weeks of your schedule based on an approximation?

When accuracy is important, for example, when signing off on a design, numerical simulation is the right tool to use. In the last five years, a whole new generation of tools has become available. These new tools have the powerful combination of being both easy to use and accurate. They can predict the characteristic impedance, cross talk, and differential impedance of any cross-section transmission line and simulate how a signal might be affected by any type of termination scheme. You don't have to be a Ph.D. to use this new generation of tools so there is no reason every engineer can't take advantage of them.

The quality of the simulation is only as good as the quality of the electrical description of the components (i.e., the equivalent circuit models). Engineers are taught about circuit models of gates that perform all the information processing, but rarely are the circuit models of the interconnects reviewed. Fifteen years ago, when interconnects looked transparent to the signals, all interconnects were considered as ideal wires—no impedance and no delay.

When these terms were added, they were lumped together as "parasitics."

Today, in a high-speed digital system with a clock frequency above about 100 MHz, it is the real wires—the wire bonds, the package leads, the pins, the circuit board traces, the connectors, and the cabling—that create signal-integrity problems and can prevent products from working correctly the first time. Understanding these "analog" effects, designing for them, specifying correct values for them, and including them in the system simulations before the design is committed to hardware, can enable moving a more robust product to market more quickly.

This book provides the tools to enable all engineers and managers involved in chip packaging and board, connector and interconnect design, to understand how these passive elements affect the electrical performance of a system and how they can be incorporated in system simulation. It illustrates how to perform engineering estimates of important electrical parameters and evaluate technology trade-offs. Examples are selected from a wide variety of common systems, including on-chip interconnects, wire bonds, flip chip attach, multilayer circuit boards, DIPs, PGAs, BGAs, QFPs, MCM connectors, and cables.

While most textbooks emphasize theoretical derivation and mathematical rigor, this book emphasizes intuitive understanding, practical tools, and engineering discipline. We use the principles of electrical engineering and physics and apply them to the world of packaging and interconnects to establish a framework of understanding and a methodology of solving problems. The tools of time- and frequency-domain measurement, two- and three-dimensional field solvers, transmission-line simulations, circuit simulators, and analytical approximations are introduced to build verified equivalent circuit models for packages and interconnects.

There are two important questions that all designers should ask of any model they use: How accurate is it? And what is the bandwidth of the model? The answers to these questions can come only from measurements. Measurements play the very important role of risk reduction.

The three generic measurement instruments, the impedance analyzer, the vector-network analyzer (VNA) and the time-domain reflectometer (TDR) are introduced and the interpretation of their data explained. Examples of measurements from real interconnects such as IC packages, printed circuit boards, cables, and connectors are included throughout this book to illustrate the principles and, by example, the value of characterization tools.

This book has been designed for use by people of all levels of expertise and training: engineers, project managers, sales and marketing managers, technology developers, and scientists. We start out with an overview of why designing the interconnects for high-speed digital systems is difficult and what major technical hurdles must be overcome to reach high-frequency operation.

We apply the tools of electrical engineering and physics to the problems of signal integrity in digital signals through the entire range of interconnects. The concept of equivalent circuit models is introduced to facilitate the quantified prediction of performance. The rest of the book describes how the circuit models of interconnects affect the electrical performance of the system in terms of the four families of noise problems: reflections, cross talk, rail collapse in the power distribution network, and EMI.

This book originated from a series of short courses and semester-long courses the author gave to packaging, circuit-board, and design engineers. It is oriented to all people who need to balance electrical performance with all other packaging and interconnect concerns in their system designs. This book provides the foundation to understand how the physical design world of geometries and material properties affects electrical performance.

If you remember nothing else about signal integrity, you should remember the following important general principles. These are summarized here and described in more detail throughout this book.

Top Ten Signal-Integrity Principles

1. The key to efficient high-speed product design is to take advantage of analysis tools that enable accurate performance prediction. Use measurements as a way of validating the design process, reducing risk, and increasing confidence in the design tools.

2. The only way to separate myth from reality is to put in the numbers using rules of thumb, approximations, numerical simulation tools, or measurements. This is the essential element of engineering discipline.

3. Each interconnect is a transmission line with a signal and a return path, regardless of its length, shape, or signal rise time. A signal sees an instantaneous impedance at each step along its way down an interconnect. Signal quality is dramatically improved if the instantaneous impedance is constant, as in a transmission line with a uniform cross section.

4. Forget the word *ground*. More problems are created than solved by using this term. Every signal has a return path. Think *return path* and

you will train your intuition to look for and treat the return path as carefully as you treat the signal path.

5. Current flows through a capacitor whenever the voltage changes. For fast edges, even the air gap between the edge of a circuit board and a dangling wire can have a low impedance through the fringe field capacitance.

6. Inductance is fundamentally related to the number of rings of magnetic-field lines completely surrounding a current. If the number of rings of field lines ever changes, for whatever reason, a voltage will be created across the conductor. This is the origin of some reflection noise, cross talk, switching noise, ground bounce, rail collapse, and some EMI.

7. Ground bounce is the voltage created on the ground return conductor due to changing currents through the total inductance of the return path. It is the primary cause of switching noise and EMI.

8. The bandwidth of a signal is the highest sine-wave frequency component that is significant, compared to an equivalent frequency square wave. The bandwidth of a model is the highest sine-wave frequency at which the model still accurately predicts the actual performance of the interconnect. Never use a model in an application where the signal bandwidth is higher than the model's bandwidth.

9. Never forget, with few exceptions, every formula used in signal integrity is either a definition or an approximation. If accuracy is important, do not use an approximation.

10. The problem caused by lossy transmission lines is the rise-time degradation. The losses increase with frequency due to skin depth and dielectric losses. If the losses were constant with frequency, the rise time would not change and lossy lines would be only a minor inconvenience.

11. The most expensive rule is the one that delays the product ship.

Register your copy of *Signal and Power Integrity–Simplified* at informit.com for convenient access to downloads, updates, and corrections as they become available. To start the registration process, go to informit.com/register and log in or create an account.* Enter the product ISBN, **9780134513416**, and click Submit. Once the process is complete, you will find any available content under "Registered Products."

*Be sure to check the box that you would like to hear from us in order to receive exclusive discounts on future editions of this product.

ACKNOWLEDGMENTS

Many colleagues, friends, and students contributed to my understanding that went into the content for both the first and second editions. Literally thousands of engineers from Intel, Cisco, Motorola, Altera, Qualcomm, Raytheon, and other companies who attended my classes provided feedback on what explanations worked and what didn't work.

In the second edition, my reviewers, Greg Edlund, Tim Swettlen, and Larry Smith, provided excellent feedback. I learned a lot from these experts. My publisher, Bernard Goodwin, was always patient and encouraging, even when I missed deadlines, and never complained when my science fiction novel was completed ahead of this second edition.

In the third edition, many of my students, both at the University of Colorado, subscribers to the Teledyne LeCroy Signal Integrity Academy, and professional engineers in the industry, encouraged me to add questions. I tried to keep track of the questions asked at public training events and added them to the appropriate chapters. To all of my students, young and old, who provided feedback, thanks for taking the time to question authority.

Thank you all for the wonderful support and encouragement.

Eric Bogatin received his B.S. in Physics from MIT in 1976 and his M.S. and Ph.D. in Physics from the University of Arizona in Tucson in 1980. For more than 30 years he has been active in the fields of signal integrity and interconnect design. He worked in senior engineering and management roles at AT&T Bell Labs, Raychem Corp, Sun Microsystems, Interconnect Devices Inc., and Teledyne LeCroy. In 2011, his company, Bogatin Enterprises, was acquired by Teledyne LeCroy.

Eric currently is a Signal Integrity Evangelist with Teledyne LeCroy, where he creates and presents educational materials related to new applications for high-performance scopes. Eric turns complexity into practical design and measurement principles, leveraging analysis techniques and measurement tools.

Since 2012, he has been an adjunct professor at the University of Colorado in Boulder, teaching graduate courses in signal integrity, interconnect design and PCB design.

He has written regular monthly columns for *PCD&F Magazine*, *Semiconductor International*, *Electronic Packaging and Production*, Altera Corporation, Mentor Graphics Corporation, *EDN*, and *EE Times*. He is currently the editor of the *Signal Integrity Journal* (www.SignalIntegrityJournal.com)

Eric is a prolific author with more than 300 publications, many posted on his website, www.beTheSignal.com, for download. He regularly presents at DesignCon, the IEEE EMC Symposium, EDI con, and at IPC's Designer Council events.

He is the coauthor of the popular Prentice Hall book, *Principles of Power Integrity for PDN Design-Simplified*, along with Larry Smith.

He was the recipient of the 2016 Engineer of the Year Award from DesignCon.

He can be reached at *eric@beTheSignal.com*.

CHAPTER 1

Signal Integrity Is in Your Future

"There are two kinds of engineers: Those who have signal-integrity problems and those who will."

—*Eric Bogatin*

Ironically, this is an era when not only are clock frequencies increasing and signal-integrity problems getting more severe, but the time design teams are available to solve these problems and designed new products are getting shorter. Product design teams have one chance to get a product to the market; the product must work successfully the first time. If identifying and eliminating signal-integrity problems isn't an active priority early in the product design cycle, chances are the product will not work.

> **TIP** As clock frequencies and data rates increase, identifying and solving signal-integrity problems becomes critical. The successful companies will be those that master signal-integrity problems and implement an efficient design process to eliminate these problems. It is by incorporating best design practices, new technologies, and new analysis techniques and tools that higher-performance designs can be implemented and meet ever-shrinking schedules.

In high-speed products, the physical and mechanical design can affect signal integrity. Figure 1-1 shows an example of how a simple 2-inch-long section of trace on a printed circuit board (PCB) can affect the signal integrity from a typical driver.

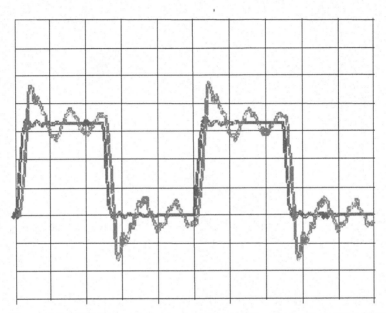

Figure 1-1   100-MHz clock waveform from a driver chip when there is no connection (smooth plot) and when there is a 2-inch length of PCB trace connected to the output (ringing). Scale is 1 v/div and 2 nsec/div, simulated with Mentor Graphics HyperLynx.

The design process is often a very intuitive and creative process. Feeding your engineering intuition about signal integrity is critically important to reaching an acceptable design as quickly as possible.

All engineers who touch a product should have an understanding of how what they do may influence the performance of the overall product. By understanding the essential principles of signal integrity at an intuitive and engineering level, every engineer involved in the design process can evaluate the impact of his or her decisions on system performance. This book is about the essential principles needed to understand signal-integrity problems and their solutions. The engineering discipline required to deal with these problems is presented at an intuitive level and a quantitative level.

1.1 What Are Signal Integrity, Power Integrity, and Electromagnetic Compatibility?

In the good old days of 10-MHz clock frequencies, the chief design challenges in circuit boards or packages were how to route all the signals in a two-layer board and how to get packages that wouldn't crack during assembly. The electrical properties of the interconnects were not important because they didn't affect system performance. In this sense, we say that the interconnects were "transparent to the signals."

A device would output a signal with a rise time of roughly 10 nsec and a clock frequency of 10 MHz, for example, and the circuits would work with the crudest of interconnects. Prototypes fabricated with wire-wrapped boards worked as well as final products with printed circuit boards and engineering change wires.

But clock frequencies have increased and rise times of signals have decreased. For most electronic products, signal-integrity effects begin to be important at clock frequencies above about 100 MHz or rise times shorter than about 1 nsec. This is sometimes called the *high-frequency* or *high-speed regime*. These terms refer to products and systems where the interconnects are no longer transparent to the signals and, if you are not careful, one or more signal-integrity problems arise.

Signal integrity refers, in its broadest sense, to all the problems that arise in high-speed products due to the interconnects. It is about how the electrical properties of the interconnects, interacting with the digital signal's voltage and current waveforms, can affect performance.

All of these problems fall into one of the following three categories, each with considerable overlap with the others:

1. **Signal integrity (SI)**, involving the distortion of signals
2. **Power integrity (PI)**, involving the noise on the interconnects and any associated components delivering power to the active devices
3. **Electromagnetic compatibility (EMC)**, the contribution to radiated emissions or susceptibility to electromagnetic interference from fields external to the product

In the design process, all three of these electrical performance issues need to be considered for a successful product.

The general area of EMC really encompasses solutions to two problems: too much radiated emission from a product into the outside world and too much interference on a product from radiation coming from the outside world. EMC is about engineering solutions for the product so it will at the same time maintain radiated emissions below the acceptable limit and not be susceptible to radiation from the external world.

When we discuss the solutions, we refer to EMC. When we discuss the problems, we refer to electromagnetic interference (EMI). Usually EMC is about passing a test for radiated emissions and another test for being robust against radiated susceptibility. This is an important perspective as some of the EMC solutions are introduced just to pass certification tests.

TIP   Generally, products can pass all their performance tests and meet all functional specs but still fail an EMC compliance test.

Designing for acceptable EMC involves good SI and PI design as well as additional considerations, especially related to cables, connectors, and enclosure design. Spread spectrum clocking (SSC), which purposely adds jitter to clocks by modulating their clock frequency, is specifically used to pass an EMC certification test.

PI is about the problems associated with the power distribution network (PDN), which includes all the interconnects from the voltage regulator modules (VRMs) to the voltage rail distributed on the die. This includes the power and ground planes in the board and in the packages, the vias in the board to the packages, the connections to the die pads, and any passive components like capacitors connected to the PDN.

While the PDN that feeds the on-die core power rail, sometimes referred to as the *Vdd rail*, is exclusively a PI issue, there are many overlapping problems between PI and SI topics. This is primarily because the return paths for the signals use the same interconnects usually associated with the PDN interconnects, and anything that affects these structures has an impact on both signal quality and power quality.

TIP   The overlap between PI and SI problems adds confusion to the industry because problems in this gray area are either owned by two different engineers or fall through the cracks as the PI engineer and the SI engineer each think it's the other's responsibility.

In the SI domain, problems generally relate to either noise issues or timing issues, each of which can cause false triggering or bit errors at the receiver.

Timing is a complicated field of study. In one cycle of a clock, a certain number of operations must happen. This short amount of time must be divided up and allocated, in a budget, to all the various operations. For example, some time is allocated for gate switching, for propagating the signal to the output gate, for waiting for the clock to get to the next gate, and for waiting for the gate to read the data at the input. Though the interconnects affect the timing budget, timing is not explicitly covered in this book. The influence on jitter from rise-time distortion, created from the interconnects, is extensively covered.

For the details on creating and managing timing budgets, we refer any interested readers to a number of other books listed Appendix C, "Selected References," for more information on this topic. This book concentrates on the effects of the interconnects on the other generic high-speed problem: too much noise.

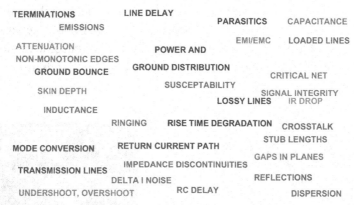

Figure 1-2   The list of signal-integrity effects seems like a random collection of terms, without any pattern.

We hear about a lot of signal-integrity noise problems, such as ringing, ground bounce, reflections, near-end cross talk, switching noise, non-monotonicity, power bounce, attenuation, and capacitive loading. All of these relate to the electrical properties of the interconnects and how the electrical properties affect the distortion of the digital signals.

It seems at first glance that there is an unlimited supply of new effects we have to take into account. This confusion is illustrated in Figure 1-2. Few digital-system designers or board designers are familiar with all of these terms as other than labels for the craters left in a previous product-design minefield. How are we to make sense of all these electrical problems? Do we just keep a growing checklist and add to it periodically?

Each of the effects listed above, is related to one of the following six unique families of problems in the fields of SI/PI/EMC:

1. Signal distortion on one net
2. Rise-time degradation from frequency-dependent losses in the interconnects
3. Cross talk between two or more nets
4. Ground and power bounce as a special case of cross talk
5. Rail collapse in the power and ground distribution
6. Electromagnetic interference and radiation from the entire system

These six families are illustrated in Figure 1-3. Once we identify the root cause of the noise associated with each of these six families of problems, and understand the essential principles behind them, the general solution for finding and fixing the problems in each family will become obvious. This is the power of being able to classify every SI/PI/EMC problem into one of these six families.

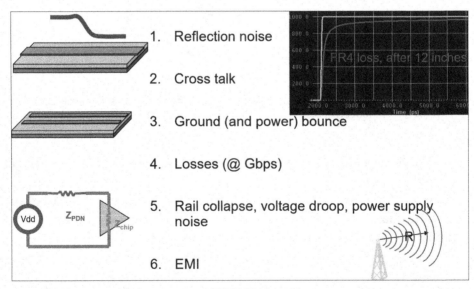

1. Reflection noise

2. Cross talk

3. Ground (and power) bounce

4. Losses (@ Gbps)

5. Rail collapse, voltage droop, power supply noise

6. EMI

Figure 1-3   The six families of SI/PI/EMC problems.

These problems play a role in all interconnects, from the smallest on-chip wire to the cables connecting racks of boards and everywhere in-between. The essential principles and effects are the same. The only differences in each physical structure are the specific geometrical feature sizes and the material properties.

1.2 Signal-Integrity Effects on One Net

A net is made up of all the metal connected together in a system. For example, there may be a trace going from a clock chip's output pin to three other chips. Each piece of metal that connects all four of these pins is considered one net. In addition, the net includes not only the signal path but also the return path for the signal current.

There are three generic problems associated with signals on a single net being distorted by the interconnect. The first is from reflections. The only thing that causes a reflection is a change in the instantaneous impedance the signal encounters. The instantaneous impedance the signal sees depends as much on the physical features of the signal trace as on the return path. An example of two different nets on a circuit board is shown in Figure 1-4.

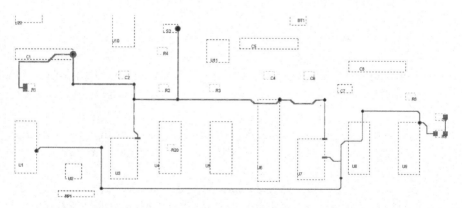

Figure 1-4 Example of two nets on a circuit board. All metal connected together is considered one net. Note: One net has a surface-mount resistor in series. Routed with Mentor Graphics HyperLynx.

When the signal leaves the output driver, the voltage and the current, which make up the signal, see the interconnect as an electrical impedance. As the signal propagates down the net, it is constantly probing and asking,

"What is the instantaneous impedance I see?" If the impedance the signal sees stays the same, the signal continues undistorted. If, however, the impedance changes, the signal will reflect from the change and continue through the rest of the interconnect distorted. The reflection itself will reflect off other discontinuities and will rattle around the interconnect, some of it making its way to the receiver, adding distortions. If there are enough impedance changes, the distortions can cause false triggering.

Any feature that changes the cross section or geometric shape of the net will change the impedance the signal sees. We call any feature that changes the impedance a *discontinuity*. Every discontinuity will, to some extent, cause the signal to be distorted from its original pristine shape. For example, some of the features that would change the impedance the signal sees include the following:

- An end of the interconnect
- A line-width change
- A layer change through a via
- A gap in return-path plane
- A connector
- A routing topology change, such as a branch, tee, or stub

These impedance discontinuities can arise from changes in the cross section, the topology of the routed traces, or the added components. The most common discontinuity is what happens at the end of a trace, which is usually either a high impedance open at the receiver or a low impedance at the output driver.

If the source of reflection noise is changes in the instantaneous impedance, then the way to fix this problem is to engineer the impedance of the interconnect to be constant.

TIP   The way to minimize the problems associated with impedance changes is to keep the instantaneous impedance the signal sees constant throughout the net.

This strategy is typically implemented by following four best design practices:

1. Use a board with constant, or "controlled," impedance traces. This usually means using uniform transmission lines.

2. To manage the reflections from the ends, use a termination strategy that controls the reflections by using a resistor to fool the signal into not seeing an impedance change.

3. Use routing rules that allow the topology to maintain a constant impedance down the trace. This generally means using point-to-point routing or minimum-length branch or short stub lengths.

4. Engineer the structures that are not uniform transmission lines to reduce their discontinuity. This means adjusting fine geometrical design features to sculpt the fringe fields.

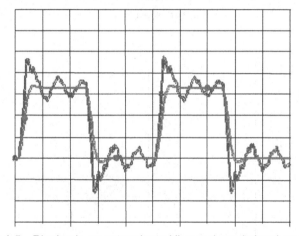

Figure 1-5 Ringing in an unterminated line and good signal quality in a source-series terminated interconnect line. The PCB trace is only 2 inches long in each case. Scale is 1 v/div and 2 nsec/div. Simulated with Mentor Graphics HyperLynx.

Figure 1-5 shows an example of poor signal quality due to impedance changes in the same net and when a terminating resistor is used to manage the impedance changes. Often, what we think of as "ringing" is really due to reflections associated with impedance changes.

Even with perfect terminations, the precise board layout can drastically affect signal quality. When a trace branches into two paths, the impedance at the junction changes, and some signal will reflect back to the source, while some will continue down the branches in a reduced and distorted form. Rerouting the trace to be a daisy chain causes the signal to see a constant impedance all down the path, and the signal quality can be restored.

The impact on a signal from any discontinuity depends on the rise time of the signal, where the discontinuity is in the circuit and other sources of reflections are in the circuit. As the rise time gets shorter, the magnitude of the distortion on the signal will increase. This means that a discontinuity that was not a problem in a 33-MHz design with a 3-nsec rise time may be a problem in a 100-MHz design with a 1-nsec rise time. This is illustrated in Figure 1-6.

The higher the frequency and the shorter the rise time, the more important it is to keep the impedance the signal sees constant. One way of achieving this is by using controlled impedance interconnects even in the packages, such as with multilayer ball grid arrays (BGAs). When the packages do not use controlled impedance, such as with lead frames, it's important to keep the leads short, such as by using chip-scale packages (CSPs).

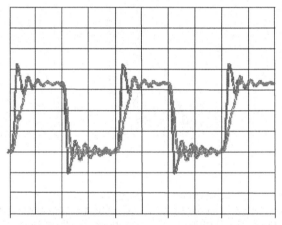

Figure 1-6   25-MHz clock waveforms with a PCB trace 6 inches long and unterminated. The slow rise time is a 3-nsec rise time. The ringing is from a rise time of 1 nsec. What may not have been a problem with one rise time can be a problem with a shorter rise time. Scale is 1 v/div and 5 nsec/div. Simulated with Mentor Graphics HyperLynx.

There are two other aspects of signal integrity associated with one net. Frequency-dependent losses in the line from the conductor and the dielectric cause higher-frequency signal components to be attenuated more than the lower-frequency components. The end result is an increase in the rise time of the signal as it propagates. When this rise-time degradation approaches the unit interval of a single bit, information from one bit will leak into the next bit and the next bit. This effect is called intersymbol interference (ISI) and is a significant source of problems in high-speed serial links, in the 1 gigabit per second (Gbps) and higher regime.

The third aspect of signal-quality problems associated with a single net is related to timing. The time delay difference between two or more signal paths is called *skew*. When a signal and clock line have a skew different than expected, false triggering and errors can result. When the skew is between the two lines that make up a differential pair, some of the differential signal will be converted into common signal, and the differential signal will be distorted. This is a special case of mode conversion, and it will result in ISI and false triggering.

While skew is a timing problem, it often arises due to the electrical properties of the interconnect. The first-order impact on skew is from the total length of the interconnects. This is easily controlled with careful layout to match lengths. However, the time delay is also related to the local dielectric constant that each signal sees and is often a much more difficult problem to fix.

TIP   Skew is the time delay difference between two or more nets. It can be controlled to first order by matching the length of the nets. In addition, the local variation in dielectric constant between the nets, as from the glass weave distribution in the laminate, will also affect the time delay and is more difficult to control.

1.3 Cross Talk

When one net carries a signal, some of this voltage and current can pass over to an adjacent quiet net, which is just sitting there, minding its own business. Even though the signal quality on the first net (the active net) is perfect, some of the signal can couple over and appear as unwanted noise on the second, quiet net.

TIP   It is the capacitive and inductive coupling between two nets that provides a path for unwanted noise from one net to the other. It can also be described in terms of the fringe electric and magnetic fields from an aggressor to a victim line.

Cross talk occurs in two different environments: when the interconnects are uniform transmission lines, as in most traces in a circuit board, and when they are not uniform transmission lines, as in connectors and packages. In controlled-impedance transmission lines where the traces have a wide uniform return path, the relative amount of capacitive coupling and inductive coupling is comparable. In this case, these two effects combine in different ways at the near end of the quiet line and at the far end of the quiet line. An example of the measured near- and far-end cross talk between two nets in a circuit board is shown in Figure 1-7.

Having a wide uniform plane as the return path is the configuration of lowest cross talk. Anything that changes the return path from a wide uniform plane will increase the amount of coupled noise between two transmission lines. Usually when this happens—for example, when the signal goes through a connector and the return paths for more than one signal path are now shared by one of the pins rather than by a plane—the inductively coupled noise increases much more than the capacitively coupled noise.

In this regime, where inductively coupled noise dominates, we usually refer to the cross talk as *switching noise, delta I noise, dI-dt noise, ground bounce, simultaneous switching noise (SSN),* or *simultaneous switching output (SSO) noise.* As we will see, this type of noise is generated by the coupling inductance, which is called *mutual inductance.* Switching noise occurs mostly in connectors, packages, and vias, where the return path conductor is not a wide, uniform plane.

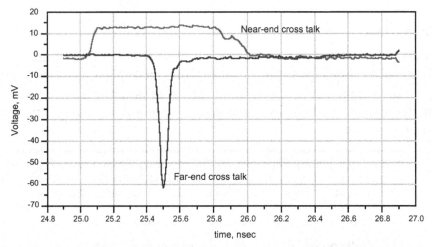

Figure 1-7 Measured voltage noise at the near end and the far end of a quiet trace when a 200-mV signal is injected in the active trace. Note the near-end noise is about 7% and the far-end noise is nearly 30% of the signal. Measurements performed with an Agilent DCA86100 with time-domain reflectometer (TDR) plug-in.

Ground bounce, reviewed later in this book, is really a special case of cross talk caused when return currents overlap in the same conductor and their mutual inductance is very high. Figure 1-8 shows an example of SSO noise from the high mutual inductance between adjacent signal and return paths in a package.

TIP   SSO noise, where the coupled or mutual-inductance dominates is the most important issue in connector and package design. It will only get worse in next-generation products. The solution lies in careful design of the geometry of paths so that mutual inductance is minimized and in the use of differential signaling.

By understanding the nature of capacitive and inductive coupling, described as either lumped elements or as fringe electric and magnetic fields, it is possible to optimize the physical design of the adjacent signal traces to minimize their coupling. This usually can be as simple as spacing the traces farther apart. In addition, the use of lower-dielectric constant material will decrease the cross talk for the same characteristic impedance lines. Some aspects of cross talk, especially switching noise, increase with the length of the interconnect and with decreasing rise time. Shorter rise-time signals will create more cross talk. Keeping interconnects short, such as by using CSPs and high-density interconnects (HDIs), helps minimize cross talk.

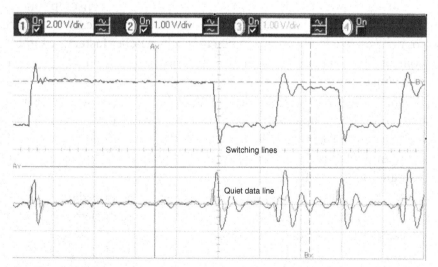

Figure 1-8   Top trace: Measured voltage on active lines in a multiline bus. Bottom trace: Measured noise on one quiet line showing the switching noise due to mutual inductance between the active and quiet nets in the package.

1.4 Rail-Collapse Noise

Noise is generated, and is a problem, in more than just the signal paths. It can also be a disaster in the power- and ground-distribution network that feeds each chip. When current through the power- and ground-path changes, as when a chip switches its outputs or core gates switch, there will be a voltage drop across the impedance of the power and ground paths. When there are reactive elements in the PDN, especially with parallel resonances, switching power currents can also result in higher voltage spikes on the power rail.

This voltage noise means a lower or higher voltage on the rail pads of the chip. A change in the voltage on the power rails can result in either voltage noise on signal lines, which can contribute to false triggering and bit errors or to enhanced jitter. Figure 1-9 shows one example of the change in the voltage across a microprocessor's power rails.

PDN noise can also cause jitter. The propagation delay for the turn-on of a gate is related to the voltage between the drain and source. When the voltage noise increases the rail voltage, gates switch faster, and clock and data edges are pulled in. When the voltage noise brings the rail voltage lower, gates switch more slowly, and clock and data edges are pushed out. This impact on the edge of clocks and signals is a dominant source of jitter.

In high-performance processors, FPGAs, and some ASICs, the trend is for lower power-supply voltage but higher power consumption. This is primarily due to more gates on a chip switching faster. In each cycle, a certain amount of energy is consumed. When the chip switches faster, the same energy is consumed in each cycle, but it is consumed more often, leading to higher power consumption.

These factors combined mean that higher currents will be switching in shorter amounts of time, and with lower signal voltage swings, the amount of noise that can be tolerated will decrease. As the drive voltage decreases and the current level increases, any voltage drops associated with rail collapse become bigger and bigger problems.

As these power rail currents travel through the impedance of the PDN interconnects, they generate a voltage drop across each element. This voltage drop is the source of power rail noise.

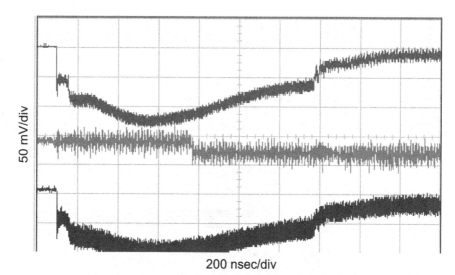

50 mV/div

200 nsec/div

Figure 1-9   Measured Vcc voltage at three pin locations on the package of a microprocessor just coming from a "stop-clock" state. Nominal voltage is supposed to be 2.5 V, but due to voltage drops in the power distribution system, the delivered voltage collapses by almost 125 mV.

T I P   To reduce the voltage noise on the power rails when power supply currents switch, the most important best design practice is to engineer the PDN as a low impedance.

In this way, even though there is current switching in the PDN, the voltage drop across a lower impedance may be kept to an acceptable level. Sun Microsystems has evaluated the requirements for the impedance of the PDN for high-end processors. Figure 1-10 shows Sun's estimate of the required impedance of the PDN. Lower impedance in the PDN is increasingly important and harder to achieve.

If we understand how the physical design of the interconnects affects their impedance, we can optimize the design of the PDN for low impedance. As we will see, designing a low-impedance PDN means including features such as:

1. Closely spaced adjacent planes for the power and ground distribution with as thin a dielectric between them as possible, near the surface of the board

2. Multiple, low-inductance decoupling capacitors

3. Multiple, very short power and ground leads in packages

4. A low-impedance VRM

5. On-package decoupling (OPD) capacitors

6. On-chip decoupling (ODC) capacitors

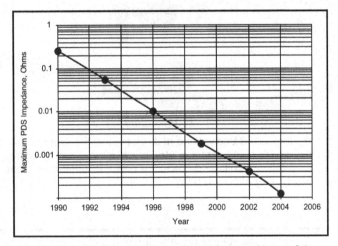

Figure 1-10   Trend in the maximum allowable impedance of the power distribution system for high-end processors. Source: Sun Microsystems.

An innovative technology to help minimize rail collapse can be seen in the new ultrathin, high-dielectric constant laminates for use between power and ground layers. One example is C-Ply from 3M Corp. This material is 8 microns thick and has a dielectric constant of 20. Used as the power and ground layers in an otherwise conventional board, the ultralow loop inductance and high distributed capacitance dramatically reduce the impedance of the power and ground distribution. Figure 1-11 shows an example of the rail-collapse noise on a small test board using conventional layers and a board with this new C-Ply.

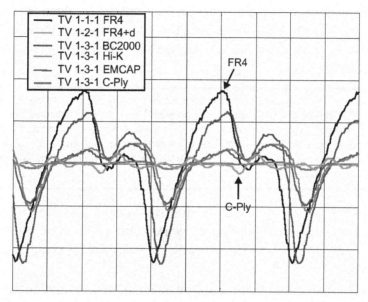

Figure 1-11 Measured rail voltage noise on a small digital board, with various methods of decoupling. The worst case is no decoupling capacitors on FR4. The best case is the 3M C-Ply material, showing virtually no voltage noise. Courtesy of National Center for Manufacturing Science. Scale is 0.5 v/div and 5 nsec/div.

1.5 Electromagnetic Interference (EMI)

With board-level clock frequencies in the 100-MHz to 500-MHz range, the first few harmonics are within the common communications bands of TV, FM radio, cell phone, and personal communications services (PCS). This means there is a very real possibility of electronic products interfering with communications unless their electromagnetic emissions are kept below acceptable levels. Unfortunately, without special design considerations, EMI will get worse at higher frequencies. The radiated far-field strength from common currents will increase linearly with frequency and from differential currents will increase with the square of the frequency. With clock frequencies increasing, the radiated emissions level will inevitably increase as well.

It takes three things to have an EMI problem: a source of noise, a pathway to a radiator, and an antenna that radiates. Every source of signal-integrity problem mentioned above will be a source of EMI. What makes EMI so challenging is that even if the noise is low enough to meet the signal-integrity and power-integrity noise budget, it may still be large enough to cause serious radiated emissions.

TIP   The two most common sources of EMI are (1) the conversion of some differential signal into a common signal, which eventually gets out on an external twisted-pair cable, and (2) ground bounce on a circuit board generating common currents on external single-ended shielded cables. Additional noise can come from internally generated radiation leaking out of the enclosure.

Most of the voltage sources that drive radiated emissions come from the power- and ground-distribution networks. Often, the same physical design features that contribute to low rail-collapse noise will also contribute to lower emissions.

Even with a voltage noise source that can drive radiation, it is possible to isolate it by grouping the high-speed sections of a board away from where they might exit the product. Shielding the box will minimize the leakage of the noise from an antenna. Many of the ills of a poorly designed board can be fixed with a good shield.

A product with a great shield will still need to have cables connecting it to the outside world for communications, for peripherals, or for interfacing. Typically, the cables extending outside the shielded enclosure act as the antennas, and radiate. The correct use of ferrites on all connected cables, especially on twisted pair, will dramatically decrease the efficiency of the cables as antennas. Figure 1-12 shows a close-up of the ferrite around a cable.

The impedance associated with the I/O connectors, especially the impedance of the return-path connections, will dramatically affect the noise voltages that can drive radiating currents. The use of shielded cables with low-impedance connections between the shields and chassis will go a long way to minimize EMI problems.

Unfortunately, for the same physical system, increasing the clock frequency generally will also increase the radiated emissions level. This means that EMI problems will be harder to solve as clock frequencies increase.

Figure 1-12 Ferrite choke around a cable, split apart. Ferrites are commonly used around cables to decrease common currents, a major source of radiated emissions. Courtesy of IM Intermark.

1.6 Two Important Signal-Integrity Generalizations

Two important generalizations should be clear from looking at the six signal-integrity problems above.

First, each of the six families of problems gets worse as rise times decrease. All the signal-integrity problems above scale with how fast the current changes or with how fast the voltage changes. This is often referred to as dI/dt or dV/dt. Shorter rise times mean higher dI/dt and dV/dt.

It is unavoidable that as rise times decrease, the noise problems will increase and be more difficult to solve. And, as a consequence of the general trends in the industry, the rise times found in *all* electronic products will continually decrease. This means that what might not have caused a problem in one design may be a killer problem in the next design with the next-generation chip sets operating with a shorter rise time. This is why it is often said that "there are two kinds of engineers: Those who have signal-integrity problems and those who will."

The second important generalization is that effective solutions to signal-integrity problems are based heavily on understanding the impedance of interconnects. If we have a clear intuitive sense of impedance and can relate the physical design of the interconnects with their impedance, many signal-integrity problems can be eliminated during the design process.

This is why an entire chapter in this book is devoted to understanding impedance from an intuitive and engineering perspective and why much of the rest of this book is about how the physical design of interconnects affects the impedance seen by signals and power and ground currents.

1.7 Trends in Electronic Products

Our insatiable thirst for ever higher performance at ever higher densities and at lower cost drives the electronics industry. This is both in the consumption of information as in video images and games, and in the processing and storage of information in industries like bioinformatics and weather or climate forecasting.

In 2015, almost 50% of Internet traffic was in the form of video images from Netflix and YouTube. The data rate of information flowing on the Internet backbone doubles almost every year.

A few trips to the local computer store over the past 10 years will have given anyone a good sense of the incredible treadmill of progress in computer performance. One measure of performance is the clock frequency of a processor chip. The trend for Intel processor chips, as illustrated in Figure 1-13, shows a doubling in clock frequency about every two years.

In some processor families, the clock frequencies have saturated, but the data rates for signals communicating chip to chip and board to board have continued to increase at a steady rate.

This trend toward ever-higher clock frequency and data rates is enabled by the same force that fuels the semiconductor revolution: photolithography. As the gate channel length of transistors is able to be manufactured at smaller size, the switching speed of the transistor increases. The electrons and holes have a shorter distance to travel and can transit the gate, effecting transitions, in a shorter time when the channel length is shorter.

When we refer to a technology generation as 0.18 microns or 0.13 microns, we are really referring to the smallest channel length that can be manufactured. The shorter switching time for smaller channel-length transistors has two important consequences for signal integrity.

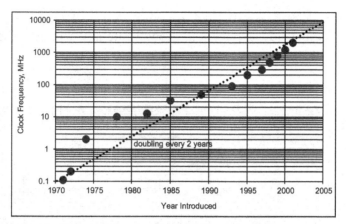

Figure 1-13   Historical trend in the clock frequency of Intel processors based on year of introduction. The trend is a doubling in clock frequency every two years. Source: Intel Corp.

The minimum time required for one clock cycle is limited by all the operations that need to be performed in one cycle. Usually, there are three main factors that contribute to this minimum time: the intrinsic time for all the gates that need to switch in series, the time for the signals to propagate through the system to all the gates that need to switch, and the setup and hold times needed for the signals at the inputs to be read by the gates.

In single-chip microprocessor-based systems, such as in personal computers, the dominant factor influencing the minimum cycle time is the switching speed of the transistors. If the switching time can be reduced, the minimum total time required for one cycle can be reduced. This is the primary reason clock frequencies have increased as feature size has been reduced.

TIP   It is inevitable that as transistor feature size continues to reduce, rise times will continue to decrease, and clock frequencies and data rates will continue to increase.

Figure 1-14 shows projections from the 2001 Semiconductor Industry Association (SIA) International Technology Roadmap for Semiconductors (ITRS) for future on-chip clock frequencies, based on projected feature size reductions, compared with the Intel processor trend. This shows the projected trend for clock frequency increasing at a slightly diminishing but still growing rate for the next 15 years as well. In the recent past and going forward, while clock frequencies have saturated at about 3 GHz, the communications data rates are on a steady, exponential growth rate and will exceed 56 Gbps per channel by 2020.

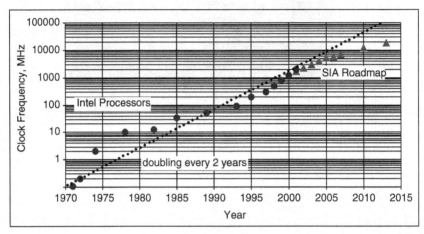

Figure 1-14   Historical trend in the clock frequency of Intel processors based on year of introduction. The trend is a doubling in clock frequency every two years. Also included is the Semiconductor Industry Association roadmap expectations. Source: Intel Corp. and SIA.

As the clock frequency increases, the rise time of the signals must also decrease. Each gate that reads either the data lines or the clock lines needs enough time with the signal being either in the high state or the low state, to read it correctly.

This means only a short time is left for the signal to be in transition. We usually measure the transition time, either the rise time or the fall time, as the time it takes to go from 10% of the final state to 90% of the final state. This is called the 10–90 rise time. Some definitions use the 20% to 80% transition points and this rise time is referred to as the 20–80 rise time. Figure 1-15 shows an example of a typical clock waveform and the time allocated for the transition. In most high-speed digital systems, the time allocated to the rise time is about 10% of the clock cycle time, or the *clock period.*

This is a simple rule of thumb, not a fundamental condition. In legacy systems with a high-end ASIC or FPGA but older peripherals, the rise time might be 1% of the clock period. In a high-speed serial link, pushing the limits to the highest data rate possible, the rise time may be 50% of the unit interval.

Based on this rule of thumb, the rise time is roughly related to the clock frequency by:

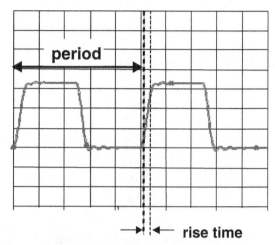

Figure 1-15   The 10–90 rise time for a typical clock waveform is roughly 10% of the period. Scale is 1 v/div and 2 nsec/div, simulated with Mentor Graphics HyperLynx.

$$RT = \frac{1}{10 \times F_{clock}} \qquad (1\text{-}1)$$

where:

RT = rise time, in nsec

F_{clock} = clock frequency, in GHz

For example, when the clock frequency is 1 GHz, the rise time of the associated signals is about 0.1 nsec, or 100 psec. When the clock frequency is 100 MHz, the rise time is roughly 1 nsec. This relationship is shown in Figure 1-16.

TIP   The treadmill-like advance of ever-increasing clock frequency means an ever-decreasing rise time and signal-integrity problems that are harder to solve.

Of course, this simple relationship between the rise time of a signal and the clock frequency is only a rough approximation and depends on the specifics of the system. The value in such simple rules of thumb is being able to get to an answer faster, without knowing all the necessary information for an accurate answer. It is based on the very important engineering principle "Sometimes an OKAY answer NOW! is more important than a good answer LATE." Always be aware of the underlying assumptions in rules of thumb.

Even if the clock frequency of a product is low, there is still the danger of shorter rise times as a direct consequence of the chip technology. A chip-fabrication factory, usually called a "fab," will try to standardize all its wafers on one process to increase the overall yield. The smaller the chip size, the more chips can be fit on a wafer and the lower the cost per chip. Even if the chip is going to be used in a slow-speed product, it may be fabricated on the same line as a leading-edge ASIC, with the same small-feature size.

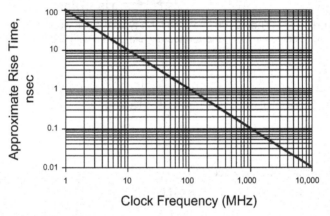

Figure 1-16   Rise time decreases as the clock frequency increases. Signal-integrity problems usually arise at rise times less than 1 nsec or at clock frequencies greater than 100 MHz.

Ironically, the lowest-cost chips will always have ever-shorter rise times, even if they don't need it for the specific application. This is an unintended, scary consequence of Moore's Law. If you have designed a chip set into your product and the rise time is 2 nsec, for example, with a 50-MHz clock, there may be no signal-integrity problems. When your chip supplier upgrades its fab line with a finer-feature process, it may provide you with lower-cost chips. You may think you are getting a good deal. However, these lower-cost chips may now have a rise time of 1 nsec. This shorter rise time may cause reflection noise, excess cross talk, and rail collapse, and the chips may fail the Federal Communications Commission (FCC) EMC certification tests. The clock frequency of your product hasn't changed, but, unknown to you, the rise time of the chips supplied to you has decreased with the newer, finer-feature manufacturing process.

> **TIP** As all fabs migrate to lower-cost, finer-feature processes, all fabricated chips will have shorter rise times, and signal-integrity problems have the potential to arise in all products, even those with clock frequencies below 50 MHz.

It's not just microprocessor-based digital products that are increasing in clock frequency and decreasing in rise times. The data rate and clock frequencies used in high-speed communications are blowing past the clock frequencies of microprocessor digital products.

One of the most common specifications for defining the speed of a high-speed serial link is the optical carrier (OC) spec. This is actually a data rate, with OC-1 corresponding to about 50 megabits per second (Mbps). OC-48, with a data rate of 2.5 gigabits per second (Gbps), is in high-volume deployment and widely implemented. OC-192, at 10 Gbps, is just ramping up. In the near future, OC-768, at 40 Gbps, will be in wide-scale deployment.

The OC designation is a specification for data rate, not for clock frequency. Depending on how a bit is encoded in the data stream, the actual clock frequency of a system can be higher or lower than the data rate.

For example, if there is one bit per clock cycle, then the actual clock frequency will be the data rate. If two data bits are encoded per clock cycle, the underlying clock frequency will be half the data rate. A 2.5-Gbps data rate can be obtained with a 1.25-GHz clock. The more bits in the data stream used for routing information or error correction and overhead, the lower the data rate, even though the clock frequency is constant.

Commonly used signaling protocols like PCIe, SATA, and Gigabit Ethernet all use a non-return-to-zero (NRZ) signaling scheme, which encodes two bits in each clock cycle. This means that the underlying clock, apparent with the highest transition density signal, such as 1010101010, will have a clock frequency half the data rate. We refer to this underlying clock frequency as the *Nyquist frequency*.

To keep signal-integrity problems at a minimum, the lowest clock frequency and longest rise time should be used. This has inspired a growing trend in data rates in excess of 28 Gbps, to encode four to eight data bits per clock cycle, using multilevel signaling and multiple signal lines in parallel. This technique of encoding bits in the amplitude is called pulse amplitude modulation (PAM).

NRZ, which encodes two bits per cycle, is considered a PAM2 signaling scheme. Using four voltage levels in one unit interval is referred to as PAM4 and is a popular technique for 56-Gbps signaling.

The trend is definitely toward higher data rates. A data rate of 10 Gbps, common for Ethernet systems, has a Nyquist clock frequency of 5 GHz. The unit interval of a 10 Gbps signal is only 100 psec. This means that the rise time must be less than 50 psec. This is a short rise time and requires extremely careful design practices.

High-speed serial links are also used for chip-to-chip and board-to-board communications in all products from cell phones and computers to large servers and mainframes. Whatever bit rate they may start at, they also have a migration path mapped to at least three generations, each one doubling in bit rate. Among these are the following:

- PCI Express, starting at 2.5 Gbps and moving to 5 Gbps, 8 Gbps, and 16 Gbps.
- InfiniBand, starting at 2.5 Gbps and moving to 5 Gbps and 10 Gbps
- Serial ATA (SATA), starting at 1.25 Gbps and moving to 2.5 Gbps, 6 Gbps, and 12 Gbps
- XAUI, starting at 3.125 Gbps and moving to 6.25 Gbps and 10 Gbps
- Gigabit Ethernet, starting at 1 Gbps and moving to 10 Gbps, 25 Gbps, 40 Gbps, and even 100 Gbps.

1.8 The Need for a New Design Methodology

We've painted a scary picture of the future. The situation analysis is as follows:

- Signal-integrity problems can prevent the correct operation of a high-speed digital product.
- These problems are a direct consequence of shorter rise time and higher clock frequencies.
- Rise times will continue their inevitable march toward shorter values, and clock frequencies will continue to increase.

- Even if we limit the clock frequency, using the lowest-cost chips means even low-speed systems will have chips with very short rise times.
- We have less time in the product-design cycle to get the product to market; it must work the first time.

What are we to do? How are we to efficiently design high-speed products in this new era? In the good old days of 10-MHz clock systems, when the interconnects were transparent, we did not have to worry about signal-integrity effects. We could get away with designing a product for functionality and ignore signal integrity. In today's products, ignoring signal integrity invites schedule slips, higher development costs, and the possibility of never being able to ship a functional product.

It will always be more profitable to pay extra to design a product right the first time than to try to fix it later. In the product life cycle, often the first six months in the market are the most profitable. If your product is late, a significant share of the life-cycle profits may be lost. Time really is money.

> **TIP**   A new product-design methodology is required which ensures that signal-integrity problems are identified and eliminated from a product as early in the design cycle as possible. To meet ever-shorter design cycle times, the product must meet performance specifications the first time.

1.9 A New Product Design Methodology

There are six key ingredients to the new product design methodology:

1. Identify which of the specific SI/PI/EMC problems will arise in your product and should be avoided, based on the six families of problems.

2. Find the root cause of each problem.

3. Apply the Youngman Principle to turn the root cause of the problem into best design practices.

4. When it is free and does not add any cost, always follow the best design practices, which we sometimes call *habits*.

5. Optimize the design for performance, cost, risk, and schedule, by evaluating design trade-offs using analysis tools: rules of thumb, approximations, and numerical simulations (as virtual prototypes.)

6. Use measurements throughout the design cycle to reduce risk and increase confidence of the quality of the predictions.

The basis of this strategy applies not just to engineering SI/PI/EMC problems out of the product but also solving all engineering design problems. The most efficient way of solving a problem is by first finding the root cause. If you have the wrong root cause for the problem, your chance of fixing the problem is based on pure luck.

TIP   Any design or debug process must have as an emphasis finding and verifying the correct root cause of the problem.

The root cause is turned into a best design practice using the Youngman Principle, named after the famous TV comedian Henny Youngman (1906–1998), known as the "king of the one-liners." One of the stories he would tell illustrates a very important engineering principle: "A guy walks into a doctor's office and tells the doctor, 'Doctor, my arm hurts when I raise it. What should I do?' The doctor says, 'Don't raise your arm.'"

As simple as this story is, it has profound significance for turning a root cause into a best design practice. If you can identify what feature in the product design causes the problem and want to eliminate the problem, eliminate that feature in your product; "don't raise your arm." For example, if reflection noise is due to impedance discontinuities, then engineer all interconnects with a constant instantaneous impedance. If ground bounce is caused by a screwed-up return path and overlapping return currents, then don't screw up the return path and don't let return currents overlap.

TIP   It is the Youngman Principle that turns the root cause of a problem into a best design practice.

If the best design practice does not add cost to the product, even if it provides only a marginal performance advantage, it should always be done. That's why these best design practices are called *habits*.

It doesn't cost any more to engineer interconnects as uniform transmission lines with a target characteristic impedance. It doesn't cost any more in the product design to optimize the via pad stack to match its impedance closer to 50 Ohms.

When it is free, it should always be done. If it costs more, it's important to answer these questions: "Is it worth it? What is the 'bang for the buck'? The way to answer these questions is to "put in the numbers" through analysis using rules of thumb, approximations, and numerical simulations. Predicting the performance ahead of time is really like constructing a "virtual prototype" to explore the product's anticipated performance.

Simulation is about predicting the performance of the system before building the hardware. It used to be that only the nets in a system that were sensitive to signal-integrity effects were simulated. These nets, called "critical nets," are typically clock lines and maybe a few high-speed bus lines. In 100-MHz clock frequency products, maybe only 5% to 10% of the nets were critical nets. In products with clock frequencies at 200 MHz and higher, more than 50% of the nets may be critical, and the entire system needs to be included in the simulation.

TIP   In all high-speed products today, system-level simulations must be performed to accurately verify that signal-integrity problems have been eliminated before the product is built.

In order to predict the electrical performance, which is typically the actual voltage and current waveforms at various nodes, we need to translate the physical design into an electrical description. This can be accomplished by one of three paths. The physical design can be converted into an equivalent circuit model and then a circuit simulator can be used to predict the voltages and currents at any node.

Alternatively, an electromagnetic simulator can be used to simulate the electric and magnetic fields everywhere in space, based on the physical design and material properties. From the electric and magnetic fields, a behavioral model of the interconnects can be generated, usually in the form of an S-parameter model, which can then be used in a circuit simulator.

Finally, a direct measurement of the S-parameter behavioral model of an interconnect can be performed using a vector-network analyzer (VNA). This measured model can be incorporated in a simulator tool, just as though it were simulated with an electromagnetic simulator.

1.10 Simulations

Three types of electrical simulation tools predict the analog effects of the interconnects on signal behavior:

1. Electromagnetic (EM) simulators, or 3D full-wave field solvers, which solve Maxwell's Equations using the design geometry as boundary conditions and the material properties to simulate the electric and magnetic fields at various locations in the time or frequency domains.

2. Circuit simulators, which solve the differential equations corresponding to various lumped circuit elements and include Kirchhoff's current and voltage relationships to predict the voltages and currents at various circuit nodes, in the time or frequency domains. These are usually SPICE-compatible simulators.

3. Numerical-simulation tools, which synthesize input waveforms, calculate the impulse response from S-parameter models of the interconnect, and then calculate, using convolution integrals or other numerical methods, the waveforms out of each port.

Blame signal integrity on Maxwell's Equations. These four equations describe how electric and magnetic fields interact with conductors and dielectrics. After all, signals are nothing more than propagating electromagnetic fields. When the electric and magnetic fields themselves are simulated, the interconnects and all passive components must be translated into the boundary conditions of conductors and dielectrics, with their associated geometries and material properties.

A device driver's signal is converted into an incident electromagnetic wave, and Maxwell's Equations are used to predict how this wave interacts with the conductors and dielectrics. The material geometries and properties define the boundary conditions in which Maxwell's Equations are solved.

Though Figure 1-17 shows the actual Maxwell's Equations, it is never necessary for any practicing engineer to solve them by hand. They are shown here only for reference and to demonstrate that there really is a set of just a few simple equations that describe absolutely everything there is to know about electromagnetic fields. How the incident electromagnetic field interacts with the geometry and materials, as calculated from these equations, can be displayed at every point in space. These fields can be simulated either in the time domain or the frequency domain.

Figure 1-18 shows an example of the simulated electric-field intensity inside a 208-pin plastic quad flat pack (PQFP) for an incident voltage sine wave on one pin at 2.0 GHz and 2.3 GHz. The different shadings show the field intensity. This simulation illustrates that if a signal has frequency components at 2.3 GHz, it will cause large field distributions inside the package. These are called *resonances* and can be disastrous for a product. These resonances will cause signal-quality degradation, enhanced cross talk, and enhanced EMC problems. Resonances will always limit the highest bandwidth for which the part can be used.

Time Domain	Frequency Domain
$\nabla \bullet \varepsilon\mathbf{E} = \rho$	$\nabla \bullet \varepsilon\mathbf{E} = \rho$
$\nabla \bullet \mathbf{B} = 0$	$\nabla \bullet \mathbf{B} = 0$
$\nabla \times \mathbf{E} = -\dfrac{\partial \mathbf{B}}{\partial t}$	$\nabla \times E = -j\omega\mathbf{B}$
$\nabla \times \dfrac{1}{\mu}\mathbf{B} = \mathbf{J} + \varepsilon\dfrac{\partial \mathbf{E}}{\partial t}$	$\nabla \times \dfrac{1}{\mu}\mathbf{B} = \mathbf{J} + j\omega\varepsilon\mathbf{E}$

Figure 1-17 Maxwell's Equations in the time and frequency domains. These equations describe how the electric and magnetic fields interact with boundary conditions and materials through time and space. They are provided here for reference.

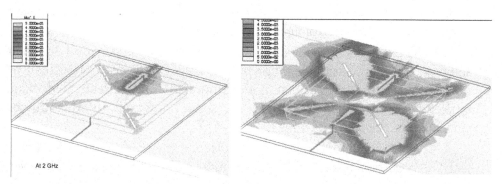

At 2 GHz

Figure 1-18 Example of an electromagnetic simulation of the electric fields in a 208-pin PQFP excited at 2.0 GHz (left) and at 2.3 GHz (right) showing a resonance. Simulation done with Ansoft's High-Frequency Structure Simulator (HFSS).

Some effects can only be simulated with an electromagnetic simulator. Usually, an EM simulator is needed when interconnects are very nonuniform and electrically long (such as traces over gaps in the return path), when electromagnetic-coupling effects dominate (such as resonances in packages and connectors), or when it's necessary to simulate EMC radiation effects.

TIP   To accurately predict the impact from interconnect structures that are not uniform, such as packages traces, connectors, vias, and discrete components mounted to pads, a 3D full-wave electromagnetic simulator is an essential tool.

Though all the physics can be taken into account with Maxwell's Equations, with the best of today's hardware and software tools, it is not practical to simulate the electromagnetic effects of other than the simplest structures. An additional limitation is that many of the current tools require a skilled user with experience in electromagnetic theory.

An alternative simulation tool, which is easier and quicker to use, is a circuit simulator. This simulation tool represents signals as voltages and currents. The various conductors and dielectrics are translated into their fundamental electrical circuit elements—resistances, capacitances, inductances, transmission lines, and their coupling. This process of turning physical structures into circuit elements is called *modeling*.

Circuit models for interconnects are approximations. When an approximation is a suitably accurate model, it will always get us to an answer faster than using an electromagnetic simulator.

Some problems in signal integrity are more easily understood and their solutions are more easily identified by using the circuit description than by using the electromagnetic analysis description. Because circuit models are inherently an approximation, there are some limitations to what can be simulated by a circuit simulator.

A circuit simulator cannot take into account electromagnetic effects such as EMC problems, resonances, and nonuniform wave propagation. However, lumped circuit models and SPICE-compatible simulations can accurately account for effects such as near-field cross talk, transmission line propagation, reflections, and switching noise. Figure 1-19 shows an example of a circuit and the resulting simulated waveforms.

A circuit diagram that contains combinations of lumped circuit elements is called a *schematic*. If you can draw a schematic composed of ideal circuit elements, a circuit simulator will be able to calculate the voltages and currents at every node.

The most popular circuit simulator is generically called SPICE (short for Simulation Program with Integrated Circuit Emphasis). The first version was created at UC Berkeley in the early 1970s as a tool to predict the performance of transistors based on their geometry and material properties.

SPICE is fundamentally a circuit simulator. From a schematic, described in a specialized text file called a netlist, the tool will solve the differential equations each circuit element represents and then calculate the voltages and currents either in the time domain, called a transient simulation, or in the frequency domain, called an AC simulation. There are more than 30 commercially available versions of SPICE, with some free student/demo versions available for download from the Web.

> **TIP**   The simplest-to-use and most versatile SPICE-compatible simulator with a very clean user interface capable of publication-quality graphics display is QUCS (Quite Universal Circuit Simulator). It is open source and available from www.QUCS.org.

Numerical simulators such as MATLAB, Python, Keysight PLTS, and the Teledyne LeCroy SI Studio are examples of simulation tools that predict output waveforms based on synthesized input waveforms and how they interact with the S-parameter interconnect models. Their chief advantage over circuit simulators is in their computation speed. Many of these simulators use proprietary simulation engines and are optimized for particular types of waveforms, such as sine waves, clocks, or NRZ data patterns.

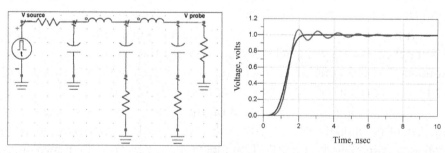

Figure 1-19   Example of a circuit model for the probe tip of a typical scope probe, approximately 5 cm long, and the resulting circuit simulation response of a clean, 1-nsec rise-time signal. The ringing is due to the excessive inductance of the probe tip.

1.11 Modeling and Models

Modeling refers to creating an electrical representation of a device or component that a simulator can interpret and use to predict voltage and current waveforms. The models for active devices, such as transistors and output drivers, are radically different from the models of passive devices, such as all interconnects and discrete components. For active devices, a model is typically either a SPICE-compatible model or an input/output buffer interface (IBIS)–compatible model.

A SPICE model of an active device will use either combinations of ideal sources and passive elements or specialized transistor models based on the geometry of the transistors. This allows easy scaling of the transistor's behavior if the process technology changes. A SPICE model contains information about the specific features and process technology of a driver. For this reason, most vendors are reluctant to give out SPICE models of their chips since they contain such valuable information.

IBIS is a format that defines the response of input or output drivers in terms of their V-I and V-t characteristics. A behavioral simulator will take the V-I and V-t curves of the active devices and simulate how these curves might change as they are affected by the transmission lines and lumped resistor (R), inductor (L), and capacitor (C) elements, which represent the interconnects. The primary advantage of an IBIS model for an active device is that an IC vendor can provide an IBIS model of its device drivers without revealing any proprietary information about the geometry of the transistors.

It is much easier to obtain an IBIS model than a SPICE model from an IC supplier. For system-level simulation, where 1000 nets and 100 ICs may be simulated at the same time, IBIS models are typically used because they are more available and typically run faster than SPICE models.

The most important limitation on the accuracy of any simulation is the quality of the models. While it is possible to get an IBIS model of a driver that compares precisely with a SPICE model and matches the actual measurement of the device perfectly, it is difficult to routinely get good models of every device.

TIP   In general, as end users, we must continually insist that our vendors supply us with some kind of verification of the quality of the models they provide for their components.

Another problem with device models is that a model that applies to one generation of chips will not match the next generation. With the next die shrink, which happens every six to nine months, the channel lengths are shorter, the rise times are shorter, and the V-I curves and transient response of the drivers change. An old model of a new part will give low estimates of the signal-integrity effects. As users, we must always insist that vendors provide current, accurate, and verified models of all drivers they supply.

> **TIP**   Though the intrinsic accuracy of all SPICE or other simulators is typically very good, the quality of the simulations is only as good as the quality of the models that are simulated. The expression garbage in, garbage out (GIGO) was invented to describe circuit simulations.

For this reason, it is critically important to verify the accuracy of the models of all device drivers, interconnects, and passive components. Only in this way can we have confidence in simulation results. Though the models for active devices are vitally important, this book deals with models for passive devices and interconnects. Other references listed Appendix C, "Selected References," discuss active-device models.

Obtaining models for all the elements that make up the system is critically important. The only way to simulate signal-integrity effects is to include models of the interconnects and passive components, such as the board-level transmission lines, package models, connector models, DC blocking capacitors, and terminating resistors.

Of course, the circuit model can only use elements that the simulator understands. For SPICE simulators, interconnects and passive components can be described by resistors, capacitors, inductors, mutual inductors, and lossless transmission lines. In some SPICE and behavioral simulators, new ideal circuit elements have been introduced that include ideal coupled transmission lines and ideal lossy transmission lines as basic ideal circuit elements.

Figure 1-20 shows an example of a physical component, two surface-mount terminating resistors, and their equivalent electrical circuit model. This model includes their inductive coupling, which gives rise to switching noise. Every electrical quality of their behavior can be described by their circuit model. This circuit model, or schematic, can accurately predict any measurable effect.

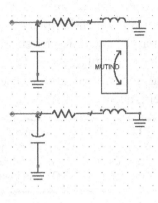

Figure 1-20   Two surface-mount 0805 resistors and their equivalent circuit model. This model has been verified up to 5 GHz.

There are two basic approaches to creating accurate circuit models of interconnects: by calculation and by measurements. We usually refer to creating a model from calculation as analysis and creating a model from a measurement as characterization.

1.12 Creating Circuit Models from Calculation

So much of life is a constant balancing act between the value received and the cost (in time, money, and expertise required). The analysis of all signal-integrity design problems, as well as problems in most other fields, is no exception. We are constantly balancing the quality of an answer, as measured by its accuracy, for example, with how long it will take and how much it will cost to get the answer.

> **TIP**   The goal in virtually all product-development programs in today's globally competitive market is to get to an acceptable design that meets the performance spec while staying within the cost, schedule, and risk budgets.

This is a tough challenge. An engineer involved in signal integrity and interconnect design can benefit from skill and versatility in selecting the best technology and establishing the optimum best design practices as early in the design cycle as possible.

The most important tool in any engineer's toolbox is the flexibility to evaluate trade-offs quickly. As Bruce Archambeault, an IBM Fellow and icon in the EMC industry, says, "Engineering is Geek for trade-off analysis."

These are really trade-offs between how choices of technology selection, material properties, and best design practices will affect system performance, cost, and risk.

TIP   The earlier in the design cycle the right trade-offs can be established, the shorter the development time and the lower the development costs.

To aid in trade-off analysis, three levels of approximation are used to predict performance or electrical properties:

1. Rules of thumb
2. Analytical approximations
3. Numerical simulations

Each approach represents a different balance between closeness to reality (i.e., accuracy) and the cost in time and effort required to get the answer. This is illustrated in Figure 1-21. Of course, these approaches are not substitutes for actual measurements. However, the correct application of the right analysis technique can sometimes shorten the design-cycle time to 10% of its original value, compared to relying on the build it/test it/redesign it approach.

Rules of thumb are simple relationships that are easy to remember and help feed our intuition. An example of a rule of thumb is "the self-inductance of a short wire is about 25 nH/inch." Based on this rule of thumb, a wire bond 0.1 inches long would have a self-inductance of about 2.5 nH.

Using a rule of thumb should be the first step in any analysis problem, if only to provide a sanity-check number to which every answer can later be compared. A rule of thumb can set an initial expectation for a simulation. This is the first of a number of consistency tests every engineer should get in the habit of performing to gain confidence in a simulation. This is the basis of Bogatin's Rule #9.

TIP   Bogatin's Rule #9 is "Never perform a measurement or simulation without first anticipating the result." If the result is not what is expected, there is a reason for it. Never proceed without verifying why it did not come out as expected. If the result did match expectations, it is a good confidence builder.

The corollary to Rule #9 is "There are so many ways of making a mistake in a measurement or simulation that you can never do too many consistency tests." An initial estimate using a rule of thumb is always the first and most important consistency test.

When you are brainstorming design/technology/cost trade-offs or looking for rough estimates, comparisons, or plausibility arguments, using rules of thumb can accelerate your progress more than tenfold. After all, the closer to the beginning of the product-development cycle the right design and technology decisions can be made, the more time and money will be saved in the project.

Rules of thumb are not meant to be accurate but to give answers quickly. They should never be used when signing off on a design. They should be used to calibrate your intuition and provide guidance to help make high-level trade-offs. Appendix B, "100 Collected Rules of Thumb to Help Estimate Signal-Integrity Effects," contains a summary of many of the important rules of thumb used in signal integrity.

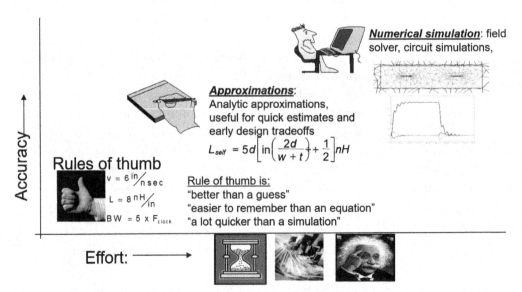

Numerical simulation: field solver, circuit simulations,

Approximations:
Analytic approximations, useful for quick estimates and early design tradeoffs

$$L_{self} = 5d \left[\ln\left(\frac{2d}{w+t}\right) + \frac{1}{2} \right] nH$$

Rules of thumb

$v = 6 \, in/nsec$

$L = 8 \, nH/in$

$BW = 5 \times F_{clock}$

Rule of thumb is:
"better than a guess"
"easier to remember than an equation"
"a lot quicker than a simulation"

Accuracy ⟶

Effort: ⟶

Figure 1-21   The balance between accuracy returned and effort required for the three levels of analysis. Each tool has an appropriate time and place.

Analytical approximations are equations or formulas. For example, an approximation for the loop self-inductance of a circular loop of wire is:

$$L_{self} = 32 \times R \times \ln\left(\frac{4R}{D}\right) nH \qquad (1-2)$$

where:

L_{self} = self-inductance, in nH

R = radius of the loop, in inches

D = diameter of the wire, in inches

For example, a round loop, ½ inch in radius or 1 inch in diameter, made from 10-mil-thick wire, would have a loop inductance of about 85 nH. Put your index finger and thumb together in a circle. If they were made of 30-gauge copper wire, the loop inductance would be about 85 nH.

Approximations have value in that they can be implemented in a spreadsheet, and they can answer what-if questions quickly. They identify the important first-order terms and how they are related. The approximation above illustrates that the inductance scales a little faster than the radius. The larger the radius of the loop, the larger the inductance of the loop. Also, the larger the wire thickness, the smaller the loop inductance, but only slightly because the loop inductance varies inversely with the natural log of the thickness, a weak function.

TIP   It is important to note that with very few exceptions, every equation you see being used in signal-integrity analysis is either a definition or an approximation.

A definition establishes an exact relationship between two or more terms. For example, the relationship between clock frequency and clock period, F = 1/T, is a definition. The relationship between voltage, current, and impedance, Z = V/I, is a definition.

You should always be concerned about the accuracy of an approximation, which may vary from 1% to 50% or more. Just because a formula allows evaluation on a calculator to five decimal places doesn't mean it is accurate to five decimal places. You can't tell the accuracy of an approximation by its complexity or its popularity.

How good are approximations? If you don't know the answer to this question for your specific case, you may not want to base a design sign-off—where a variance of more than 5% might mean the part won't work in the application—on an approximation with unknown quality. The first question you should always answer of every approximation is: How accurate is it?

One way of verifying the accuracy of an approximation is to build well-characterized test vehicles and perform measurements that can be compared to the calculation. Figure 1-22 shows the very good agreement between the approximation for loop inductance offered above and the measured values of loop inductance based on building loops and measuring them with an impedance analyzer. The agreement is shown to be better than 2%.

Approximations are extremely important when exploring design space or performing tolerance analysis. They are wonderful when balancing trade-offs. However, when being off by 20% will cost a significant amount of time, money, or resources, you should never rely on approximations whose accuracy is uncertain.

There is a more accurate method for calculating the parameter values of the electrical circuit elements of interconnects from the geometry and material properties. It is based on the numerical calculation of Maxwell's Equations. These tools are called *field solvers*, in that they solve for the electric and magnetic fields based on Maxwell's Equations, using as boundary conditions the distribution of conductors and dielectrics. A type of field solver that converts the calculated fields into the actual parameter values of the equivalent circuit-model elements, such as the R, L, or C values, is called a *parasitic extraction tool*.

When the geometry of an interconnect is uniform down its length, it can be described by its cross section, and a 2D field solver can be used to extract its transmission line properties. An example of a typical cross section of a microstrip transmission line and the simulated electric field lines and equipotentials is shown in Figure 1-23. For this structure, the extracted parameter values were $Z_0 = 50.3$ Ohms and TD = 142 psec.

When the cross section is nonuniform, such as in a connector or an IC package, a 3D field solver is needed for the most accurate results.

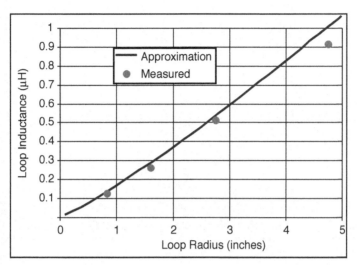

Figure 1-22 Comparison of the measured and calculated loop inductance of various circular wire loops as measured with an impedance analyzer. The accuracy of the approximation is seen to be about 2%.

Before relying on any numerical-simulation tool, it is always important to have its results and the process flow used to set up and solve problems verified for some test cases similar to the end application for which you will be solving. Every user should insist on vendor verification that the tool is suitably accurate for the typical application. In this way, you gain confidence in the quality of your results. The accuracy of some field solvers has been verified to better than 1%. Obviously, not all field solvers are always this accurate.

When accuracy is important, as, for example, in any design sign-off, a numerical-simulation tool, such as a parasitic extraction tool, should be used. It may take longer to construct a model using a numerical-simulation tool than it would using a rule of thumb or even using an analytic approximation. More effort in time and in expertise is required. But such models offer greater accuracy and a higher confidence that the as-manufactured part will match the predicted performance. As new commercially available numerical-simulation tools advance, market pressures will drive them to be easier to use.

Using a combination of these three analysis techniques, the trade-offs between the costs in time, money, and risk can be balanced to make a very good prediction of the possible performance gain.

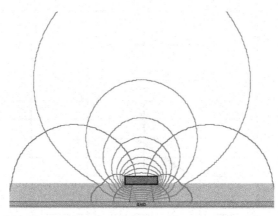

Figure 1-23   The results of a 2D field solver used to calculate the electric fields in a microstrip transmission line. The parasitic extraction tool used was HyperLynx from Mentor Graphics. The accuracy of this tool has been independently verified to be better than 2%.

1.13 Three Types of Measurements

TIP   Though calculations play a critical role of offering a prediction of performance before a product is built, measurements play a critical role of risk reduction. The ultimate test of any calculation result is a measurement. Measurements can be anchors to reality.

When it comes to measuring passive interconnects, as distinct from active devices, the measuring instrument must create a precision reference signal, apply it to the device under test, and measure the response. Ultimately, this response is related to the spatial variation of the impedance of the device. In active devices, which create their own signals, the measurement instrument can be passive, merely measuring the created voltages or currents. Three primary instruments are used to perform measurements on passive components:

1. Impedance analyzer
2. Vector-network analyzer (VNA)
3. Time-domain reflectometer (TDR)

An impedance analyzer, typically a four-terminal instrument, operates in the frequency domain. One pair of terminals is used to generate a constant-current sine wave through the device under test (DUT). The second pair of terminals is used to measure the sine-wave voltage across the DUT.

The ratio of the measured voltage to the measured current is the impedance. The frequency is typically stepped from the low 100-Hz range to about 40 MHz. The magnitude and phase of the impedance at each frequency point is measured based on the definition of impedance.

The vector-network analyzer also operates in the frequency domain. Each terminal or port emits a sine-wave voltage at frequencies that can range in the low kHz to over 50 GHz. At each frequency, both the incident-voltage amplitude and phase and the reflected amplitude and phase are measured. These measurements are in the form of scattering parameters (S-parameters). This topic is extensively covered in Chapter 12, "S-Parameters for Signal Integrity Applications."

The reflected signal will depend on the incident signal and the impedance change in going from the VNA to the DUT. The output impedance of a VNA is typically 50 Ohms. By measuring the reflected signal, the impedance of the device under test can be determined at each frequency point. The relationship between the reflected signal and the impedance of the DUT is:

$$\frac{V_{\text{reflected}}}{V_{\text{incident}}} = \frac{Z_{\text{DUT}} - 50\,\Omega}{Z_{\text{DUT}} + 50\,\Omega} \tag{1-3}$$

where:

$V_{\text{reflected}}$ = amplitude and phase of the reflected sine-wave voltage
V_{incident} = amplitude and phase of the incident sine-wave voltage
Z_{DUT} = impedance of the DUT
$50\,\Omega$ = impedance of the VNA

The ratio of the reflected to the incident voltage, at each frequency, is often referred to as one of S-parameters, signified as S_{11}. Measuring S_{11} and knowing that the source impedance is 50 Ohms allow us to extract the impedance of the device under test at any frequency. An example of the measured impedance of a short-length transmission line is shown in Figure 1-24.

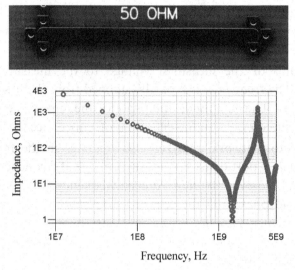

Figure 1-24 Measured impedance of a 1-inch length of transmission line. The network analyzer measured the reflected sine-wave signal between the front of the line and a via to the plane below the trace. This reflected signal was converted into the magnitude of the impedance. The phase of the impedance was measured, but is not displayed here. The frequency range is from 12 MHz to 5 GHz. Measured with a GigaTest Labs Probe Station.

A time-domain reflectometer (TDR) is similar to a VNA but operates in the time domain. It emits a fast rise-time step signal, typically 35 psec to 150 psec, and measures the reflected transient voltage signal. Again, using the reflected voltage, the impedance of the DUT can be extracted. In the time domain, the impedance measured represents the instantaneous impedance of the DUT. For an interconnect that is electrically long, such as a transmission line, a TDR can map the impedance profile. Figure 1-25 is an example of a TDR profile from a 4-inch-long transmission line with a small gap in the return plane showing a higher impedance where the gap occurs.

The principles of how a TDR works and how to interpret the measured voltages from a TDR are described in Chapter 8, "Transmission Lines and Reflections." It is also possible to take the measured frequency-domain S_{11} response from a VNA and mathematically transform it into the step response in the time domain. Whether measured in the time or frequency domain, the measured response can be displayed in the time or frequency. This important principle is detailed in Chapter 12.

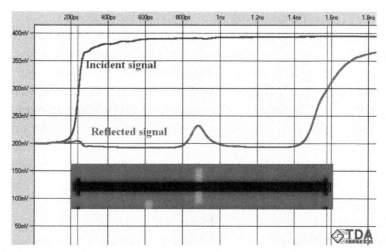

Figure 1-25   Measured TDR profile of a 4-inch-long uniform transmission line with a gap in the return path near the middle. The far end of the line is open. Measured with an Agilent 86100 DCA TDR and recorded with TDA Systems IConnect software, using a GigaTest Labs Probe Station.

T I P   Though the same impedance of a DUT can be displayed in the frequency domain or the time domain, it is really a different impedance in each case. When displayed in the frequency domain, the impedance is the total, integrated input impedance of the entire DUT at each frequency. When displayed in the time domain, it is the instantaneous impedance at each spatially distinct point on the DUT that is displayed.

1.14 The Role of Measurements

If it is possible to calculate the expected electrical performance of a component or system, why bother with a measurement? Why not just rely on the modeling and simulation tool? Measurements can only be performed on real devices. Isn't the goal to avoid the build-it/measure-it/redesign-it iteration loop that takes so much time?

Measurements such as the ones illustrated above play six critically important basic roles, at various stages in the product life cycle, all related to risk reduction and establishing higher confidence in the accuracy of the simulations. Measurements allow designers to do the following:

1. Verify the accuracy of the design/modeling/simulation process before significant resources are expended using an unverified process.

2. Verify an as-fabricated component meets the performance spec.

3. Extract the material properties of the as-fabricated structure for use as an input parameter to the simulation tools.
4. Create an equivalent electrical model for a component at any stage of the design cycle as parts are made available or acquired from a vendor.
5. Emulate system performance of a component as a quick way of determining expected performance without building an electrical model at any stage of the design cycle as parts are made available or acquired from a vendor.
6. Debug a functional part or system at any stage of the design cycle as parts are made available or acquired from a vendor.

An example from Delphi Electronics illustrates the incredible power of the combination of a design/modeling/simulation process that has been verified using measurements. One of the products Delphi Electronics makes is a custom flexible connector for connecting two circuit boards for high-speed signals (see Figure 1-26). These connectors are used in servers, computers, and switching systems. The electrical performance of such a connector is critical to the correct function of the system.

A customer comes to Delphi with a set of performance specifications, and Delphi is to deliver a part that meets those specs. The old way of designing the product was to make a best guess, manufacture the part, bring it back to the lab and perform measurements on it, compare to the specs, and redesign it. This is the old build-it/test-it/redesign-it approach. In this old process, one iteration took almost nine weeks because of the long CAD and manufacturing cycle. Sometimes, the first design did not meet the customer specs and a second cycle was required. This meant an 18-week development cycle.

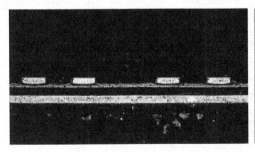 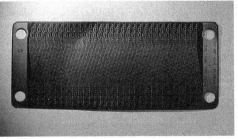

Figure 1-26 Gold Dot connector from Delphi Automotive. Left: Cross section of the two-metal-layer flex substrate. Right: Top view of the hundreds of conductors connecting one board to the other, each with controlled electrical performance. Photo courtesy of Laurie Taira-Griffin, Delphi Electronics.

To shorten this design-cycle time, Delphi implemented a 2D modeling tool that allows Delphi to predict the electrical properties of the connector based on the geometry and material properties. Through a few cycles of experimenting, using measurements with a TDR and VNA as the verification process, Delphi was able to fine-tune its modeling process to ensure accurate predictability of its product before the part was sent to manufacturing. An example of the final agreement between what its modeling tool is able to predict and what was actually measured for a connector, compared with the customer's original specification, is shown in Figure 1-27.

Parameter	Simulation	Measured	Goal
Single-Ended Impedance	52.1 Ohms	53 Ohms	50 +/- 10% Ohms
Differential Impedance	95.2 Ohms	98 Ohms	100 +/- 10% Ohms
Attenuation (5GHz)	<.44 dB/inch	<.44 dB/inch	<.5 dB/inch
Propagation Delay	152 ps/inch	158 ps/inch	170 ps/inch
Single-Ended NEXT	<4.5%	<4.5%	<5%
Differential NEXT	<.3%	<.3%	<.5%
Data Rate	>5 Gbps	>5 Gbps	5 Gbps

Figure 1-27 Summary of the predicted and measured electrical specifications of the connector compared with the requirements for a particular connector. After the modeling/simulation process was optimized, the ability to predict performance was excellent.

Once the modeling/simulation process was in place and Delphi had confidence in its ability to accurately predict the final performance of the manufactured connectors, Delphi was able to reduce the design-cycle time to less than four hours. Nine weeks to four hours is a reduction of more than 100x. Measurements provided the critical verification step for this process.

1.15 The Bottom Line

1. Signal-integrity problems relate to how the physical interconnects screw up pristine signals coming from the integrated circuits.

2. There are six general families of signal-integrity problems: signal quality on one net, rise-time degradation from losses, cross talk between adjacent nets, ground and power bounce, rail collapse, and EMI.

3. Each of these problems gets worse and harder to solve as rise times decrease or clock frequencies increase.

4. It is inevitable that as transistors get smaller, their rise times will get shorter, and signal integrity will be a greater problem.

5. To find, fix, and prevent signal-integrity problems, it is essential to be able to convert a physical design into its equivalent electrical circuit model and to use that model to simulate waveforms so as to predict performance before the product is built.

6. Three levels of analysis can be used for calculating electrical effects—rules of thumb, approximations, and numerical tools—and can be applied to modeling and simulation.

7. Three general instruments can be used to measure electrical properties of passives and interconnects: an impedance analyzer, a network analyzer, and a time-domain reflectometer.

8. These instruments play the important role of reducing the risk and increasing the confidence level in the accuracy of the modeling and simulation process.

9. Understanding the six signal-integrity problems leads to the most important best design practices to engineer them out of the product. Figure 1-28 summarizes the general best design practices for the six families of signal-integrity problems.

Noise Category	Design Principle
Signal quality	Engineer a constant instantaneous impedance by using controlled impedance interconnects terminating the ends of the lines and as close to a point-to-point routing topology as possible.
Rise-time degradation from losses	Reduce frequency-dependent losses by using shorter interconnects, wider traces, and lower-dissipation-factor laminates.
Cross talk	Keep spacing of traces greater than a minimum value.
Ground and power bounce	Don't screw up return paths and don't share return currents; minimize mutual inductance of non-ideal returns.
Rail collapse	Minimize the impedance of the power/ground path and the delta I.
EMI	Minimize bandwidth, minimize ground impedance, and minimize common currents on external cables and shield.

Figure 1-28 Summary of the six families of signal-integrity problems and the general design guidelines to minimize these problems. Even if these guidelines are followed, it is still essential to model and simulate the system to evaluate whether the design will meet the performance requirements.

The rest of this book discusses the expected SI/PI/EMC problems, the best design practices, the essential principles required to understand these problems, the best design practices and the specific techniques to apply them in your product design.

End-of-Chapter Review Questions

The answers to the following review questions can be found in Appendix D, "Review Questions and Answers."

1.1 Name one problem that is just a signal-integrity problem.

1.2 Name one problem that is just a power-integrity problem.

1.3 Name one problem that is just an electromagnetic compliance problem.

1.4 Name one problem that is considered both an SI and PI problem.

1.5 What causes an impedance discontinuity?

1.6 What happens to a propagating signal when the interconnect has frequency-dependent losses?

1.7 What are the two mechanisms that cause cross talk?

1.8 For lowest cross talk, what should the return path for adjacent signal paths look like?

1.9 A low impedance PDN reduces power-integrity problems. List three design features for a low-impedance PDN.

1.10 List two design features that contribute to reduced EMI.

1.11 When is it a good idea to use a rule of thumb? When is it not a good idea to use a rule of thumb?

1.12 What is the most important feature of a signal that influences whether it might have a signal-integrity problem?

1.13 What is the most important piece of information you need to know to fix a problem?

1.14 Best design practices are good habits to follow. Give three examples of a best design practice for circuit board interconnects.

1.15 What is the difference between a model and a simulation?

1.16 What are the three important types of analysis tools.

1.17 What is Bogatin's rule #9?

1.18 What are three important reasons to integrate measurements somewhere in your design flow?

1.19 A clock signal is 2 GHz. What is the period? What is a reasonable estimate for its rise time?

1.20 What is the difference between a SPICE and an IBIS model?

1.21 What do Maxwell's Equations describe?

1.22 If the underlying clock frequency is 2 GHz, and the data is clocked at a double data rate, what is the bit rate of the signal?

CHAPTER 2

Time and Frequency Domains

In this chapter, we explore the basic properties of signals in preparation for looking at how they interact with interconnects. There are multiple ways of looking at a signal, each providing a different perspective. The quickest path to an answer may not be the most obvious or direct path. The different perspectives we will use to look at signals are called *domains*. In particular, we'll use the time domain and the frequency domain.

While people may generally be more familiar with the time domain, the frequency domain can provide valuable insight to understand and master many signal-integrity effects, such as impedance, lossy lines, the power-distribution network, measurements, and models.

After introducing the two domains, we will look at how to translate between the two for some special cases. We will apply what we learn to relate two important quantities: rise time and bandwidth. The first is a time-domain term and the second a frequency-domain term. However, as we will see, they are intimately related.

Finally, we'll apply this concept of bandwidth to interconnects, models, and measurements.

2.1 The Time Domain

We use the term —*the time domain* a lot. But what do we really mean when we say this? What is the time domain? What special features of the time domain that make it useful? These are surprisingly difficult questions to answer because they seem so obvious, and we rarely think about what we really mean by *time domain*.

TIP   The time domain is the real world. It is the only domain that actually exists.

We take this domain for granted because from the moment we are born, our experiences are developed and calibrated in the time domain. We are used to seeing events happen with a time stamp and ordered sequentially.

The time domain is the world of our experiences and is the domain in which high-speed digital products perform. When evaluating the behavior of a digital product, we typically do the analysis in the time domain because that's where performance is ultimately measured.

For example, two important properties of a clock waveform are clock period and rise time. Figure 2-1 illustrates these features.

The clock period is the time interval to complete one clock cycle, usually measured in nanoseconds (nsec). The clock frequency, F_{clock}, or how many cycles per second the clock goes through, is the inverse of the clock period, T_{clock}:

$$F_{clock} = \frac{1}{T_{clock}} \tag{2-1}$$

where:

F_{clock} = clock frequency, in GHz

T_{clock} = clock period, in nsec

For example, a clock with a period of 10 nsec will have a clock frequency of 1/10 nsec = 0.1 GHz or 100 MHz.

The rise time is related to how long it takes for the signal to transition from a low value to a high value. There are two popular definitions of rise time.

The 10–90 rise time is how long it takes for the signal to transition from 10% of its final value to 90% of its final value. This is usually the default meaning of rise time. It can be read directly off the time-domain plot of a waveform.

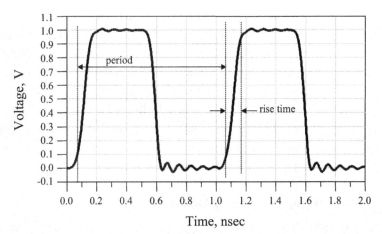

Figure 2-1 Typical clock waveform showing the clock period and the 10–90 rise time for a 1-GHz clock. The fall time is typically slightly shorter than the rise time and sometimes creates more noise.

The second definition is the 20–80 rise time. This is the time it takes for the signal to transition from 20% of its final value to 80% of its final value. Of course, for the same waveform the 20–80 rise time is shorter than the 10–90 rise time. When dealing with signals transmitted down lossy interconnects, the rising or falling edge shape is distorted, and the signal has a long tail. The 10–90 rise time has little significance, and the 20–80 rise time might be a better figure of merit.

Some IBIS models of real devices use the 20–80 definition of rise time. This makes it confusing. To remove ambiguity, it's often good practice to refer explicitly to the 10–90 rise time or the 20–80 rise time.

There is a corresponding value for the fall time of a time-domain waveform. Depending on the logic family, the fall time is usually slightly shorter than the rise time. This is due to the design of typical CMOS output drivers. In a typical output driver, a p and an n transistor are in series between the V_{CC} (+) and the V_{SS} (–) power rails. The output is connected to the center, between them. Only one transistor is on at any one time, depending on whether the output is a low or a high.

When the driver switches from a low to a high (i.e., rising edge), the n transistor turns off, and the p transistor turns on. The rise time is related to how fast the p transistor can turn on. When switching from the high to the low state (i.e., a falling edge), the p transistor turns off, and the n transistor turns on.

In general, for the same feature-size transistor, an n transistor can turn on faster than a p transistor. This means that when switching from high to low, the falling edge will be shorter than the rising edge. In general, signal-integrity problems are more likely to occur when switching from a high to low level than when switching from a low to high level. By making the n channel transistor longer than the p channel, the rising and falling edges can be closely matched.

Having established an awareness of the time domain as a distinct way of looking at events, we can turn our attention to one of a number of alternative ways of analyzing the world—the frequency domain.

2.2 Sine Waves in the Frequency Domain

We hear the term *frequency domain* quite a bit, especially when it involves radio frequency (rf) or communications systems. We also encounter the frequency domain in high-speed digital applications. There are few engineers who have not heard of and used the term multiple times. Yet, what do we really mean by the *frequency domain*? What is the frequency domain, and what makes it special and useful?

TIP   The most important quality of the frequency domain is that it is not real. It is a mathematical construct. The only reality is the time domain. The frequency domain is a mathematical world where very specific rules are followed.

The most important rule in the frequency domain is that the only kind of waveforms that exist are sine waves. Sine waves are the language of the frequency domain.

There are other domains that use other special functions. For example, the JPEG picture-compression algorithm takes advantage of special waveforms that are called *wavelets*. The wavelet transform takes the space domain, with a lot of x–y amplitude information content, and translates it into a different mathematical description that is able to use less than 10% of the memory to describe the same information. It is an approximation—but a very good one.

It's common for engineers to think that we use sine waves in the frequency domain because we can build any time-domain waveform from combinations of sine waves. This is a very important property of sine waves. However, there are many other waveforms with this property. It is not a property that is unique to sine waves.

In fact, four properties make sine waves very useful for describing any other waveform:

1. Any waveform in the time domain can be completely and uniquely described by combinations of sine waves.
2. Any two sine waves with different frequencies are orthogonal to each other. If you multiply them together and integrate over all time, they integrate to zero. This means you can separate each component from every other.
3. They are well defined mathematically.
4. They have a value everywhere, with no infinities, and they have derivatives that have no infinities anywhere. This means they can be used to describe real-world waveforms because there are no infinities in the real world.

All of these properties are vitally important but are not unique to sine waves. There is a whole class of functions called *orthonormal functions*, or sometimes called *eigenfunctions* or *basis functions*, which could be used to describe any time-domain waveform. Other orthonormal functions are Hermite polynomials, Legendre polynomials, Laguerre polynomials, and Bessel functions.

Why did we choose sine waves as our functions in the frequency domain? What's so special about sine waves? The real answer is that by using sine waves, some problems related to the electrical effects of interconnects will be easier to understand and solve using sine waves. If we switch to the frequency domain and use sine-wave descriptions, we can sometimes get to an answer faster than if we stay in the time domain.

TIP After all, if the time domain is the real world, we would never leave it unless the frequency domain provides a faster route to an acceptable answer.

Sine waves can sometimes provide a faster path to an acceptable answer because of the types of electrical problems we often encounter in signal integrity. If we look at the circuits that describe interconnects, we find that they often include combinations of resistors (R), inductors (L), and capacitors (C).

These elements in a circuit can be described by a second-order linear differential equation. The solution to this type of differential equation is a sine wave. In these circuits, the naturally occurring waveforms will be combinations of the waveforms that are solutions to the differential equation.

We find that in the real world, if we build circuits that contain Rs, Ls, and Cs and send any arbitrary waveform in, more often than not, we get waveforms out that look like sine waves and can more simply be described by a combination of a few sine waves. Figure 2-2 shows an example of this.

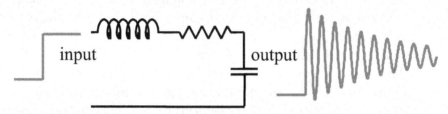

Figure 2-2   Time-domain behavior of a fast edge interacting with an ideal RLC circuit. Sine waves are naturally occurring when digital signals interact with interconnects, which can often be described as combinations of ideal RLC circuit elements.

2.3 Shorter Time to a Solution in the Frequency Domain

> **TIP**   The only reason we would ever want to move to another domain is to get to an acceptable answer faster.

In some situations, if we use the naturally occurring sine waves in the frequency domain rather than in the time domain, we may arrive at a simpler description to a problem and get to a solution faster.

It is important to keep in mind that there is fundamentally no new information in the frequency domain. The time- and frequency-domain descriptions of the same waveforms will each have exactly the same information content.

However, some problems are easier to understand and describe in the frequency domain than in the time domain. For example, the concept of bandwidth is intrinsically a frequency-domain idea. We use this term to describe the most significant sine-wave frequency components associated with a signal, a measurement, a model, or an interconnect.

Impedance is defined in both the time and the frequency domains. However, it is far easier to understand, to use, and to apply the concepts of impedance in the frequency domain. We need to understand impedance in both domains, but we will often get to an answer faster by solving an impedance problem in the frequency domain first.

Looking at the impedance of the power and ground distribution in the frequency domain will allow a simpler explanation and solution to rail-collapse problems. As we shall see, the design goal for the power-distribution system is to keep its impedance below a target value from direct current (DC) up to the bandwidth of the typical signals.

When dealing with EMI issues, both the FCC specifications and the methods of measuring the electromagnetic compliance of a product are more easily performed in the frequency domain.

With today's current capabilities of hardware and software tools, the quality of the measurements and the computation speed of the numerical-simulation tools can sometimes be better in the frequency domain.

A high signal-to-noise ratio (SNR) means higher-quality measurements. The SNR of a vector-network analyzer (VNA), which operates in the frequency domain, is constant over its entire frequency range, which can be -130 dB from 10 MHz up to 50 GHz and more. For a time-domain reflectometer (TDR), the effective bandwidth may be as high as 20 GHz, but the SNR starts at -70 dB at low frequency and drops to as low as -30 dB at 20 GHz.

Many of the effects related to lossy transmission lines are more easily analyzed, measured, and simulated by using the frequency domain. The series resistance of a transmission line increases with the square root of frequency, and the shunt AC leakage current in the dielectric increases linearly with frequency. The transient (time-domain) performance of lossy transmission lines is often more easily obtained by first transforming the signal into the frequency domain, looking at how the transmission line affects each frequency component separately, and then transforming the sine-wave components back to the time domain.

T I P   Because the frequency domain is so useful, developing skill in thinking in both the time and frequency domains is valuable. The successful engineer should become bilingual—fluent in both the time and frequency domains.

2.4 Sine-Wave Features

As we now know, by definition, the only waveforms that exist in the frequency domain are sine waves. We should also be familiar with the description of a sine wave in the time domain. It is a well-defined mathematical curve that has three terms that fully characterize absolutely everything you could ever ask about it (see Figure 2-3):

- Frequency
- Amplitude
- Phase

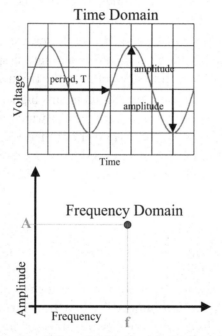

Figure 2-3   Top: Description of a sine wave in the time domain. It is composed of more than 1000 voltage-versus-time data points. Bottom: Description of a sine wave in the frequency domain. Only three terms define a sine wave, which is a single point in the frequency domain.

The frequency, usually identified using a small f, is the number of complete cycles per second made by the sine wave, in Hertz. Angular frequency is measured in radians per second. Radians are like degrees, describing a fraction of a cycle. There are $2 \times \pi$ radians in one complete cycle. The Greek letter omega, ω is often used to refer to the angular frequency, measured in radians per second. The sine-wave frequency and the angular frequency are related by:

$$\omega = 2\pi \times f \qquad (2\text{-}2)$$

where:

ω = angular frequency, in radians/sec

π = constant, 3.14159...

f = sine-wave frequency, in Hz

For example, if the frequency of a sine wave is 100 MHz, the angular frequency is $2 \times 3.14159 \times 100$ MHz $\sim 6.3 \times 10^8$ radians/sec.

The amplitude is the maximum value of the peak height above the center value. The wave peak goes below the horizontal just as much as it goes above. For an ideal sine wave, the DC or average value is always 0. This is an important observation. It means no combination of sine waves will produce a signal with other than an average of 0. To describe a signal with a DC value other than 0, we need to explicitly add the DC or offset value. This is usually stored in the 0 Hz frequency component.

The phase is more complicated and identifies where the wave is in its cycle at the beginning of the time axis. The units of phase are in cycles, radians, or degrees, with 360 degrees in one cycle. While phase is important in mathematical analysis, we will minimize the use of phase in most of our discussion to concentrate on the more important aspects of sine waves.

In the time domain, describing a sine wave requires plotting a lot of voltage-versus-time data points to draw the complete sine-wave curve. However, in the frequency domain, describing a sine wave is much simpler.

In the frequency domain, we already know that the only waveforms we can talk about are sine waves, so all we have to identify are the amplitude, frequency, and phase. If there is only one sine wave we are describing, all we need are these three values, and we have identified a complete description of the sine wave.

Since we are going to ignore phase for right now, we really only need two terms to completely describe a sine wave: its amplitude and its frequency. These two values are plotted with the frequency as one axis and the amplitude as the other axis, as shown in Figure 2-3. Of course, if we were including phase, we'd have a third axis.

A sine wave, plotted in the frequency domain, is just one single data point. This is the key reason we will go into the frequency domain. What might have been 1000 voltage-versus-time data points in the time domain is converted to a single amplitude-versus-frequency data point in the frequency domain.

When we have multiple frequency values, the collection of amplitudes is called the *spectrum*. As we will see, every time-domain waveform has a particular pattern to its spectrum. The only way to calculate the spectrum of a waveform in the time domain is with the Fourier transform.

2.5 The Fourier Transform

The starting place for using the frequency domain is being able to convert a waveform from the time domain into a waveform in the frequency domain. We do this with the Fourier transform. There are three types of Fourier transforms:

- Fourier integral (FI)
- Discrete Fourier transform (DFT)
- Fast Fourier transform (FFT)

The Fourier integral (FI) is a mathematical technique of transforming an ideal mathematical expression in the time domain into a description in the frequency domain. For example, if the entire waveform in the time domain were just a short pulse, and nothing else, the Fourier integral would be used to transform to the frequency domain.

This is done with an integral over all time from − infinity to + infinity. The result is a frequency-domain function that is also continuous from 0 to + infinity frequencies. There is a value for the amplitude at every continuous frequency value in this range.

For real-world waveforms, the time-domain waveform is actually composed of a series of discrete points, measured over a finite time, T. For example, a clock waveform may be a signal from 0 v to 1 v and have a period of 1 nsec and a repeat frequency of 1 GHz. To represent one cycle of the clock, there might be as many as 1000 discrete data points, taken at 1-psec intervals. Figure 2-4 shows an example of a 1-GHz clock wave in the time domain.

To transform this waveform into the frequency domain, the discrete Fourier transform (DFT) would be used. The basic assumption is that the original time-domain waveform is periodic and repeats every T seconds. Rather than integrals, just summations are used so any arbitrary set of data can be converted to the frequency domain using simple numerical techniques.

Finally, the fast Fourier transform (FFT) is exactly the same as a DFT except that the actual algorithm used to calculate the amplitude values at each frequency point uses a trick of very fast matrix algebra. This trick works only if the number of time-domain data points is a power of two—for example 256 points, or 512 points, or 1,024 points. The result is a DFT but calculated 100–10,000 times faster than the general DFT algorithm, depending on the number of voltage points.

In general, it is common in the industry to use all three terms, FI, DFT, and FFT, synonymously. We now know there is a difference between them, but they have the same purpose: to translate a time-domain waveform into its frequency-domain spectrum.

TIP   Once in the frequency domain, the description of a waveform is a collection of sine-wave frequency values. Each frequency component has an amplitude and a phase associated with it. We call the entire set of frequency values and their amplitudes the *spectrum* of the waveform.

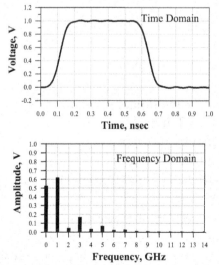

Figure 2-4   One cycle of a 1-GHz clock signal in the time domain (top) and frequency domain (bottom).

An example of a simple time-domain waveform and its associated spectrum, calculated by using a DFT, is shown in Figure 2-4.

At least once in his or her life, every serious engineer should calculate a Fourier integral by hand, just to see the details. After this, we never again need to do the calculation manually. We can always get to an answer faster by using one of the many commercially available software tools that calculate Fourier transforms for us.

There are a number of relatively easy-to-use, commercially available software tools that calculate the DFT or FFT of any waveform entered. Every version of SPICE has a function called the .FOUR command that will generate the amplitude of the first nine frequency components for any waveform. Most versions of the more advanced SPICE tools will also compute the complete set of amplitude and frequency values using a DFT. Microsoft Excel has an FFT function, usually found in the engineering add-ins.

2.6 The Spectrum of a Repetitive Signal

In practice, the DFT or FFT is used to translate a real waveform from the time domain to the frequency domain. It is possible to take a DFT of any arbitrary, measured waveform. A key requirement of the waveform is that it be repetitive. We usually designate the repeat frequency of the time-domain waveform with the capital letter F.

For example, an ideal square wave might go from 0 v to 1 v, with a repeat time of 1 nsec and a 50% duty cycle. As an ideal square wave, the rise time to transition from 0 v to 1 v is precisely 0 sec. The repeat frequency would be 1/1 nsec = 1 GHz.

If a signal in the time domain is some arbitrary waveform over a time interval from t = 0 to t = T, it may not look repetitive. However, it can be turned into a repetitive signal by just repeating the interval every T seconds. In this case, the repeat frequency would be F = 1/T. Any arbitrary waveform can be made repetitive and the DFT used to convert it to the frequency domain. This is illustrated in Figure 2-5.

When we turn a section of a waveform into a repetitive waveform, there is the possibility of an artifact. When the ends of the waveforms do not begin and end on the same value, the artificial transition when the sections are concatenated can create an artifact in the DFT. To avoid this problem, "windowing" filters are often used to guarantee that the voltages on each end have the same value. Filters that do this are the Hamming and Hanning filters, for example.

For a DFT, only certain frequency values exist in the spectrum. These values are determined by the choice of the time interval or the repeat frequency. When using an automated DFT tool, such as in SPICE, it is recommended to choose a value for the period equal to the clock period. This will simplify the interpretation of the results.

The only sine-wave frequency values that will exist in the spectrum will be multiples of the repeat frequency. If the clock frequency is 1 GHz, for example, the DFT will only have sine-wave components at 1 GHz, 2 GHz, 3 GHz, and so on.

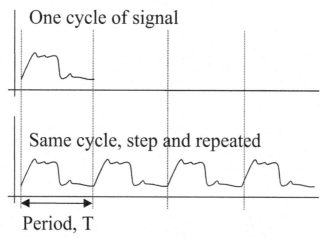

Figure 2-5 Any arbitrary waveform can be made to look repetitive. A DFT can be performed only on a repetitive waveform. When the voltages on the ends do not match, a windowing filter can be used to minimize artifacts in the spectrum.

The first sine-wave frequency is called the *first harmonic*. The second sine-wave frequency is called the *second harmonic*, and so on. Each harmonic will have a different amplitude and phase associated with it. The collection of all the harmonics and their amplitudes is called the *spectrum*.

The actual amplitudes of each harmonic will be determined by the values calculated by the DFT. Every specific waveform will have its own spectrum.

2.7 The Spectrum of an Ideal Square Wave

An ideal square wave has a zero rise time, by definition. It is not a real waveform; it is an approximation to the real world. However, useful insight can be gained by looking at the spectrum of an ideal square wave and using this to evaluate real waveforms later. An ideal square wave has a 50% duty cycle, is symmetrical, and has a peak voltage of 1 v. This is illustrated in Figure 2-6.

If the ideal square-wave repeat frequency is 1 GHz, the sine-wave frequency values in its spectrum will be multiples of 1 GHz. We expect to see components at f = 1 GHz, 2 GHz, 3 GHz, and so on. But what are the amplitudes of each sine wave? The only way to determine this is to perform a DFT on the ideal square wave. Luckily, it is possible to calculate the DFT exactly for this special case of an ideal square wave. The result is relatively simple.

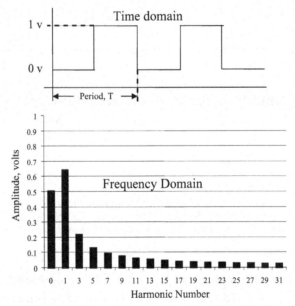

Figure 2-6   Time- and frequency-domain views of an ideal square wave.

The amplitudes of all the even harmonics (e.g., 2 GHz, 4 GHz, 6 GHz) are all zero. It is only odd harmonics that have nonzero values. This is a feature of any waveform in which the second half of the waveform is exactly the negative of the first half. We call these waveforms *anti-symmetric*, or *odd*, waveforms.

The amplitudes, A_n, of the odd harmonics are given by:

$$A_n = \frac{2}{\pi \times n}$$

(2.3)

where:

A_n = amplitude of the nth harmonic

π = constant, 3.14159...

n = harmonic number, only odd allowed

For example, an ideal square wave with 50% duty-cycle and 0 v to 1 v transition has a first harmonic amplitude of 0.63 v. The amplitude of the third harmonic is 0.21 v. We can even calculate the amplitude of the 1001st harmonic: 0.00063 v. It is important to note that the amplitudes of higher sine-wave-frequency components decrease with 1/f.

If the transition-voltage range of the ideal square wave were to double to 0 v to 2 v, the amplitudes of each harmonic would double as well.

There is one other special frequency value, 0 Hz. Since sine waves are all centered about zero, any combination of sine waves can only describe waveforms in the time domain that are centered about zero. To allow a DC offset, or a nonzero average value, the DC component is stored in the zero-frequency value. This is sometimes called the zeroth harmonic. Its amplitude is equal to the average value of the signal. In the case of the 50% duty-cycle square wave, the zeroth harmonic is 0.5 v.

When the square wave has a 0 v DC value, it transitions from –0.5 v to +0.5 v. Its amplitude is 0.5 v, and its peak to peak value is 1 v. In its spectrum, the zeroth harmonic is 0 v, since there is no DC value, and the first harmonic is 0.63 v. This is rather startling. The amplitude of the first harmonic component buried in the square wave is *larger* than the amplitude of the ideal square wave itself! This is an important property of the Fourier transform.

If you were to filter all the higher harmonics from an ideal square wave, the amplitude of the signal would be *larger* than the original signal! It is counterintuitive, but taking away frequency components leaves you with a larger signal. This has significance in signal processing and when analyzing signals received by an oscilloscope when the signals are near the bandwidth limit of the scope.

To summarize:

- The collection of sine-wave-frequency components and their amplitudes is called the *spectrum*. Each component is called a *harmonic*.
- The zeroth harmonic is the DC value.
- For the special case of a 50% duty-cycle ideal square wave, the even harmonics have an amplitude of zero.
- The amplitude of any harmonic can be calculated as $2/(\pi \times n)$.

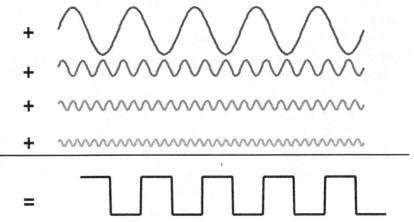

Figure 2-7   Convert the frequency-domain spectrum into the time-domain waveform by adding up each sine-wave component.

2.8 From the Frequency Domain to the Time Domain

The spectrum in the frequency domain represents all the sine-wave-frequency amplitudes of the time-domain waveform. If we have a spectrum and want to look at the time-domain waveform, we simply take each frequency component, convert it into its time-domain sine wave, and add it to all the rest. This process is called the inverse Fourier transform, and it is illustrated in Figure 2-7.

Each component in the frequency domain is a sine wave in the time domain, defined from $t = -$ infinity to $t = +$ infinity. To re-create the time-domain waveform, we take each of the sine waves described in the spectrum and add them up in the time domain at each time-interval point. We start at the low-frequency end and add each harmonic based on the spectrum.

For a 1-GHz ideal-square-wave spectrum, the first term in the frequency domain is the zeroth harmonic, with amplitude of 0.5 v. This component describes a constant DC value in the time domain.

The next component is the first harmonic, which is a sine wave in the time domain with a frequency of 1 GHz and an amplitude of 0.63 v. When this is added to the previous term, the result in the time domain is a sine wave, offset to 0.5 v. It is not a very good approximation to the ideal square wave. This is shown in Figure 2-8.

The next term is the third harmonic. The amplitude of the 3-GHz sine-wave-frequency component is 0.21 v. When we add this to the existing time-domain waveform, we see that it changes the shape of the new waveform slightly. The top is a bit flatter, better approximating a square wave, and the rise time is a little sharper. As we go through this process, adding each successive higher harmonic to re-create the ideal square wave, the resulting waveform begins to look more and more like a square wave. In particular, the rise time of the resulting time-domain waveform changes as we add higher harmonics.

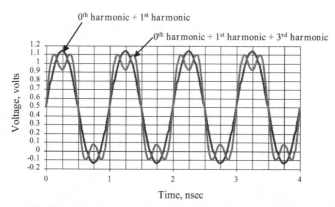

Figure 2-8 The time-domain waveform is created by adding together the zeroth harmonic and first harmonic and then the third harmonic, for a 1-GHz ideal square wave.

To illustrate this in more detail, we can zoom in on the rise time of the waveform, centered about the beginning of a cycle. As we add all the harmonics up to the seventh harmonic, and then all the way up to the nineteenth, and finally, all the way up to the thirty-first harmonic, we see that the rise time of the resulting waveform in the time domain continually gets shorter. This is shown in Figure 2-9.

Depending on how the DFT was set up, there could be more than 100 different harmonics listed in the spectrum. Logical questions come up at this point: Do we have to include all of them, or can we still re-create a "good enough" representation of the original time-domain waveform with just a limited number of harmonics? What really is the impact of limiting the highest harmonic included in the re-created time-domain waveform? Is there a highest sine-wave-frequency component at which we can stop?

2.9 Effect of Bandwidth on Rise Time

The term *bandwidth* is used for the highest sine-wave-frequency component that is significant in the spectrum. This is the highest sine-wave frequency we need to include to adequately approximate the important features of the time-domain waveform. All frequency components of higher frequency than the bandwidth can be ignored. In particular, as we will see, the bandwidth we choose will have a direct effect on the shortest rise time of the signal we are able to describe in the time domain.

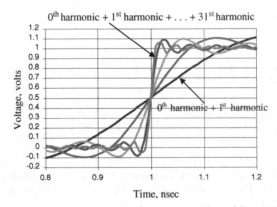

Figure 2-9 The time-domain waveform created by adding together the zeroth harmonic and first harmonic, then the third harmonic and then up to the seventh harmonic, then up to the nineteenth harmonic, and then all harmonics up to the thirty-first harmonic, for a 1-GHz ideal square wave.

The term *bandwidth* historically is used in the rf world to refer to the range of frequencies in a signal. In rf applications, a carrier frequency is typically modulated with some amplitude or phase pattern. The spectrum of frequency components in the signal falls within a band. The range of frequencies in the rf signal is called the *bandwidth*. Typical rf signals might have a carrier frequency of 1.8 GHz with a bandwidth about this frequency of 100 MHz. The bandwidth of an rf signal defines how dense different communications channels can fit.

With digital signals, bandwidth also refers to the range of frequencies in the signal's spectrum. It's just that for digital signals, the low-frequency range starts at DC and extends to the highest frequency component. In the world of digital signals, since the lowest frequency will always be DC, bandwidth will always be a measure of the highest sine-wave-frequency component that is significant.

This difference in the bandwidth of signals between rf and high-speed digital applications is one of the most important differences between these applications.

TIP   In rf products, it's important to engineer the interconnects to have controlled impedance properties in a relatively narrow frequency range. In high-speed digital applications, it's important to engineer the impedance of interconnects over a very wide frequency range. This generally is more difficult.

When we created a time-domain waveform from just the zeroth, the first, and the third harmonics included in Figure 2-8, the bandwidth of the resulting waveform was just up to the third harmonic, or 3 GHz in this case. By design, the highest sine-wave-frequency component in this waveform is 3 GHz. The amplitude of all other sine-wave components in this time-domain waveform is exactly 0.

When we added higher harmonics to create the waveforms in Figure 2-9, we designed their bandwidths to be 7 GHz, 19 GHz, and 31 GHz. If we were to take the shortest rise-time waveform in Figure 2-9 and transform it back into the frequency domain, its spectrum would look exactly like that shown in Figure 2-6. It would have components from the zeroth to the thirty-first harmonics. Beyond the thirty-first harmonic, all the components would be zero. The highest sine-wave-frequency component that is significant in this waveform is the thirty-first harmonic, or the waveform has a bandwidth of 31 GHz.

In each case, we created a waveform with a higher bandwidth, using the ideal-square-wave's spectrum as the starting place. And, in each case, the higher-bandwidth waveform also had a shorter 10–90 rise time. The higher the bandwidth, the shorter the rise time and the more closely the waveform approximates an ideal square wave. Likewise, if we do something to a short rise-time signal to decrease its bandwidth (i.e., eliminate high-frequency components), its rise time will increase.

For example, it is initially difficult to evaluate the time-domain response of a signal propagating down a lossy transmission line in FR4. As we will see, there are two loss mechanisms: conductor loss and dielectric loss. If each of these processes were to attenuate low-frequency components the same as they do high-frequency components, there would simply be less signal at the far end, but the pattern of the spectrum would look the same coming out as it does going in. There would be no impact on the rise time of the waveform.

However, both conductor loss and dielectric loss will attenuate the higher-frequency components more than the low-frequency components. By the time the signal has traveled through even 4 inches of trace, the high-frequency components, above about 8 GHz, can have lost more than 50% of their power, leaving the low-frequency terms less affected. In Figure 2-10 (top), we show the measured attenuation of sine-wave-frequency components through a 4-inch length of transmission line in FR4. This transmission line happens to have a 50-Ohm characteristic impedance and was measured with a network analyzer. Frequency components below 2 GHz are not attenuated more than −1 dB, while components at 10 GHz are attenuated by −4 dB.

This preferential attenuation of higher frequencies has the impact of decreasing the bandwidth of a signal that would propagate through the interconnect. Figure 2-10 (bottom) is an example of the measured rise time of a 50-psec signal entering a 36-inch-long trace in FR4 and this same waveform when it exits the trace. The rise time has been increased from 50 psec to nearly 1.5 nsec due to the higher attenuation of the high-frequency components. Thirty-six inches is a typical length for a trace that travels over two 6-inch-long daughter cards and 24 inches of backplane. This rise-time degradation is the chief limitation to the use of FR4 laminate in high-speed serial links above 1 GHz.

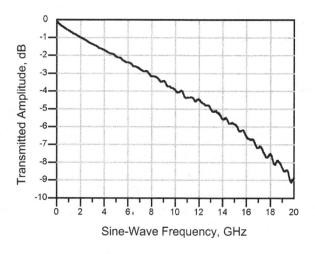

Sine-Wave Frequency, GHz

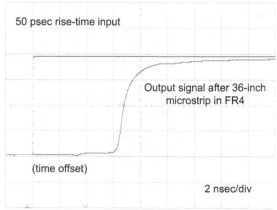

Figure 2-10   Top: The measured attenuation through a 4-inch length of 50-Ohm transmission line in FR4 showing the higher attenuation at higher frequencies. Bottom: The measured input and transmitted signal through a 36-inch 50-Ohm transmission line in FR4, showing the rise time to have degraded from 50 psec to more than 1.5 nsec.

TIP   In general, a shorter rise-time waveform in the time domain will have a higher bandwidth in the frequency domain. If something is done to the spectrum to decrease the bandwidth of a waveform, the rise time of the waveform will be increased.

The connection between the highest sine-wave-frequency component that is significant in a spectrum and the corresponding rise time of the waveform in the time domain is a very important property.

2.10 Bandwidth and Rise Time

The relationship between rise time and bandwidth for a re-created ideal square wave can be quantified. In each synthesized waveform in the previous example re-creating an ideal square wave, the bandwidth is explicitly known because each waveform was artificially created by including sine-wave-frequency components only up to a specified value. The rise time, defined as the time from the 10% point to the 90% point, can be measured from time-domain plots.

When we plot the measured 10–90 rise time and the known bandwidth for each waveform, we see that empirically there is a simple relationship. This is a fundamental relationship for all signals and is shown in Figure 2-11.

For the special case of a re-created square wave with only some of the higher harmonics included, the bandwidth is inversely related to the rise time. We can fit a straight-line approximation through the points and find the relationship between bandwidth and rise time as:

$$BW = \frac{0.35}{RT} \qquad\qquad (2\text{-}4)$$

where:

BW = bandwidth, in GHz

RT = 10–90 rise time, in nsec

For example, if the rise time of a signal is 1 nsec, the bandwidth is about 0.35 GHz or 350 MHz. Likewise, if the bandwidth of a signal is 3 GHz, the rise time of the signal will be about 0.1 nsec. A signal with a rise time of 0.25 nsec, such as might be seen in a DDR3-based system, has a bandwidth of 0.35/0.25 nsec = 1.4 GHz.

There are other ways of deriving this relationship for other waveforms, such as with Gaussian or exponential edges. The approach we took here for square waves is purely empirical and makes no assumptions. It is one of the most useful rules of thumb in our toolbox.

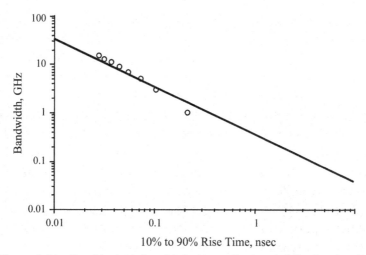

Figure 2-11 Empirical relationship between the bandwidth of a signal and its 10–90 rise time, as measured from a re-created ideal square wave with each harmonic added one at a time. Circles are the values extracted from the data; line is the approximation of BW = 0.35/rise time.

It is important to keep the units straight. When rise time is in microseconds, the bandwidth is in MHz. For example, a very long rise time of 10 microsec has a bandwidth of about 0.35/10 microsec = 0.035 MHz. This is equivalent to 35 kHz.

When the rise time is in nanoseconds, the bandwidth is in GHz. A 10-nsec rise time, typical of a 10-MHz clock frequency, has a bandwidth of about 0.35/10 nsec = 0.035 GHz or 35 MHz.

2.11 What Does *Significant* Mean?

We defined the bandwidth of a signal as the highest sine-wave-frequency component that is significant. In the example where we started with an ideal square wave and limited the high-frequency components, there was absolutely no ambiguity about what *significant* meant. We explicitly cut off all higher frequency sine-wave components in the frequency domain so that the highest significant component was the last harmonic in the spectrum.

We simply showed that if we include 100% of all the frequency components of an ideal square wave, up to the bandwidth, we would be able to re-create a square wave with a limited rise time, where the relationship of rise time = 0.35/BW. But what is the impact from adding only a fraction of the next component?

For example, if we take an ideal-square-wave clock signal with clock frequency of 1 GHz, its first harmonic will be a 1-GHz sine-wave frequency. If we were to include 100% of every component up to the twenty-first harmonic, the bandwidth would be 21 GHz and the resulting rise time of the re-created signal would be 0.35/21 GHz = 0.0167 nsec or 16.7 psec.

How would the rise time change if we added the twenty-third harmonic? The rise time would be 0.35/23 GHz = 0.0152 nsec, or 15.2 psec. That is the rise time would drop by 1.5 psec. This is about 10% of the rise time, which is consistent, because we increased the bandwidth by 10%. The magnitude of the component we added was just 0.028 v, compared with the first harmonic of 0.63 v. Even though this amplitude is a small amount, less than 5% of the first harmonic amplitude and less than 3% of the peak value of the original square wave, it had the impact of dropping the rise time by 10%.

The spectrum of an ideal square wave has components that extend to infinite frequency. In order to achieve the zero rise time of an ideal square wave, each of these components is needed and is significant.

For a real time-domain waveform, the spectral components will almost always drop off in frequency faster than those of an ideal square wave of the same repeat frequency. The question of significance is really about the frequency at which amplitudes of the higher harmonics become small compared to the corresponding amplitudes of an ideal square wave.

By "small," we usually mean when the power in the component is less than 50% of the power in an ideal square wave's amplitude. A drop of 50% in power is the same as a drop to 70% in amplitude. This is really the definition of significant. Significant is when the amplitude is still above 70% of an ideal square wave's amplitude of the same harmonic.

> **TIP**   For any real waveform that has a finite rise time, *significant* refers to the point at which its harmonics are still more than 70% of the amplitude of an equivalent repeat-frequency ideal square wave's.

In a slightly different view, we can define *significant* as the frequency at which the harmonic components of the real waveform begin to drop off faster than 1/f. The frequency at which this happens is sometimes referred to as the *knee frequency*. The harmonic amplitudes of an ideal square wave will initially drop off similarly to 1/f. The frequency at which the harmonic amplitudes of a real waveform begin to significantly deviate from an ideal square wave is the knee frequency.

To evaluate the bandwidth of a time-domain waveform, we are really asking what is the highest-frequency component that is just barely above 70% of the same harmonic of an equivalent ideal square wave. When the harmonic amplitudes of the real waveform are significantly less than an ideal square wave's, these lower-amplitude harmonics will not contribute significantly to decreasing the rise time, and we can ignore them.

For example, we can compare, in the time-domain waveform, two clock waves with a repeat frequency of 1 GHz: an ideal square wave and an ideal trapezoidal waveform, which is a non-ideal square wave with a long rise time. In this example, the 10–90 rise time is about 0.08 nsec, which is a rise time of about 8% of the period, typical of many clock waveforms. These two waveforms are shown in Figure 2-12.

If we compare the frequency components of these waveforms, at what frequency will the trapezoid's spectrum start to differ significantly from the ideal square wave's? We would expect the trapezoid's higher frequency components to begin to become insignificant at about 0.35/0.08 nsec = about 5 GHz. This is the fifth harmonic. After all, we could create a non-ideal square wave with this rise time if we were to take the ideal-square-wave spectrum and drop all components above the fifth harmonic, as we saw earlier.

When we look at the actual spectrum of the trapezoid compared to the square wave, we see that the first and third harmonics are about the same for each. The trapezoid's fifth harmonic is about 70% of the square wave's, which is still a large fraction. However, the trapezoid's seventh harmonic is only about 30% of the ideal square wave's. This is illustrated in Figure 2-12.

We would conclude, by simply looking at the spectrum of the trapezoid, that harmonics above the fifth harmonic (i.e., the seventh and beyond) are contributing only a very small fraction of the amount of voltage as in the ideal square wave. Thus, their ability to further affect the rise time is going to be minimal. From the spectrum, we would say that the highest sine-wave-frequency component that is significant in the trapezoid, *compared to that in the ideal square wave*, is the fifth harmonic, which is what our approximation gave us.

There are higher harmonics in the trapezoid's spectrum than the fifth harmonic. However, the largest amplitude is 30% of the square wave's and then only a few percent after this. Their magnitude is such a small fraction of the amplitude of the ideal square wave's that they will contribute very little to the decrease of the rise time and can be ignored.

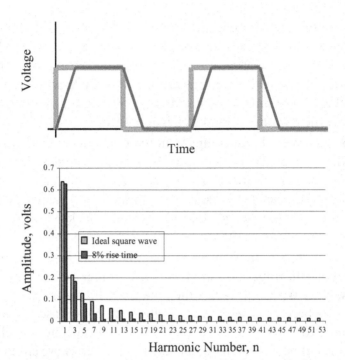

Figure 2-12 Top: Time-domain waveforms of 1-GHz repeat frequency: an ideal square wave and an ideal trapezoidal wave with 0.08-nsec rise time. Bottom: Frequency-domain spectra of these waveforms showing the drop-off of the trapezoidal wave's higher harmonics, compared to the square wave's.

The bandwidth of any waveform is always the highest sine-wave-frequency component in its spectrum that is comparable in magnitude to a corresponding ideal square wave. We can find out the bandwidth of any waveform by using a DFT to calculate its spectrum and compare it to an ideal square wave. We identify the frequency component of the waveform that is less than 70% of the ideal square wave, or we can use the rule of thumb developed earlier, that the BW is 0.35/rise time.

> **TIP**   It is important to note that this concept of bandwidth is inherently an approximation. It is really a rule of thumb, identifying roughly where the amplitude of frequency components in a real waveform begin to drop off faster than in an ideal square wave.

If you have a problem where it is important to know whether the bandwidth of a waveform is 900 MHz or 950 MHz, you should not use this value of *bandwidth*. Rather, you should use the *whole spectrum*. The entire spectrum is always an accurate representation of the time-domain waveform.

2.12 Bandwidth of Real Signals

Other than the approximation for the bandwidth of a waveform based on its rise time, there is little calculation we can do by hand. Fourier transforms of arbitrary waveforms can only be done using numerical simulation.

For example, the spectrum of a good-quality, nearly square wave signal has a simple behavior. If a transmission line circuit is poorly terminated, the signal may develop ringing. The resulting spectrum will have peaks at the ringing frequency. The amplitudes of the ringing frequency can be more than a factor of 10 greater than the amplitudes of the signal without ringing. This is shown in Figure 2-13.

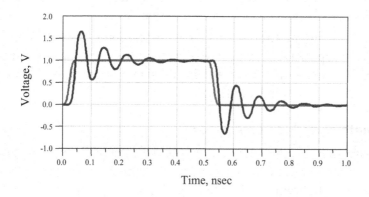

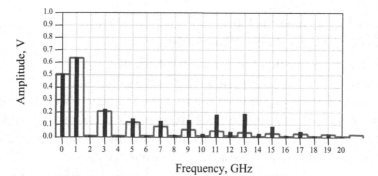

Figure 2-13 Top: The time-domain waveform of a nearly square wave and one that has significant ringing due to poor termination. Bottom: The resulting DFT spectrum of these two waves, showing the effect of the ringing on the spectrum. The wide bars are for the ideal waveform, while the narrow bars are for the ringing waveform.

The bandwidth of a waveform with ringing is clearly higher than one without. When ringing is present in a waveform, the bandwidth is better approximated by the ringing frequency. Just using the bandwidth to characterize a ringing signal, though, may be misleading. Rather, the whole spectrum needs to be considered.

EMI arises from each frequency component of the currents radiating. For the worst offender, the common currents, the amount of radiated emissions will increase linearly with the frequency. This means that if the current had an ideal-square-wave behavior, though the amplitude of each harmonic drops off at a rate of 1/f, the ability to radiate would increase at the rate of f, so all harmonics contribute equally to EMI. To minimize EMI, the design goal is to use absolutely the lowest bandwidth possible in all signals. Above the bandwidth, the harmonic amplitudes drop off faster than 1/f and would contribute to less radiated emissions. By keeping the bandwidth low, the radiated emissions will be kept to a minimum.

Any ringing in the circuits may increase the amplitudes of higher-frequency components and increase the magnitude of radiated emissions by a factor of 10. This is one reason solving all signal-integrity problems is usually a starting place in minimizing EMI problems.

2.13 Bandwidth and Clock Frequency

As we have seen, bandwidth relates to the rise time of a signal. It is possible to have two different waveforms with exactly the same clock frequency but different rise times and different bandwidths. Just knowing the clock frequency cannot tell us what the bandwidth is. Figure 2-14 shows four different waveforms, each with exactly the same clock frequency of 1 GHz. However, they have different rise times and hence different bandwidths.

Sometimes, we don't know the rise time of a signal but need an idea of its bandwidth anyway. Using a simplifying assumption, we can estimate the bandwidth of a clock wave from just its clock frequency. Still, it is important to keep in mind that it is not the clock frequency that determines the bandwidth but the rise time. If all we know about the waveform is the clock frequency, we can't know the bandwidth for sure; we can only guess.

To evaluate the bandwidth of a signal from just its clock frequency, we have to make a very important assumption: We need to estimate what a typical rise time might be for a clock wave.

How is the rise time related to the clock period in a real clock waveform? In principle, the only relationship is that the rise time must be less than 50%

of the period. Other than this, there is no restriction, and the rise time can be any arbitrary fraction of the period. It could be 25% of the period, as in cases where the clock frequency is pushing the limits of the device technology, such as in 1-GHz clocks. It could be 10% of the period, which is typical of many microprocessor-based products. It could be 5% of the period, which is found in high-end FPGAs driving external low-clock-frequency memory buses. It could even be 1% if the board-level bus is a legacy system.

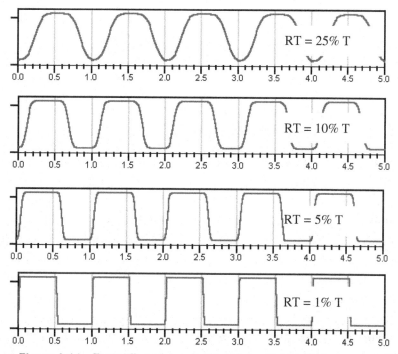

Figure 2-14   Four different waveforms, each with exactly the same 1-GHz clock frequency. Each of them has a different rise time, as a fraction of the period, and hence they have different bandwidths.

If we don't know what fraction of the period the rise time is, a reasonable generalization is that the rise time is 7% of the clock period. This approximates many typical microprocessor-based boards and ASICs driving board-level buses. From this, we can estimate the bandwidth of the clock waveform.

It should be kept in mind that this assumption of the rise time being 7% of the period is a bit aggressive. Most systems are probably closer to 10%, so we are assuming a rise time slightly shorter than might typically be found. Likewise, if we are underestimating the rise time, we will be overestimating the bandwidth, which is safer than underestimating it.

If the rise time is 7% of the period, then the period is 1/0.07, or 15 times the rise time. We have an approximation for the bandwidth as 0.35/rise time. We can relate the clock frequency to the clock period because they are each the inverse of the other. Replacing the clock period for the clock frequency results in the final relationship; the bandwidth is five times the clock frequency:

$$BW_{clock} = 5 \times F_{clock} \qquad\qquad (2\text{-}5)$$

where:

BW_{clock} = approximate bandwidth of the clock, in GHz

F_{clock} = clock repeat frequency, in GHz

For example, if the clock frequency is 100 MHz, the bandwidth of the signal is about 500 MHz. If the clock frequency is 1 GHz, the bandwidth of the signal is about 5 GHz.

This is a generalization and an approximation, based on the assumption that the rise time is 7% of the clock period. Given this assumption, it is a very powerful rule of thumb, which can give an estimate of bandwidth with very little effort. It says that the highest sine-wave-frequency component in a clock wave is typically the fifth harmonic!

It's obvious but bears repeating that we always want to use the rise time to evaluate the bandwidth. Unfortunately, we do not always have the luxury of knowing the rise time for a waveform. And yet, we need an answer *now*!

TIP   Sometimes getting an OKAY answer NOW is more important than getting a BETTER answer LATER.

2.14 Bandwidth of a Measurement

So far, we have been using the term *bandwidth* to refer to signals, or clock waveforms. We have said that the bandwidth is the highest significant sine-wave-frequency component in the waveform's spectrum. And, for signals, we said *significant* is based on comparing the amplitude of the signal's harmonic to the amplitude of an equivalent repeat frequency ideal square wave's.

We also use this term *bandwidth* to refer to other quantities. In particular, it can relate to the bandwidth of a measurement, the bandwidth of a model, and the bandwidth of an interconnect. In each case, it refers to the highest sine-wave-frequency component that is significant, but the definition of *significant* varies per application.

The bandwidth of a measurement is the highest sine-wave-frequency component that has significant accuracy. When the measurement is done in the frequency domain, using an impedance analyzer or a network analyzer, the bandwidth of the measurement is very easy to determine. It is simply the highest sine-wave frequency in the measurement.

The measured impedance of a decoupling capacitor, from 1 MHz up to 1 GHz, shows that below about 10 MHz, the impedance behaves like an ideal capacitor, but above 10 MHz, it looks like an ideal inductor. Such a measurement is shown in Figure 2-15. There is good, accurate data up to the full range of the network analyzer—in this case, up to 1 GHz. The bandwidth of the measurement is 1 GHz in this example. The measurement bandwidth is not the same as the useful application bandwidth of the device.

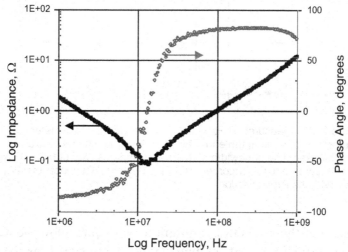

Figure 2-15 Measured impedance of a small 1,206 ceramic decoupling capacitor. The measurement bandwidth for this data is 1 GHz.

When the measuring instrument works in the time domain, such as a time-domain reflectometer (TDR), the bandwidth of the measurement can be found by the rise time of the fastest signal that can be launched into the DUT. After all, this is a rough measure of when the higher-frequency components are small.

In a typical TDR, a fast step edge is created and its change due to interaction with the DUT is measured. A typical rise time entering the DUT is 35 psec to 70 psec, depending on the probes and cables used. Figure 2-16 shows the measured rise time of a TDR as about 52 psec. The bandwidth of the edge is 0.35/52 psec = 0.007 THz or 7 GHz. This is the bandwidth of the signal coming out of the TDR and is a good first order measure of the bandwidth of the measurement.

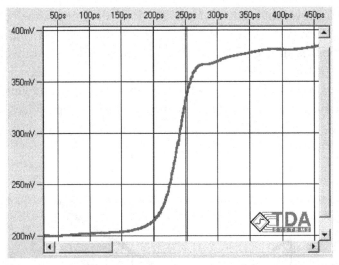

Figure 2-16   Measured TDR profile from the output of a 1-meter cable and microprobe tip, open at the end. The TDR rise time after the cable and probe is about 52 psec. The bandwidth of the measurement is about 0.35/ 52 psec = 7 GHz. The measurement was recorded with TDA Systems IConnect software, using a GigaTest Labs Probe Station.

In state-of-the-art TDRs, calibration techniques allow the bandwidth of the measurement to exceed the bandwidth of the signal, as defined by our simple rule of thumb. The bandwidth of the measurement is set by when the signal-to-noise ratio of a frequency component is below a reasonable value, like

10, not based on when the amplitude of a harmonic is blow 70% of the ideal square wave's. The bandwidth of the measurement of some TDRs can exceed the signal's bandwidth by a factor of three to five, making the bandwidth of a TDR's measurement as high as 30 GHz.

2.15 Bandwidth of a Model

> **T I P**   When we refer to the *bandwidth of a model*, we are referring to the highest sine-wave-frequency component where the model will accurately predict the actual behavior of the structure it is representing. There are a few tricks that can be used to determine this, but in general, only a comparison to a measurement will give a confident measure of a model's bandwidth.

The simplest starting equivalent circuit model to represent a wire bond is an inductor. Up to what bandwidth might this be a good model? The only way to really tell is to compare a measurement with the prediction of this model. Of course, it will be different for different wire bonds.

As an example, we take the case of a very long wire bond, 300 mils long, connecting two pads over a return-path plane 10 mils below. This is diagrammed in Figure 2-17. A simple starting circuit model is a single ideal inductor and ideal resistor in series, as shown in Figure 2-18. The best values for the L and R give a prediction for the impedance that closely matches the measured impedance up to 2 GHz. The bandwidth of this simple model is 2 GHz. This is shown in Figure 2-18.

We could confidently use this simple model to predict performance of this physical structure in applications that had signal bandwidths of 2 GHz. It is surprising that for a wire bond this long, the simplest model—that of a constant ideal inductor and resistor—works so well up to 2 GHz. This is probably higher than the useful bandwidth of the wire bond, but the model is still accurate up to a frequency this high.

Suppose we wanted a model with an even higher bandwidth that would predict the actual impedance of this real wire bond to higher frequency. We might add the effect of the pad capacitance. Building a new model, a second-order model, and finding the best values for the ideal R, L, and C elements result in a simulated impedance that matches the actual impedance to almost 4 GHz. This is shown in Figure 2-18.

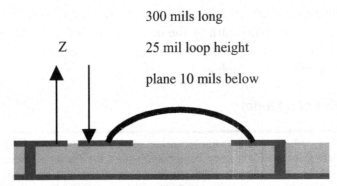

300 mils long

Z   25 mil loop height

plane 10 mils below

Figure 2-17   Diagram of a wire-bond loop between two pads, with a return path about 10 mils beneath the wire bond.

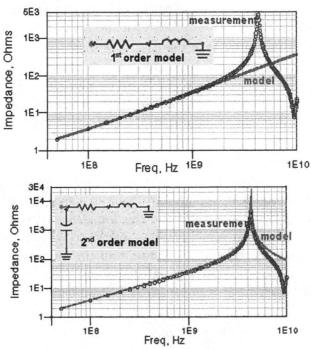

Figure 2-18   Top: Comparison of the measured impedance and the simulation based on the first-order model. The agreement is good up to a bandwidth of about 2 GHz. Bottom: Comparison of the measured impedance and the simulation based on the second-order model. The agreement is good up to a bandwidth of about 4 GHz. The bandwidth of the measurement is 10 GHz, measured with a GigaTest Labs Probe Station.

2.16 Bandwidth of an Interconnect

The *bandwidth of an interconnect* refers to the highest sine-wave-frequency component that can be transmitted by the interconnect without significant loss. What does *significant* mean? In some applications, a transmitted signal that is within 95% of the incident signal is considered too small to be useful. In other cases, a transmitted signal that is less than 10% of the incident signal is considered usable. In long-distance cable-TV systems, the receivers can use signals that have only 1% of the original power. Obviously, the notion of how much transmitted signal is significant is very dependent on the application and the particular specification. In reality, the bandwidth of an interconnect is the highest sine-wave frequency at which the interconnect still meets the performance specification for the application.

TIP   In practice, *significant* means the transmitted frequency-component amplitude is reduced by –3 dB, which means that its amplitude is reduced to 70% of the incident value. This is often referred to as the 3-dB bandwidth of an interconnect.

The bandwidth of an interconnect can be measured in either the time domain or the frequency domain. In general, we have to be careful interpreting the results if the source impedance is different than the characteristic impedance of the line, due to the complication of multiple reflections.

Measuring the bandwidth of an interconnect in the frequency domain is very straightforward. A network analyzer is used to generate sine waves of various frequencies. It injects the sine waves in the front of the interconnect and measures how much of each sine wave comes out at the far end. It is basically measuring the transfer function of the interconnect, and the interconnect is acting like a filter. This is also sometimes referred to as the *insertion loss* of the interconnect. The interpretation is simple when the interconnect is 50 Ohms, matched to the network analyzer's impedance.

For example, Figure 2-19 shows the measured transmitted amplitude of sine waves through a 4-inch length of a 50-Ohm transmission line in FR4. The measurement bandwidth is 20 GHz in this case. The 3-dB bandwidth of the interconnect is seen to be about 8 GHz. This means that if we send in a sine wave at 8 GHz, at least 70% of the amplitude of the 8-GHz sine wave would appear at the far end. More than likely, if the interconnect bandwidth were 8 GHz, nearly 100% of a 1-GHz sine wave would be transmitted to the far end of the same interconnect.

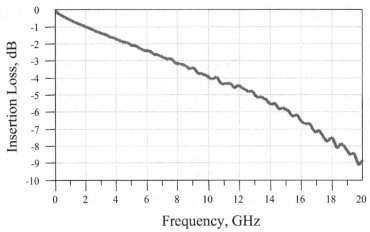

Figure 2-19 Measured transmitted amplitude of different sine-wave signals through a 4-inch-long transmission line made in FR4. The 3 dB bandwidth is seen to be about 8 GHz for this cross section and material properties. Measured with a GigaTest Labs Probe Station.

The interpretation of the bandwidth of an interconnect is the approximation that if an ideal square wave were transmitted through this interconnect, each sine-wave component would be transmitted, with those components lower than 8 GHz having roughly the same amplitude coming out as they did going in. But the amplitude of those components above 8 GHz would be reduced to insignificance.

A signal that might have a rise time of 1 psec going into the interconnect would have a rise time of 0.35/8 GHz = 0.043 nsec, or 43 psec when it came out. The interconnect will degrade the rise time.

TIP   The bandwidth of the interconnect is a direct measure of the minimum rise-time signal an interconnect can transmit.

If the bandwidth of an interconnect is 1 GHz, the fastest edge it can transmit is 350 psec. This is sometimes referred to as its *intrinsic* rise time. If a signal with a 350-psec edge enters the interconnect, what will be the rise time coming out? This is a subtle question. The rise time exiting the interconnect can be approximated by:

$$RT_{out}^2 = RT_{in}^2 + RT_{interconnect}^2 \qquad (2\text{-}6)$$

where:

RT_{out} = 10–90 rise time of the output signal
RT_{in} = 10–90 rise time of the input signal
$RT_{interconnect}$ = intrinsic 10–90 rise time of the interconnect

This assumes that both the incident spectra and the response of the interconnect correspond to a Gaussian-shaped rise time.

For example, in the case of this 4-inch-long interconnect, if a signal with a rise time of 50 psec were input, the rise time of the transmitted signal would be:

$$\text{Sqrt}\left(50 \text{ psec}^2 + 43 \text{ psec}^2\right) = 67 \text{ psec} \qquad (2\text{-}7)$$

This is an increase of about 17 psec in the rise time of the transmitted waveform compared to the incident rise time.

In Figure 2-20, we show the measured time-domain response of the same 4-inch-long, 50-Ohm interconnect that was measured in the frequency domain above. The input waveform has been time shifted to lie directly at the start of the measured output waveform.

The rise time of the waveform going into the PCB trace is 50 psec. The measured 10–90 rise time of the output waveform is about 80 psec. However, this is somewhat distorted by the long roll to stabilize at the top, characteristic of the behavior of lossy lines. The extra delay at the 70% point is about 15 psec, which is very close to what our approximation above predicted.

If a 1-nsec rise-time signal enters an interconnect with an intrinsic rise time of 0.1 nsec, the rise time of the signal transmitted would be about sqrt(1 nsec² + 0.1 nsec²), or 1.005 nsec, which is still basically 1 nsec. The interconnect would not affect the rise time. However, if the interconnect intrinsic rise time were 0.5 nsec, the output rise time would be 1.1 nsec, and would start to have a significant impact.

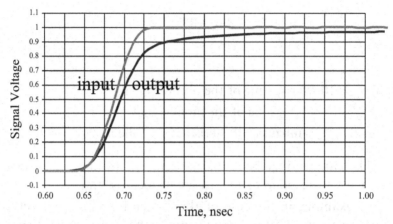

Figure 2-20 Measured input and transmitted signal through a 4-inch-long, 50-Ohm transmission line in FR4 showing the rise-time degradation. The input rise time is 50 psec. The predicted output rise time is 67 psec, based on the measured bandwidth of the interconnect. Measured with a GigaTest Labs Probe Station.

TIP   As a simple rule of thumb, in order for the rise time of the signal to be increased by the interconnect less than 10%, the intrinsic rise time of the interconnect should be shorter than 50% of the rise time of the signal.

TIP   In the frequency-domain perspective, to support the transmission of a 1-GHz bandwidth signal, we want the bandwidth of the interconnect to be at least twice as high, or 2 GHz.

It is important to keep in mind that this is a rule of thumb, and it should not be used for design sign-off. It should be used only for a rough estimate or to identify a goal. If the bandwidth of an interconnect is within a factor of two of the bandwidth of the signal, it would probably be important to perform an analysis of how the interconnect affected the entire signal's spectrum.

2.17 The Bottom Line

1. The time domain is the real world and is typically where high-speed digital performance is measured.

2. The frequency domain is a mathematical construct where very specific, specialized rules apply.

3. The only reason to ever leave the time domain and use the frequency domain is to get to an answer faster.

4. The rise time of a digital signal is commonly measured from 10% of the final value to 90% of the final value.

5. Sine waves are the only waveforms that can exist in the frequency domain.

6. The Fourier transform converts a time-domain waveform into its spectrum of sine-wave-frequency components.

7. The spectrum of an ideal square wave has amplitudes that drop off at a rate of 1/f.

8. If the higher-frequency components are removed in the square wave, the rise time will increase.

9. The bandwidth of a signal is the highest sine-wave-frequency component that is significant, compared to the same harmonics in an ideal square wave with the same repeat frequency.

10. A good rule of thumb is that the bandwidth of a signal is 0.35/rise time of the signal.

11. Anything that decreases the bandwidth of a signal will increase its rise time.

12. The bandwidth of a measurement is the highest sine-wave frequency where the measurement has good accuracy.

13. The bandwidth of a model is the highest sine-wave frequency where the predictions of the model give good agreement with the actual performance of the interconnect.

14. The bandwidth of an interconnect is the highest sine-wave frequency where the performance of the interconnect still meets specifications.

15. The 3-dB bandwidth of an interconnect is the highest sine-wave frequency where the attenuation of a signal is less than −3 dB.

End-of-Chapter Review Questions

The answers to the following review questions can be found in Appendix D, "Review Questions and Answers."

2.1 What is the difference between the time domain and the frequency domain?

2.2 What is the special feature of the frequency domain, and why it is so important for signal analysis on interconnects?

2.3 What is the only reason you would ever leave the real world of the time domain to go into the frequency domain?

2.4 What feature is required in a signal for the even harmonics to be nearly zero?

2.5 What is bandwidth? Why is it only an approximate term?

2.6 In order to perform a DFT, what is the most important property the signal has to have?

2.7 Why is designing interconnects for high-speed digital applications more difficult than designing interconnects for rf applications?

2.8 What feature in a signal changes if its bandwidth is decreased?

2.9 If there is −10 dB attenuation in an interconnect, but it is flat with frequency, what will happen to the rise time of the signal as it propagates through the interconnect?

2.10 When describing the bandwidth as the highest significant frequency component, what does the word significant mean?

2.11 Some published rules of thumb suggest the bandwidth of a signal is actually 0.5/RT. Is it 0.35/RT or 0.5/RT?

2.12 What is meant by the bandwidth of a measurement?

2.13 What is meant by the bandwidth of a model?

2.14 What is meant by the bandwidth of an interconnect?

2.15 When measuring the bandwidth of an interconnect, why should the source impedance and the receiver impedance be matched to the characteristic impedance of the interconnect?

2.16 If a higher bandwidth scope will distort the signal less than a lower bandwidth scope, why shouldn't you just buy a scope with a bandwidth 20 times the signal bandwidth?

2.17 In high-speed serial links, the −10 dB interconnect bandwidth is the frequency where the first harmonic has −10 dB attenuation. How much attenuation will the third harmonic have? How small an amplitude is this?

2.18 What is the potential danger of using a model with a bandwidth lower than the signal bandwidth?

2.19 To measure an interconnect's model bandwidth using a VNA, what should the bandwidth of the VNA instrument be?

2.20 The clock frequency is 2.5 GHz. What is the period? What would you estimate the 10–90 rise time to be?

2.21 A repetitive signal has a period of 500 MHz. What are the frequencies of the first three harmonics?

2.22 An ideal 50% duty cycle square wave has a peak-to-peak value of 1 V. What is the peak-to-peak value of the first harmonic? What stands out as startling about this result?

2.23 In the spectrum of an ideal square wave, the amplitude of the first harmonic is 0.63 times the peak-to-peak value of the square wave. What harmonic has an amplitude 3 dB lower than the first harmonic?

2.24 What is the rise time of an ideal square wave? What is the amplitude of the 1,001st harmonic compared to the first harmonic? If it is so small, do you really need to include it?

2.25 The 10–90 rise time of a signal is 1 nsec. What is its bandwidth? If the 20–80 rise time was 1 nsec, would this increase, decrease, or have no effect on the signal's bandwidth?

2.26 A signal has a clock frequency of 3 GHz. Without knowing the rise time of the signal, what would be your estimate of its bandwidth? What is the underlying assumption in your estimate?

2.27 A signal rise time is 100 psec. What is the minimum bandwidth scope you should use to measure it?

2.28 An interconnect's bandwidth is 5 GHz. What is the shortest rise time you would ever expect to see coming out of this interconnect?

2.29 A clock signal is 2.5 GHz. What is the lowest bandwidth scope you need to use to measure it? What is the lowest bandwidth interconnect you could use to transmit it and what is the lowest bandwidth model you should use for the interconnects to simulate it?

CHAPTER 3

Impedance and Electrical Models

In high-speed digital systems, where signal integrity plays a significant role, we often refer to signals as either changing voltages or changing currents. All the effects that we lump in the general category of signal integrity are due to how analog signals (those changing voltages and currents) interact with the electrical properties of the interconnects. The key electrical property with which signals interact is the impedance of the interconnects.

Impedance is defined as the ratio of the voltage to the current. We usually use the letter Z to represent impedance. The definition, which is *always* true, is $Z = V/I$. The manner in which these fundamental quantities, voltage, and current, interact with the impedance of the interconnects determines all signal-integrity effects. As a signal propagates down an interconnect, it is constantly probing the impedance of the interconnect and reacting based on the answer.

TIP   If we know the impedance of an interconnect, we can accurately predict how the signals will be distorted and whether a design will meet the performance specification before we build it.

Likewise, if we have a target spec for performance and know what the signals will be, we can sometimes specify an impedance requirement for the interconnects. If we understand how the geometry and material properties affect the impedance of the interconnects, then we will be able to design the cross section, the topology, and the materials and select the other components so they will meet the impedance spec and result in a product that works the first time.

TIP   Impedance is the key term that describes every important electrical property of an interconnect. Knowing the impedance and propagation delay of an interconnect means knowing almost everything about it electrically.

3.1 Describing Signal-Integrity Solutions in Terms of Impedance

Each of the six basic families of signal-integrity problems can be described based on impedance:

1. Signal-quality problems arise because voltage signals reflect and are distorted whenever the impedance the signal sees changes. If the impedance the signal sees is always constant, there will be no reflection, and the signal will continue undistorted. Attenuation effects are due to series and shunt-resistive impedances.
2. Cross talk arises from the electric and magnetic fields coupling between two adjacent signal traces (and, of course, their return paths). The mutual capacitance and mutual inductance between the traces establishes an impedance, which determines the amount of coupled current and voltage.
3. Ground bounce arises because of the higher mutual inductance between signal lines. While the mutual capacitance between signal lines increases, the mutual inductance dramatically increases when the return path is screwed up.
4. Rise time is degraded due to the impedance of the series resistance, which is frequency dependent, and the shunt conductance, which is frequency dependent. These loss mechanisms increase attenuation with frequency and cause the rise time to increase.
5. Rail collapse of the voltage supply is really about the impedance in the power distribution network (PDN). A certain amount of current must flow to feed all the ICs in the system. Because of the impedance of

the power and ground distribution, a voltage drop will occur as the IC current switches. This voltage drop means the power and ground rails have collapsed from their nominal values.

6. The greatest source of EMI is common currents, driven by voltages in the ground planes, through external cables. The higher the impedance of the return current paths in the ground planes, the greater the voltage drop, or ground bounce, that will drive the radiating currents. The most common fix for EMI from cables is the use of a ferrite choke around the cable. This works by increasing the impedance the common currents see, thereby reducing the amount of common current.

There are a number of design rules, or guidelines, that establish constraints on the physical features of the interconnects. For example, "keep the spacing between adjacent signal traces greater than 10 mils" is a design rule to minimize cross talk. "Use power and ground planes on adjacent layers separated by less than 5 mils" is a design rule for the power and ground distribution. Such guidelines are designed to engineer the impedance of interconnects.

TIP   Not only are the problems associated with signal integrity best described by the use of impedance, but the solutions and the design methodology for good signal integrity are also based on the use of impedance.

These rules establish a specific impedance for the physical interconnects. This impedance provides a specific environment for the signals, resulting in a desired performance. For example, keeping the power and ground planes closely spaced will result in a low impedance for the power distribution system and hence a lower voltage drop for a given power and ground current. This helps minimize rail collapse and EMI.

If we understand how the physical design of the interconnects affects their impedance, we will be able to interpret how they will interact with signals and what performance they might have.

TIP   Impedance is the Rosetta stone that links physical design and electrical performance. Our strategy is to translate system-performance needs into an impedance requirement and physical design into an impedance property.

Impedance is at the heart of the methodology we will use to solve signal-integrity problems. Once we have designed the physical system as we think it should be for optimal performance, we will translate the physical structure into its equivalent electrical circuit model. This process is called *modeling*.

It is the impedance of the resulting circuit model that determines how the interconnects will affect the voltage and current signals. Once we have the circuit model, we will use a circuit simulator, such as SPICE, to predict the new waveforms as the voltage sources are affected by the impedances of the interconnects. Alternatively, behavioral models of the drivers or interconnects can be used where the interaction of the signals with the impedance, described by the behavioral model, will predict performance. This process is called *simulation*.

Finally, the predicted waveforms will be analyzed to determine if they meet the timing and distortion or noise specs and are acceptable or if the physical design has to be modified. This process flow for a new design is illustrated in Figure 3-1.

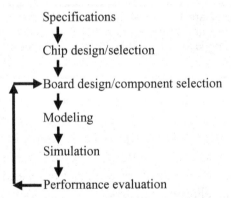

Figure 3-1 Process flow for hardware design. The modeling, simulation, and evaluation steps should be implemented as early and often in the design cycle as possible.

The two key processes, modeling and simulation, are based on converting electrical properties into an impedance and analyzing the impact of the impedance on the signals.

If we understand the impedance of each of the circuit elements used in a schematic and how the impedance is calculated for a combination of circuit elements, the electrical behavior of *any* model and *any* interconnect can be evaluated. This concept of impedance is absolutely critical in all aspects of signal-integrity analysis.

3.2 What Is Impedance?

We use the term *impedance* in everyday language and often confuse the electrical definition with the common usage definition. As we saw earlier, the electrical term *impedance* has a very precise definition based on the relationship between the current through a device and the voltage across it: $Z = V/I$. This basic definition applies to any two-terminal device, such as a surface-mount resistor, a decoupling capacitor, a lead in a package, or the front connections to a printed circuit-board trace and its return path. When there are more than two terminals, such as in coupled conductors, or between the front and back ends of a transmission line, the definition of impedance is the same, it's just more complex to take into account the additional terminals.

For two-terminal devices, the definition of impedance, as illustrated in Figure 3-2, is simply:

$$Z = \frac{V}{I} \qquad\qquad (3\text{-}1)$$

where:
Z = impedance, measured in Ohms
V = voltage across the device, in units of volts
I = current through the device, in units of Amps

For example, if the voltage across a terminating resistor is 5 v and the current through it is 0.1 A, then the impedance of the device must be 5 v/0.1 A = 50 Ohms. No matter what type of device the impedance is referring to, in both the time and the frequency domain, the units of impedance are always in Ohms.

TIP   This definition of impedance applies to absolutely all situations, whether in the time domain or the frequency domain, whether for real devices that are measured or for ideal devices that are calculated.

If we always go back to this basic definition, we will never go wrong and will often eliminate many sources of confusion. One aspect of impedance that is often confusing is to think of it only in terms of resistance. As we will see, the impedance of an ideal resistor-circuit element, with resistance, R, is, in fact, Z = R.

Impedance is the general term, applying to all circuit elements in the time and frequency domains. Resistance is the specific figure of merit for an ideal circuit element we call a *resistor element*. While impedance is sometimes referred to as *AC resistance*, it applies to all circuit elements, not just resistors.

Our intuition of the impedance of a resistor is that a higher impedance means less current flows for a fixed voltage. Likewise, a lower impedance means a lot of current can flow for the same voltage. This is consistent with the definition I = V/Z and applies just as well when the voltage or current is not DC.

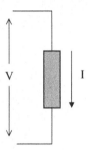

Figure 3-2 The definition of impedance for any two-terminal device, showing the current through the component and the voltage across the leads.

In addition to the notion of the impedance of a resistor, the concept of impedance can apply to an ideal capacitor, an ideal inductor, a real-wire bond, a printed circuit trace, or even a pair of connector pins.

There are two special extreme cases of impedance. For a device that is an open, there will be no current flow. If the current through the device for any voltage applied is zero, the impedance is Z = 1 v / 0 A = infinite Ohms. The impedance of an open device is very, very large. When the device is a short, there will be no voltage across it, no matter what the current through it. The impedance of a short is Z = 0 v / 1 A = 0 Ohms. The impedance of a short is always 0 Ohms.

3.3 Real Versus Ideal Circuit Elements

There are two types of electrical devices, real and ideal. Real devices can be measured; they physically exist. They are the actual interconnects or components that make up the hardware of a real system. Examples of real devices are traces on a board, leads in a package, or discrete decoupling capacitors mounted to a board.

Ideal devices are mathematical descriptions of specialized circuit elements that have precise, specific definitions. Each ideal circuit element has a very specific model or defined behavior associated with it. Models are the language that simulators understand. We build circuits by using combinations of ideal circuit elements.

Generally, simulators can only simulate circuits described by ideal circuit elements. The formalism and power of circuit theory applies only to ideal devices. Models are composed of combinations of ideal devices.

It is remarkable how well simulations of combinations of ideal circuit elements can match the measurements of real components.

In recent years, some simulators have been able to incorporate measured, behavioral models of components based on their measured S-parameters. This topic is covered in Chapter 12, "S-Parameters for Signal Integrity Applications."

Most simulators can only simulate the performance of ideal devices. It is very important to keep separate real versus ideal circuit elements. The impedance of any real, physical interconnect, or passive component can be measured. However, when the impedance of a passive interconnect is calculated, it is only the impedance of four very-well-defined, ideal passive circuit elements that can be considered. We cannot measure ideal circuit elements, nor can we calculate the impedance of any circuit elements other than ideal ones, except in the case of a measured behavioral S-parameter model. This is why it is important to make the distinction between real components and ideal circuit elements. This distinction is illustrated in Figure 3-3.

TIP   Ultimately, our goal is to create an equivalent circuit model composed of combinations of ideal circuit elements whose simulated impedance closely approximates the actual, measured impedance of a real component.

A circuit model will always be an approximation of the real-world structure. However, it is possible to construct an ideal model with a simulated impedance that accurately matches the measured impedance of a real device.

For example, Figure 3-4 shows the measured impedance of a real decoupling capacitor and the simulated impedance based on an RLC circuit model. These are the component and model in Figure 3-3. The agreement is excellent even up to 5 GHz, the bandwidth of the measurement.

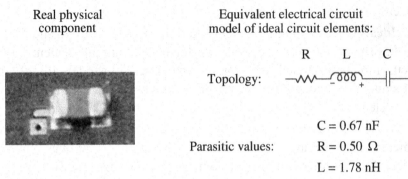

Figure 3-3 The two worldviews of a component, in this case a 1,206 decoupling capacitor mounted to a circuit board and an equivalent circuit model composed of combinations of ideal circuit elements.

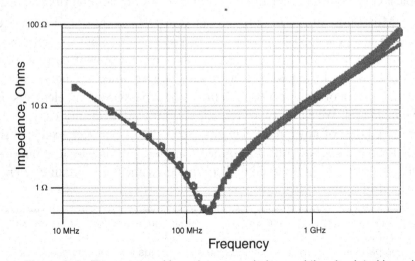

Figure 3-4 The measured impedance, as circles, and the simulated impedance, as the line, for a nominal 1-nF decoupling capacitor. The measurement was performed with a network analyzer and a GigaTest Labs Probe Station.

There are four ideal, two-terminal, circuit elements that we will use in combination as building blocks to describe any real interconnect:

1. An ideal resistor
2. An ideal capacitor
3. An ideal inductor
4. An ideal transmission line

We usually group the first three elements in a category called *lumped circuit elements*, in the sense that their properties can be lumped into a single point. This is different from the properties of an ideal transmission line, which are "distributed" along its length.

As ideal circuit elements, these elements have precise definitions that describe how they interact with currents and voltages. It is very important to keep in mind that ideal elements are different from real components, such as real resistors, real capacitors, or real inductors. One is a physical component, the other an ideal element.

It is unfortunate that we chose to call the ideal circuit elements and real circuit components by the same name. An ideal resistor is not the same as an ideal resistor electrical model. Their behaviors can be very similar, but they are not the same.

To minimize the confusion, it is a good habit to use the preface ideal or real when describing the components. Doing so minimizes the confusion to others and helps train our own intuition to keep track of which type of component we are referring.

While an ideal resistor model starts with a resistance that is constant with frequency, for example, it is possible to also add complexity to the ideal model, as long as the simulation tool understands what to do with the model. If the model were to include a resistance term that increased proportional to the square root of frequency, for example, the model would still be an ideal model; it would just be a second-order model and a better approximation to the real world.

Generally, the simulator tool defines what ideal models are available in its toolbox. This is an important differentiator for different simulators. Keysight Technology's Advanced System Designer (ADS), for example, is a sophisticated simulator with very complex models of all the ideal circuit elements.

The properties of a transmission line are initially seen as so confusing and nonintuitive, yet so important, that we devote the entire Chapter 7 to transmission lines and their impedance. In this chapter, we will concentrate on the impedance of just R, L, and C elements.

TIP   Only real devices can be measured, and only ideal elements can be calculated or simulated.

An equivalent electrical circuit model is an idealized electrical description of a real structure. It is an approximation, based on using combinations of ideal circuit elements. A good model will have a calculated impedance that closely matches the measured impedance of the real device. The better we can match the simulated impedance of an interconnect to its measurement, the better we can predict how a signal will interact with it.

When dealing with some high-frequency effects, such as lossy lines, we will need to invent new ideal circuit elements to get a closer match between simulated and measured performance. It is always astonishing how close a simulation of combinations of ideal circuit elements can be to the measured performance of a real interconnect.

3.4 Impedance of an Ideal Resistor in the Time Domain

Each of the four basic circuit elements above has a definition of how voltage and current interact with it. This is not the same as their impedance, though their impedance can be derived from this definition.

The relationship between the voltage across and the current through an ideal resistor is:

$$V = I \times R \tag{3-2}$$

where:

V = voltage across the ends of the resistor

I = current through the resistor

R = resistance of the resistor, in Ohms

An ideal resistor has a voltage across it that increases with the current through it. This definition of the I-V properties of an ideal resistor applies in both the time domain and the frequency domain. This is Ohm's Law, which applies to resistor elements. In fact, we define a component that obeys Ohm's Law as a resistor component.

In the time domain, we can apply the definition of the impedance and, using the definition of the ideal element, calculate the impedance of an ideal resistor:

$$Z = \frac{V}{I} = \frac{I \times R}{I} = R \qquad (3\text{-}3)$$

This basically says that the impedance is constant and independent of the current or voltage across a resistor. The impedance of an ideal resistor is pretty boring.

3.5 Impedance of an Ideal Capacitor in the Time Domain

In an ideal capacitor, there is a relationship between the charge stored between the two leads and the voltage across the leads. The capacitance of an ideal capacitor is defined as:

$$C = \frac{Q}{V} \qquad (3\text{-}4)$$

where:

C = capacitance, in Farads

V = voltage across the leads, in volts

Q = charge stored between the leads, in Coulombs

The value of the capacitance of a capacitor describes its capacity to store charge at the expense of voltage. A large capacitance means the ability to store a lot of charge at a low voltage across the terminals.

Capacitance is a measure of the *efficiency* of two conductors to store charge, at the expense of the voltage across it. A pair of conductors that are very efficient at storing charge and store a lot of change for a little voltage have a high capacitance.

The impedance of a capacitor can only be calculated based on the current through it and the voltage across its terminals. In order to relate the voltage across the terminals with the current through it, we need to know how the current flows through a capacitor. A real capacitor is made from two conductors separated by a dielectric. How does current get from one conductor to the other, when it has an insulating dielectric between them? This fundamental question will pop up over and over in signal integrity applications. The answer is that real current probably doesn't really flow through a capacitor; it just acts as though it does when the voltage across the capacitor changes.

Suppose the voltage across a capacitor were to increase. This means that some positive charge had to be added to the top conductor and some negative charge had to be added to the bottom conductor. Adding negative charge to the bottom conductor is the same as pushing positive charge out; it is as though positive charges were added to the top terminal and positive charges were pushed out of the bottom terminal. This is illustrated in Figure 3-5. The capacitor behaves as though current flows through it—but only when the voltage across it changes.

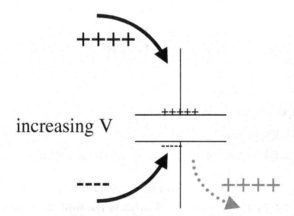

Figure 3-5   Increasing the voltage across a capacitor adds positive charge to one conductor and negative charge to the other. Adding negative charge to one conductor is the same as taking positive charges from it. It looks like positive charge enters one terminal and comes out of the other.

James Clerk Maxwell, now known as the father of electromagnetism, realized that if only conduction currents were considered, the *continuity* of current through a capacitor would be violated. Conduction current flows into and out of a capacitor, but because of the insulating properties of the dielectric between the plates of a capacitor, there could not be conduction current through it. There had to be a new sort of current to keep the current continuous through the insulating space between the plates.

Maxwell envisioned the missing current to be the motion of the *bound* charges due to the change in the *polarization* of the dielectric between the plates. As the voltage across the plates changed, the electric field changed, and the material inside the capacitor became more polarized. The bound charges in the material would *displace* due to the applied electric field. He called this current of bound charges *displacement* current, as the displacement of the bound changes in the polarized dielectric.

This displacement current only flowed when the bound charges' polarization increased or decreased, which happened when the electric field changed.

While this model made sense when there was a polarizable dielectric in the space between the conductors, what about when the gap between the plates was filled with the vacuum of free space? Maxwell assumed there was some polarization to free space, which, in his time in the mid-1880s, he envisioned to be filled with *ether*. The displacement current was the displacement of the bound charges of the ether.

Of course, today we have no evidence to support the presence of an ether. What is the displacement current, if there are no bound charges to "displace"? Rather than describing displacement current as a displacement of bound charges, we assign displacement current to a new property of electric fields— that a changing electric field has equivalent properties of a current, which we still refer to today as a displacement current.

A cornerstone of Maxwell's equations is the realization that, built into the very fabric of spacetime, a changing electric field has the properties of a current. To distinguish this current from the conduction current of free charges, or the changing polarization of a material, we call this current due to a changing electric field *displacement current*. It is just as real as a conduction current, just arising from a different property of electric fields.

By taking derivatives of both sides of the previous equation, a new definition of the I-V behavior of a capacitor can be developed:

$$I = \frac{dQ}{dt} = C\frac{dV}{dt} \qquad (3\text{-}5)$$

where:

I = current through the capacitor

Q = charge on one conductor of the capacitor

C = capacitance of the capacitor

V = voltage across the capacitor

This relationship points out, as we saw previously, that the only way current flows through a capacitor is when the voltage across it changes. If the voltage is constant, the current through a capacitor is zero. We also saw that for a resistor, the current through it doubles if the voltage across it doubles. However, in the case of a capacitor, the current through it doubles if the rate of change of the voltage across it doubles.

This definition is consistent with our intuition. If the voltage changes rapidly, the current through a capacitor is large. If the voltage is nearly constant, the current through a capacitor is near zero. Using this relationship, we can calculate the impedance of an ideal capacitor in the time domain:

$$Z = \frac{V}{I} = \frac{V}{C\dfrac{dV}{dt}} \qquad (3\text{-}6)$$

where:

V = voltage across the capacitor

C = capacitance of the capacitor

I = current through the capacitor

This is a complicated expression. It says that the impedance of a capacitor depends on the precise shape of the voltage waveform across it. If the slope of the waveform is large (i.e., if the voltage changes very fast), the current through it is high and the impedance is small. It also says that a large capacitor will have a lower impedance than a small capacitor for the same rate of change of the voltage signal.

However, the precise value of the impedance of a capacitor is more complicated. It is hard to generalize what the impedance of a capacitor is other than it depends on the shape of the voltage waveform. The impedance of a capacitor is not an easy term to use in the time domain.

3.6 Impedance of an Ideal Inductor in the Time Domain

The behavior of an ideal inductor is defined by:

$$V = L \frac{dI}{dt} \tag{3-7}$$

where:
V = voltage across the inductor
L = inductance of the inductor
I = current through the inductor

This says that the voltage across an inductor depends on how fast the current through it changes. It also says that the current change through an inductor depends on the voltage difference across it. Which one is the cause and which one is the response depends on which is the driving force.

If the current is constant, the voltage across the inductor will be zero. Likewise, if the current changes rapidly through an inductor, there will be a large voltage drop across it. The inductance is the proportionality constant that says how sensitive the voltage generated is to a changing current. A large inductance means that a small changing current produces a large voltage.

There is often confusion about the direction of the voltage drop that is generated across an inductor. If the direction of the changing current reverses, the polarity of the induced voltage will reverse. An easy way of remembering the polarity of the voltage is to base it on the voltage drop of a resistor.

In a resistive element, DC *always* flows from the positive side to the negative side of the component. The terminal the current goes into is the positive side, and the terminal the current flows out of is the negative side. Likewise, with an inductor, the terminal the current increases into is the positive side, and the other is the negative side for the induced voltage. This is illustrated in Figure 3-6.

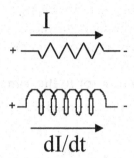

Figure 3-6   The direction of voltage drop across an inductor for a changing current is in the same direction as the voltage drop across a resistor for a DC.

Using this basic definition of inductance, we can calculate the impedance of an inductor. This is, by definition, the ratio of the voltage to the current through an inductor:

$$Z = \frac{V}{I} = L\frac{\frac{dI}{dt}}{I} \tag{3-8}$$

where:

V = voltage across the inductor

L = inductance of the inductor

I = current through the inductor

Again, we see the impedance of an inductor, though well defined, is awkward to use in the time domain. The general features are easy to discern. If the current through an inductor increases rapidly, the impedance of the inductor is large. An inductor will have a high impedance when current through it changes. If the current through an inductor changes only slightly, its impedance will be very small. For DC current the impedance of an inductor is nearly zero. But, other than these simple generalities, the actual impedance of an inductor depends very strongly on the precise waveform of the current through it.

TIP   For both the capacitor and the inductor, the impedance, in the time domain, is not a simple function at all. Impedance in the time domain is a very complicated way of describing these basic building-block ideal circuit elements. It is not wrong; it is just complicated.

This is one of the important occasions where moving to the frequency domain will make the analysis of a problem much simpler.

3.7 Impedance in the Frequency Domain

The important feature of the frequency domain is that the only waveforms that can exist are sine waves. We can only describe the behavior of ideal circuit elements in the frequency domain by how they interact with sine waves: sine waves of current and sine waves of voltage. These sine waves have three and only three features: the frequency, the amplitude, and the phase associated with each wave.

Rather than describe the phase in cycles or degrees, it is more common to use radians. There are $2 \times \pi$ radians in one cycle, so a radian is about 57 degrees. The frequency in radians per second is referred to as the *angular frequency*. The Greek letter omega (ω) is used to denote the angular frequency. ω is related to the frequency, by:

$$\omega = 2\pi \times f \tag{3-9}$$

where:

ω = angular frequency, in radians/sec

f = sine-wave frequency, in Hertz

We can apply sine-wave voltages across a circuit element and look at the sine waves of current through it. When we do this, we will still use the same basic definition of impedance (that is, the ratio of the voltage to the current), except that we will be taking the ratio of two sine waves: a voltage sine wave and a current sine wave.

It is important to keep in mind that all the basic building-block circuit elements and all the interconnects are linear devices. If a voltage sine wave of 1 MHz, for example, is applied across any of the four ideal circuit elements, the only sine-wave-frequency components that will be present in the current waveform will be a sine wave at 1 MHz. The amplitude of the current sine wave will be some number of Amps, and it will have some phase shift with respect to the voltage wave, but it will have exactly the same frequency. This is illustrated in Figure 3-7.

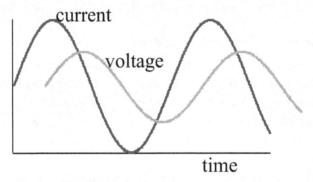

Figure 3-7   The sine-wave current through and voltage across an ideal circuit element will have exactly the same frequency but different amplitudes and some phase shift.

T I P   When we take the ratio of two sine waves, we need to account for the ratio of the amplitudes and the phase shift between the two waves.

What does it mean to take the ratio of two sine waves: the voltage and the current? The ratio of two sine waves is not a sine wave. It is a pair of numbers that contains information about the ratio of the amplitudes and the phase shift, at each frequency value. The magnitude of the ratio is just the ratio of the amplitudes of the two sine waves:

$$|Z| = \frac{|V|}{|I|}$$

(3-10)

The ratio of the voltage amplitude to the current amplitude will have units of Ohms. We refer to this ratio as the *magnitude of the impedance*. The phase of the ratio is the phase shift between the two waves. The phase shift has units of degrees or radians. In the frequency domain, the impedance of a circuit element or combination of circuit elements would be of the form: at 20 MHz, the magnitude of the impedance is 15 Ohms and the phase of the impedance is 25 degrees. This means the impedance is 15 Ohms and the voltage wave is leading the current wave by 25 degrees.

The impedance of any circuit element is two numbers, a magnitude and a phase, at every frequency value. Both the magnitude of the impedance and the phase of the impedance may be frequency dependent. The ratio of the amplitudes may vary with frequency or the phase may vary with frequency. When we describe the impedance, we need to specify at what frequency we are describing the impedance.

In the frequency domain, impedance can also be described with complex numbers. For example, the impedance of a circuit can be described as having a real component and an imaginary component. The use of real and imaginary components allows the powerful formalism of complex numbers to be applied, which dramatically simplifies the calculations of impedance in large circuits. Exactly the same information is contained in the magnitude and the phase information. These are two different and equivalent ways of describing impedance.

With this new idea of working in the frequency domain, and dealing only with sine waves of current and voltage, we can take another look at impedance.

We apply a sine wave of current through a resistor and we get a sine wave of voltage across it that is simply R times the current wave:

$$V = I_0 \sin(\omega t) \times R \qquad (3\text{-}11)$$

We can describe the sine wave of current in terms of sine and cosine waves or in terms of complex exponential notation.

When we take the ratio of the voltage to the current for a resistor, we find that it is simply the value of the resistance:

$$|Z| = \frac{|V|}{|I|} = \frac{I_0 \times R}{I_0} = R \qquad (3\text{-}12)$$

The impedance is independent of frequency, and the phase shift is zero. The impedance of an ideal resistor is flat with frequency. This is basically the same result we saw in the time domain—and is still pretty boring.

When we look at an ideal capacitor in the frequency domain, we will apply a sine-wave voltage across the ends. The current through the capacitor is the derivative of the voltage, which is a cosine wave:

$$I = C \times \frac{d}{dt} V_0 \sin(\omega t) = C \times \omega V_0 \cos(\omega t) \qquad (3\text{-}13)$$

This says the current amplitude will increase with frequency, even if the voltage amplitude stays constant. The higher the frequency, the larger the amplitude of the current through the capacitor. This suggests the impedance of a capacitor will decrease with increasing frequency. The magnitude of the impedance of a capacitor is calculated from:

$$|Z| = \frac{|V|}{|I|} = \frac{V_0}{C \times \omega V_0} = \frac{1}{\omega C} \qquad (3\text{-}14)$$

Here is where it gets confusing. This ratio is easily described using complex math, but most of the insight can also be gained from sine and cosine waves. The magnitude of the impedance of a capacitor is just $1/\omega C$. All the important information is here.

As the angular frequency increases, the impedance of a capacitor decreases. This says that even though the value of the capacitance is constant with frequency, the impedance gets smaller with higher frequency. We see this is reasonable because the current through the capacitor will increase with higher frequency and hence its impedance will be less.

The phase of the impedance is the phase shift between a sine and cosine wave, which is −90 degrees. When described in complex notation, the −90 degree phase shift is represented by the complex number, −i. In complex notation, the impedance of a capacitor is −i/ωC. For most of the following discussion, the phase adds more confusion than value and will generally be ignored.

A real decoupling capacitor has a capacitance of 10 nF. What is its impedance at 1 kHz? First, we assume that this capacitor is an ideal capacitor. A 10-nF ideal capacitor will have an impedance of $1/(2\,\pi \times 1$ kHz $\times 10$ nF) $=$ $1/(6 \times 10^3 \times 10 \times 10^{-9}) = 1/60 \sim 16$ kOhm. Of course, at lower frequency, its impedance would be higher. At 1 Hz, its impedance would be about 16 megaOhms.

Let's use this same frequency-domain analysis with an inductor. When we apply a sine wave current through an inductor, the voltage generated is:

$$V = L \times \frac{d}{dt} I_0 \sin(\omega t) = L \times \omega I_0 \cos(\omega t) \tag{3-15}$$

This says that for a fixed current amplitude, the voltage across an inductor gets larger at higher frequency. It takes a higher voltage to push the same current amplitude through an inductor. This would hint that the impedance of an inductor increases with frequency.

Using the basic definition of impedance, the magnitude of the impedance of an inductor in the frequency domain can be derived as:

$$|Z| = \frac{|V|}{|I|} = \frac{L \times \omega I_0}{I_0} = \omega L \tag{3-16}$$

The magnitude of the impedance increases with frequency, even though the value of the inductance is constant with frequency. It is a natural consequence of the behavior of an inductor that it is harder to shove AC current through it with increasing frequency.

The phase of the impedance of an inductor is the phase shift between the voltage and the current, which is +90 degrees. In complex notation, a +90 degree phase shift is i. The complex impedance of an inductor is $Z = i\omega L$.

In a real decoupling capacitor, there is inductance associated with the intrinsic shape of the capacitor and its board-attach footprint. A rough estimate for this intrinsic inductance is 2 nH. We really have to work hard to get it any lower than this. What is the impedance of just the series inductance of the real capacitor that we will model as an ideal inductor of 2 nH, at a frequency of 1 GHz?

The impedance is $Z = 2 \times \pi \times 1$ GHz $\times 2$ nH = 12 Ohms. When it is in series with the power and ground distribution and we want a low impedance, for example less than 0.1 Ohm, 12 Ohms is a lot. How does this compare with the impedance of the ideal-capacitor component of the real decoupling capacitor? The impedance of an ideal 10 nF capacitor element at 1 GHz is 0.01 Ohm. The impedance of the ideal inductor component is more than 1000 times higher than this and will clearly dominate the high-frequency behavior of a real capacitor.

We see that for both the ideal capacitor and inductor, the impedance in the frequency domain has a very simple form and is easily described. This is one of the powers of the frequency domain and why we will often turn to it to help solve problems.

The values of the resistance, capacitance, and inductance of ideal resistors, capacitors, and inductors are all constant with frequency. For the case of an ideal resistor, the impedance is also constant with frequency. However, for a capacitor, its impedance will decrease with frequency, and for an inductor, its impedance will increase with frequency.

TIP   It is important to keep straight that for an ideal capacitor or inductor, even though its value of capacitance and inductance is absolutely constant with frequency, the impedance will vary with frequency.

3.8 Equivalent Electrical Circuit Models

The impedance behavior of real interconnects can be closely approximated by combinations of these ideal elements. A combination of ideal circuit elements is called an *equivalent electrical circuit model*, or just a *model*. The drawing of the circuit model is often referred to as a *schematic*.

An equivalent circuit model has two features: It identifies how the circuit elements are connected together (called the *circuit topology*), and it identifies the value of each circuit element (referred to as the *parameter values*, or *parasitic values*).

Chip designers, who like to think they produce drivers with perfect, pristine waveforms, view all interconnects as *parasitics* in that they can only screw up their wonderful waveforms. To the chip designer, the process of determining the parameter values of the interconnects is really *parasitic extraction*, and the term has stuck in general use.

TIP It is important to keep in mind that whenever we draw circuit elements, they are always ideal circuit elements. We will have to use combinations of ideal elements to approximate the actual performance of real interconnects.

There will always be a limit to how well we can accurately predict the actual impedance behavior of real interconnects, using an ideal equivalent circuit model. This limit can often be found only by measuring the actual impedance of an interconnect and comparing it to the predictions based on the simulations of circuits containing these ideal circuit elements.

There are always two important questions to answer of every model: How accurate is it? and What is its bandwidth? Remember, its bandwidth is the highest sine-wave frequency at which we get good agreement between the measured impedance and the predicted impedance. As a general rule, the closer we would like the predictions of a circuit model to be to the actual measured performance, the more complex the model may have to be.

TIP It is good practice to always start the process of modeling with the simplest model possible and grow in complexity from there.

Take, for example, a real decoupling capacitor and its impedance, as measured from one of the capacitor pads, through a via and a plane below it, coming back up to the start of the capacitor. This is the example shown previously in Figure 3-3. We might expect that this real device could be modeled as a simple ideal capacitor. But, at how high a frequency will this simple ideal capacitor circuit model still behave like the real capacitor? The measured impedance of this real device, from 10 MHz to 5 GHz, is shown in Figure 3-8, with the impedance predicted for an ideal capacitor superimposed.

It is clear that this simple model works really well at low frequency. This simple model of an ideal capacitor with a value of 0.67 nF is a very good approximation of the real capacitor. It's just that it gives good agreement only up to about 70 MHz. The bandwidth of this model is 70 MHz.

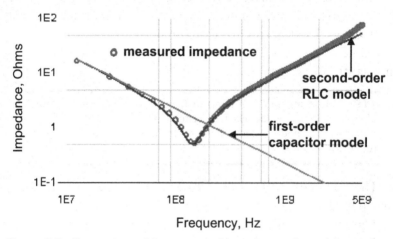

Figure 3-8 Comparison of the measured impedance of a real decoupling capacitor and the predicted impedance of a simple first-order model using a single C element and a second-order model using an RLC circuit model. Measured with a GigaTest Labs Probe Station.

If we expend a little more effort, we can create a more accurate circuit model with a higher bandwidth. A more accurate model for a real decoupling capacitor is an ideal capacitor, inductor, and resistor in series. Choosing the best parameter values, we see in Figure 3-8 that the agreement between the predicted impedance of this model and the measured impedance of the real device is excellent, all the way up to the bandwidth of the measurement, 5 GHz in this case.

We often refer to the simplest model we create as a *first-order model*, as it is the first starting place. As we increase the complexity, and hopefully, better agreement with the real device, we refer to each successive model as the *second-order model*, *third-order model*, and so on.

Using the second-order model for a real capacitor would let us accurately predict every important electrical feature of this real capacitor as it would behave in a system with application bandwidths at least up to 5 GHz.

TIP   It is remarkable that the relatively complex behavior of real components can be very accurately approximated, to very high bandwidths, by combinations of ideal circuit elements.

3.9 Circuit Theory and SPICE

There is a well-defined and relatively straightforward formalism to describe the impedance of combinations of ideal circuit elements. This is usually referred to as *circuit theory*. The important rule in circuit theory is that when two or more elements are in series—that is, connected end-to-end—the impedance of the combination, from one end terminal to the other end terminal, is the sum of the impedances of each element. What makes it a little complicated is that when in the frequency domain, the impedances that are summed are complex and must obey complex algebra.

In the previous section, we saw that it is possible to calculate the impedance of each individual circuit element by hand. When there are combinations of circuit elements, it gets more complicated. For example, the impedance of an RLC model approximating a real capacitor is given by:

$$Z(\omega) = R + i\left(\omega L - \frac{1}{\omega C}\right) \tag{3-17}$$

We could use this analytic expression for the impedance of the RLC circuit to plot the impedance versus frequency for any chosen values of R, L, and C. It can conveniently be used in a spreadsheet and each element changed. When there are five or ten elements in the circuit model, the resulting impedance can be calculated by hand, but it can be very complicated and tedious.

However, there is a commonly available tool that is much more versatile in calculating and plotting the impedance of any arbitrary circuit. It is so common and so easy to use, every engineer who cares about impedance or circuits in general, should have access to it on their desktop. It is SPICE.

SPICE stands for *Simulation Program with Integrated Circuit Emphasis.* It was developed in the early 1970s at UC Berkeley as a tool to predict the behavior of transistors based on the as-fabricated dimensions. It is basically a circuit simulator. Any circuit we can draw with R, L, C, and T elements can be simulated for a variety of voltage or current-exciting waveforms. It has evolved and diversified over the past 50 years, with over 30 vendors each adding their own special features and capabilities. There are a few either free versions or student versions for less than $100 that can be downloaded from the Web. Some of the free versions have limited capability but are excellent tools for learning about circuits.

A very powerful, simple-to-use and free SPICE tool is QUCS (Quite Universal Circuit Simulator). It is open source and available from www.QUCS. org. Another version is available from Linear Technologies, under the name LTSpice. Either tool should be on the desktop of every engineer.

In SPICE, only ideal circuit elements are used and every circuit element has a well-defined, precise behavior. There are three basic types of elements: active sources, passive elements, and nonlinear elements.

The active elements are the signal sources, current, or voltage waveforms. The passive elements are the R, L, C, and T elements. The nonlinear elements are all the semiconductor elements, such as the diodes and transistor-level models.

One of the distinctions between the various forms of SPICE is the variety of ideal circuit elements they provide. Every version of SPICE includes at least the R, L, C, and T (transmission-line) elements.

SPICE simulators allow the prediction of the voltage or current at every point in a circuit, simulated either in the time domain or the frequency domain. A time-domain simulation is called a *transient simulation*, and a frequency-domain simulation is called an *AC simulation*. Some versions of SPICE, such as QUCS, also allow an S-parameter simulation. SPICE is an incredibly powerful tool.

For example, a driver connected to two receivers located very close together can be modeled with a simple voltage source and an RLC circuit. The R is the impedance of the driver, typically about 10 Ohms. The C is the capacitance of the interconnect traces and the input capacitance of the two receivers, typically about 5 pF total. The L is the total loop inductance of the package leads and the interconnect traces, typically about 7 nH.

Figure 3-9 shows the setup of this circuit in SPICE and the resulting time-domain waveform, with the ringing that might be found in the actual circuit.

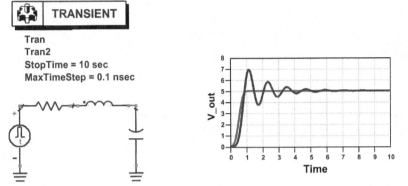

Figure 3-9 Simple equivalent circuit model to represent a driver and receiver fanout of two, including the packaging and interconnects, as set up in Keysight's Advanced Design System (ADS), a version of SPICE, and the resulting simulation of the internal-voltage waveform and the voltage at the input of the receivers. The rise time simulated is 0.5 nsec. The lead and interconnect inductance plus the input-gate capacitance dominate the source of the ringing.

TIP   If the circuit schematic can be drawn, SPICE can simulate the voltage and current waveforms. This is the real power of SPICE for general electrical engineering analysis.

SPICE can be used to calculate and plot the impedance of any circuit in the frequency domain. Normally, it plots only the voltage or current waveforms at every connection point, but a trick can be used to convert this into impedance.

One of the source circuit elements SPICE has in its toolbox for AC simulation is a constant-current sine-wave-current source. This current source will output a sine wave of current, with a constant amplitude, at a predetermined frequency. When running an AC analysis, the SPICE engine will step the frequency of the sine-wave-current source from the start frequency value to the stop frequency value with a number of intermediate frequency points.

The source generates the constant-current amplitude by outputting a sine wave with some voltage amplitude. The amplitude of the voltage wave is automatically adjusted to result in the specified constant amplitude of current.

To build an impedance analyzer in SPICE, we set the current source to have a constant amplitude of 1 Amp. No matter what circuit elements are connected to the current source, SPICE will adjust the voltage amplitude to result in a 1-Amp current amplitude through the circuit. If the constant-current source is connected to a circuit that has some impedance associated with it, $Z(\omega)$, then to keep the amplitude of the current constant, the voltage it applies will have to adjust.

The voltage applied to the circuit, from the constant-current source, with a 1-Amp current amplitude, is $V(\omega) = Z(\omega) \times 1$ Amp. The voltage across the current source, in volts, is numerically equal to the impedance of the circuit attached, in Ohms.

For example, if we attach a 1-Ohm resistor across the terminals, in order to maintain the constant current of 1 Amp, the voltage amplitude generated must be $V = 1$ Ohm $\times 1A = 1$ v. If we attach a capacitor with capacitance C, the voltage amplitude at any frequency will be $V = 1/\omega C$. Effectively, this circuit will emulate an impedance analyzer. Plotting the voltage versus the frequency is a measure of the magnitude of the impedance versus frequency for any circuit. The phase of the voltage is also a measure of the phase of the impedance.

To use SPICE to plot an impedance profile, we construct an AC constant-current source with amplitude of 1 A and connect the circuit under test across the terminals. The voltage measured across the current source is a direct measure of the impedance of the circuit. An example of a simple circuit is shown in Figure 3-10. As a trivial example, we connect a few different circuit elements to the impedance analyzer and plot their impedance profiles.

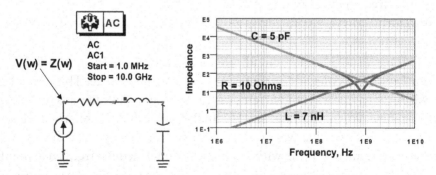

Figure 3-10   Left: An impedance analyzer in SPICE. The voltage across the constant-current source is a direct measure of the impedance of the circuit connected to it. Right: An example of the magnitude of the impedance of various circuit elements, calculated with the impedance analyzer in SPICE.

We can use this impedance analyzer to plot the impedance of any circuit model. Impedance is complex. It has not only magnitude information but also phase information. We can plot each of these separately in SPICE. The phase is also available in an AC simulation in SPICE. In Figure 3-11, we illustrate using the impedance analyzer to simulate the impedance of an RLC circuit model, approximating a real capacitor, plotting the magnitude and phase of the impedance across a wide frequency range.

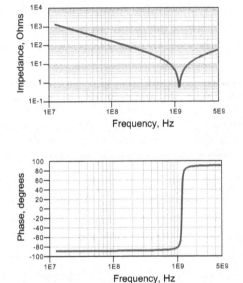

Figure 3-11   Simulated magnitude and phase of an ideal RLC circuit. The phase shows the capacitive behavior at low frequency and the inductive behavior at high frequency.

It is exactly as expected. At low frequency, the phase of the impedance is −90 degrees, suggesting capacitive behavior. At high frequency the phase of the impedance is +90 degrees, suggesting inductive behavior.

3.10 Introduction to Measurement-Based Modeling

As pointed out in Chapter 1, "Signal Integrity Is in Your Future," equivalent circuit models for interconnects and passive devices can be created based either on measurements or on calculations. In either case, the starting place is always some assumed topology for the circuit model. How do we pick the right topology? How do we know what is the best circuit schematic with which to start?

The strategy for building models of interconnects or other structures is to follow the principle that Albert Einstein articulated when he said, "Everything should be made as simple as possible, but not simpler." Always start with the simplest model first, and build in complexity from there.

Building models is a constant balancing act between the accuracy and bandwidth of the model required and the amount of time and effort we are willing to expend in getting the result. In general, the more accuracy required, the more expensive the cost in time, effort, and dollars. This is illustrated in Figure 3-12.

TIP   When constructing models for interconnects, it is always important to keep in mind that sometimes an OKAY answer NOW! is more important than a good answer LATE. This is why Einstein's advice should be followed: Start with the simplest model first and build in complexity from there.

It is always a good idea to start with an ideal transmission line as the first-order model for any interconnect. An ideal transmission line is a good low-frequency and high-frequency model for an interconnect, which is why we devote the entire Chapter 7 to this ideal circuit element.

However, sometimes the question you want answered is phrased in terms of lumped circuit elements, such as What is the inductance of a package lead? or What is the capacitance of an interconnect trace? When the interconnect structure is electrically short, or at low frequency, interconnects can be approximated by ideal R, L, or C elements. This concept of electrical length is described in Chapter 7.

The simplest lumped circuit model is just a single R, L, or C circuit element. The next simplest are combinations of two of them, and then three of them, and so on. The key factor that determines when we need to increase the complexity of a model is the bandwidth of the model required. As a general trend, the higher the bandwidth, the more complex the model. However, every high-bandwidth model must still give good agreement at low frequency; otherwise, it will not be accurate for transient simulations that can have low-frequency components in the signals.

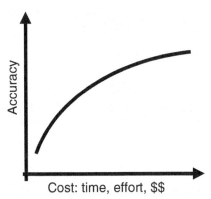

Figure 3-12   Fundamental trade-off between the accuracy of a model and how much effort is required to achieve it. This is a fundamental relationship for most issues in general.

For discrete passive devices, such as surface-mount technology (SMT) terminating resistors, decoupling capacitors, and filter inductors, the low-bandwidth and high-bandwidth ideal circuit model topologies are illustrated in Figure 3-13. As we saw earlier for the case of decoupling capacitors, the single-element circuit model worked very well at low frequency. The higher-bandwidth model for a real decoupling capacitor worked even up to 5 GHz for the specific component measured. The bandwidth of a circuit model for a real component is not easy to estimate except from a measurement.

For many interconnects that are electrically short, simple circuit models can also be used. The simplest starting place for a printed-circuit trace over a return plane in the board, which might be used to connect one driver to another, is a single capacitor. Figure 3-14 shows an example of the measured impedance of a 1-inch interconnect and the simulated impedance of a first-order model consisting of a single C element model. In this case, the agreement is excellent up to about 1 GHz. If the application bandwidth was less than 1 GHz, a simple ideal capacitor could be used to accurately model this 1-inch-long interconnect.

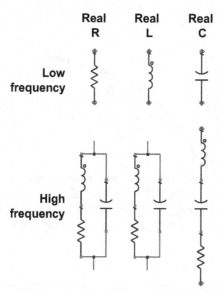

Figure 3-13   Simplest starting models for real components or interconnect elements, at low frequency and for higher bandwidth.

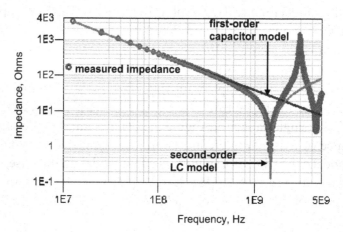

Figure 3-14   Measured impedance of a 1-inch-long microstrip trace and the simulated impedance of first- and second-order models. The first-order model is a single C element and has a bandwidth of about 1 GHz. The second-order model uses a series LC circuit and has a bandwidth of about 2 GHz.

For a higher-bandwidth model, a second-order model consisting of an inductor in series with the capacitor can be used. The agreement of this higher-bandwidth model is up to about 2 GHz.

As we show in Chapter 7, the best model for any uniform interconnect is an ideal transmission line model. The T element works at low frequency and at high frequency. Figure 3-15 illustrates the excellent agreement between the measured impedance and the simulated impedance of an ideal T element across the entire bandwidth of the measurement.

An ideal resistor-circuit element can model the actual behavior of real resistor devices up to surprisingly high bandwidth. There are three general technologies for resistor components, such as those used for terminating resistors: axial lead, SMT, and integrated passive devices (IPDs). The measured impedance of a representative of each technology is shown in Figure 3-16.

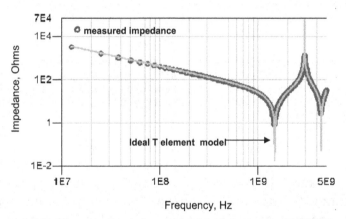

Figure 3-15   Measured impedance of a 1-inch-long microstrip trace and the simulated impedance of an ideal T element model. The agreement is excellent up to the full bandwidth of the measurement. Agreement is also excellent at low frequency.

An ideal resistor will have an impedance that is constant with frequency. As can be seen, the IPD resistors match the ideal resistor-element behavior up to the full-measurement bandwidth of 5 GHz. SMT resistors are well approximated by an ideal resistor up to about 2 GHz, depending on the mounting geometry and board stack-up, and axial-lead resistors can be approximated to about 500 MHz by an ideal resistor. In general, the primary effect that arises at higher frequency is the impact from the inductive properties of the real resistors. A higher-bandwidth model would have to include inductor elements and maybe also capacitor elements.

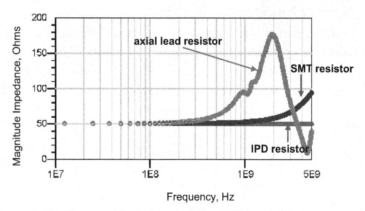

Figure 3-16   Measured impedance of three different resistor components, axial lead, surface-mount technology (SMT), and integrated passive device (IPD). An ideal resistor element has an impedance constant with frequency. This simple model matches each real resistor at low frequency but has limited bandwidth depending on the resistor technology.

Having the circuit-model topology is only half of the solution. The other half is to extract the parameter values, either from a measurement or with a calculation. Starting with the circuit topology, we can use rules of thumb, analytic approximations, and numerical-simulation tools to calculate the parameter values from the geometry and material properties for each of the circuit elements. This is detailed in the next chapters.

3.11 The Bottom Line

1. Impedance is a powerful concept to describe all signal-integrity problems and solutions.
2. Impedance describes how voltages and currents are related in an interconnect or a component. It is fundamentally the ratio of the voltage across a device to the current through it.

3. Real components that make up the actual hardware are not to be confused with ideal circuit elements that are the mathematical description of an approximation to the real world.

4. Our goal is to create an ideal circuit model that adequately approximates the impedance of the real physical interconnect or component. There will always be a bandwidth beyond which the model is no longer an accurate description, but simple models can work to surprisingly high bandwidth.

5. The resistance of an ideal resistor, the capacitance of an ideal capacitor, and the inductance of an ideal inductor are all constant with frequency.

6. Though impedance has the same definition in the time and frequency domains, the description is simpler and easier to generalize for C and L components in the frequency domain.

7. The impedance of an ideal R is constant with frequency. The impedance of an ideal capacitor varies as $1/\omega C$, and the impedance of an ideal inductor varies as ωL.

8. SPICE is a very powerful tool to simulate the impedance of any circuit or the voltage and current waveforms expected in both the time and frequency domains. All engineers who deal with impedance should have a version of SPICE available to them on their desktop.

9. When building equivalent circuit models for real interconnects, it is always important to start with the simplest model possible and build in complexity from there. The simplest starting models are single R, L, C, or T elements. Higher-bandwidth models use combinations of these ideal circuit elements.

10. Real components can have very simple equivalent circuit models with bandwidths in the GHz range. The only way to know what the bandwidth of a model is, however, is to compare a measurement of the real device to the simulation of the impedance using the ideal circuit model.

End-of-Chapter Review Questions

The answers to the following review questions can be found in Appendix D, "Review Questions and Answers."

3.1 What is the most important electrical property of an interconnect?

3.2 How would you describe the origin of reflection noise in terms of impedance?

3.3 How would you describe the origin of cross talk in terms of impedance?

3.4 What is the difference between modeling and simulation?

3.5 What is impedance?

3.6 What is the difference between a real capacitor and an ideal capacitor?

3.7 What is meant by the bandwidth of an ideal circuit model used to describe a real component?

3.8 What are the four ideal passive circuit elements used to build interconnect models?

3.9 What are two differences between the behavior you might expect between an ideal inductance described by a simple L element and a real inductor?

3.10 Give two examples of an interconnect structure that could be modeled as an ideal inductor.

3.11 What is displacement current, and where do you find it?

3.12 What happens to the capacitance of an ideal capacitor as frequency increases?

3.13 If you attach an open to the output of an impedance analyzer in SPICE, what impedance will you simulate?

3.14 What is the simplest starting model for an interconnect?

3.15 What is the simplest circuit topology to model a real capacitor? How could this model be improved at higher frequency?

3.16 What is the simplest circuit topology to model a real resistor? How could this model be improved at higher frequency?

3.17 Up to what bandwidth might a real axial lead resistor match the behavior of a simple ideal resistor element?

3.18 In which domain is it easiest to evaluate the bandwidth of a model?

3.19 What is the impedance of an ideal resistor with a resistance of 253 Ohms at 1 kHz and at 1 MHz?

3.20 What is the impedance of an ideal 100 nF capacitor at 1 MHz and at 1 GHz? Why is it unlikely a real capacitor will have such a low impedance at 1 GHz?

3.21 The voltage on a power rail on-die may drop by 50 mV very quickly. What will be the dI/dt driven through a 1 nH package lead?

3.22 To get the largest dI/dt through the package lead, do you want a large lead inductance or a small lead inductance?

3.23 In a series RLC circuit with R = 0.12 Ohms, C = 10 nF, and L = 2 nH, what is the minimum impedance?

3.24 In the circuit in Question 3.23, what is the impedance at 1 Hz? At 1 GHz?

3.25 If an ideal transmission line matches the behavior of a real interconnect really well, what is the impedance of an ideal transmission line at low frequency? High or low?

3.26 What is the SPICE circuit for an impedance analyzer?

The Physical Basis of Resistance

The electrical description of every interconnect and passive device is based on using just three ideal lumped circuit elements (resistors, capacitors, and inductors) and one distributed element (a transmission line). The electrical properties of the interconnects are all due to the precise layout of the conductors and dielectrics and how they interact with the electric and magnetic fields of the signals.

Understanding the connection between geometry and electrical properties will give us insight into how signals are affected by the physical design of interconnects and feed our intuition about manipulating signal-integrity performance by design.

TIP   The key to optimizing the physical design of a system for good signal integrity is to be able to accurately predict the electrical performance from the physical design and to efficiently optimize the physical design for a target electrical performance.

All the electrical properties of interconnects can be completely described by the application of Maxwell's Equations. These four equations describe how electric and magnetic fields interact with with the boundary conditions: conductors and dielectrics in some geometry. In principle, with optimized software and a powerful enough computing platform, we should be able to input the precise layout of a circuit board and all the various initial voltages coming out of the devices, push a button, and see the evolution of all the electric and magnetic fields. In principle. After all, there are no new physics or unknown, mysterious effects involved in signal propagation. It's all described by Maxwell's Equations.

There are some software tools available that, running on even a PC, will allow small problems to be completely simulated by Maxwell's Equations. However, there is no way of simulating an entire board directly with Maxwell's Equations—yet. Even if there were, we would be able to perform only a final system verification, which could tell us only whether this specific board met or failed the performance spec. Just being able to solve for all the time-varying electric and magnetic fields doesn't give us insight into what should be changed or done differently in the next design.

> **TIP**   The design process is a very intuitive process. New ideas come from imagination and creativity. These are fed not by numerically solving a set of equations but by understanding, at an intuitive level, the meaning of the equations and what they tell us.

4.1 Translating Physical Design into Electrical Performance

As we saw in Chapter 3, "Impedance and Electrical Models," the simplest starting place in thinking about the electrical performance of an interconnect is with its equivalent electrical circuit model. Every model has two parts: the circuit topology and the parameter values of each circuit element. And the simplest starting place in modeling any interconnect is using some combination of the three ideal lumped circuit elements (resistors, capacitors, and inductors) or the distributed element (an ideal transmission line circuit element).

Modeling is the process of translating the physical design of line widths, lengths, thicknesses, and material properties into the electrical view of R, L, and C elements. Figure 4-1 shows this relationship between the physical view and the electrical view for the special case of a generic RLC model.

Once we have established the topology of the circuit model for an interconnect, the next step is to extract the parameter values. This is sometimes called *parasitic extraction*. The task then is to take the geometry and material properties and determine how they translate into the equivalent parameter values of the ideal R, C, L, or T elements. We will use rules of thumb, analytic approximations, and numerical simulation tools to do this.

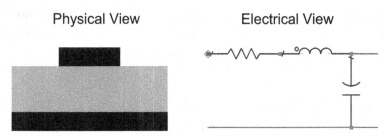

Figure 4-1   The physical worldview and the electrical worldview of the special case of a microstrip interconnect, illustrating the two different perspectives.

In this chapter, we look at how resistance is determined by the geometry and material properties. In the next three chapters, we look at the physical basis of capacitance, inductance, and transmission lines.

4.2 The Only Good Approximation for the Resistance of Interconnects

When we take the two ends of any conductor, such as a copper trace on a board, and apply a voltage across the ends, we get a current through the conductor. Double the voltage, and the current doubles. The impedance across the ends of the real copper trace can be modeled as an ideal resistor. It has an impedance that is constant in time and frequency.

> **TIP**   When we extract the resistance of an interconnect, what we are really doing is first implicitly assuming that we will model the interconnect as an ideal resistor.

Once we've established the circuit topology to be an ideal resistor element, we apply one of the three analysis techniques to extract the parameter value based on the specific geometry of the interconnect. The initial accuracy of the model will depend on how well we can translate the actual geometry into one of the standard patterns for which we have good approximations or how well we can apply a numerical simulation tool. When we want just a rough, approximate, ball-park number, we can apply a rule of thumb.

There is only one good analytical approximation for the resistance of an interconnect. This approximation is for a conductor that has a uniform cross section down its length. For example, a wire bond, a hook-up wire, and a trace on a circuit board have the same diameter or cross section all down their length. This approximation will be a good match in these cases. Figure 4-2 illustrates the geometrical features for this approximation.

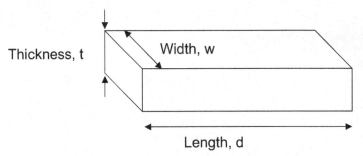

Figure 4-2 Description of the geometrical features for an interconnect that will be modeled by an ideal R element. The resistance is between the two end faces, spaced a distance, d, apart. We can also use the term Len to label the distance between the ends.

For the special case of a conductor with the same cross section down its length, the resistance can be approximated by:

$$R = \rho \frac{\text{Len}}{A} \qquad (4\text{-}1)$$

where:

R = resistance, in Ohms

ρ = bulk resistivity of the conductor, in Ohm-cm

Len = distance between the ends of the interconnect, in cm

A = cross-section area, in cm^2

For example, if a wire bond has a length of 0.2 cm, or about 80 mils, and a diameter of 0.0025 cm, or 1 mil, and is composed of gold with a resistivity of 2.5 microOhm-cm, then the resistance from one end to the other will be:

$$R = \rho \frac{\text{Len}}{A} = 2.5 \times 10^{-6} \times \frac{0.2}{(\pi/4)0.0025^2} = 0.1\Omega \qquad (4\text{-}2)$$

TIP   This is a good rule of thumb to remember: The resistance of a 1-mil-diameter wire bond that is 80 mils long is about 0.1 Ohm. This is pretty close to 1 Ohm/inch, an easy number to remember.

This approximation says that the value of the resistance will increase linearly with length. Double the length of the interconnect, and the resistance will double. The value of the resistance also varies inversely with the cross-sectional area. If we make the cross section larger, the resistance decreases. This matches what we know about water flowing through a pipe: A wider pipe means less resistance to water flow, and a longer pipe means more resistance.

The parameter value of the resistance of the equivalent ideal resistor depends on the geometry of the structure and its material property (that is, its bulk resistivity). If we change the shape of the wire, the equivalent resistance will change. If the cross section of the conductor changes down its length, as it does in a lead frame of a plastic quad flat pack (PQFP), we must find a way to approximate the actual cross section in terms of a constant cross section, or we cannot use this approximation.

Consider a lead in a 25-mil pitch, 208-pin PQFP. The lead has a total length of 0.5 inch but changes shape and does not have a constant cross section. It always has a thickness of 3 mils, but its width starts out at 10 mils and widens to 20 mils on the outside edge. How are we to estimate the resistance from one end to the other? The key term in this question is *estimate*. If we needed the most accurate result, we would probably want to take a precise profile of the shape of the conductor and use a 3D-modeling tool to calculate the resistance of each section, taking into account the changes in width.

The only way we can use the approximation above is if we have a structure with a constant cross section. We must approximate the real variable-width PQFP lead with a geometry that has a constant cross section. One way of doing this is to assume that the lead tapers uniformly down its length. If it is

10 mils wide on one end and 20 mils wide at the other, the average width is 15 mils. As a first-pass approximation, we can assume a constant cross section of 3 mils thick and 15 mils wide. Then, using the resistivity of copper, the resistance of the lead is:

$$R = \rho \frac{\text{Len}}{A} = 1.8 \times 10^{-6} \text{Ohm} - \text{cm} \times \frac{0.5 \text{ inch}}{0.003 \text{ inch} \times 0.015 \text{ inch}} \times \frac{1 \text{ inch}}{2.54 \text{ cm}} = 8 \text{ m}\Omega \quad (4\text{-}3)$$

It is important to be careful with the units and always be consistent. Resistance will always be measured in Ohms.

4.3 Bulk Resistivity

Bulk resistivity is a fundamental material property that all conductors have. It has units of Ohms-length, such as Ohms-inches or Ohms-cm. This is very confusing. We might expect resistivity to have units of Ohms/cm, and in fact, we often see the bulk resistivity incorrectly reported with these units. However, do not confuse the intrinsic material property with the extrinsic *resistance* of a piece of interconnect.

Bulk resistivity must have the units it does so that the resistance of an interconnect has units of Ohms. For resistivity × length / (length × length) to equal Ohms, resistivity must have units of Ohms-length.

TIP   Bulk resistivity is not a property of the structure or object made from a material; it is an intrinsic property of the material, independent of the size of the sample.

Bulk resistivity is an intrinsic material property, independent of the size of the chunk of material we look at. It is a measure of the intrinsic resistance to current flow of a material. The copper in a chunk 1 mil on a side will have the same bulk resistivity as the copper in a chunk 10 inches on a side.

The worse the conductor, the higher the resistivity. Usually, the Greek letter ρ is used to represent the bulk resistivity of a material. There is another term, *conductivity*, that is sometimes used to describe the electrical resistivity of a material. Usually, the Greek letter σ is used for the conductivity of a material. It is not surprising that a more conductive material will have a higher conductivity. Numerically, resistivity and conductivity are inversely related to each other:

$$\rho = \frac{1}{\sigma} \qquad (4\text{-}4)$$

While the units of resistivity are Ohms-m, for example, the units for conductivity are 1/(Ohms-m). The unit of 1/Ohms is given the special name *Siemens*. The units of conductivity are Siemens/meter. Figure 4-3 lists the values of many common conductors used in interconnects and their resistivity. It is important to note that the bulk resistivity of most interconnect metals will vary as much as 50% due to different processing conditions. For example, the bulk resistivity of copper is reported as between 1.8 and 4.5 μOhms-cm, depending on whether it is electroplated, electrolessly deposited, sputtered, rolled, extruded, or annealed. The more porous it is, the higher the resistivity. If it is important to know the bulk resistivity of the conductor to better than 10%, it should be measured for the specific sample.

Material	Resistivity μ Ohms-cm
Silver	1.47
Copper	1.58
Gold	2.01
Aluminum	2.61
Molybdenum	5.3
Tungsten	5.3
Nickel	6.2
Silver-filled glass	~10
Tin	10.1
Eutectic Pb/Sn solder	15
Lead	19.3
Kovar	49
Alloy42	57
Silver-filled epoxy	~300

Figure 4-3 Typical bulk resistivities of common interconnect materials.

We sometimes use the terms *bulk resistivity* and *volume resistivity* to refer to this intrinsic material property. This is to distinguish it from two other resistance-related terms: *resistance per length* and *sheet resistance*.

4.4 Resistance per Length

When the cross section of the conductor is uniform down the length, such as in any wire or even in a trace on a circuit board, the resistance of the interconnect will be directly proportional to the length. Using the approximation above, we can see that for a uniform cross-section conductor, the resistance per length is constant and given by:

$$R_L = \frac{R}{Len} = \frac{\rho}{A} \qquad (4\text{-}5)$$

where:

R_L = resistance per length

Len = interconnect length

ρ = bulk resistivity

A = cross-section area current travels through

For example, for a wire bond with a diameter of 1 mil, the cross section is constant down the length and the cross-sectional area is $A = \pi/4 \times 1$ mil^2 = 0.8×10^{-6} inches^2. With a bulk resistivity of gold of roughly 1 μOhm-inch, the resistance per length is calculated as $R_L = 1$ μOhm-inch/0.8×10^{-6} inches^2 = 0.8 Ohms/inch ~ 1 Ohm/inch.

This is an important number to keep in mind and is often called a rule of thumb: The resistance per length of a wire bond is about 1 Ohm/inch. The typical length of a wire bond is about 0.1 inch, so the typical resistance is about 1 Ohm/inch × 0.1 inch = 0.1 Ohm. A wire bond 0.05 inches long would have a resistance of 1 Ohm/inch × 0.05 inch = 0.05 Ohm, or 50 mOhms.

The diameter of wires is measured by a standard reference number referred to as the American wire gauge (AWG). Figure 4-4 lists some gauge values and the equivalent diameters. From the diameter and assuming copper wire, we can estimate the resistance per length. For example, 22-gauge wire, which is typical in many personal computer boxes, has a diameter of 25 mils. Its resistance per length is $R_L = 1.58$ μOhm-cm/(2.54 cm/inch)/($\pi/4 \times 25$ mil^2) = 1.2×10^{-3} Ohms/inch, or about 15×10^{-3} Ohms/foot, or 15 Ohms per 1000 ft.

AWG wire size	Diameter (inches)	Resistance per 1,000 ft (Ohms) (assumes $\rho = 1.74$ μOhms-cm)
24	0.0201	25.67
22	0.0254	16.14
20	0.0320	10.15
18	0.0403	6.385
16	0.0508	4.016
14	0.0640	2.525
12	0.0808	1.588
10	0.1019	0.999

Figure 4-4   AWG, diameter, and resistance per length.

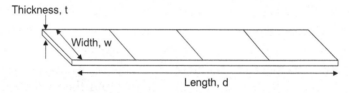

Figure 4-5   A uniform trace cut from a sheet can be divided into a number of squares, n = Len/w.

4.5 Sheet Resistance

Many interconnect substrates, such as printed circuit boards, cofired ceramic substrates, and thin film substrates, are fabricated with uniform sheets of conductor that are patterned into traces. All the conductors on each layer have exactly the same thickness. For the special case where the width of a trace is uniform, as illustrated in Figure 4-5, the resistance of the trace is given by:

$$R = \rho \frac{Len}{t \times w} = \left(\frac{\rho}{t} \right) \times \left(\frac{Len}{w} \right) \qquad (4\text{-}6)$$

The first term, (ρ/t), is constant for every trace built on the layer with thickness t. After all, every trace on the same layer will have the same bulk resistivity and the same thickness. This term is given the special name *sheet resistance* and is designated by R_{sq}.

The second term, (Len/w), is the ratio of the length to the width for a specific trace. This is the number of squares that can be drawn down the trace. It is referred to as n and is a dimensionless number. The resistance of a rectangular trace can be rewritten as:

$$R = R_{sq} \times n \qquad\qquad (4\text{-}7)$$

where:

R_{sq} = sheet resistance

n = number of squares down the trace

Interestingly, the units of sheet resistance are just Ohms. Sheet resistance has units of resistance, but what resistance does *sheet resistance* refer to? The simplest way of thinking about sheet resistance is to consider the resistance between the two ends of a section of sheet that is one square in shape (i.e., the length equals the width). In this case, n = 1, and the resistance between the ends of the square trace is just the sheet resistance. Sheet resistance refers to the resistance of one square of conductor.

Surprisingly, whether the square is 10 mils on a side or 10 inches on a side, the resistance across opposite ends of the square is constant. If the length of the square is doubled, we might expect the resistance to double. However, the width would also double, and the resistance would be cut in half. These two effects cancel, and the net resistance is constant as we change the size of a square.

TIP   All square pieces cut from the same sheet of conductor have the same resistance between opposite ends, and we call this resistance the sheet resistance; it is measured in Ohms and often referred to as Ohms per square.

The sheet resistance will depend on the bulk resistivity of the conductor and the thickness of the sheet. In typical printed circuit boards fabricated with layers of copper conductor, the thickness of copper is described by the weight of copper per square foot. This is a holdover from the days when the plating thickness was measured by weighing a 1-square-foot panel. A 1-ounce copper sheet gives 1 ounce of weight of copper per square foot of board. The thickness of 1-ounce copper is about 1.4 mils, or 35 microns. A 1/2-ounce copper sheet has a thickness of 0.7 mil, or 17.5 microns. Based on the thickness and the bulk resistivity of copper, the sheet resistance of 1-ounce copper is $R_{sq} = 1.6 \times 10^{-6}$ Ohm-cm /35 $\times$ 10^{-4} cm = 0.5 mOhm per sq.

If the copper thickness is cut in half to 1/2-ounce copper, the sheet resistance doubles. Half-ounce copper has a sheet resistance of 1 mOhm per square.

TIP   A simple rule of thumb to remember is that the sheet resistance of 1/2-ounce copper is 1 mOhm/sq. A trace 5 mil wide and 5 inches long has 1,000 squares in series and a resistance of 1 Ohm with 1/2-ounce copper sheet.

Sheet resistance is an important characteristic of the metallization of a layer. If the thickness and the sheet resistance are measured, then the bulk resistivity of the deposited metal can be found. Sheet resistance is measured by using a specially designed four-point probe. The four tips are usually mounted to a rigid fixture so they are held in a line with equal spacing. These four probes are placed in contact with the sheet being measured and connected to a four-point impedance analyzer, or Ohmmeter. When a constant current is applied to the outer two points and the voltage is measured between the two inner points, the resistance is measured as R_{meas} = V/I. Figure 4-6 illustrates the alignment of the probe points and their connection.

Figure 4-6   Four probes all in a row can be used to measure sheet resistance.

As long as the probes are far from the edge (i.e., at least four probe spacings from any edge), the measured resistance is completely independent of the actual spacing of the probe points. The sheet resistance, R_{sq}, can be calculated from the measured resistance using:

$$R_{sq} = 4.53 \times R_{meas}$$ (4-8)

It is interesting to note that for 1-ounce copper, a sheet resistance of 0.5 mOhms/sq, the measured resistance of the four-point probe would be 0.1 mOhm. This is an incredibly small resistance and requires a special microOhmmeter to measure. If we want the sheet resistance to 1% precision, we must be able to resolve 1 microOhm of resistance in the measurement.

If we know the sheet resistance of the sheet, we can use this to calculate the resistance per length and the total resistance of any conductor made in the sheet. Typically, a trace will be defined by a line width, w, and a length, Len. The resistance per length of a trace is given by:

$$R_L = \frac{R}{Len} = R_{sq} \times \frac{1}{w}$$ (4-9)

where:
R_L = resistance per length
R = trace resistance
R_{sq} = sheet resistance
w = line width
Len = length of the trace

Figure 4-7 illustrates the resistance per length of different line widths for 1-ounce and 1/2-ounce copper traces. The wider the line, the lower the resistance per length, as expected. For a 5-mil-wide trace, typical of many backplane applications, a 1/2-ounce copper trace would have a resistance per length of 0.2 Ohm/inch. A 10-inch long trace would have a resistance of 0.2 Ohm/inch × 10 inches = 2 Ohms.

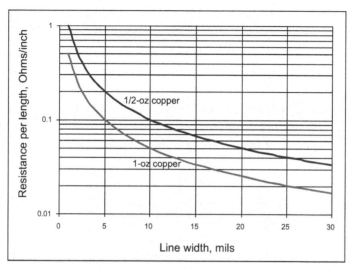

Figure 4-7   The resistance per length for traces with different line widths in 1-ounce copper and 1/2-ounce copper.

It is important to keep in mind that these resistances calculated so far are all resistances at DC, or at least at low frequency. As we show in Chapter 6, the resistance of a trace will increase with frequency due to skin-depth-related effects. The bulk resistivity of the conductor does not change; the current distribution through the conductor changes. Higher-frequency signal components will travel through a thin layer near the surface, decreasing the cross-sectional area. For 1-ounce copper traces, the resistance begins to increase at about 20 MHz and will increase roughly with the square root of frequency. It's all related to inductance.

4.6 The Bottom Line

1. Translating physical features into an electrical model is a key step in optimizing system electrical performance.
2. The first step in calculating the resistance of an interconnect is assuming that the equivalent circuit model is a simple ideal resistor.
3. The most useful approximation for the end-to-end resistance of an interconnect is R = ρ × length/cross-sectional area.

4. Bulk resistivity is an intrinsic material property, independent of the amount of material.

5. If the structure is not uniform in cross section, either it must be approximated as uniform or a field solver should be used to calculate its resistance.

6. Resistance per length of a uniform trace is constant. A 10-mil-wide trace in 1/2-ounce copper has a resistance per length of 0.1 Ohm/inch.

7. Every square cut from the same sheet will have the same edge-to-edge resistance.

8. Sheet resistance is a measure of the edge-to-edge resistance of one square of conductor cut from the sheet.

9. For 1/2-ounce copper, the sheet resistance is 1 mOhm/square.

10. The resistance of a conductor will increase at higher frequency due to skin-depth effects. For 1-ounce copper, this begins above 20 MHz.

End-of-Chapter Review Questions

The answers to the following review questions can be found in Appendix D, "Review Questions and Answers."

4.1 What three terms influence the resistance of an interconnect?

4.2 While almost every resistance problem can be calculated using a 3D field solver, what is the downside of using a 3D field solver as the first step to approaching all problems?

4.3 What is Bogatin's rule #9, and why should this always be followed?

4.4 What are the units for bulk resistivity, and why do they have such strange units?

4.5 What is the difference between resistivity and conductivity?

4.6 What is the difference between bulk resistivity and sheet resistivity?

4.7 If the length of an interconnect increases, what happens to the bulk resistivity of the conductor? What happens to the sheet resistance of the conductor?

4.8 What metal has the lowest resistivity?

4.9 How does the bulk resistivity of a conductor vary with frequency?

4.10 Generally, will the resistance of an interconnect trace increase or decrease with frequency? What causes this?

4.11 If gold has a higher resistivity than copper, why is it used in so many interconnect applications?

4.12 What is the sheet resistance of ½-ounce copper?

4.13 A 5-mil wide trace in ½-ounce copper is 10 inches long. What is its total DC resistance?

4.14 Why does every square cut out of the same sheet of conductor have the same edge to edge resistance?

4.15 When you calculate the edge-to-edge resistance of a square of metal, what is the fundamental assumption you are making about the current distribution in the square?

4.16 What is the resistance per length of a signal line 5 mils wide in ½-ounce copper?

4.17 Surface traces are often plated up to 2-ounce copper thickness. What is the resistance per length of a 5-mil wide trace on the surface compared to on a stripline layer where it is ½-ounce thick?

4.18 To measure the sheet resistance of ½-ounce copper using a 4-point probe to 1%, you are resolving a resistance of 1 uOhm. If you use a current of 100 mA, what is the voltage you have to resolve to see such a small resistance?

4.19 Which has higher resistance: a copper wire 10 mil in diameter and 100 inches long, or a copper wire 20 mils in diameter but only 50 inches long? What if the second wire were made of tungsten?

4.20 What is a good rule of thumb for the resistance per length of a wirebond?

4.21 Estimate the resistance of a solder ball used in a chip attach application in the shape of a cylinder, 0.15 mm in diameter and 0.15 mm long with a bulk resistivity of 15 uOhm-cm. How does this compare to a wire bond?

4.22 The bulk resistivity of copper is 1.6 uOhms-cm. What is the resistance between opposite faces of a cube of copper 1 cm on a side? What if it is 10 cm on a side?

4.23 Generally, a resistance less than 1 Ohms is not significant in the signal path. If the line width of ½-ounce copper is 5 mils, how long could a trace be before its DC resistance is > 1 Ohms?

4.24 The drilled diameter of a via is typically 10 mils. After plating it is coated with a layer of copper equivalent to about ½-ounce copper. If the via is 64 mils long, what is the resistance of the copper cylinder inside the via?

4.25 Sometimes, it is recommended to fill the via with silver filled epoxy, with a bulk resistivity of 300 uOhm-cm. What is the resistance of the fillet of silver filled epoxy inside a through via? How does this compare with the copper resistance? What might be an advantage of a filled via?

4.26 Engineering change wires on the surface of a board sometimes use 24 AWG wire. If the wire is 4 inches long, what is the resistance of the wire?

CHAPTER 5

The Physical Basis of Capacitance

A capacitor is physically made up of two conductors, and between every two conductors there is some capacitance.

TIP   The capacitance between any two conductors is basically a measure of their capacity to store charge, at the cost of a voltage between them.

If we take two conductors and add positive charge to one of them and negative charge to the other, as illustrated in Figure 5-1, there will be a voltage between them. The capacitance of the pair of conductors is the ratio of the amount of charge stored on each conductor per voltage between them:

$$C = \frac{Q}{V} \tag{5-1}$$

where:

C = capacitance, in Farads

Q = total amount of charge, in Coulombs

V = voltage between the conductors, in volts

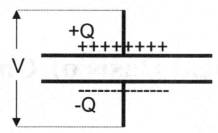

Figure 5-1 Capacitance is a measure of the capacity to store charge for a given voltage between the conductors.

Voltage is the price paid to store charge. The more charge that can be stored for a fixed voltage, the higher the capacitance of the pair of conductors.

Capacitance is also a measure of the efficiency with which two conductors can store charge, at the cost of voltage. A higher capacitance means the conductors are more efficient at storing more charge for the same cost of voltage. If the voltage across the conductors that make up a capacitor increases, the charge stored increases, but the capacitance does not change, and the efficiency of storing charge does not change.

The actual capacitance between two conductors is determined by their geometry and the material properties of any dielectrics nearby. It is completely independent of the voltage applied. If the geometry of the conductors changes, the capacitance will change. The more closely the conductors are brought together, or the more their areas overlap, the greater their capacitance.

Capacitance plays a key role in describing how signals interact with interconnects and is one of the four fundamental ideal circuit elements used to model interconnects.

Even when two conductors have no DC path between them, they can have capacitance. Their impedance will decrease with frequency and could result in a very low impedance between them at high frequency. Because of the potential fringe electric fields between any two conductors, in signal integrity applications, there is no such thing as an "open."

TIP   What makes capacitance so subtle is that even though there may be no direct wire connection between two conductors, which might be two different signal traces, there will always be some capacitance between them. This capacitance will allow a current flow in some cases, especially at higher frequency, which can contribute to cross talk and other signal-integrity problems.

By understanding the physical nature of capacitance, we will be able to see with our mind's eye this sneak path for current flow.

5.1 Current Flow in Capacitors

There is no DC path between the two conductors that are separated by a dielectric material in an ideal capacitor. Normally, we would think there could not be any current flow through a real capacitor. After all, there is insulating dielectric between the conductors. How could we get current flow through the insulating dielectric? As we showed in Chapter 3, "Impedance and Electrical Models," it is possible to get current through a capacitor, but only in the special case when the voltage between the conductors changes.

The current through a capacitor is related by:

$$I = \frac{\Delta Q}{\Delta t} = C \frac{dV}{dt} \qquad (5\text{-}2)$$

where:

I = current through the capacitor

ΔQ = change in charge on the capacitor

Δt = time it takes the charge to change

C = capacitance

dV = voltage change between the conductors

dt = time period for the changing voltage

TIP   Capacitance is also a measure of how much current we can get through a pair of conductors when the voltage between them changes.

If the capacitance is large, we get a lot of current through it for a fixed dV/dt. The impedance, in the time domain, would be low if the capacitance were high.

How does the current flow through the empty space between the conductors? There is an insulating dielectric. Of course, we can't have any real conduction current through the insulating dielectric, but it sure looks that way.

Rather, there is apparent current flow. If we increase the voltage on the conductors, for example, we must add + charges to one conductor and push out + charges from the other. It looks like we add the charges to one conductor, and they come out of the other. Current effectively flows through a capacitor when the voltage across the conductors changes.

We often refer to the current that effectively flows through the empty space of the capacitor as *displacement current*. This is a term we inherited from James Clerk Maxwell, the father of electromagnetism. In his mind, the current through the empty space between the conductors, when the voltage changed, was due to the increasing separation of charges, or polarization, in the *ether*.

To his 1880s mind, the vacuum was not empty. Filling the vacuum, and the medium for light to propagate, was a tenuous medium called ether. When the voltage between conductors changed, charges in the ether were pulled apart slightly, or displaced. It was their motion, while they were being displaced, that he imagined to be the current, and he termed the displacement *current*.

As distinct from conduction current, which is the motion of free charges in a conductor, *polarization current* is the motion of bound changes in a dielectric when its polarization changes, as when the electric field inside the material changes. Displacement current is the special case of current flow in a vacuum when electric field changes. This is a fundamental property of spacetime in which the properties of electric fields was frozen in about 1 nanosecond after the Big Bang.

TIP   Whenever you see fringe electric fields between conductors, imagine that displacement current will flow along those field lines when they change.

5.2 The Capacitance of a Sphere

The actual amount of capacitance between two conductors is related to how many electric field lines would connect the two conductors. The closer the spacing, the greater the area of overlap, the more field lines would connect the conductors, and the larger the capacity and efficiency to store charge.

Each specific geometrical configuration of conductors will have a different relationship between the dimensions and the resulting capacitance. With few exceptions, most of the formulae that relate geometry to capacitance are approximations. In general, we can use a field solver to accurately calculate the capacitance between every pair of conductors in any arbitrary collection of conductors, such as with multiple pins in a connector. There are a few special geometries where Maxwell's Equations can be solved exactly and where exact analytical expressions exist. One of these is for the capacitance between two concentric spheres, one inside the other.

The capacitance between the two spheres is:

$$C = 4\pi\varepsilon_0 \frac{rr_b}{r_b - r} \tag{5-3}$$

where:

C = capacitance, in pF

ε_0 = permittivity of free space = 0.089 pF/cm, or 0.225 pF/inch

r = radius of the inner sphere, in inches or cm

r_b = radius of the bigger, outer sphere, in inches or cm

When the outer sphere's radius is more than 10 times the inner sphere's, the capacitance of the sphere is approximated by:

$$C \approx 4\pi \times \varepsilon_0 \times r \tag{5-4}$$

where:

C = capacitance, in pF

ε_0 = permittivity of free space = 0.089 pF/cm, or 0.225 pF/inch

r = radius of the sphere, in inches or cm

For example, a sphere 0.5 inch in radius, or 1 inch in diameter, has a capacitance of C = $4\pi \times$ 0.225 pF/in $\times$ 0.5 in = 1.8 pF. As a rough rule of thumb, a sphere with a 1-inch diameter has a capacitance of about 2 pF.

TIP   This subtle relationship says that any conductor, just sitting, isolated in space, has some capacitance with respect to even the earth's surface. It does not get smaller and smaller. It has a minimum amount of capacitance, related to its diameter. The closer to a nearby surface, the higher the capacitance might be above this minimum.

This means that if a small wire pigtail, even only a few inches in length, were sticking outside a box, it could have a stray capacitance of at least 2 pF. At 1 GHz, the impedance to the earth, or the chassis, would be about 100 Ohms. This is an example of how subtle capacitance can be and how it can create significant sneak current paths, especially at high frequency.

5.3 Parallel Plate Approximation

A very common approximation is the parallel plate approximation. For the case of two flat plates, as shown in Figure 5-2, separated by a distance, h, with total area, A, with just air between the plates, the capacitance is given by:

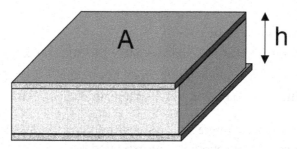

Figure 5-2   Most common approximation for the capacitance and geometry is for a pair of parallel plates with plate area, A, and separation, h.

$$C = \varepsilon_0 \frac{A}{h} \tag{5-5}$$

where:

C = capacitance, in pF

ε_0 = permittivity of free space = 0.089 pF/cm, or 0.225 pF/inch

A = area of the plates

h = separation between the plates

For example, a pair of plates that look like the faces of a penny, about 1 cm^2 in area, separated by 1 mm of air, has a capacitance of C = 0.089 pF/cm × 1 cm^2/0.1 cm = 0.9 pF. This is a good way of getting a feel for 1 pF of capacitance. It is roughly the capacitance associated with plates of comparable size to the faces of a penny.

TIP   This relationship points out the important geometrical features for all capacitors. The farther apart the conductors, the lower the capacitance, and the larger the area of overlap, the greater the capacitance.

It is always important to keep in mind that with the exception of only a few equations, every equation used in signal integrity is either a definition or an approximation. The parallel plate approximation is an approximation. It assumes that the fringe fields around the perimeter of the plates are negligible. The thinner the spacing or the wider the plates, the better this approximation. For the case of plates that are square and of dimension w on a side, the approximation gets better as w/h gets larger.

In general, the parallel plate approximation underestimates the capacitance. It only accounts for the field lines that are vertical between the two conductors. It does not include the fringe fields along the sides of the conductors.

The actual capacitance is larger than the approximation because of the contribution of the fringe fields from the edge. As a rough rule of thumb, when the distance between the plates is equal to a lateral dimension, so the plates look like a cube, the actual capacitance between the two plates is roughly twice what the parallel plate approximation predicts. In other words, the fringe fields from the edge contribute an equal amount of capacitance as the parallel plate approximation predicts, when the spacing is comparable to the width of a plate.

5.4 Dielectric Constant

The presence of an insulating material between the conductors will increase the capacitance between them. The special material property that causes the capacitance to increase is called the *relative dielectric constant*. We usually use the Greek letter epsilon plus a subscript r (ε_r) to describe the relative dielectric constant of a material. Alternatively, we use the abbreviation Dk to designate the dielectric constant of a material. It is the dielectric constant, relative to air, which has a dielectric constant of 1. As a ratio, there are no units for relative dielectric constant. We often leave off the term *relative*.

Dielectric constant is an intrinsic bulk property of an insulator. A small piece of epoxy will have the same dielectric constant as a large chunk of the same material. The way to measure the dielectric constant of an insulator is to compare the capacitance of a pair of conductors when they are surrounded by air, C_0, and when they are completely surrounded by the material, C. The dielectric constant of the material is defined as:

$$\varepsilon_r = \frac{C}{C_0} \tag{5-6}$$

where:

ε_r = relative dielectric constant of the material

C = capacitance when conductors are completely surrounded by the material

C_0 = capacitance when air completely surrounds the conductors

The higher the dielectric constant, the more the capacitance between the fixed conductors is increased by the material. The dielectric constant will increase the capacitance of any two conductors, completely independently of their shape (whether they are shaped like a parallel plate, two rods, or one wire near a wide plane), provided that all space in the vicinity of the conductors is uniformly filled with the material.

Figure 5-3 lists the dielectric constants of many common insulators used in interconnects. For most polymers, the dielectric constant is about 3.5 to 4.5. This says that the capacitance between two electrodes is increased by a factor of roughly four by the addition of a polymer material. It is important to note that the dielectric constant of most polymers will vary due to processing conditions, degree of cure, and any filler materials used. There may be some frequency dependence. If it is important to know the dielectric constant to better than 10%, the dielectric constant of the sample should be measured.

The dielectric constant of a material is roughly related to the number of dipoles and their size. A material having molecules with a lot of dipoles, such as water, will have a high dielectric constant (over 80). A material with very few dipoles, such as air, will have a low dielectric constant (1). The lowest dielectric constant of any solid, homogeneous material is about 2, which is for Teflon.

The dielectric constant can be decreased by adding air to the material. Foams have a dielectric constant that can approach 1. At the other extreme, some ceramics, such as barium titanate, have dielectric constants as large as 5000.

Material	Dielectric Constant
Air	1
Teflon	2.1
Polyethylene	2.3
BCB	2.6
PTFE	2.8
Polyimide	3.4
GETEK	3.6–4.2
BT/Glass	3.7–3.9
Quartz	3.8
Kapton	4
FR4	4–4.5
Glass-Ceramic	5
Diamond	5.7
Alumina	9–10
Barium Titanate	5,000

Figure 5-3 Dielectric constants of commonly used interconnect dielectric materials.

The dielectric constant will sometimes vary with frequency. For example, from 1 kHz to 10 MHz, the dielectric constant of FR4 can vary from 4.8 to 4.4. However, from 1 GHz to 10 GHz, the dielectric constant of FR4 can be very constant. The exact value of the dielectric constant for FR4 varies depending on the relative amount of epoxy resin and glass weave. To remove the ambiguity, it is important to specify the frequency at which the dielectric constant is measured.

When the frequency dependence to the dielectric constant is important, we can describe the frequency-dependent behavior with *causal models*. These models usually assume a Dk that varies with the log of frequency, and a *dissipation factor*, Df, that is a measure of the slope of the frequency variation. A higher dissipation factor means more frequency dependence to the Dk and a higher loss materials.

When using a causal model to describe the frequency dependence of the dielectric constant, it is only necessary to specify the Dk and Df at one frequency, usually 1 GHz, and their value at all other frequencies can be calculated.

5.5 Power and Ground Planes and Decoupling Capacitance

One of the most important applications of the parallel plate approximation is to analyze the capacitance between the power and ground planes in an IC package or in a multilayer printed circuit board.

As we will show later, in order to reduce the voltage rail collapse in the power-distribution system, it is important to have a lot of decoupling capacitance between the power and ground return. A capacitance, C, will prevent the droop in the power voltage for a certain amount of time, δt.

One metric for the current draw on a power rail is the power dissipation since this is the current draw multiplied by the voltage of the rail. If the power dissipation of the chip is P, the current is I = P/V. The time until the voltage droop increases to 5% of the supply voltage because of the decoupling capacitance, is approximately:

$$\delta t = C \times 0.05 \times \frac{V^2}{P} \qquad (5\text{-}7)$$

where:

δt = time in seconds before droop exceeds 5%

C = decoupling capacitance, in Farads

0.05 = 5% voltage droop allowed

P = average power dissipation of the chip, in watts

V = supply voltage, in volts

For example, if the chip power is 1 watt, and the capacitance available for decoupling is 1 nF, with a supply voltage of 3.3 v, the time the capacitance will provide decoupling is δt = 1 nF × 0.05 × 3.3²/1 = 0.5 nsec. This is not a very long period of time compared to what is required.

More typically, enough decoupling capacitance is required to provide decoupling for at least 5 μsec, until the power supply regulator can provide adequate current. In this example, we would actually need more than 10,000 times this decoupling capacitance, or 10 µF, to provide adequate decoupling.

It is often erroneously assumed that the capacitance found in the power and ground planes of the circuit board will provide a significant amount of decoupling. By using the parallel plate approximation, we can put in the numbers and evaluate just how much decoupling capacitance they provide and how long the planes can decouple a chip.

In a multilayer circuit board, with the power plane on an adjacent layer to the ground plane, we can estimate the capacitance between the layers, per square inch of area. The capacitance is given by:

$$C = \varepsilon_0 \varepsilon_r \frac{A}{h} \tag{5-8}$$

where:

C = capacitance, in pF

ε_0 = permittivity of free space = 0.089 pF/cm, or 0.225 pF/inch

ε_r = relative dielectric constant of the FR4, typically ~ 4

A = area of the planes

h = separation between the planes

For the case of FR4, having a dielectric constant of 4, the capacitance for a 1 in² of planes is C = 0.225 pF/inch × 4 × 1 in²/h ~ 1000 pF/h, with h in mils. This is 1 Nf/mil of thickness per square in of board area. This is a simple rule of thumb to remember.

If the dielectric spacing is 10 mils, a very common thickness, the capacitance between the power and ground planes is only 100 pF for 1 in² of planes.

If there are 4 in² of board area allocated for the ASIC, then the total plane-to-plane decoupling capacitance available in the power and ground planes of the board is only 0.4 nF. This is more than four orders of magnitude below the required 10-µF capacitance.

How long will this amount of capacitance provide decoupling? Using the relationship above, the amount of time the 0.4-nF capacitance in the planes would provide decoupling is 0.2 nsec. This is not a very significant amount of time. Furthermore, the chip would have to see this small capacitance through its package leads. The impedance of the package leads would make this 0.4-nF plane capacitance almost invisible. In addition, the capacitance already integrated on a chip is typically more than 100 times this plane-to-plane capacitance.

> **TIP**   Though there is plane-to-plane capacitance in a multilayer circuit board, in general, it is too small to play a significant role in power management. As we show in Chapter 6, "The Physical Basis of Inductance," the real role of the power and ground planes is to provide a low-inductance path between the chip and the bulk decoupling capacitors. It is not to provide decoupling capacitance.

What could be done to dramatically increase the capacitance in the power and ground planes? The parallel plate approximation points out only two knobs that affect the capacitance: the dielectric thickness and the dielectric constant. The thinnest layer of FR4 in commercial production is 2 mils. This results in a capacitance per area of about $1000 \text{ pF/in}^2/2 = 500 \text{ pF/in}^2$. For this example, if the area allocated for decoupling a chip is two inches on a side, the total capacitance would be $500 \text{ pF/in}^2 \times 4 \text{ in}^2$, or 2 nF. The time this would keep the voltage from collapsing is about 1 nsec—still not very significant.

However, if the dielectric thickness can be made thin enough and the dielectric constant high enough, a significant amount of capacitance can be designed into the power and ground planes. Figure 5-4 shows the capacitance per square inch as the dielectric thickness changes, for four different dielectric constants of 1, 4, 10, and 20. Obviously, if the goal is to increase the capacitance per area, the way to do this is use thin dielectric layers and high dielectric constant.

An example of such a material is in development by 3M, under the trade name C-Ply. It is composed of ground-up barium titanate in a polymer matrix. The dielectric constant is 20, and the layer thickness is 8 microns, or 0.33 mil. With ½-ounce copper layers laminated to each side, the capacitance per area of one layer is $C/A = 0.225 \text{ pF/inch} \times 20 / 0.33 \text{ mil} \sim 14 \text{ nF/in2}$. This is about 30 times greater than the best alternative.

A 4-in 2 region of a board with a layer of C-Ply would have about 56 nF of decoupling capacitance. For the 1-watt chip, this would provide decoupling for about 28 nsec, a significant amount of time.

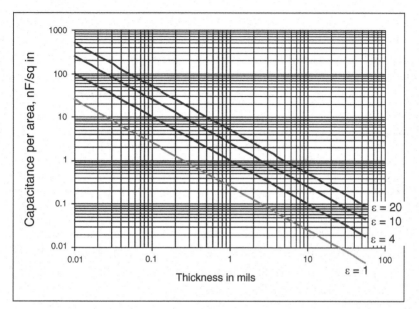

Figure 5-4 Capacitance per area for power and ground planes with four different dielectric constants and different plane-to-plane thicknesses.

5.6 Capacitance per Length

Most uniform interconnects have a signal path and a return path with a fixed cross section. In this case, the capacitance between the signal trace and its return path scales with the length of the interconnect. If the interconnect length doubles, the total capacitance of the trace will double. It is convenient to describe the capacitance of the line by the capacitance per length. As long as the cross section remains uniform, the capacitance per length will be constant.

In the special case of uniform cross-section interconnects, the total capacitance between the signal and return path is related to:

$$C = C_L \times Len \tag{5-9}$$

where:

C = total capacitance of the interconnect

C_L = capacitance per length

Len = length of the interconnect

There are three cross sections for which Maxwell's Equations can be solved in cylindrical coordinates exactly (see Figure 5-5). For these structures, the capacitance per length can be calculated exactly based on the cross section. They are good calibration structures to test any field solver. In addition, there are many other approximations for other geometries, but they are approximations.

TIP   In general, when the cross section is uniform, a 2D field solver can be used to very accurately calculate the capacitance per length of any arbitrary shape.

A coax cable is an interconnect with a central round conductor surrounded by dielectric material and then enclosed by an outer round conductor. The center conductor is usually termed the *signal path*, and the outer conductor is termed the *return path*. The capacitance per length between the inner conductor and the outer conductor is given exactly by:

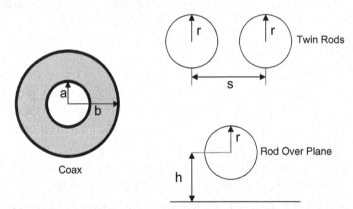

Figure 5-5   The three cross-section geometries for which there are very good approximations for the capacitance per length: coax, twin rods, and rod over plane.

$$C_L = \frac{2\pi\varepsilon_0\varepsilon_r}{\ln\left(\dfrac{b}{a}\right)}$$ (5-10)

where:

C_L = capacitance per length

ε_0 = permittivity of free space = 0.089 pF/cm, or 0.225 pF/inch

ε_r = relative dielectric constant of the insulation

a = inner radius of the signal conductor

b = outer radius of the return conductor

For example, in an RG58 coax cable (the most common type of coax cable, typically with BNC connectors on the ends), the ratio of the outer to the inner diameters is (1.62 mm/0.54 mm) = 3, with a dielectric constant of 2.3 for polyethylene, the capacitance per length is:

$$C_L = \frac{2\pi \times 0.225 \times 2.3}{\ln(3)} = 2.9 \frac{pF}{inch} \tag{5-11}$$

A second exact relationship is for the capacitance between two parallel rods. It is given by:

$$C_L = \frac{\pi \varepsilon_0 \varepsilon_r}{\ln\left\{ \frac{s}{2r} \left[1 + \sqrt{1 - \left(\frac{2r}{s}\right)^2} \right] \right\}} \tag{5-12}$$

where:

C_L = capacitance per length

ε_0 = permittivity of free space = 0.089 pF/cm, or 0.225 pF/inch

ε_r = relative dielectric constant of the insulation

s = center-to-center separation of the rods

r = radius of the two rods

If the spacing between the rods is large compared with their radius (i.e., s >> r), this relatively complex relationship can be approximated by:

$$C_L = \frac{\pi \varepsilon_0 \varepsilon_r}{\ln\left\{\frac{s}{r}\right\}} \tag{5-13}$$

Both cases assume that the dielectric material surrounding the two rods is uniform everywhere. Unfortunately, this is not often the case, and this approximation is not very useful except in special cases, such as wire bonds in air. For example, two parallel wire bonds, both of radius 0.5-mil and 5-mil center-to-center separation, have a capacitance per length of about:

$$C_L = \frac{\pi \varepsilon_0 \varepsilon_r}{\ln\left\{\frac{s}{r}\right\}} = \frac{3.14 \times 0.225 \times 1}{\ln\left\{\frac{5}{0.5}\right\}} = 0.3 \frac{pF}{inch} \tag{5-14}$$

If they are 40 mils long, the total capacitance is $0.3 \times 0.04 = 0.012$ pF.

The third exact relationship is for a rod over a plane. The capacitance, when the rod is far from the plane (i.e., h >> r), is approximately:

$$C_L = \frac{2\pi \varepsilon_0 \varepsilon_r}{\ln\left\{\frac{2h}{r}\right\}} \tag{5-15}$$

where:

C_L = capacitance per length

ε_0 = permittivity of free space = 0.089 pF/cm, or 0.225 pF/inch

ε_r = relative dielectric constant of the insulation

h = center of the rod to surface of the plane

r = radius of the rod

Two other useful approximations for cross sections are commonly found in circuit-board interconnects. These are for microstrip and stripline interconnects, as illustrated in Figure 5-6.

In microstrip interconnects, a signal trace rests on top of a dielectric layer that has a plane below. This is the common geometry for surface traces in a multilayer board. In a stripline, two planes provide the return path on either side of the signal trace. Whether or not the two planes actually have a DC connection between them, for high-frequency signals, they are effectively shorted together and can be considered connected. A signal trace is symmetrically spaced between them. A dielectric material, the board laminate, completely surrounds the signal conductors. In both cases, the capacitance per length between the signal trace and return path is calculated.

Though many approximations exist in the literature, the two offered here are recommended by the IPC, the industry association for the printed circuit board industry. The capacitance per length of a microstrip is given by:

$$C_L = \frac{0.67(1.41+\varepsilon_r)}{\ln\left\{\dfrac{5.98\times h}{0.8\times w+t}\right\}} \cong \frac{0.67(1.41+\varepsilon_r)}{\ln\left\{7.5\left(\dfrac{h}{w}\right)\right\}} \tag{5-16}$$

where:

C_L = capacitance per length, in pF/inch

ε_r = relative dielectric constant of the insulation

h = dielectric thickness, in mils

w = line width, in mils

t = thickness of the conductor, in mils

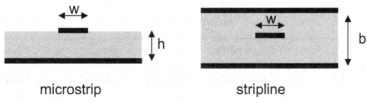

microstrip   stripline

Figure 5-6   The cross-section geometries for microstrip and stripline interconnects illustrating the important geometrical features.

It is important to keep in mind that though there is a parameter to include the thickness of the trace, if a problem requires the level of accuracy where the impact from the trace thickness is important, this approximation should not be used. Rather, a 2D field solver should be used. For all practical purposes, the accuracy of this tool will not be affected if the trace thickness is assumed to be 0.

If the line width is twice the dielectric thickness (i.e., $w = 2 \times h$), and the dielectric constant is 4, the capacitance per length is about $C_L = 2.7$ pF/inch. These are the dimensions of a microstrip that is approximately a 50-Ohm transmission line.

The capacitance per length of a stripline, as illustrated in Figure 5-6, is approximated by:

$$C_L = \frac{1.4\varepsilon_r}{\ln\left\{\dfrac{1.9 \times b}{0.8 \times w + t}\right\}} \cong \frac{1.4\varepsilon_r}{\ln\left\{2.4\left(\dfrac{b}{w}\right)\right\}} \tag{5-17}$$

where:

C_L = capacitance per length in pF/inch

ε_r = relative dielectric constant of the insulation

b = total dielectric thickness, in mils

w = line width, in mils

t = thickness of the conductor, in mils

For example, if the total dielectric thickness, b, is twice the line width, $b = 2w$, corresponding to roughly a 50-Ohm line, the capacitance per length is $C_L = 3.8$ pF/inch.

In both geometries, we see that the capacitance per length of a 50-Ohm line is about 3.5 pF/inch. This is a good rule of thumb to keep in mind.

TIP   As a rough rule of thumb, the capacitance per length of a 50-Ohm transmission line in FR4 is about 3.5 pF/inch.

For example, in a multilayer BGA package, signal traces are designed as microstrip geometries and are roughly 50-Ohm characteristic impedance. The dielectric material bismaleimide triazine (BT), has a dielectric constant of about 3.9. The capacitance per length of a signal trace is roughly 3.5 pF/inch. A trace that is 0.5 inch long will have a capacitance of about 3.5 pF/inch × 0.5 inch = 1.7 pF. The capacitive load of a receiver would be approximated as the roughly 2 pF of input-gate capacitance and the 1.7 pF of the lead capacitance, or about 3.7 pF.

TIP   It is important to keep in mind that these approximations are *approximations*. If it is important to have confidence in the accuracy, the capacitance per length should be calculated with a 2D field solver.

5.7 2D Field Solvers

When accuracy is important, the best numerical tool to use to calculate the capacitance per length between any two conductors is a 2D field solver. These tools assume that the cross section of the conductors is constant down their length. When this is the case, the capacitance per length is also constant down the length.

A 2D field solver will solve LaPlace's Equation, one of Maxwell's Equations, using the geometry of the conductors as the boundary conditions. The process that most tools use, without the user really needing to know, is to set the voltage of one conductor to be 1 volt and solving for the electric fields everywhere in the space. The charge on the conductors is then calculated from the electric fields. The capacitance per length between the two conductors is directly calculated as the ratio of the charge on the conductors per the 1 v applied.

One way of evaluating the accuracy of a 2D field solver is to use the tool to calculate a geometry for which there is an exact expression, such as a coax or twin-rod structure. For example, Figure 5-7 shows the comparison of the capacitance per length calculated by a field solver for the twin-rod structure, compared to the exact expression above. The agreement is seen to be excellent. To quantify the residual error, the relative difference between the field-solver result and the analytic result is plotted. The worst-case error is seen to be less than 1% for this specific tool.

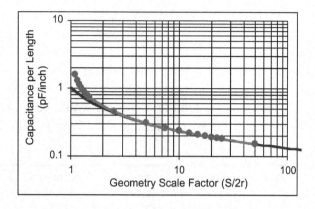

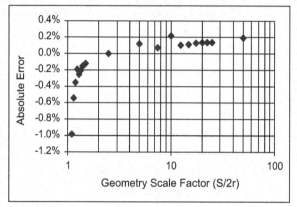

Figure 5-7 Calculated capacitance per length of twin rods, comparing Ansoft's 2D Extractor field solver and the exact formula and the approximation. Top: The points are from the field solver; the line through them is the exact expression, and the other line is the approximation. For s > 4r, the approximation is excellent. Bottom: Absolute error of the Ansoft 2D field solver is seen to be better than 1%.

TIP   For any arbitrary geometry, it is possible to get better than 1% absolute error by using a field solver to calculate the capacitance per length.

Using a field solver whose accuracy has been verified, it is interesting to evaluate the accuracy of some of the popular approximations, such as the capacitance per length for a microstrip or stripline structure.

Figure 5-8 compares the approximation above for a microstrip and the field-solver result. In some cases, the approximation can be as good as 5%, but in other cases, the results differ by more than 20%. No approximation should be relied on for better than 10%–20% unless previously verified.

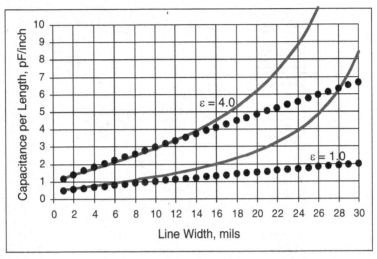

Figure 5-8 Calculated capacitance per length of a microstrip with 5-mil-thick dielectric, and dielectric constants of 4 and 1, comparing Ansoft's SI2D field solver and the IPC approximation above, as the line width is increased. Circles are the field-solver results; the lines are the IPC approximation.

An additional advantage of using a field solver is the ability to take into account second-order effects. One important effect is the impact from increasing trace thickness on the capacitance per length for a microstrip.

Before we blindly use the field solver to get a result, we should always apply rule 9 and anticipate what we expect to see.

As the metal thickness increases, from very thin to very thick, the fringe fields between the signal line and the return plane will increase. In fact, as shown in Figure 5-9, the capacitance per length does increase with increasing trace thickness, but not very much. It increases only about 3% from very thin to 3 mils thick, or 2-ounce copper.

There are no approximations that can accurately predict these sorts of effects. For comparison, we show how the IPC approximation for the capacitance matches the more accurate field-solver results. The poor agreement is the reason the approximation should not be used to evaluate the impact of second-order effects, such as trace thickness.

A 2D field solver is a very important tool for calculating the electrical properties of any uniform cross-section interconnect. It is especially important when the dielectric material is not homogeneously distributed around the conductors.

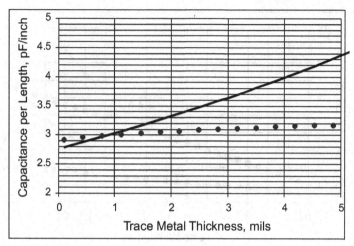

Figure 5-9   Effect on capacitance per length of a microstrip as the trace thickness is increased from 0.1 mil to 5 mils with 5-mil-thick dielectric and 10-mil-wide conductor. Circles are the field-solver results; the line is the IPC approximation.

5.8 Effective Dielectric Constant

In any cross section where the dielectric material completely surrounds the conductors, all the field lines between the conductors will see the same dielectric constant. This is the case for stripline, for example. However, if the material is not uniformly distributed around the conductors, such as with a microstrip or twisted pair or coplanar, some field lines may encounter air, while other field lines encounter the dielectric. An example of the field lines in microstrip is shown in Figure 5-10.

The presence of the insulating material will increase the capacitance between the conductors compared to if the material was not there. When the material is homogeneously distributed between and around the conductors, such as in stripline, the material will increase the capacitance by a factor equal to the dielectric constant of the material.

In microstrip, some of the field lines encounter air, and some encounter the dielectric constant of the laminate. The capacitance between the signal path and the plane will increase due to the material between them—but by how much?

The combination of air and partial filling of the dielectric creates an *effective dielectric constant*. Just as the dielectric constant is the ratio of the filled capacitance to the empty capacitance, the effective dielectric constant is the ratio of the capacitance when the material is in place with whatever distribution is really there, compared to the capacitance between the conductors when there is only air surrounding them.

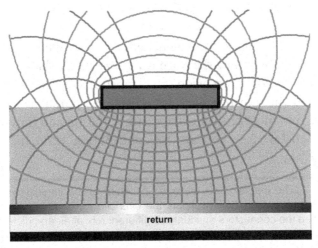

Figure 5-10   Field lines in a microstrip showing some in air and some in the bulk material. The effective dielectric constant is a combination of the dielectric constant of air and the bulk dielectric constant. Field lines calculated using Mentor Graphics HyperLynx.

The first step in calculating the effective dielectric constant, ε_{eff}, is to calculate the capacitance per length between the two conductors with empty space between them, C_0. The second step is to add the material as it is distributed and calculate, using the 2D field solver, the resulting capacitance per length, C_{filled}. The effective dielectric constant is just the ratio:

$$\varepsilon_{eff} = \frac{C_{filled}}{C_0}$$

(5-18)

where:

C_0 = empty space capacitance

C_{filled} = capacitance with the actual material distribution

ε_{eff} = effective dielectric constant

Both the empty and filled capacitance can be accurately calculated using a 2D field solver. The only accurate way of calculating the effective dielectric constant of a transmission line is by using a 2D field solver. As we show in Chapter 7, "The Physical Basis of Transmission Lines," the effective dielectric constant is a very important performance term, as it directly determines the speed of a signal in a transmission line.

Figure 5-11 shows the calculated effective dielectric constant for a microstrip as the width of the trace is increased. The bulk dielectric constant in this case is 4. For a very wide trace, most of the field lines are in the bulk material, and the effective dielectric constant approaches 4. When the line width is narrow, most of the field lines are in the air, and the effective dielectric constant is below 3, reflecting the contribution from the lower dielectric constant of air.

> **TIP**   The intrinsic bulk dielectric constant of the laminate material is not changing. Only how it affects the capacitance changes as the fields between the conductors encounter a different mix of air and dielectric.

If dielectric material is added to the top surface of the microstrip, the fringe field lines that were in air will see a higher dielectric constant, and the capacitance of the microstrip will increase. With dielectric above, we call this an *embedded microstrip*. When only some of the field lines are in the material, it is called a *partially embedded microstrip*. This is the case for a soldermask coating, for example. When all the field lines are covered by dielectric, it is called a *fully embedded microstrip*. Figure 5-12 shows the field line distribution for three different degrees of embedded microstrip.

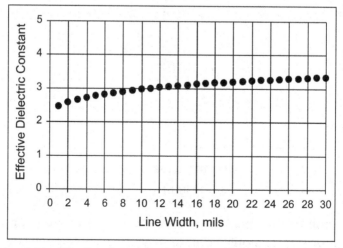

Figure 5-11   Effective dielectric constant of a microstrip with a 5-mil-thick dielectric, as the line width is increased. The dielectric constant of the bulk material is 4. Results are calculated with the Ansoft 2D Extractor.

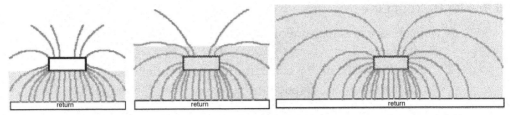

Figure 5-12   The electric field distribution around a microstrip with different thickness covering. For a thick enough layer, all the fields are confined in the bulk material, and the capacitance will be independent of more thickness. Simulations performed with Mentor Graphics HyperLynx.

How much material would have to be added to a microstrip to completely cover all the field lines and have the bulk dielectric constant match the effective dielectric constant? This is an easy problem to solve with a 2D field solver. Figure 5-13 shows the calculated capacitance per length for a microstrip with 5-mil-thick dielectric and 10-mil-wide trace as a top layer of dielectric is added with the same bulk dielectric constant value as the laminate, 4.

In this example, a thickness of dielectric on top of the trace about equal to the line width is required to fully cover all the fringe fields.

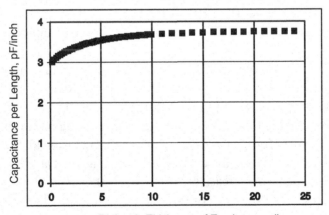

Dielectric Thickness of Top Layer, mils

Figure 5-13   Capacitance per length of microstrip as top dielectric thickness is increased. Calculated using Ansoft's 2D Extractor.

5.9 The Bottom Line

1. Capacitance is a measure of the capacity to store charge between two conductors.
2. Current flows through a capacitor when the voltage between the conductors changes. Capacitance is a measure of how much current will flow.
3. With few exceptions, every formula relating the capacitance of two conductors is an approximation. If better than 10%–20% accuracy is required, approximations should not be used.
4. The only three exact expressions are for coax, twin rods, and rod over ground.
5. In general, the farther apart the conductors, the lower the capacitance. The greater the overlap of their areas, the higher the capacitance.
6. Dielectric constant is an intrinsic bulk-material property that relates how much the capacitance is increased due to the presence of the material.
7. The power and ground planes in a circuit board have capacitance. However, the amount is so small as to be negligible. The value the planes provide is not for their decoupling capacitance but for their low loop inductance.
8. The IPC approximations for microstrip and stripline should not be used if accuracy better than 10% is required.
9. A 2D field solver, once verified, can be used to calculate the capacitance per length of a uniform transmission line structure to better than 1%.
10. Increasing conductor thickness in a microstrip will increase the capacitance per length—but only slightly. A change from very thin to 2-ounce copper only increases the capacitance by 3%.
11. Increasing the thickness of the dielectric coating on top of a microstrip will increase the capacitance. Completely enclosing all the fringe field lines happens when the coating is as thick as the trace is wide and the capacitance can increase by as much as 20%.
12. The effective dielectric constant is the composite dielectric constant when the material is not homogeneously distributed and some field lines see different materials, such as in a microstrip. It can easily be calculated with a 2D field solver.

End-of-Chapter Review Questions

The answers to the following review questions can be found in Appendix D, "Review Questions and Answers."

5.1 What is capacitance?

5.2 Give one example where capacitance is an important performance metric.

5.3 What are two different interpretations of what the capacitance between two conductors measures.

5.4 A small piece of metal might have a capacitance of 1 pF to the nearest metal, inches away. There is no DC connection between these pieces of metal. At 1 GHz, what is the impedance between these conductors?

5.5 How does conduction current flow through the insulating dielectric of a capacitor?

5.6 What is the origin of displacement current, and where will it flow?

5.7 If you wanted to engineer a higher capacitance between the power and ground planes in a board, what three design features would you change?

5.8 What primary property about the chemistry of a dielectric most strongly influences its dielectric constant?

5.9 What happens to the capacitance between two conductors when the voltage between them increases?

5.10 In a coax geometry, what happens to the capacitance if the outer radius is increased?

5.11 What happens to the capacitance per length in a microstrip if the signal path is moved away from the return path?

5.12 Is there any geometry in which capacitance increases when the conductors are moved farther apart?

5.13 Why does the effective dielectric constant increase as the thickness of the dielectric coating increases in microstrip?

5.14 What is the lowest dielectric constant of a solid, homogenous material? What material is that?

5.15 What could you do to a material to dramatically reduce its dielectric constant?

5.16 What happens to the capacitance per length of a microstrip if solder mask is added to the top surface?

5.17 What happens to the capacitance per length of a microstrip if the conductor thickness increases?

5.18 What happens to the capacitance per length of a stripline if the line width increases?

5.19 What happens to the capacitance per length of a stripline if the trace thickness increases?

5.20 For the same line width and dielectric thickness per layer, which will have more capacitance per length: a microstrip or a stripline?

5.21 What is theoretically the lowest dielectric constant any material could have?

5.22 Why is the capacitance per length constant in a uniform cross section interconnect?

5.23 On die, the dielectric thickness between the power and ground rails can be as thin as 0.1 micron. If the SIO2 dielectric constant is also 4, how does the on-die capacitance per square inch compare to the on-board capacitance per square inch if the power and ground plane separation is 10 mils?

5.24 What is the capacitance between the faces of a penny if they are separated with air?

5.25 What is the minimum capacitance between a sphere 2 cm in diameter suspended a meter above the floor? How does this capacitance change as the sphere is raised higher above the floor?

5.26 What is the capacitance between the power and ground planes in a circuit board if the planes are 10 inches on a side, 10 mils separation, and filled with FR4?

5.27 Derive the capacitance per length of the rod over a plane from the capacitance per length of the twin rod geometry, in air.

CHAPTER 6

The Physical Basis of Inductance

Inductance is a critically important electrical property because it affects virtually all signal-integrity problems. Inductance plays a role in signal propagation for uniform transmission lines as a discontinuity, in the coupling between two signal lines, in the power distribution network, and in EMI.

In many cases, the goal will be to decrease inductance, such as the mutual inductance between signal paths for reduced switching noise, the loop inductance in the power distribution network, and the effective inductance of return planes for EMI. In other cases, the goal may be to optimize the inductance, as in achieving a target characteristic impedance.

By understanding the basic types of inductance and how the physical design influences the magnitude of the inductance, we will see how to optimize the physical design for acceptable signal integrity.

6.1 What Is Inductance?

There is not a single person involved with signal integrity and interconnect design who has not worried about inductance at one time or another. Yet, very few engineers use the term correctly. This is fundamentally due to the way we all learned about inductance in high school or college physics or electrical engineering.

Typically, we were taught about inductance and how it relates to flux lines in coils. We were introduced to the inductance of a coil with figures of coiled wires, or solenoids, with flux lines through them. Or, we were told inductance was mathematically an integral of magnetic-field density through surfaces. For example, a commonly used definition of inductance, L, is:

$$L = \frac{1}{I} \int_{area} \vec{B} \bullet \hat{n} da \qquad (6\text{-}1)$$

While all of these explanations may be perfectly true, they don't help us on a practical level. Where are the coils in a signal-return path? What does an integral of magnetic-field density really mean? We have not been trained to apply the concepts of inductance to the applications we face daily— applications related to the interaction of signals on interconnects as packages, connectors, or boards.

While all this is true, we still need to understand inductance in a more fundamental way. Our understanding of inductance needs to feed our engineering intuition and give us the tools we need to solve real interconnect problems. A practical approach to inductance is based on just three fundamental principles.

6.2 Inductance Principle 1: There Are Circular Rings of Magnetic-Field Lines Around All Currents

There is a new fundamental entity, called *magnetic-field lines*, that surrounds every current. If we have a straight wire, as shown in Figure 6-1, for example, and send a current of 1 Amp through it, there will be concentric circular magnetic-field lines in the shape of rings created around the wire. These rings exist up and down the length of the wire. Imagine walking along the wire and counting the specific number of field line rings that completely surround it; the farther from the surface of the current, the fewer the number of field line rings we will encounter. If we move far enough away from the surface, we will count very few magnetic-field line rings.

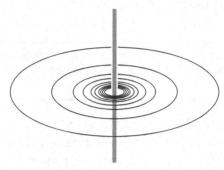

Figure 6-1   Some of the circular magnetic-field line rings around a current. The rings exist up and down the length of the wire.

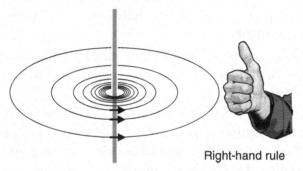

Right-hand rule

Figure 6-2   The direction of circulation of the magnetic-field line rings is based on the right-hand rule.

Where do these magnetic-field lines come from? Why are they created by moving charges in a wire? Like the properties of electric fields coming from charges, the properties of magnetic fields and their interaction with moving charges is built into the very fabric of spacetime. Electric and magnetic fields and their interactions with charges are intrinsic properties of spacetime that seems to have been frozen into the structure of spacetime about 1 nanosecond after the Big Bang. Whatever their origin, their properties seem to be incredibly well described by Maxwell's Equations.

The magnetic-field line rings around a current have a specific direction, as though they were circulating around the wire. To determine their direction, use the familiar right-hand rule: Point the thumb of your right hand in the direction of the positive current and your fingers curl in the direction the field line rings circulate. This is illustrated in Figure 6-2.

TIP   Magnetic-field line rings are always complete circles and always enclose some current. There must be some current encircled by the field line rings.

As we walk along the wire with a current, we can imagine counting each specific magnetic-field line ring we encounter. When we count or measure the magnetic-field lines, it is only the number of rings we count.

In what units do we count the field line rings? We count pens in units of gross. There are 144 pens in a gross. Paper is counted in units of reams, with 500 sheets to each ream. Apples are counted by the bushel. How many apples are there in a bushel? It's not clear exactly how many apples are in a bushel, but there is some number.

Likewise, we count the number of magnetic-field line rings around a current in units of *Webers*. There are some number of magnetic-field line rings in a Weber of field lines.

The number of Webers of field line rings we count around the current in the wire is influenced by a number of factors. First is the amount of current in the wire. If we double the current in the wire, we will double the number of Webers of magnetic-field line rings around it.

Second, the length of the wire around which the field line rings appear will affect the number of field line rings we count. The longer the wire, the more field line rings we will count. Third, the cross section of the wire affects the total number of field line rings surrounding the current. This is a second-order effect and is more subtle. As we will see, if the cross-sectional area is increased—for example, if the wire diameter is increased—the number of field line rings will decrease a little bit.

Fourth, the presence of other currents nearby will affect the total number of field line rings around the first current. A special current to watch for is the return current. As the return current of the first wire is brought closer, some of its field line rings will be so big as to also enclose the first current.

When we count the total number of magnetic-field line rings around the first wire, we will count its own rings, plus the rings from its return current. And, since the magnetic-field line rings around the first current circulate in the opposite direction, the total number of rings around the first wire will be reduced due to the proximity of the return current.

On the other hand, the presence of dielectric materials does not affect the number of field line rings around the current.

> **TIP**   Magnetic fields do not interact with dielectric materials at all. There is no change in the number of magnetic-field line rings around a current if it is surrounded by Teflon or by barium titanate.

The fifth factor that may influence the number of magnetic-field line rings around a current is the metal of which the wire is composed. Only three metals affect the magnetic-field lines around a current: iron, nickel, and cobalt (and combinations of them).

These three metals are called *ferromagnetic metals*. They, or alloys containing them, have a permeability greater than 1. If any magnetic-field line rings are completely contained within one of these metals, the metal will have the effect of dramatically increasing the number of field line rings. However, it is only those field line rings that are circulating internally in the conductor that will be affected. Two common interconnect alloys, Alloy 42 and Kovar, are ferromagnetic because they both contain iron, nickel, and cobalt.

The composition of a wire made of any other metal, such as copper, silver, tin, aluminum, gold, lead, or even carbon, will have absolutely no effect on the number of field line rings around the current.

6.3 Inductance Principle 2: Inductance Is the Number of Webers of Field Line Rings Around a Conductor per Amp of Current Through It

Inductance is fundamentally related to the number of magnetic-field line rings around a conductor, per Amp of current through it.

> **TIP**   Inductance is about the number of rings of magnetic-field lines enclosing a current per amp of the current, not about the absolute value of the magnetic-field density at any one point. Don't worry about magnetic-field concentration, or density; worry about the number of magnetic-field line rings per Amp of current.

The units we use to measure inductance are Webers of field line rings per Amp of current. One Weber/Amp is given the special name, Henry. The inductance of most interconnect structures is typically such a small fraction of a Henry, it is more common to use nanoHenry units. A nanoHenry, abbreviated nH, is a measure of how many Webers of field line rings we would count around a conductor per Amp of current through it:

$$L = \frac{N}{I} \qquad\qquad (6\text{-}2)$$

where:

L = inductance, in Henrys

N = number of magnetic-field line rings around the conductor, in Webers

I = current through the conductor, in Amps

If the current through a conductor doubles, the number of field line rings doubles, but the ratio stays the same. This ratio is completely independent of how much current is going through the conductor. A conductor has the same inductance if 0 Amps are flowing through it or 100 Amps. Yes, the number of field line rings changes, but the ratio doesn't, and inductance is the ratio.

TIP   This means that inductance is really related to the geometry of the conductors. The only thing that influences inductance is the distribution of the conductors and, in the case of ferromagnetic metals, their permeability.

In this respect, inductance is a measure of the efficiency of the conductors to create magnetic-field line rings. A conductor that is not very efficient at generating magnetic-field line rings has a low inductance. A conductor has an efficiency whether or not it has current through it. The amount of current in the conductor does not in any way affect its efficiency. Inductance is only about a conductor's geometry.

Inductance is not about the current in a wire. Inductance is not about how much magnetic field is around a wire. Inductance is not about how much concentration there is in the magnetic-field density around a current. Inductance is not about the energy stored in the magnetic field. Inductance is only about the efficiency of how good the conductor geometry is in creating magnetic-field line rings per amp of current.

We can apply this simple definition to *all* cases involving inductance. What makes it complicated and confusing is having to keep track of how much of the current loop we are counting the field line rings around and which other currents are present, creating field line rings. This gives rise to many qualifiers for inductance.

To keep track of the source of the magnetic-field line rings, we will use the terms *self-inductance* and *mutual inductance*. To keep track of how much of the current loop around which we are counting field line rings, we will use the terms *loop inductance* and *partial inductance*. Finally, when referring to the magnetic-field line rings around just a section of an interconnect, while the current is flowing through the entire loop, we will use the terms *total inductance*, *net inductance*, and *effective inductance*. These last three terms are used interchangeably in the industry.

Just using the term *inductance* is ambiguous. We must develop the discipline to always use the qualifier of the exact type of inductance to which we are referring. The most common source of confusion about inductance arises from mixing up the various types of inductance.

6.4 Self-Inductance and Mutual Inductance

If the only current that existed in the universe were the current in a single wire, the number of field line rings around it would be easy to count. However, when there are other currents nearby, their magnetic-field line rings can encircle multiple currents. Consider two adjacent wires, labeled a and b, as shown in Figure 6-3. If there were current in only one wire, a, it would have some number of field line rings around it, per amp of current through it, and some inductance.

Suppose there were some current in the second wire, b. It would have some field line rings around it, per amp of current through it, and hence, it would have some inductance. Some of the field line rings from this wire, b, would also encircle the first wire, a. Around the first wire, a, there are now some field line rings from its own current and some field line rings from the adjacent (second) current, b.

When we count field line rings around one wire, we need a way of keeping track of the source of the field line rings. We do this by labeling those field line rings from a wire's own currents as the self-field line rings and those from an adjacent current's as the mutual-field line rings.

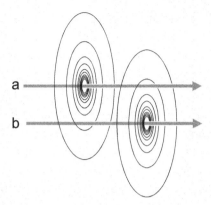

Figure 6-3 Magnetic-field line rings around one conductor can arise from its own currents and from another current.

TIP   The self-field line rings are those field line rings around a wire that arise from its own currents *only*. The mutual-field line rings are those magnetic-field line rings completely surrounding a wire that arise from another wire's current.

Every field line that is created by current in wire b and also goes around wire a must be around both of them. In this way, we say the mutual-field line rings "link," or are around, both of the two conductors.

If we have two adjacent wires and we put current in the second wire, it will have some of its field line rings around the first wire. As we might imagine, if we move the second wire farther away, the number of mutual-field line rings that are around both wires will decrease. Move them closer, and the number of mutual-field line rings will increase.

But, what happens to the total number of field line rings around the first wire? If there is current through both wires, they will each have some self-field line rings. If the currents are in the same direction, the circulation direction of their self-field line rings will be the same. To count the total number of field line rings around the first wire, we would count its self-field line rings plus the mutual-field line rings from the second wire, because they are both circulating in the same direction.

However, if the direction of their currents is opposite, the circulation direction of the self- and mutual-field line rings around the first wire will be opposite. The mutual-field line rings will subtract from the self-field line rings.

The total number of field line rings around the first wire will be decreased due to the presence of the adjacent opposite current.

Given this new perspective of keeping track of the source of the field line rings, we can make inductance more specific.

T I P   We use the term *self-inductance* to refer to the number of field line rings around a wire per Amp of current in its own wire. What we normally think of as *inductance* we now see is really specifically the self-inductance of a wire.

The self-inductance of a wire will be independent of the presence of another conductor's current. If we bring a second current near the first wire, the total number of field line rings around the first may change, but the number of field line rings from its own currents will not.

T I P   Likewise, we use the term *mutual inductance* to refer to the number of field line rings around one wire per Amp of current in another wire.

As we bring the two wires close together, their mutual inductance increases. Pull them farther apart, and their mutual inductance decreases. Of course, the units we use to measure mutual inductance are also nH, since it is the ratio of a number of field line rings per Amp of current.

Mutual inductance has two very unusual and subtle properties. Mutual inductance is symmetric. Whether we send 1 Amp of current in one wire and count the number of field line rings around the other wire or send 1 Amp of current through the other and count the field line rings it produces around the first, we get exactly the same ratio of field line rings per Amp of current. In this respect, mutual inductance is related to the field line rings that link two conductors and is tied to both of them equally. We sometimes refer to the "mutual inductance between two conductors" because it is a property shared equally by the two conductors. This is true no matter what the size or shape of each individual wire. One can be a narrow strip and the other a wide plane. The number of field line rings around one conductor per Amp of current in the other conductor is the same whether we send the current in the wide conductor or the narrow one.

The second property is that the mutual inductance between any two conductors can never be greater than the self-inductance of either one. After all, every mutual-field line ring is coming from one conductor and must also be a self-field line ring around that conductor. Since the mutual inductance between two conductors is independent of which one has the source current, it must always be less than the smallest of the two conductors' self-inductance.

6.5 Inductance Principle 3: When the Number of Field Line Rings Around a Conductor Changes, There Will Be a Voltage Induced Across the Ends of the Conductor

So far, we have discussed what inductance is. Now we will look at the "so what": Why do we care about inductance? A special property of magnetic-field line rings is that when the actual, total number of field line rings around a section of a wire changes, for whatever reason, there will be a voltage created across the length of the conductor. This is illustrated in Figure 6-4. The voltage created is directly related to how fast the total number of field line rings changes:

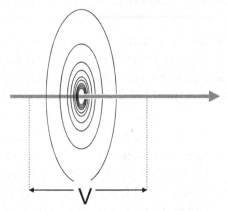

Figure 6-4 Voltage induced across a conductor due to the changing number of magnetic-field line rings around it.

$$V = \frac{\Delta N}{\Delta t} \tag{6-3}$$

where:

V = voltage induced across the ends of a conductor

ΔN = number of field line rings that change

Δt = time in which they change

If the current in a wire changes, the number of self-field line rings around it will change, and there will be a voltage generated across the ends of the wire. The number of field line rings around the wire is $N = L \times I$, where L in this example is the self-inductance of the section of the wire. The voltage created, or induced, across a wire can be related to the inductance of the wire and how fast the current in it changes:

$$V = \frac{\Delta N}{\Delta t} = \frac{\Delta LI}{\Delta t} = L\frac{dI}{dt} \qquad (6\text{-}4)$$

TIP   Induced voltage is the fundamental reason inductance plays such an important role in signal integrity. If there were no induced voltage when a current changed, there would be no impact from inductance on a signal. This induced voltage from a changing current gives rise to transmission line effects, discontinuities, cross talk, switching noise, rail collapse, ground bounce, and most sources of EMI.

This relationship is, after all, the definition of an inductor. If we change the current through an inductor, we get a voltage generated across it. We sometimes describe this voltage as an *induced electromotive force* (EMF). This is a new voltage source, and it will act like a battery to pump current from the minus side to the plus side. The voltage polarity is created to drive an induced current that would counteract the change in current. This is why we say that "an inductor resists a change in current."

If we happened to have an adjacent current in another wire, near the first wire, some of the field line rings from this second wire may also go around the first wire. If the current in the second wire changes, the number of its field line rings around the first wire will change. This changing number of field line rings will cause a voltage to be created across the first wire. This is illustrated in Figure 6-5. These changes in the mutual-field line rings contribute to the induced voltage across the first wire. When it is another conductor in which the current changes, we typically use the term *cross talk* to describe the induced voltage in the adjacent conductor. This induced voltage is a source of *noise*, an unwanted signal. In this case, the voltage noise generated is:

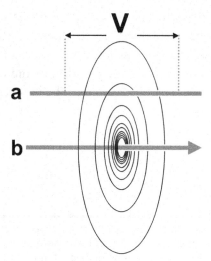

Figure 6-5 Induced voltage on one conductor due to a changing current in another and the subsequent changing mutual-field lines between them are a form of cross talk.

$$V_{noise} = M \frac{dI}{dt} \qquad (6\text{-}5)$$

where:

V_{noise} = voltage noise induced in the first, quiet wire

M = mutual inductance between the two wires, in Webers

I = current in the second wire

Because the voltage induced depends on how fast the current changes, we sometimes use the term *switching noise* or *delta I noise* when describing the noise created when the current switches through a mutual inductance.

To be able to analyze real-world problems that involve multiple conductors, we need to be able to keep track of all the various currents that are the sources of field line rings. The effects are the same; it's just more complicated when there are many conductors, each with possible currents and magnetic-field line rings.

6.6 Partial Inductance

Of course, real currents *only* flow in complete current loops. In the previous examples, we have been looking at just a part of a circuit, where the only current that exists is the specific current in the part of wire we had drawn. As we counted field line rings, we ignored the rings from current in the rest of the current loop to which this wire segment belonged. When we look at the field line rings around part of a loop from currents in only part of the loop, we call this type of inductance the *partial inductance* of the wire.

It is important to keep in mind that when we speak of partial inductance, it is as though the *rest of the loop does not exist*. In the view of partial inductance, no other currents exist, except in the specific part of the conductor in which we are looking. The concept of partial inductance is a mathematical construct. It can never be measured, since an isolated current can never exist.

> **TIP**   In reality, we can never have a partial current; we must always have current loops. However, the concept of partial inductance is a very powerful tool for understanding and calculating the other flavors of inductance, especially if we don't know what the rest of the loop looks like yet.

Partial inductance has two flavors: partial self-inductance and partial mutual inductance. What we have been discussing above has really been the partial inductance of parts of two wires. More often than not, when referring to the inductance of a lead in a package, or a connector pin, or a surface trace, we are really referring to the partial self-inductance of this interconnect element.

The precise definition of partial self- and partial-mutual inductance is based on a mathematical calculation of the number of field line rings around a section of a wire. Take a fixed-length section of conductor that might be part of a current loop. Rip it out of the loop so it is isolated in space but maintains its original geometry. On the ends, put large planes that are perpendicular to the length of the conductor. Now imagine injecting 1 A of current, appearing suddenly in one end of the wire, traveling through the wire, coming out the other end, and disappearing back into nothing when it exits.

The only current that exists in the universe is in the section of wire between the end planes. From this small section of current, count the number of field line rings that fit between the two end-cap planes. The number of field line rings counted, per the 1 A of current in the wire section, is the partial self-inductance of that section of the conductor. Obviously, if we make the part of the wire longer, the total number of field line rings surrounding it will increase and its partial self-inductance will increase.

Now bring another short section of interconnect near this first section. Inject current from nowhere into the second wire and have it disappear out the opposite end. This partial current will create magnetic-field line rings throughout space, some of which will fit within the plane end caps of the first partial section of wire, completely enclosing the first wire segment. The number of field line rings around the first conductor segment, per Amp of current in the second wire, is the partial mutual inductance between the two sections.

Obviously, in the real world, we can't create from nothing a current into the end of the wire without it coming from some other part of a circuit. However, we can perform this operation mathematically. The term *partial inductance* is a very well-defined quantity; it just can't be measured. As we shall see, it is a very powerful concept to facilitate optimizing the design for reduced ground bounce and calculating the other, measurable terms of inductance.

There are only a few geometries of conductors with a reasonably good approximation for their partial self-inductance. The partial self-inductance of a straight, round conductor, illustrated in Figure 6-6, can be calculated to better than a few percent accuracy using a simple approximation. It is given by:

$$L = 5 \text{Len} \left\{ \ln \left(\frac{2 \text{Len}}{r} \right) - \frac{3}{4} \right\} \qquad (6\text{-}6)$$

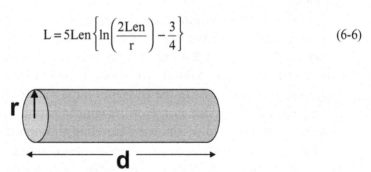

Figure 6-6 Geometry for the approximation of the partial self-inductance of a round rod. The length of the wire, d, is also represented by Len.

where:

L = partial self-inductance of the wire, in nH

r = radius of the wire, in inches

Len = length of the wire, in inches

For example, an engineering change wire is typically 30-gauge wire, or roughly 10 mils in diameter. For a length that is 1 inch long, the partial self-inductance is:

$$L = 5 \times 1 \left\{ \ln\left(\frac{2 \times 1}{0.005} \right) - \frac{3}{4} \right\} = 26 \, \text{nH} \qquad (6\text{-}7)$$

T I P   This gives rise to an important rule of thumb: The partial self-inductance of a wire is about 25 nH/inch, or 1 nH/mm. Always keep in mind that this is a rule of thumb and sacrifices accuracy for ease of use.

We see that the partial self-inductance increases as we increase the length of the conductor. But, surprisingly, it increases faster than just linearly. If we double the length of the conductor, the partial self-inductance increases by more than a factor of two. This is because as we increase the length of the wire, there are more field line rings around this newly created section of the wire from the current in the new section, and some field line rings from the current in the other part of the wire are also around this new section.

The partial self-inductance decreases as the cross-sectional area increases. If we make the radius of the wire larger, the current will spread out more, and the partial self-inductance will decrease. As we spread out the current distribution, the total number of field line rings will decrease a little bit.

T I P   This points out a very important property of partial self-inductance: The more spread out the current distribution, the lower the partial self-inductance. The more dense we make the current distribution, the higher the partial self-inductance.

In this geometry of a round rod, the partial self-inductance varies only with the natural log of the radius, so it is only weakly dependent on the cross-sectional area. Other cross sections, such as wide planes, will have a partial self-inductance that is more sensitive to spreading out the current distribution.

Using the rule of thumb above, we can estimate the partial self-inductance of a number of interconnects. A surface trace from a capacitor to a via, 50 mils long, has a partial self-inductance of about 25 nH/inch × 0.05 inch = 1.2 nH. A via through a board, 064 mils thick, has a partial self-inductance of about 25 nH/inch × 0.064 inch = 1.6 nH.

Both the approximation and the rule of thumb are very good estimates of the partial self-inductance of a narrow rod. Figure 6-7 compares the estimated partial self-inductance for a 1-mil-diameter wire bond, based on the rule of thumb and the approximation given above, with the calculation from a 3D field solver. For lengths of typical wire bonds, about 100 mils long, the agreement is very good.

The partial mutual inductance between two conductor segments is the number of field line rings from one conductor that completely surround the other conductor's segment. In general, the partial mutual inductance between two wires is a small fraction of the partial self-inductance of either one and drops off very rapidly as the wires are pulled apart. The partial mutual inductance between two straight, round wires can be approximated by:

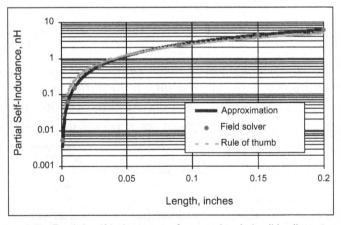

Figure 6-7 Partial self-inductance of a round rod, 1 mil in diameter, comparing the rule of thumb, the approximation, and the results from the Ansoft Q3D field solver.

$$M = 5\text{Len}\left\{\ln\left(\frac{2\text{Len}}{s}\right) - 1 + \frac{s}{\text{Len}} - \left(\frac{s}{2\text{Len}}\right)^2\right\} \tag{6-8}$$

where:

M = partial mutual inductance, in nH

Len = length of the two rods, in inches

s = center-to-center separation, in inches

This formidable approximation takes into account second-order effects and is often referred to as a second-order model. It can be simplified by a further approximation when the separation is small compared to the length of the rods (s << Len) as:

$$M = 5\text{Len}\left\{\ln\left(\frac{2\text{Len}}{s}\right) - 1\right\} \tag{6-9}$$

This is a first-order model, as it ignores some of the details of the long-range coupling between the rods. What it sacrifices in accuracy is made up for by its slight advantage in ease of use. Figure 6-8 shows the predictions for the mutual inductance between two rods, as the rods are moved farther apart, and how they compare with the predictions from a 3D field solver. We see that when the partial mutual inductance is greater than about 20% of the partial self-inductance (i.e., when mutual inductance is significant), the first-order approximation is pretty good. It is a good, practical approximation.

For example, two wire bonds may each have a partial self-inductance of 2.5 nH if they are 100 mils long. If they are on a 5 mil pitch, they will have a partial mutual inductance of 1.3 nH. This means that if there were 1 A of current in one wire bond, there would be 1.3 nH × 1 Amp = 1.3 nanoWebers of field line rings around the second wire bond. The ratio of the partial mutual inductance between the two wire bonds to the partial self-inductance of either one is about 50% with this separation.

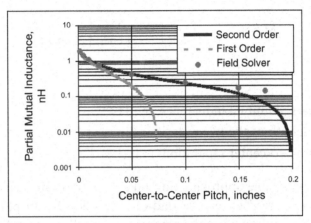

Figure 6-8 Partial mutual inductance between two round rods, 0.1-inch long, as their center-to-center separation increases, comparing the accurate approximation, the simplified approximation, and the result from Ansoft's Q3D field solver.

> **T I P**   From the graph of the mutual inductance and spacing, a good rule of thumb can be identified: If the spacing between two conductor segments is farther apart than their length, their partial mutual inductance is less than 10% of the partial self-inductance of either one and can often be ignored.

This says that the coupling between two sections of an interconnect is not important if the sections are farther apart than their length. For example, two vias 20 mils long have virtually no coupling between them if they are spaced more than 20 mils apart, center to center.

The concept of partial inductance is really the fundamental basis for all aspects of inductance. All the other forms of inductance can be described in terms of partial inductance. Package and connector models are really based on partial inductances. The output result of 3D static field solvers, when they calculate inductance values, is really in partial-inductance terms. SPICE models really use partial-inductance terms.

> **T I P**   If we can identify the performance design goals in terms of how we need to optimize each type of inductance, we can use our understanding of how the physical design will affect the partial self- and mutual inductances of a collection of conductors to optimize the physical design.

6.7 Effective, Total, or Net Inductance and Ground Bounce

Consider a wire that is straight for some length and then loops back on itself, as shown in Figure 6-9, making a complete loop. This sort of configuration is very common for all interconnects, including signal and return paths and power paths and their "ground"-return paths. For example, it is common to have adjacent power- and ground-return wire bonds in a package. The pair could be adjacent signal and return leads in an IC package, or they could be an adjacent signal and a return plane pair in a circuit board.

When current flows in the loop, magnetic-field line rings are created from each of the two legs. If the current in the loop changes, the number of field line rings around each half of the wire would change. Likewise, there would be a voltage created across each leg that would depend on how fast the total number of field line rings around each leg was changing.

The voltage noise created across one leg of the current loop depends on how fast the *total* number of field line rings around the leg changes, while it is part of this complete current loop.

The total number of field line rings around one leg arises from the current in that leg (the partial self-field line rings) and the field line rings coming from the other leg (the partial mutual-field line rings). But, the field line rings from the two currents circulate around the leg in opposite directions, so the total number of field line rings around this section of the loop is the difference between the self- and mutual-field line rings around it. The total number of field line rings per Amp of current around this leg is given the special name *effective inductance*, *total inductance*, or *net inductance*.

> **TIP**   Effective, net, or total inductance of a section of a loop is the total number of field line rings around just this section, per Amp of current in the loop. This includes the contribution of field line rings from *all* the current segments in the complete loop.

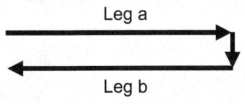

Figure 6-9 A current loop with two legs: an initial current and its return current.

We can calculate the effective inductance of one leg based on the partial inductances of the two legs. Each of the two legs of the loop has a partial self-inductance associated with it, which we label L_a and L_b. There is a partial mutual inductance between the two legs, which we label L_{ab}. We label the current in the loop, I, which is the same current in each leg but, of course, going in opposite directions.

If we could distinguish the separate sources of the field line rings around leg b, we would see that some of the field line rings come from the current in leg b and are self-field lines. Around leg b, there would be $N_b = I \times L_b$ field line rings from its own current. At the same time, some of the field line rings around leg b are mutual-field line rings and come from the current in leg a. The number of mutual-field line rings coming from leg a, surrounding leg b, is $N_{ab} = L_{ab} \times I$.

What is the total number of field line rings around leg b? Since the current in leg a is moving in the opposite direction of leg b, the field line rings from leg a, around leg b, will be in the opposite direction of the field line rings from the current in leg b. When we count the total number of field line rings around leg b, we will be subtracting both sets of field line rings. The total number of field line rings around leg b is:

$$N_{total} = N_b - N_{ab} = (L_b - L_{ab}) \times I \qquad (6\text{-}10)$$

We call $(L_b - L_{ab})$ the total, net, or effective inductance of leg b. It is the total number of field line rings around leg b, per Amp of current in the loop, including the effects of all the current segments in the entire loop. When the adjacent current is in the opposite direction, as when the two legs are part of the same loop and one leg has the return current for the other leg, the effective inductance will determine how much voltage is created across one leg when the current in the loop changes. When the second leg is the return path, we call this voltage generated across the return path *ground bounce*.

The ground-bounce voltage drop across the return path is:

$$V_{gb} = L_{total} \times \frac{dI}{dt} = (L_b - L_{ab}) \times \frac{dI}{dt} \qquad (6\text{-}11)$$

where:

V_{gb} = ground-bounce voltage

L_{total} = total inductance of just the return path

I = current in the loop

L_b = partial self-inductance of the return path leg

L_{ab} = partial mutual inductance between the return path and the initial path

If the goal is to minimize the voltage drop in the return path (i.e., the ground-bounce voltage), there are only two approaches. First, we can do everything possible to decrease the rate of current change in the loop. This means slow down the edges, limit the number of signal paths that use the same, shared return path, and use differential signaling for the signals. Rarely do we have the luxury of affecting these terms much. However, we should always ask.

Second, we must find every way possible to decrease L_{total}. There are only two knobs to tweak to decrease the total inductance of the return path: *Decrease* the partial self-inductance of the leg and *increase* the partial mutual inductance between the two legs. Decreasing the partial self-inductance of the leg means making the return path as short and wide as possible (i.e., using planes). Increasing the mutual inductance between the return path and the initial path means doing everything possible to bring the first leg and its return path as close together as possible.

> **T I P**   Ground bounce is the voltage between two points in the return path due to a changing current in a loop. Ground bounce is the primary cause of switching noise and EMI. It is primarily related to the total inductance of the return path and shared return current paths. To decrease ground-bounce voltage noise, there are two significant features to change: Decrease the partial self-inductance of the return path by using short lengths and wide interconnects and increase the mutual inductance of the two legs by bringing the current and its return path closer together.

Surprisingly, decreasing ground bounce on the return path requires more than just doing something to the return path. It also requires thought about the placement of the signal current path and the resulting partial mutual inductance with the return path.

We can evaluate how much the total inductance of one wire bond can be reduced by bringing an adjacent wire bond closer, using the approximations above. Suppose one wire bond carries the power current and the other carries the ground-return current. They would have equal and opposite currents. In this case, the partial mutual inductance between them will act to decrease the total inductance of one of them: $L_{total} = L_a - L_{ab}$. The closer we bring the wires, the greater the partial mutual inductance between them and the greater the reduction in the total inductance of one of the wire bonds.

If each wire bond were 1 mil in diameter and 100 mils in length, the partial self-inductance of each would be about 2.5 nH. Using the approximation above for the partial self-inductance and the partial mutual inductance, we can estimate the net or effective inductance of one wire bond as we change the center-to-center spacing, s.

When the spacing is greater than 100 mils, the partial mutual inductance is less than 10% of the partial self-inductance. The effective inductance of one wire bond is nearly the same as its isolated partial self-inductance. But, when we bring them as close as 5 mils center-to-center pitch, the mutual inductance increases considerably, and we can reduce the effective inductance of one wire bond down to 1.3 nH. This is a reduction of more than 50%. The lower the effective inductance, the lower the voltage drop across this wire bond and the less ground-bounce voltage noise the chip will experience.

If the effective inductance of one wire bond, when the other current is far away, is 2.5 nH and there is 100 mA of current that switches in 1 nsec (typical of what goes into a transmission line), the ground-bounce voltage generated across the wire bond is V_{gb} = 2.5 nH × 0.1A/1 nsec = 250 mV. This is a lot of voltage noise. When the two wire bonds are routed close together, with a center pitch of 5 mils, the ground-bounce noise is reduced to V_{gb} = 1.3 nH × 0.1A/1 nsec = 130 mV—considerably less.

TIP   This demonstrates a very important design rule: To decrease effective inductance, bring the return current as close as possible to the signal current.

Suppose we have the opposite case: Suppose that both wires will be carrying power current. This is often the case in many IC packages because multiple leads are used to carry power and carry the ground return. If we look at the net inductance of one of the power wires and bring an adjacent power wire nearby, what happens?

In this case, both currents are in the same direction. The mutual-field line rings will be in the same direction as the self-field line rings. The mutual-field line rings from the second wire will add its field line rings to the field line rings already around the first wire. The net inductance of one of the power wires will be $L_{total} = L_a + L_{ab}$.

If the goal is to minimize the net inductance of the power lead, our design goal, as always, is to do everything possible to decrease the partial self-inductance of the lead. However, in this case, because the field line rings from the adjacent wire are in the same direction, we must do everything possible to *decrease* their mutual inductance. This means spacing the wires out as much as possible.

We can estimate the net or effective inductance of one wire bond when the adjacent one carries the same current, for the two cases of current in the same direction or in opposite directions, as we pull them farther apart. This is plotted in Figure 6-10.

As long as the two wires are farther apart than their length, the net inductance is not much different from the partial self-inductance of one. As they are brought closer together, the net inductance decreases when the currents are in opposite directions but increases when the currents are in the same direction.

TIP A good general design rule to minimize the net inductance of either leg in the power distribution system is to keep similar parallel currents as far apart as their lengths.

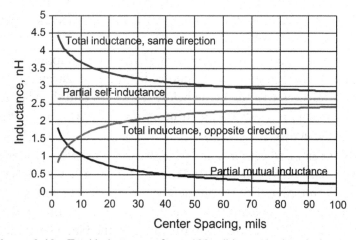

Figure 6-10 Total inductance of one 100-mil-long wire bond when an adjacent wire bond carries the same current, for the two cases of the currents in the same direction and in opposite directions. The wires are pulled apart, comparing the total inductances with the partial self- and mutual inductances.

In other words, if there are two adjacent wire bonds, each 100 mils long and both carrying power, they should be at least 100-mils center to center. Any closer, and their mutual inductance will increase the effective inductance of each leg and increase the switching noise on one of the wires. This is not to say that there is no benefit for parallel currents in close proximity. It's just that their effectiveness is reduced over the maximum possible gain.

A common practice in high-power chips is double bonding—that is, using two wire bonds between the same pad on the die and the same pad on the package. The series resistance between the two pads will be decreased because of the two wires in parallel, and the equivalent inductance of the two wire bonds will be reduced compared to that when using only one wire bond. The closer the wires, the higher the mutual inductance, and the larger the effective inductance. But, since there are two in parallel, the equivalent inductance is half the total inductance of either one.

When double bonding, the wire bond loops should be created to keep the wires as far apart as possible. This is in keeping with the general principle that total inductance is decreased if we can spread out the same direction current paths. If they are 50 mils long and can be kept 5 mils apart, the partial self-inductance of either wire bond would be about 1.25 nH, and their partial mutual inductance would be about 0.5 nH. Their effective inductance would be 1.75 nH. With the two in parallel, their equivalent inductance would be $1/2 \times 1.75$ nH = 0.88 nH. This is reduced from the 1.25 nH expected with just one wire bond. Double bonding can, in fact, decrease the equivalent inductance between two pads.

As another example, consider the vias from a decoupling capacitor's pads to the power- and ground-return planes below, as illustrated in Figure 6-11. Suppose the distance to the plane below is 20 mils, and the vias are 10 mils in diameter. Is there any advantage in using multiple vias in parallel from each pad of the capacitor?

If the center-to-center spacing, s, between the vias is greater than the length of a via, 20 mils, their partial mutual inductance will be very small, and they will not interact. The net inductance of either one will be just its partial self-inductance. Having multiple vias in parallel from one pad to the plane below will reduce the equivalent inductance to the plane below inversely with the number of vias. The more vias we have in parallel, the lower the equivalent inductance. In Figure 6-11, this means s_2 must be roughly at least as large as the distance to the planes, 20 mils. Likewise, if it is possible to bring vias with opposite currents closer together than their length, the effective inductance of each via will be reduced. If it is possible to place the vias with s_1 less than 20 mils apart, the net inductance of each via will be reduced, the equivalent inductance from the pad to the plane below will be reduced, and the rail-collapse voltage will be reduced.

Another important advantage of multiple vias in the same pad is that the spreading inductance into the power and ground planes will be reduced due to a larger contact area into the planes. This can sometimes be a larger advantage than the reduced inductance of the vias.

TIP Consider the following design rule for minimizing the total inductance of either path: Keep the center-to-center spacing between vias of the same current direction at least as far apart as the length of the via; keep the center-to-center spacing between vias with opposite-direction current much closer than the length of the vias.

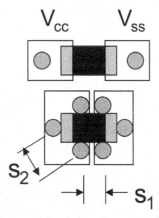

Figure 6-11 Via placement for decoupling capacitor pads between V_{cc} and V_{ss} planes. Top: Conventional placement. Bottom: Optimized for low total inductance and lowest voltage-collapse noise: $s_2 >$ via length, $s_1 <$ via length.

6.8 Loop Self- and Mutual Inductance

The general definition of inductance is the number of field line rings around a conductor per Amp of current through it. In the real world, current always flows in complete loops. When we describe the total inductance of the complete current loop, we call it the *loop inductance*. This loop inductance is really the self-inductance of the entire current loop, or the loop self-inductance.

TIP   The loop self-inductance of a current loop is the total number of field line rings surrounding the entire loop per Amp of current in the loop. With 1 Amp of current in the loop, we start at one end of the loop, walk along the wire, and count the total number of field line rings we encounter from all the current in the loop. This includes the contribution from the current distribution of each section of the wire to every other section.

Let's look at the loop self-inductance of the wire loop with two straight legs, as in Figure 6-9. Leg a is like a signal path, and leg b is like a return path. As we walk along leg a, counting field line rings, we see the field line rings coming from the current in leg a (the partial self-inductance of leg a), and we see the field line rings around leg a coming from the current in leg b (or the partial mutual inductance between leg a and leg b).

Moving down leg a, the total number of field line rings we count is really the total inductance of leg a. When we move down leg b and count the total number of field line rings around leg b, we count the total inductance of leg b. The combination of these two is the loop self-inductance of the entire loop:

$$L_{loop} = L_a - L_{ab} + L_b - L_{ab} = L_a + L_b - 2L_{ab} \qquad (6\text{-}12)$$

where:

L_{loop} = loop self-inductance of the twin-lead loop

L_a = partial self-inductance of leg a

L_b = partial self-inductance of leg b

L_{ab} = partial mutual inductance between legs a and b

This may look familiar, as it appears in many textbooks. What is often not explicitly stated and what often makes inductance confusing is that the self- and mutual inductances in this relationship are really the partial self- and mutual inductances.

The relationship shown in Equation 6-12 says that as we bring the two legs of current closer together, the loop inductance will decrease. The partial self-inductances will stay the same; it is the partial mutual inductance that increases. In this case, the larger mutual inductance between the two legs will act to decrease the total number of field line rings around each leg and to reduce the loop self-inductance.

> **TIP**   It is sometimes stated that the loop self-inductance depends on the "area of the loop." While this is roughly true in general, it doesn't help feed our intuition much. As we have seen, it is not so much the area that is important as the total number of field line rings encircling each leg that is important.

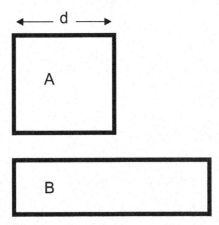

Figure 6-12   Two loops with exactly the same area but very different loop inductances. Bringing the return section of the loop closer to the other leg reduces the loop inductance by increasing the partial mutual inductance.

For example, in Figure 6-12 we show two different current loops, each with exactly the same area. They will have different loop inductances since the partial mutual inductances are so different. The closer together we bring the legs with opposite current, the larger their partial mutual inductances and the smaller the resulting loop inductance.

There is a reason to think the loop inductance scales with the area of the loop. When we count the total number of magnetic-field line rings surrounding the loop, we have to note that every one of these rings passes through the center of the loop. Counting the total number of rings is really the same as integrating the magnetic-field density over the entire area of the loop.

While the area over which the integral is performed obviously scales with the area, the value of the magnetic-field density inside the loop that we integrate up strongly depends on the shape of the loop and the current distribution.

As we have seen, the underlying mechanism for decreasing loop self-inductance is the increase in the partial mutual inductance between the signal and return paths when the return path is brought closer to the other leg and the loop is made smaller.

There are three important special-case geometries that have good approximations for loop inductance: a circular loop; two long, parallel rods; and two wide planes.

For a circular loop, the loop inductance is approximated by:

$$L_{loop} = 32 \times R \times \ln\left(\frac{4R}{D}\right) nH \qquad (6-13)$$

where:

L_{loop} = loop inductance, in nH

R = radius of the loop, in inches

D = diameter of the wire making up the loop, in inches

For example, a 30-gauge wire, about 10 mils thick, that is bent in a circle with a 1-inch diameter, has a loop inductance of:

$$L_{loop} = 32 \times 0.5 \times \ln\left(\frac{4 \times 0.5}{0.01}\right) nH = 85 \; nH \qquad (6-14)$$

TIP   This is a good rule of thumb to remember: Hold your index finger and thumb in a circle; a loop of 30-gauge wire this size has a loop inductance of about 85 nH.

This approximation illustrates that loop inductance is not really proportional to the area of the loop nor to the circumference. It is proportional to the radius times ln (radius). The larger the circumference, the larger the partial self-inductance of each section, but also the farther away the opposite-direction currents in the loop and the lower their mutual inductance.

However, to first order, the loop inductance is roughly proportional to the radius. If the circumference is increased, the loop inductance will increase. For the 1-inch loop, the circumference is 1 inch × 3.14, or about 3.14 inches.

This corresponds to a loop inductance per inch of circumference of 85 nH/ 3.14 inches ~ 25 nH/in. As a good rule of thumb, we see again that the loop inductance per length is about 25 nH/inch for loops near 1 inch in diameter.

Figure 6-13 compares the loop inductance predicted by this approximation with the actual measured loop inductance of small copper wires. The accuracy is good to a few percent.

The loop inductance of two adjacent, straight-round wires, assuming that one carries the return current of the other, is approximately:

$$L_{loop} = 10 \times Len \times \ln\left(\frac{s}{r}\right) nH \qquad (6\text{-}15)$$

where:

L_{loop} = loop inductance, in nH

Len = length of the rods, in inches

r = radius of the rods, in mils

s = center-to-center separation of the rods, in mils

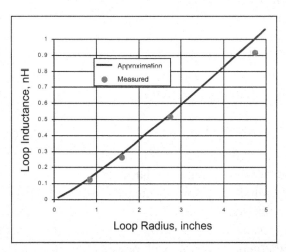

Figure 6-13   Comparison of the measured loop inductance of small loops made from 25-mil-thick wire with different loop radius and the predictions of the approximation. The approximation is seen to be good to a few percent.

For example, the loop inductance of two 1-mil-diameter wire bonds, 100 mils long and separated by 5 mils, is:

$$L_{loop} = 10 \times 0.1 \times \ln\left(\frac{5}{0.5}\right) nH = 2.3 \ nH \qquad (6\text{-}16)$$

This relationship points out that the loop inductance of two parallel wires is directly proportional to the length of the wires. It scales directly with the natural log of the separation. Increase the separation and the loop inductance increases, but only slowly, as the natural log of the separation.

The loop inductance of long, straight, parallel rods is directly proportional to the length of the rods. For example, if the rods represent the conductors in a ribbon cable, with a radius of 10 mils and center-to-center spacing of 50 mils, the loop inductance for adjacent wires carrying equal and opposite current is about 16 nH for a 1-inch-long section.

In the special case where the signal and return paths have a constant cross section down their length, the loop inductance scales directly with the length, and we refer to the loop inductance per length of the interconnect. Signal and return paths in ribbon cable have a constant loop inductance per length. In the example above, the loop inductance per length of the ribbon-cable wires is about 16 nH/inch. In the case of two adjacent wire bonds, the loop inductance per length is 2.3 nH/0.1 inch, or 23 nH/inch.

As we shall see, any controlled-impedance interconnect will have a constant loop inductance per length.

6.9 The Power Distribution Network (PDN) and Loop Inductance

When we think "signal integrity," we usually think about reflection problems and cross talk between signal nets. Though these are important problems, they represent only some of the signal-integrity problems. Another set of problems is due not to the signal paths but to the power and ground paths. We call this the *power distribution network* (*PDN*), and the design of the PDN we lump under the general heading of power integrity. Though power integrity is covered in more depth in Chapter 13, "The Power Distribution Network (PDN)," some of the relevant aspects of the role inductance plays in the PDN are introduced here.

The purpose of the PDN is to deliver a constant voltage across the power and ground pads of each chip. Depending on the device technology, this voltage difference is typically 5 v to 0.8 v. Most noise budgets will allocate no more than 5% ripple on top of this. Isn't the regulator supposed to keep the voltage constant? If ripple is too high, why not just use a heftier regulator?

The expression "there's many a slip twixt cup and lip" pretty much summarizes the problem. Between the regulator and the chip are a lot of interconnects in the PDN—vias, planes, package leads, and wire bonds, for example. When current going to the chip changes (e.g., when the microcode causes more or fewer gates to switch, or at clock edges, where most gates tend to switch), the changing current, passing through the impedance of the PDN interconnects, will cause a voltage drop, called either *rail droop* or *rail collapse*.

To minimize this voltage noise from the changing current, the design goal is to keep the impedance of the PDN below a target value. There will still be changing current, but if the impedance can be kept low enough, the voltage changes across this impedance can be kept below the 5% allowed ripple.

T I P   The impedance of the PDN is kept low by two design features: the addition of low-loop inductance decoupling capacitors to keep the impedance low at lower frequencies and minimized loop inductance between the decoupling capacitors and the chip's pads to keep the impedance low at higher frequencies.

How much decoupling capacitance is really needed? We can estimate roughly how much total capacitance is needed by assuming that the decoupling capacitor provides all the charge that must flow for some time period, Δt.

During this time, the voltage across the capacitor, C, will drop as its charge is depleted by ΔQ, which has flowed through the chip. The voltage drop, ΔV, is given by:

$$\Delta V = \frac{\Delta Q}{C} \qquad (6\text{-}17)$$

where:

ΔV = change in voltage across the capacitor

ΔQ = charge depletion from the capacitor

C = capacitance

How much current, I, flows through the chip? Obviously this depends strongly on the specific chip and will change a lot depending on the code running through it. However, we can get a rough estimate by assuming that the power dissipation of the chip, P, is related to the voltage, V, across it, and the average current, I, through it. Given the chip's average power dissipation, the average current through the chip is:

$$I = \frac{P}{V} \qquad (6\text{-}18)$$

The total amount of decoupling needed for the chip to be able to decouple for a time, Δt, is related by:

$$\frac{P}{V} = \frac{\Delta Q}{\Delta t} = \frac{C\Delta V}{\Delta t} \qquad (6\text{-}19)$$

From this relationship, the time a capacitor will decouple can be found as:

$$C = 0.05 \times C \times \frac{V^2}{P} \qquad (6\text{-}20)$$

or the capacitance required to decouple for a given time can be found from:

$$C = \frac{1}{0.05} \times \frac{P}{V^2} \times \Delta t \qquad (6\text{-}21)$$

where:

Δt = time during which the charge flows from the capacitor, in sec

0.05 = 5% voltage droop allowed

C = capacitance of the decoupling capacitor, in Farads

V = voltage of the rail, in Volts

P = power dissipation of the chip, in watts

For example, if the chip power is 1 watt, typical for a memory chip or small ASIC running at 3.3 v and with a 5% ripple allowed, the amount of total decoupling capacitance needed is about:

$$C = \frac{1}{0.05} \times \frac{1}{3.3^2} \times \Delta t = 2 \times \Delta t \qquad (6\text{-}22)$$

If the regulator cannot respond to a voltage change in less time than 10 microseconds, for example, then we need to provide at least 2×10 microsec = 20 microFarad capacitance for decoupling. Less than this, and the voltage droop across the capacitor will exceed the 5% allowed ripple.

Why not just use a single 20-microFarad capacitor to provide all the decoupling required? The impedance of an ideal capacitor will decrease with increasing frequency. At first glance, it would seem that if a capacitor has low enough impedance (e.g., in the 1-MHz range) where the regulator cannot respond, then it should have even lower impedance at higher frequency.

Unfortunately, in real capacitors, there is a loop associated with the connection between the terminals of the capacitor and the rest of the connections to the pads on the chip. This loop inductance, in series with the ideal capacitance of the component, causes the impedance of a real capacitor to increase with increasing frequency.

Figure 6-14 is a plot of the measured impedance of an 0603 decoupling capacitor. This is the measured loop impedance between one end of the capacitor and the other, through a plane below the component. At low frequency, the impedance decreases, exactly as it does in an ideal capacitor. However, as the frequency increases, we reach a point where the series loop inductance of the real capacitor begins to dominate the impedance. Above this frequency, called the *self-resonant frequency*, the impedance begins to increase. Above the self-resonant frequency, the impedance of the real capacitor is completely independent of its capacitance. It is only related to its associated loop inductance. If we want to decrease the impedance of decoupling capacitors at the higher frequency end, we need to decrease their associated loop inductance, not increase their capacitance.

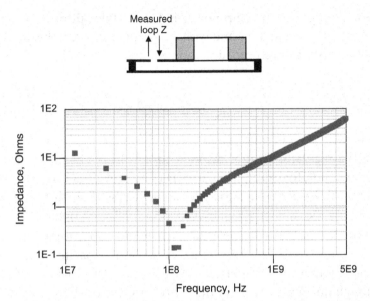

Figure 6-14 Measured loop impedance of a 1 nF 0603 decoupling capacitor, with a current loop configured as shown, measured with a GigaTest Labs Probe Station.

TIP   A key feature of decoupling capacitors is that at high frequency, the impedance is solely related to their loop inductance, referred to as the equivalent series inductance (ESL). Decreasing the impedance of a decoupling capacitor at high frequency is all about decreasing the loop inductance of the complete path from the chip's pads to the decoupling capacitor.

The measured loop impedance of six 0603 decoupling capacitors with different values is shown in Figure 6-15. Their impedance at low frequency is radically different since they have orders-of-magnitude different capacitance. However, at high frequency, their impedances are identical because they have the same mounting geometry on the test board.

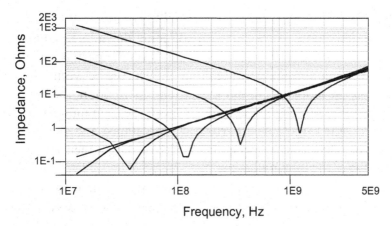

Figure 6-15 Measured loop impedance of six different 0603 capacitors with capacitances varying from 10 pF to 1 µF, but all with the same mounting geometry, measured with GigaTest Labs Probe Station.

TIP   The only way of decreasing the impedance of a decoupling capacitor at high frequency is by decreasing its loop self-inductance.

The best ways of decreasing the loop inductance of a decoupling capacitor are as follows:

- Keep vias short by assigning the power and ground planes close to the surface.
- Use small-body-size capacitors.
- Use very short connections between the capacitor pads and the vias to the underlying planes.
- Use multiple capacitors in parallel.

If the loop inductance associated with one decoupling capacitor and its mounting is 2 nH and the maximum allowed inductance is 0.1 nH, then there must be at least 20 capacitors in parallel for the equivalent loop inductance to meet the requirement.

From the decoupling capacitors to the chips' pads, the interconnect should be designed for the lowest loop inductance. In addition to short surface pads and short vias, planes are the interconnect geometry with the lowest loop inductance.

6.10 Loop Inductance per Square of Planes

The loop inductance for a current path going down one plane and back the other, as illustrated in Figure 6-16, depends on the partial self-inductance of each plane path and the partial mutual inductance between them. The wider the planes, the more spread out the current distribution, the lower the partial self-inductance of each plane, and the lower the loop inductance. The longer the planes, the larger their partial self-inductance and the larger the loop inductance. The closer we bring the planes, the larger their mutual inductance and the lower the loop inductance.

For the case of wide conductors, where the width, w, is much larger than their spacing, h, or w >> h, the loop inductance between two planes is to a very good approximation given by:

$$L_{loop} = \mu_0 h \frac{Len}{w} \qquad (6\text{-}23)$$

where:

L_{loop} = loop inductance, in pH

μ_0 = permeability of free space = 32 pH/mil

h = spacing between the planes, in mils

Len = length of the planes, in mils

w = width of the planes, in mils

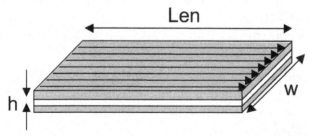

Figure 6-16   Geometrical configuration for the current flow in two planes forming a loop. The return current is in the opposite direction in the bottom plane.

This assumes the current flows uniformly from one edge and back to the other edge.

If the section is in the shape of a square, the length equals the width, and the ratio is always 1, independent of the length of a side. It's startling that the loop inductance of a square section of a pair of planes that is 100 mils on a side is exactly the same as for a square section 1 inch on a side. Any square section of a pair of planes will have the same edge-to-edge loop inductance. This is why we often use the term *loop inductance per square* of the planes, or the shortened version *inductance per square* or *sheet inductance* of the board. It makes life more confusing, but now we know this really refers to the loop inductance between two edges of a square section of planes, when the other edges are shorted together.

For the thinnest dielectric spacing currently in volume production, 2 mils, our approximation above estimates the loop inductance per square as about $L_{loop} = 32$ pH/mil $\times$ 2 mil $= 64$ pH. As the dielectric thickness increases, the loop inductance per square increases. A dielectric spacing of 5 mils has a loop inductance per square of $L_{loop} = 32$ pH/mil $\times$ 5 mil $= 160$ pH.

As the spacing between adjacent planes increases, the partial mutual inductance will decrease, and there won't be as many field line rings from one plane around the other to cancel out the total number of field line rings. With greater dielectric spacing, the loop inductance increases, and the rail-collapse noise will increase. This will make the PDN noise worse and will also increase the ground-bounce noise that drives common currents on external cables and causes EMI problems.

TIP Spacing the power and ground planes as close together as possible will decrease the loop inductance in the planes, decreasing rail collapse, ground bounce in the planes, and EMI.

6.11 Loop Inductance of Planes and Via Contacts

Current doesn't flow from one edge to another edge in planes. Between a discrete decoupling capacitor and the package leads, the connection to the planes is more like point contacts. In the above analysis, we assumed the current was flowing uniformly down the plane. However, in actual practice, the current is not uniform. If the current is restricted due to point contacts, the loop inductance will increase.

The only reason we assumed the current was uniform was because this was the only case for which we had a simple approximation to help estimate the loop inductance. When balancing accuracy versus little effort to get an answer, we chose the path of little effort. The only way to get a better estimate of the loop inductance associated between two planes with real contacts is by using a 3D field solver.

We can gain some insight into how the geometry affects the associated loop inductance between two contact points between the planes. By using a 3D field solver, we can calculate the specific current distribution between the contact points and from the specific current distributions, the resulting loop inductance.

The good news about field solvers is that the accuracy can be very good, and they can include many real-world effects for which there are no good approximations. The bad news is that we can't generalize an answer from a field solver. It can run only one specific problem at a time.

As an example of the impact of the via contacts on the loop inductance between two planes, two special cases are compared. In both cases, a 1-inch-square section of two planes, 2 mils apart, was set up. In the first case, one edge of the top plane and the adjacent edge of the bottom plane were used as the current source and sink. The far ends of the planes were shorted together.

In the second case, two small via contacts were used as the current source and sink at one end, and a similar pair of contacts shorted the planes together at the other end. The contacts were 10 mils in diameter, spaced on 25-mil centers, similar to how a pair of vias would contact the planes in an actual board.

Figure 6-17 shows the current distribution in one of the planes for each case. When the edge contact is used, the current distribution is uniform, as expected. The loop inductance between the two planes is extracted as 62 pH. The approximation above, for the same geometry, was 64 pH. We see that for this special case, the approximation is pretty good.

When the current flows from one via contact, down the board to a second via contact, through the via to the bottom plane, back through the board and up through the end vias—as would be the case in a real board—the loop inductance extracted by the field solver is 252 pH. This is an increase by about a factor of four. The increase in loop inductance between the planes is due to the higher current density where the current is restricted to flow up the via.

The more constricted the current flow, the higher the partial self-inductance and loop inductance. This increase in loop inductance is sometimes referred to as *spreading inductance*. If the contact area were increased, the current density would decrease, and the spreading inductance would decrease.

 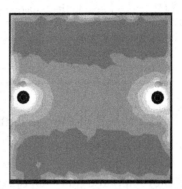

Figure 6-17 Current distribution in the top plane of a pair of planes with edge contact and a pair with point contacts. The lighter shade corresponds to higher current density. With edge contacts, the current distribution is uniform. With via point contacts, the current is crowded near the contact points. The higher current density creates higher inductance. Simulation with Ansoft's Q3D field solver, courtesy of Charles Grasso.

The loop inductance between the planes, even with the spreading inductance, will still scale with the plane-to-plane separation. The thinner the dielectric between the power and ground planes, the lower the sheet inductance and the lower the spreading inductance. Likewise, when the dielectric spacing between planes is large, the spreading inductance is large.

> **TIP**   The spreading inductance associated with the via contact points to a pair of power and ground planes is usually larger than the associated sheet loop inductance and must be taken into account to accurately estimate the loop inductance of the planes.

When many pairs of vias contribute current to the planes connecting many capacitors and many package leads, closely spaced planes will minimize the voltage drop from all the simultaneous dI/dts.

The loop inductance in the planes associated with the decoupling capacitor is dominated by the spreading inductance, not by the distance between the chip and the capacitor. To a first order, the total loop inductance of the decoupling capacitor is only weakly dependent on its proximity to the chip. However, the closer the capacitor is to the chip, the more the high-frequency power and return currents will be confined to the proximity of the chip and the lower the ground-bounce voltage will be on the return plane.

TIP   By keeping decoupling capacitors close to the high-power chips, the high-frequency currents in the return plane can be localized to the chip and kept away from I/O regions of the board. This will minimize the ground-bounce voltage noise that might drive common currents on external cables and contribute to EMI.

6.12 Loop Inductance of Planes with a Field of Clearance Holes

A field solver is a useful tool for exploring the impact of a field of via clearance holes on the loop inductance between two planes. Arrays of vias occur all the time. They happen under BGA packages, under connectors, and in very high-density regions of the board.

Often, there will be clearance holes in the power and ground planes of the via field. What impact will the holes have on the loop inductance of the planes? To a first order, we would expect the inductance to increase. But by how much? We often hear that an array of clearance holes under a package—the Swiss cheese effect—will dramatically increase the inductance of the planes. The only way to know by how much is to put in the numbers—that is, use a field solver.

Two identical pairs of planes were created, each 0.25 inch on a side and separated by 2 mils. Two via contacts were connected to each end of the planes. At one end, the vias were shorted together. At the other end, current was injected in one via and taken out from the other. This simulates one end connected to a decoupling capacitor and the other connected to the power and ground leads of a package through vias to the top surface.

In one pair of planes, a field of clearance holes was created in each plane. Each hole was 20 mils in diameter on 25 mil centers. This is an open area of about 50%. For each of the two cases, the current distribution was calculated and the loop inductance extracted with a 3D static field solver. Figure 6-18 shows the current distribution with and without the field of holes. The clearance holes constrict the current to the narrow channels between the holes, so we would expect to see the loop inductance increase.

The field solver calculates a loop inductance of 192 pH with no holes and 243 pH with this high density of holes. This is an increase of about 25% in loop inductance for an open area of 50%. We see that holes do increase the loop inductance of planes but not nearly as dramatically as we might have expected. To minimize the impact from the holes, it is important to make the holes as small as possible. Of course, bringing the planes closer together will also decrease the loop inductance of the planes, with or without clearance holes.

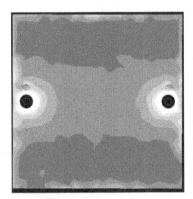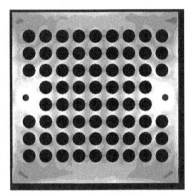

Figure 6-18   Current distribution on closely spaced planes with contact via points with and without a field of clearance holes. The lighter colors correspond to higher current densities. The holes cause the current to constrict, increasing the loop inductance. Simulated with Ansoft's Q3D field solver, courtesy of Charles Grasso.

It is important to note that while a field of clearance holes in the power and ground planes will increase the loop inductance, the loop inductance can be kept well below a factor of two increase. It is not nearly as much of a disaster as commonly believed.

TIP   The optimum power and ground interconnects, for lowest loop inductance, are planes as wide as possible and as closely spaced as possible. When we use a very thin dielectric between the planes, we reduce the loop inductance between the bulk decoupling capacitors and the chip's pads. This will decrease the rail collapse and EMI.

Both rail collapse and EMI will get worse as rise times decrease. As we move into the future of higher and higher clock frequencies, thin dielectrics in the power distribution network will play an increasingly important role.

6.13 Loop Mutual Inductance

If there are two independent current loops, there will be a mutual inductance between them. The loop mutual inductance is the number of field line rings from the current in one loop that completely surrounds the current of the second loop per Amp of current in the first loop.

When current in one loop changes, it will change the number of field line rings around the second current loop and induce noise in the second current loop. The amount of noise created is:

$$V_{noise} = L_m \frac{dI}{dt} \tag{6-24}$$

where:

V_{noise} = voltage noise induced on one loop

L_m = loop mutual inductance between the two rings

dI/dt = how fast the current in the second loop changes

The noise in the quiet loop will arise only when there is a dI/dt in the active loop, which is only during the switching transitions. This is why this sort of noise is often called *switching noise, simultaneous switching noise (SSN),* or *delta I noise.*

TIP   The most important way of reducing switching noise is reducing the mutual inductance between the signal- and return-path loops. This can be accomplished by moving the loops farther apart from each other. Since the mutual inductance between two loops can never be greater than the self-inductance of the smallest loop, another way of decreasing the loop mutual inductance is to decrease the loop self-inductance of both loops.

Loop mutual inductance also contributes to the cross talk between two uniform transmission lines and is discussed in Chapter 11, "Differential Pairs and Differential Impedance."

6.14 Equivalent Inductance of Multiple Inductors

So far, we have been considering just the partial inductance associated with a single interconnect element that has two terminals, one on either end, and then the resulting loop inductance consisting of the two elements in series.

For two separate interconnect elements, there are two connection topologies: end to end (in series) or with each end together (in parallel). These two circuit configurations are illustrated in Figure 6-19.

In either configuration, the resulting combination has two terminals, and there is an equivalent inductance for the combination. We are used to thinking about the equivalent inductance of a series combination of two inductors simply as the sum of each individual partial self-inductance. But what about the impact from their mutual inductance?

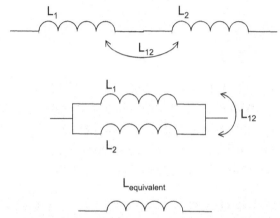

Figure 6-19   Circuit topologies for combining partial inductors in series (top) and in parallel (middle) into an equivalent inductance (bottom).

The inclusion of the mutual inductance between the interconnect elements makes the equivalent inductance a bit more complicated. We first must evaluate the total inductance of each element from its self-inductance and the contribution of mutual inductance from the other element. Then we combine total inductances either in series or in parallel.

Since the currents in each element are in the same direction, the direction of circulation of all the magnetic-field line rings is in the same direction, and the mutual partial inductances add to the partial self-inductances. In each case, the total inductances are larger than the partial self-inductances, and the equivalent circuit inductances are larger including the mutual inductances than not including them.

For the series combination of two partial inductances, the resulting equivalent partial self-inductance of the series combination is:

$$L_{series} = L_1 + L_2 + 2L_{12} \qquad (6\text{-}25)$$

The equivalent partial self-inductance when the elements are connected in parallel is given by:

$$L_{parallel} = \frac{L_1 L_2 + L_{12}(L_1 + L_2) + L_{12}^2}{L_1 + L_2 + 2L_{12}} \qquad (6\text{-}26)$$

where:

L_{series} = equivalent partial self-inductance of the series combination
$L_{parallel}$ = equivalent partial self-inductance of the parallel combination
L_1 = partial self-inductance of one element
L_2 = partial self-inductance of the other element
L_{12} = partial mutual-inductance between the two elements

When the partial mutual inductance is zero, and the partial self-inductances are the same, both of these models reduce to the familiar expressions of the series and parallel combination. The series combination is just twice the self-inductance of one of them, and the equivalent parallel inductance is just half the partial self-inductance of either of them.

In the special case when the partial self-inductance of the two inductors is the same, the series combination of the two inductors is just twice the sum of the self- and mutual inductance. For the parallel combination, when the two partial self-inductances are equal, the equivalent inductance is:

$$L_{parallel} = \frac{1}{2}(L + M) \qquad (6\text{-}27)$$

where:

$L_{parallel}$ = equivalent partial self-inductance of the parallel combination
L = partial self-inductance of either element
M = partial mutual inductance of either element

This says that if the goal is to reduce the equivalent inductance of two current paths in parallel, we do get a reduction, as long as the mutual inductance between the elements is kept small.

6.15 Summary of Inductance

All the various flavors of inductance are directly related to the number of magnetic-field line rings around a conductor per Amp of current. The importance of inductance is due to the induced voltage across a conductor when currents change. Just referring to an "inductance" can be ambiguous.

To be unambiguous, we need to specify the source of the currents, as in the self-inductance or mutual inductance. Then we need to specify whether we are referring to part of the circuit using partial inductance or the whole circuit using loop inductance. When we look at the voltage noise generated across part of the circuit, because this depends on all the field line rings and how they change, we need to specify the total inductance of just that section of the circuit. Finally, if we have multiple inductors in some combination, such as multiple parallel leads in a package or vias in parallel, we need to use the equivalent inductance.

The greatest source of confusion arises when we misuse the term *inductance*. As long as we include the correct qualifier, we will never go wrong. The various flavors of inductance are listed here:

- **Inductance:** The number of magnetic-field line rings around a conductor per Amp of current through it
- **Self-inductance:** The number of field line rings around the conductor per Amp of current through the same conductor
- **Mutual inductance:** The number of field line rings around a conductor per Amp of current through another conductor
- **Loop inductance:** The total number of field line rings around the complete current loop per Amp of current
- **Loop self-inductance:** The total number of field line rings around the complete current loop per Amp of current in the same loop
- **Loop mutual-inductance:** The total number of field line rings around a complete current loop per Amp of current in another loop
- **Partial inductance:** The number of field line rings around a section of a conductor as though no other currents exist anywhere

- **Partial self-inductance:** The number of field line rings around a section of a conductor per Amp of current in that section as though no other currents exist anywhere
- **Partial mutual-inductance:** The number of field line rings around a section of a conductor per Amp of current in another section as though no other currents exist anywhere
- **Effective, net, or total inductance:** The total number of field line rings around a section of a conductor per Amp of current in the entire loop, taking into account the presence of field line rings from current in every part of the loop
- **Equivalent inductance:** The single self-inductance corresponding to the series or parallel combination of multiple inductors including the effect of their mutual inductance

6.16 Current Distributions and Skin Depth

In evaluating the resistance and inductance of conductors, we have been assuming that the currents are uniformly distributed through the conductors. While this is the case for DC currents, it is not always true when currents change. AC currents can have a radically different current distribution, which can dramatically affect the resistance and to some extent the inductance of conductors.

It is easiest to calculate the current distributions in the frequency domain where currents are sine waves. This is a situation where moving to the frequency domain gets us to an answer faster than staying in the time domain.

At DC frequency, the current distribution in a solid rod is uniform throughout the rod. When we counted magnetic-field line rings previously, we focused on the field line rings that were around the outside of the conductor. In fact, there are some magnetic-field line rings inside the conductor as well. They are also counted in the self-inductance. This is illustrated in Figure 6-20.

To distinguish the magnetic-field line rings contributing to the conductor's self-inductance that are enclosed inside the conductor from those that are outside the conductor, we separate the self-inductance into internal self-inductance and external self-inductance.

The internal magnetic-field line rings are the only field line rings that actually see the conductor metal and can be affected by the metal. For a round wire, the external-field line rings never see any conductor and they will never change with frequency. However, the internal-field line rings do see the conductor, and they can change as the current distribution inside the conductor changes with frequency.

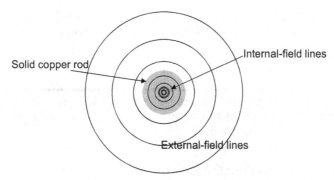

Figure 6-20 Magnetic-field line rings from the DC current in a uniform solid rod of copper. Some of the field line rings are internal, and some are external.

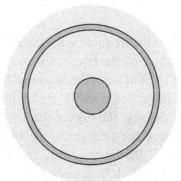

Figure 6-21 Two annuluses of current, traveling into the paper, singled out in a solid rod of copper, both of exactly the same cross-sectional area.

Consider two annuluses of current with exactly the same cross-sectional area in a solid rod of copper, as shown in Figure 6-21. If the cross-sectional area of each is exactly the same, each annulus will carry the same current. Which annulus of current has more field line rings around it?

The outer annulus and the inner annulus have exactly the same number of field line rings in the region outside the outer cylinder. The number of field line rings outside a current is only related to the total amount of current enclosed by the field line rings. There are no field line rings from currents in the outer annulus inside this annulus, since field line rings must encircle a current.

The current from the inner annulus has more internal self-field line rings, since it has more distance from its current to the outer wall of the rod. The closer to the center of the rod the current is located, the greater the total number of field line rings around that current.

TIP   Currents closer to the center of the rod will have more field line rings per Amp of current and a higher self-inductance than annuluses of current toward the outside of the conductor.

Now, we turn on the AC current. Currents are sine waves. Each frequency component will travel the path of lowest impedance. The current paths that have the highest inductance will have the highest impedance. As the sine-wave frequency increases, the impedance of the higher-inductance paths get even larger. The higher the frequency, the greater the tendency for current to want to take the lower-inductance path. This is the path toward the outer surface of the rod.

In general, the higher the frequency, the greater the tendency for current to travel on the outside surface of the conductor. At a given frequency, there will be some distribution of current from the center to the outside surface. This will depend on the relative amount of resistive impedance and inductive impedance. The higher the current density, the higher the voltage drop from the resistive impedance. But the higher the frequency, the bigger the difference between the inductive impedance of the inner and outer paths. This balance means the current distribution changes with frequency, tending toward all the current in a thin layer around the outer surface, at high frequency.

TIP   As the frequency of the sine waves of current through the rod increases, the current will redistribute, so most of the current is taking the path of lowest impedance—that is, toward the outside of the conductor. At high frequency, it appears as though all the current is traveling in a thin shell at the conductor's surface.

There are only a few simple geometries where there is a good approximation for the actual current distribution of the current in a conductor. One of these is for a round cylinder, assuming that the return path is coaxial. At each frequency, the current distribution will drop off exponentially toward the center of the conductor. This is illustrated in Figure 6-22.

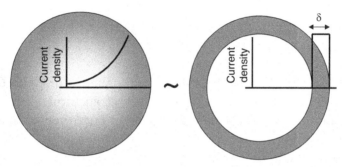

Figure 6-22 Left: Current distribution in a solid rod of copper at some frequency, showing the concentration of current near the outer surface. Darker color is high-current density. Right: Approximating the current in the rod in terms of a uniform current distribution with a thickness equal to the skin depth.

In this geometry, we can approximate the shell of current as a uniform distribution with a fixed thickness, a distance δ, from the outer surface. We call this equivalent thickness of the current shell the *skin depth*. It depends on the frequency, the conductivity of the metal, and the permeability of the metal:

$$\delta = \sqrt{\frac{1}{\sigma \pi \mu_0 \mu_r f}} \qquad (6\text{-}28)$$

where:

δ = skin depth, in meters

σ = conductivity of conductor, in Siemens/m

μ_0 = permeability of free space, $4 \times \pi \times 10^{-7}$ H/m

μ_r = relative permeability of the conductor

f = sine-wave frequency, in Hz

For the case of copper with a conductivity of 5.6×10^7 Siemens/m and relative permeability of 1, the skin depth is approximately:

$$\delta = 66 \text{ microns}\sqrt{\frac{1}{f}} \qquad (6\text{-}29)$$

where:

δ = skin depth, in microns

f = sine-wave frequency, in MHz

At 1 MHz, the skin depth for copper is 66 microns. In Figure 6-23, the skin depth for copper is plotted and compared to the geometrical thickness of 1-ounce and 1/2-ounce copper. This points out that for 1-ounce copper traces, when the sine-wave frequency of the current is higher than about 10 MHz, the current distribution is determined by skin depth, not the geometrical cross section. Below 10 MHz, the current distribution will be uniform and independent of frequency. When the skin depth is thinner than the geometrical cross section, the current distribution, the resistance, and the loop inductance will be frequency dependent.

This is a handy rule of thumb to keep in mind, and it means that if we have a board with 1-ounce copper traces or a geometrical thickness of 34 microns, current that is moving through the conductor at 10-MHz or higher sine-wave frequencies would not use the entire cross section of the trace. It would be dominated by these skin-depth effects.

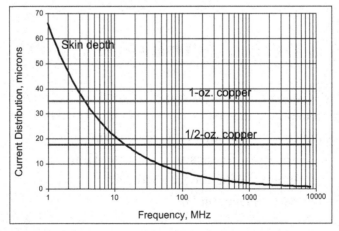

Figure 6-23 Current distribution in copper when skin depth is limited and compared with the geometrical thickness of 1-ounce and 1/2-ounce copper.

In real interconnects, there are always a signal path and a return path. As the current loop propagates down the signal path and the return path, it is the loop self-inductance that influences the impedance the current sees. As frequency increases, the impedance from the loop self-inductance will increase, and the current in both conductors will redistribute to take the path of lowest impedance or lowest loop self-inductance. What is the current distribution for lowest loop self-inductance?

There are two ways the loop self-inductance can be decreased: by spreading out the current within each conductor and by bringing the return current closer to the signal current. This will decrease the partial self-inductance of each conductor and increase the partial mutual inductance between them. Both effects will happen: Current will spread out within each conductor, and the current distribution in both conductors will redistribute so the two currents get closer to each other. The precise current distribution in each conductor is determined by the balance of two forces. Within each conductor, the current wants to spread apart to decrease its partial self-inductance. At the same time, the current in one conductor wants to get as close as possible to the current in the return path to increase their partial mutual inductance. The resulting current distribution can only be calculated with a 2D field solver. Figure 6-24 shows the simulated current distribution in a pair of 20-mil- (500-micron-) diameter ribbon wires. At low frequency, the skin depth is large compared with the geometrical cross section, and the current is uniformly distributed in each conductor. At 100 kHz, the skin depth in copper is about 10 mils (250 microns) and is comparable with the cross section, so it will begin to redistribute. At 1 MHz, the skin depth is 2.5 mils (66 microns), small compared to the diameter. The current distribution will be dominated by the skin depth. As the frequency increases, the current redistributes to minimize the loop impedance.

Also shown is the current distribution for a 1-ounce copper microstrip. At 1 MHz, the current is mostly uniformly distributed. At 10 MHz, it begins to redistribute. Above 10 MHz, the skin depth is much smaller than the cross section and will dominate the current distribution. In both examples, the current redistributes at higher frequency to minimize the impedance.

As the frequency increases, the bulk resistivity of the conductor does not change. For copper, the bulk resistivity doesn't begin to change until above 100 GHz. However, if the current travels through a thinner cross section due to skin-depth effects, the resistance of the interconnect will increase.

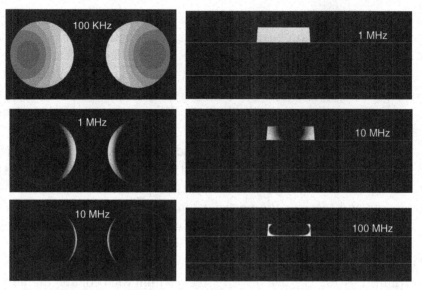

Figure 6-24 Current distribution in 20-mil-diameter wire at three different frequencies and 1-ounce copper microstrip. The lighter the color, the higher the current density. Simulated with Ansoft's 2D Extractor.

TIP   In the skin-depth regime, when the skin depth is thinner than the geometrical cross section, as frequency increases, the cross-sectional area the current travels through decreases proportionally to the square root of the frequency. This will cause the resistance per length of the line to increase with the square root of the frequency as well.

Consider the case of a simple microstrip line, made from 1-ounce copper, 5 mils wide. The resistance per length of the signal path at DC frequency is:

$$R_{DC} = \frac{\rho}{wt}$$
(6-30)

where:

R_{DC} = resistance per length for DC currents

ρ = bulk resistivity of copper

w = line width of the signal trace

t = geometrical thickness of the signal trace

At frequencies above about 10 MHz, the current is skin-depth limited, and the resistance is frequency dependent. Above this frequency, the thickness of the conductor in which the current flows is roughly the skin depth, so the high-frequency resistance, assuming current flows in one surface only, is really:

$$R_{HF} = \frac{\rho}{w\delta}$$

(6-31)

where:

R_{HF} = resistance per length for high-frequency currents

ρ = bulk resistivity of copper

w = line width of the signal trace

δ = skin depth of copper at the high frequency

The ratio of resistance at high frequency to the resistance at DC is roughly $R_{HF}/R_{DC} = t/\delta$. At 1 GHz, where the skin depth of copper is 2 microns, the high-frequency resistance will be 30 microns/2 microns = 15 × higher than the low-frequency resistance for 1-ounce copper. The series resistance of a signal trace will only get larger at higher frequencies.

Figure 6-25 shows the measured resistance of a small loop of 22-gauge copper wire, 25 mils in thickness. The loop was about 1 inch in diameter. The skin depth is comparable to the geometrical thickness at 10 kHz. At higher frequencies, the resistance should increase with roughly the square root of frequency.

One consequence of the frequency-dependent current distribution is the frequency-dependent resistance. In addition, the inductance will change. Since the driving force for the current redistribution is decreasing loop self-inductance, the loop self-inductance will decrease with higher frequency.

At DC frequency, the self-inductance of a wire will be composed of the external self-inductance and the internal self-inductance. The external self-inductance will not change as the current redistributes within the conductor, but the internal self-inductance will decrease as more current moves to the outside. At frequencies far above where the skin depth is comparable to the geometrical thickness, and the current distribution is skin-depth dominated, there will be very little current inside the conductor, and there will be no internal self-inductance.

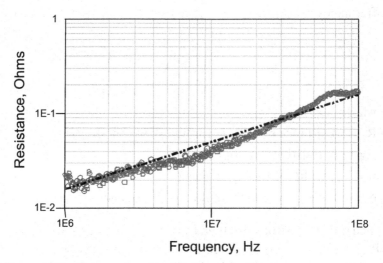

Figure 6-25   Measured series resistance of a 22-gauge copper wire loop, 1 inch in diameter, showing the resistance increasing with the square root of frequency. Circles are measured resistance; the line is resistance increasing with the square root of frequency.

We would expect the self-inductance of a wire to be frequency dependent. At low frequency, it will be $L_{internal} + L_{external}$. At high frequency, it should be simply $L_{external}$. The transition should start about where the skin depth is comparable to the geometrical thickness and reach a steady value above a frequency where the skin depth is just a small fraction of the geometrical thickness.

The precise current distribution and the contribution of internal and external self-inductance are hard to estimate analytically, especially for rectangular cross sections. However, they can be calculated very easily using a 2D field solver.

Figure 6-26 shows an example of the loop self-inductance per length for a microstrip as the current redistributes itself. This illustrates that due to skin-depth effects, the low-frequency inductance is higher than the high-frequency inductance by an amount equal to the internal self-inductance. Above about 100 MHz, the current travels in a thin shell, and the inductance is constant with further increase in frequency.

TIP   When we refer to the loop self-inductance of a microstrip, for example, we are more often referring to this high-frequency limit, assuming that all the current is in the outside surface. *High frequency* refers to above the skin-depth limit, where current is close to the surface and not dependent on the geometrical thickness.

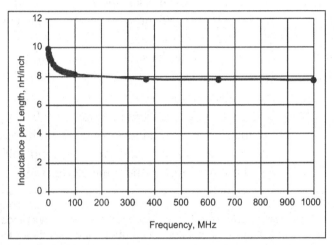

Figure 6-26   Loop self-inductance of a microstrip as the current redistributes due to skin-depth effects, calculated with Ansoft's 2D Extractor.

6.17 High-Permeability Materials

There is an important term affecting skin depth that applies to only a few special materials. This term is the *permeability of the conductor*. Permeability refers to how the conductor interacts with magnetic-field line rings. Most metals have a permeability of 1, and they do not interact with magnetic-field line rings.

However, when the permeability is greater than 1, the number of field line rings in the metal is amplified over what would be there if the permeability were 1. There are only three metals that have a permeability other than 1. These are the ferromagnetic metals: iron, nickel, and cobalt. Most alloys that contain any combination of these metals have a permeability much greater than 1. We are most familiar with ferrites, which usually contain iron and cobalt and can have a permeability of over 1,000. Two important interconnect metals are ferromagnetic: Alloy 42 and Kovar. The permeability of these metals can be 100–500. This high permeability can have a significant impact on the frequency dependence of the resistance and inductance of an interconnect made from these materials.

For a ferromagnetic wire, at DC frequency, the self-inductance of the wire will be related to the internal and external self-inductance. All the field line rings that contribute to the external self-inductance see only air, which has a permeability of 1. The external self-inductance of a ferromagnetic wire will be exactly the same as if the wire were made of copper. After all, for the same current in the wire, there would be the same external-field line rings per amp of current.

However, the internal-field line rings in the ferromagnetic wire will see a high permeability, and these magnetic-field line rings will be amplified. At low frequency, the inductance of a ferromagnetic wire is high, but above about 1 MHz, all the field line rings are external, and the loop self-inductance is comparable to that of a copper loop of the same dimension.

> **TIP**   The loop inductance that high-speed signals would encounter in a ferromagnetic conductor is comparable to the loop inductance if the conductor were made from copper, since above the skin-depth limit the loop inductance is composed almost solely of external-field line rings.

The skin depth for a ferromagnetic conductor can be much smaller than for a copper conductor due to its high permeability. For example, nickel, having a bulk conductivity of about 1.4×10^7 Siemens/m and permeability of about 100, has a skin depth of roughly:

$$\delta = 13 \text{ microns} \sqrt{\frac{1}{f\,[\text{MHz}]}} \tag{6-32}$$

At the same frequency, the cross section for current in a nickel conductor will be much thinner than for an equivalent-geometry copper conductor. In addition, the bulk resistivity is higher. This means the series resistance will be much higher. Figure 6-27 shows the measured series resistance of 1-inch-diameter rings of copper wire and nickel wire of roughly comparable wire diameter. The resistance of the nickel wire is more than 10 times higher than that of the copper wire and is clearly increasing with the square root of frequency, which is characteristic of a skin-depth-limited current distribution. This is why the high-frequency resistance of Kovar or Alloy 42 leads can be very high compared to that of nonferromagnetic leads.

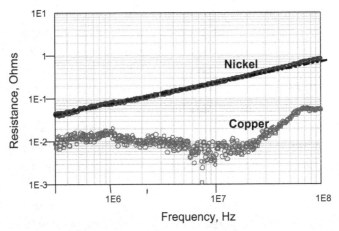

Figure 6-27   Measured resistance of a 1-inch-diameter loop of copper wire and nickel wire, of roughly the same cross section, showing much higher resistance of the nickel conductor due to skin-depth effects. The resistance increases with the square root of frequency, shown by the superimposed line. The noise floor of the measurement was about 10 milliOhms.

This is why a plating of silver is sometimes applied to Alloy 42 leads to limit their resistance at high frequency. It provides a nonferromagnetic conductor on the outer surface for high-frequency currents to travel. The highest-frequency components will experience a larger skin depth and the higher-conductivity material.

The precise resistance of a conductor will depend on the frequency-dependent current distribution, which, for arbitrary shapes, may be difficult to calculate. This is one of the values of using a good 2D field solver that allows the calculation of frequency-dependent current distributions and the resulting inductance and resistance.

6.18 Eddy Currents

As mentioned previously, if there are two conductors, and the current in one changes, there will be a voltage created across the second due to the changing mutual magnetic-field line rings around it. This voltage induced in the second conductor can drive currents in the second conductor. In other words, changing the current in one conductor can induce current in the second conductor. We call the induced currents in the second conductor *eddy currents*.

There is an important geometry where eddy currents can significantly affect the partial self-inductance of a conductor and the loop self-inductance of a current loop. This geometry occurs when a loop is near a large conducting surface, such as a plane in a circuit board or the sides of a metal enclosure.

As a simple example, consider a round wire above a metal plane. It is important to keep in mind that this metal plane can be any conductor and can float at any voltage. It does not matter what its voltage is or to what else it is connected. All that is important is that it is conductive and that it is continuous.

When there is current in the wire, some of the field line rings will pass through the conducting plane. There will be some mutual inductance between the wire and the plane. When current in the wire changes, some of the magnetic-field line rings going through the plane will change, and voltages will be induced in the plane. These voltages will drive eddy currents in the plane. These eddy currents will, in turn, produce their own magnetic fields.

By solving Maxwell's Equations, it can be shown that the pattern of the magnetic-field line rings generated by the eddy currents looks exactly like the magnetic-field lines from another current that would be located below the surface of the plane (i.e., a distance below equal to the height above the plane of the real current). This is illustrated in Figure 6-28. This fictitious current is called an *image current*. The direction of the image current is opposite the direction of the real current that induced it. The net magnetic-field lines from the real current and the induced eddy currents will have the same field line distribution as from the real current and the image current, as though the plane were not there. To understand the actual magnetic-field line rings present with the combination of the real current and the induced eddy currents, we can throw out the plane and the real eddy currents and replace them with the image currents. The field line rings from the real current and the image current will be exactly the same as the field line rings from real current and the eddy currents.

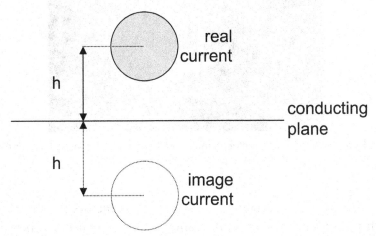

Figure 6-28 Generation of an image current in a plane exactly opposite from the initial current.

The current in the image current is exactly the same magnitude as the current in the real current but in the opposite direction. Some of the field line rings from the image current will go around the real current. But, because the direction of the image current is opposite the direction of the source current, the field line rings from the image current will subtract from the field line rings around the real current.

The mutual-field line rings from the eddy currents (the image current) will reduce the total inductance of the wire, which is really the partial self-inductance of the wire. If a current loop is over the surface of a floating, conducting plane, which has absolutely no electrical contact to the loop, the mere presence of the plane, and the eddy currents induced in it, will decrease the loop inductance of the loop. The closer the wire is to the plane, the closer is the image current, and the more mutual-field line rings there are from the image current, and the lower the partial self-inductance of the real current. The closer the floating plane below, the larger the induced eddy currents in the plane and the greater the reduction of the self-inductance of the signal path. Figure 6-29 shows the eddy current distributed in an adjacent plane when the signal path is in proximity to the floating plane.

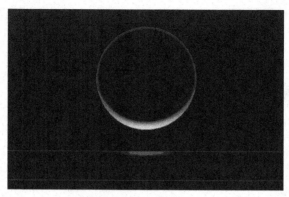

Figure 6-29 Current distribution at 1 MHz in a round conductor near a floating plane showing the eddy currents induced in the plane.

When two long, rectangular coplanar conductors make up a signal and return path loop, they will have some loop self-inductance per length. If a uniform, floating conducting plane is brought in proximity, their loop self-inductance will decrease due to the field line rings from the eddy currents in the plane below. The closer the plane, the lower the loop inductance. Figure 6-30 shows this reduction in loop self-inductance per length for a simple case.

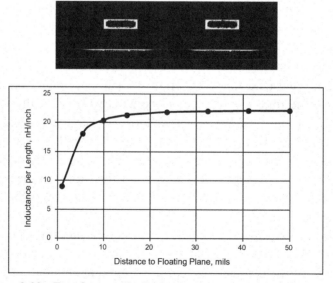

Figure 6-30 Top: Current distribution in the two legs of a long rectangular coplanar loop, showing the induced eddy currents in the floating plane. Bottom: The change in loop self-inductance per length as the distance to the floating plane changes, calculated with Ansoft's 2D Extractor.

In this example, the line width is 5 mils, and the spacing is 10 mils. The span between the outer edges of the conductors is 20 mils. As a rough rule of thumb, the induced eddy currents will play a role if the spacing to the floating plane is closer than the total span of the conductors.

TIP   Induced eddy currents will be created in any plane conductor when a current loop is as close as the span of the conductors. The presence of adjacent planes will always decrease the loop self-inductance of an interconnect.

6.19 The Bottom Line

1. *Inductance* is a critically important term that affects all aspects of signal integrity.
2. The basic definition of *inductance* is that it is the number of magnetic-field line rings around a conductor per Amp of current through the conductor.
3. All the various types of inductance are special qualifiers that specify from which conductor the field line rings arise (self and mutual), around how much of the conductor the field line rings are counted (partial and loop), and if all the field line rings are included, even those from the rest of the loop (total).
4. The only reason inductance is important is because of induced voltage: If the number of field line rings around a conductor changes, a voltage is induced across the ends of the conductor related to how fast the field line rings change.
5. Ground bounce is the voltage induced between different parts of the ground-return path due to a total inductance of the return path and a dI/dt through it.
6. Reducing ground bounce is about reducing the total inductance of the return path: wide conductors, short lengths, and the signal path as close as possible to the return path.
7. The lowest rail-collapse noise is obtained when the loop inductance from the chip pad to the decoupling capacitor is as low as possible. The lowest loop inductance interconnect is obtained with two wide planes as close together as possible.

8. The loop inductance between two planes is increased by the presence of a field of via holes. For the case of about 50% open area, the loop inductance is increased by about 25%.

9. As the sine-wave frequency components of currents increase, they will take the path of lowest impedance, which translates to a distribution toward the outside surface of the conductors and signal and return currents as close together as possible. This causes the inductance to be slightly frequency dependent, decreasing toward higher frequency, and the resistance to be strongly frequency dependent, increasing with the square root of the frequency.

10. When a current is in proximity to a uniform plane, even if it is floating, induced eddy currents will cause the self-inductance of the current to decrease.

End-of-Chapter Review Questions

The answers to the following review questions can be found in Appendix D, "Review Questions and Answers."

6.1 What is inductance?

6.2 What are the units we use to count magnetic-field lines?

6.3 List three properties of magnetic-field lines around currents.

6.4 How many field line rings are around a conductor when there is no current through it?

6.5 If the current in a wire increases, what happens to the number of rings of magnetic-field lines?

6.6 If the current in a wire increases, what happens to the inductance of the wire?

6.7 What is the difference between self-inductance and mutual-inductance?

6.8 What happens to the mutual-inductance between two conductors when the spacing between then increases? Why?

6.9 What two geometrical features influence the self-inductance of a conductor?

6.10 Why does self-inductance increase when the length of the conductor increases?

6.11 What influences the induced voltage on a conductor?

6.12 What is the difference between partial and loop inductance?

6.13 Why does the mutual-inductance subtract from the self-inductance to give the total inductance when the other conductor is the return path?

6.14 In what cases should the mutual-inductance add to the self-inductance to give the total inductance?

6.15 What three design features will decrease the loop inductance of a current loop?

6.16 When estimating the magnitude of ground bounce, what type of inductance should be calculated?

6.17 If you want to reduce the ground bounce in a leaded package, which leads should be selected as the return leads?

6.18 If you want to reduce the loop inductance in the power and ground paths in a connector, what are two important design features when selecting the power and ground pins?

6.19 There are 24 Webers of field line rings around a conductor with 2 A of current. What happens to the number of field lines when the current increases to 6 A?

6.20 A conductor has 0.1 A of current and generates 1 microWeber of field lines. What is the inductance of the conductor?

6.21 A current in a conductor generates 100 microWebers of rings of field lines. The current turns off in 1 nsec. What is the voltage induced across the ends of the conductor?

6.22 The return lead in a package has 5 nH of total inductance. When the 20 mA of current through the lead turns off in 1 nsec, what is the voltage noise induced across the lead?

6.23 What if four signals use the lead in Question 6.22 as its return path. What is the total ground bounce noise generated?

6.24 What is the skin depth of copper at 1 GHz?

6.25 If current flows in both the top and bottom surfaces of a signal trace on a circuit board, how much does the resistance increase at 1 GHz compared to DC?

6.26 Based on the simulation results in Figure 6-26, what is the percentage of decrease in inductance from DC to 1 GHz?

6.27 When the spacing between two loops doubles, does the loop mutual-inductance increase or decrease?

6.28 What is the loop inductance of a loop composed of 10-mil diameter wire, in a circular loop 2 inches in diameter?

6.29 What is the loop inductance per length of two rods 100 mils in diameter, and spaced by 1 inch? What is the total inductance per inch of each leg?

6.30 A typical dielectric thickness between the power and ground planes in a 4-layer board is 40 mils. What is the sheet inductance of the power and ground planes? How does 1 square of sheet inductance compare to the typically 2 nH of mounting inductance of a decoupling capacitor?

The Physical Basis of Transmission Lines

We hear the term *transmission line* all the time, and we probably use it every day, yet what really is a transmission line? A coax cable is a transmission line. A PCB trace and the adjacent plane beneath it in a multilayer board is a transmission line.

TIP   Fundamentally, a transmission line is composed of any two conductors that have length. This is all it takes to make a transmission line.

As we will see, a transmission line is used to transport a signal from one point to another. Figure 7-1 illustrates the general features of all transmission lines. To distinguish the two conductors, we refer to one as the *signal path* and the other as the *return path*.

A transmission line is also a new ideal circuit element with very different properties than the three previously introduced ideal circuit elements: resistors, capacitors, and inductors. It has two very important parameters: a characteristic impedance and a time delay. The way a signal interacts with an ideal transmission line is radically different from the way it interacts with the other three ideal elements.

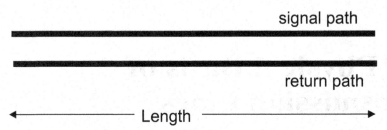

Figure 7-1 A transmission line is any two conductors with length. We label one of the conductors the signal path and the other the return path.

Though in some cases we can approximate the electrical properties of an ideal transmission line with combinations of Ls and Cs, the behavior of an ideal transmission line matches the actual, measured behavior of real interconnects more accurately and to much higher bandwidth than does an LC approximation. Adding an ideal transmission line circuit element to our toolbox will dramatically increase our ability to describe the interactions of signals and interconnects.

7.1 Forget the Word *Ground*

Generally, the term *ground* is reserved for the conductor with the lowest voltage in the circuit compared to any other node in the circuit. Only a voltage difference is ever measured. When the ground node is selected as the reference point, all other circuit nodes are at a higher voltage. There is nothing special about this conductor labeled as ground other than that it is the lowest voltage in the circuit, and all other nodes are at a higher voltage. This type of ground is referred to as the *circuit* ground.

This is distinct from the *chassis* ground and from *earth* ground. Chassis ground is a special conductor. It refers to the connection to the product's metal enclosure. This conductor is unique.

Earth ground is also a special connection. Ultimately, any conductor connected to earth ground can trace its path to a copper rod driven at least 4 feet into the ground. Many building codes specify the details of how the rod is placed to be considered an earth ground.

The round socket of three-prong AC power plugs is connected to earth ground. Usually, as a safety precaution, chassis ground is connected to earth ground. This is required by Underwriters Laboratory (UL) specifications, for example.

In bipolar circuits, the ground node is often the node that has a voltage halfway between +Vcc and –Vcc.

Too often, the other line in a transmission line is referred to as the *ground* line. Using the term ground to refer to the return path is a very bad habit and should be avoided.

TIP   Far more problems are created than solved by referring to the second line as *ground*. It is a good habit to use the term *return path* instead.

One of the most common ways of getting into trouble with signal-integrity design is overusing the term *ground*. It is much healthier to get into the habit of calling the other conductor the *return path*.

Many of the problems related to signal integrity are due to poorly designed return paths. If we are always consciously aware that the other path plays the important role of being the return path for the signal current, we will take as much care in designing its geometry as we take for the signal path.

When we label the other path as ground, we typically think it is a universal sink for current. Return current goes into this connection and comes out wherever there is another ground connection. *This is totally wrong.* The return current will closely follow the signal current. As we saw in Chapter 6, "The Physical Basis of Inductance," at higher frequencies, the loop inductance of the signal and return paths will be minimized, which means the return path will distribute as close to the signal path as the conductors will allow.

Figure 7-2   Forget the word *ground*, and more problems will be avoided than created.

Further, the return current has no idea what the absolute voltage level is of the return conductor. The actual return conductor may in fact be a voltage plane such as the Vcc or Vdd plane. Other times, it might be a low-voltage plane. That it is labeled as a *ground* connection in the schematic is totally irrelevant to the signal, which is propagating on the transmission line. Start calling the return path the return path, and problems will be reduced in the future. This guideline is illustrated in Figure 7-2.

7.2 The Signal

As we will see, when a signal moves down a transmission line, it simultaneously uses the signal path and the return path. Both conductors are equally important in determining how the signal interacts with the interconnect.

When both lines look the same, as in a twisted pair, it is inconsequential which we call the signal path and which we call the return path. When one is different from the other, such as in a microstrip, we usually refer to the narrow conductor as the signal path and the plane as the return path.

When a signal is launched into a transmission line, it propagates down the line at the speed of light in the material. After the signal is launched in the transmission line, we can freeze time a moment later, move along the line, and measure the signal. The signal is always the voltage difference between two adjacent points on the signal and return paths. This is illustrated in Figure 7-3.

TIP   If we know the impedance the signal sees, we can always calculate the current associated with the signal voltage. In this respect, a signal is equally well defined as a voltage or a current.

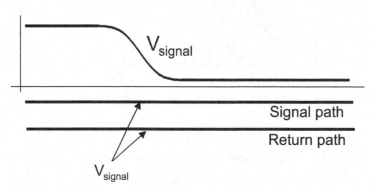

Figure 7-3 Map of the signal, frozen in time on a transmission line. The signal is the voltage between two adjacent points on the signal and the return paths.

It is important to distinguish, and pay attention to, the voltage on the signal line, which is what would be measured by a scope probe, and the propagating signal. The propagating signal is the voltage pattern that is moving down the transmission line and is dynamic.

When the signal passes by a point on the transmission line, the voltage a scope probe would measure is the same magnitude as the signal. However, if there are multiple signals on the transmission line propagating in opposite directions, the scope probe cannot separate them. The voltage that would be measured is *not* the same as the propagating signal.

These general principles apply to all transmission lines—single ended and, as we will see, differential transmission lines.

7.3 Uniform Transmission Lines

We classify transmission lines by their geometry. The two general features of geometry that strongly determine the electrical properties of a transmission line are the uniformity of the cross section down its length and how identical are each of its two conductors.

When the cross section is the same down the length, as in a coax cable, the transmission line is called *uniform*. Examples of various uniform transmission lines are illustrated in Figure 7-4.

As we will see, uniform transmission lines are also called *controlled impedance lines*. There are a great variety of uniform transmission lines, such as twin leads, microstrips, striplines, and coplanar lines.

TIP   Reflections will be minimized and signal quality optimized if the transmission lines are uniform or are controlled impedance. All high-speed interconnects should be designed as uniform transmission lines.

Nonuniform transmission lines exist when some geometry or material property changes as we move down the length of the line. For example, if the spacing between two wires is not controlled but varies, this is a nonuniform line. A pair of leads in a dual in-line package (DIP) or quad flat pack (QFP) are nonuniform lines.

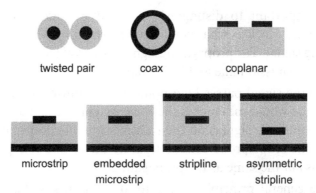

Figure 7-4 Examples of cross sections of uniform transmission lines commonly used as interconnects.

Adjacent traces in a connector are often nonuniform transmission lines. Traces on a PCB that do not have a return-path plane are often nonuniform lines. Nonuniform transmission lines will lead to signal-integrity problems and should be avoided unless they are kept short.

TIP   One of the goals in designing for optimized signal integrity is to design all interconnects as uniform transmission lines and to minimize the length of all nonuniform transmission lines.

Another quality of the geometry affecting transmission lines is how similar the two conductors are. When each conductor is the same shape and size as the other one, there is symmetry, and we call this line a *balanced line*. A pair of twisted wires is symmetric since each conductor looks identical. A coplanar line has two narrow strips side by side on the same layer and is balanced.

A coax cable is unbalanced since the center conductor is much smaller than the outer conductor. A microstrip is unbalanced, since one wire is a narrow trace, while the second is a wide trace. A stripline is unbalanced for the same reason.

TIP   In general, for most transmission lines, the signal quality and cross-talk effects will be completely unaffected by whether the line is balanced or unbalanced. However, ground-bounce and EMI issues will be strongly affected by the specific geometry of the return path.

Whether the transmission line is uniform or nonuniform, balanced or unbalanced, it has just one role to play: to transmit a signal from one end to the other with an acceptable level of distortion.

7.4 The Speed of Electrons in Copper

How fast do signals travel down a transmission line? If it is often erroneously believed that the speed of a signal down a transmission line depends on the speed of the electrons in the wire. With this false intuition, we might imagine that reducing the resistance of the interconnect will increase the speed of a signal. In fact, the speed of the electrons in a typical copper wire is actually about 10 billion times slower than the speed of the signal.

It is easy to estimate the speed of an electron in a copper wire. Suppose we have a roughly 18-gauge round wire, 1 mm in diameter, with 1 Amp of current. We can calculate the speed of the electrons in the wire based on how many electrons pass by one section of the wire per second, the density of the electrons in the wire, and the cross-sectional area of the wire. This is illustrated in Figure 7-5. The current in the wire is related to:

$$I = \frac{\Delta Q}{\Delta t} = \frac{q \times n \times A \times v \times \Delta t}{\Delta t} = q \times n \times A \times v \qquad (7\text{-}1)$$

from which we can calculate the velocity of the electrons as:

$$v = \frac{I}{q \times n \times A} \qquad (7\text{-}2)$$

where:

I = current passing one point, in Amps

ΔQ = charge flowing in a time interval, in Coulombs

Δt = time interval

q = charge of one electron = 1.6×10^{-19} Coulombs

n = density of free electrons, in #/m^3

A = cross-sectional area of the wire, in m^2

v = speed of the electrons in the wire, in m/sec

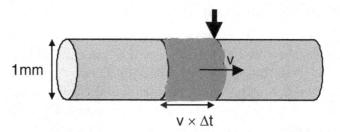

Figure 7-5 The electrons moving down a wire. The number passing the arrow per second is the current and is related to the velocity and number density of the electrons.

Each copper atom contributes roughly two free electrons that can move through the wire. Atoms of copper are about 1 nm apart. This makes the density of free electrons, n, about n ~ 10^{27}/m^3.

For a wire that is 1 mm in diameter, the cross-sectional area is about A ~ 10^{-6} m^2. Combining these terms and using a 1-Amp current in the wire, the speed of an electron in the wire can be estimated to be roughly:

$$v = \frac{I}{q \times n \times A} = \frac{1\,Amp}{10^{-19} \times 10^{27} \times 10^{-6}} = 10^{-2}\,\frac{m}{sec} = 1\frac{cm}{sec} \qquad (7\text{-}3)$$

TIP   An electron travels at a speed of about 1 cm/sec. This is about as fast as an ant scurries on the ground.

With this simple analysis, we see that the speed of an electron in a wire is incredibly slow compared to the speed of light in air. The speed of an electron in a wire really has virtually nothing to do with the speed of a signal. Likewise, as we will see, the resistance of the wire has only a very small, almost irrelevant effect on the speed of a signal in a transmission line. It is only in extreme cases that the resistance of an interconnect affects the signal speed—and even then the effect is only very slight. We must recalibrate our intuition from the erroneous notion that lower resistance will mean faster signals.

But how do we reconcile the speed of a signal with the incredibly slow speed of the electrons in a wire? How does the signal get from one end of the wire to the other in a much shorter amount of time than it takes an electron to get from one end to the other? The answer lies in the interactions between the electrons.

If we have a pipe filled with marbles and push one marble in on one end, another marble comes out the other end almost instantaneously. The interaction between marbles—the force of one marble on the next one—is transmitted from marble to marble much faster than the actual speed of the marbles.

When a train is stopped at a crossing and starts up again, the couplings between the cars engage down the train much faster than the locomotive inches forward. In the same way, when one electron in a wire is jiggled by the source, its interaction with the adjacent electron through the electric field between them jiggles. This *kink* in the electric field propagates to the next electron at the speed of light allowed by the changing field.

When an electron on one end of the wire moves, the kink in the electric field propagates to the next electron, and that electron moves, creating a kink in its electric field to the next electron, and all the way down the chain until the last electron moves out the wire at the other end. It's the speed of the interactions—the kinks in the electric field between the electrons—that determines how fast this signal propagates, not the speed of the electrons themselves.

7.5 The Speed of a Signal in a Transmission Line

If the speed of the electrons isn't what determines the speed of the signal, what does?

TIP   The speed of a signal depends on the materials that surround the conductors and how quickly the changing electric and magnetic fields associated with the signal can build up and propagate in the space around the transmission line conductors.

Figure 7-6 illustrates the simplest way to think of a signal propagating down a transmission line. The signal, after all, is a propagating voltage difference between the signal path and the return path. As the signal propagates, a voltage difference must be created between the two conductors. Accompanying the voltage difference is an electric field between the conductors.

In addition to the voltage, a current must be flowing in the signal conductor and in the return conductor to provide the charge that charges up the conductors that generates the voltage difference that creates the electric field. This current loop moving through the conductors will produce a magnetic field.

A signal can be launched into a transmission line simply by touching the leads of a battery to the signal and return paths. The sudden voltage change creates a sudden electric and magnetic-field change. This *kink of field* will propagate through the dielectric material surrounding the transmission line at the speed of a changing electric and magnetic field, which is the speed of light in the material.

We usually think of light as the electromagnetic radiation we can see. However, all changing electromagnetic fields are exactly the same and are described by exactly the same set of equations, Maxwell's Equations. The only difference is the frequency of the waves. For visible light, the frequency is about 1,000,000 GHz. For the signals typically found in high-speed digital products, the frequency is about 1–10 GHz.

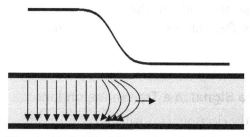

Figure 7-6   The electric field building up in a transmission line as the signal propagates down the line. The speed of the signal depends on how fast the changing electric and magnetic fields can build up and propagate in the materials surrounding the signal- and return-path conductors.

How quickly the electric and magnetic fields can build up is what really determines the speed of the signal. The propagation and interaction of these fields is described by Maxwell's Equations, which say that if the electric and magnetic fields ever change, the kink they make will propagate outward at a speed that depends on some constants and material properties.

The speed of the change, or the kink, v, is given by:

$$v = \frac{1}{\sqrt{\varepsilon_0 \varepsilon_r \mu_0 \mu_r}}$$ (7-4)

where:

ε_0 = permittivity of free space = 8.89×10^{-12} F/m

ε_r = relative dielectric constant of the material

μ_0 = permeability of free space = $4\pi \times 10^{-7}$ H/m

μ_r = relative permeability of the material

Putting in the numbers, we find:

$$v = \frac{2.99 \times 10^8}{\sqrt{\varepsilon_r \mu_r}} \frac{m}{sec\ s} = \frac{(11.8)}{\sqrt{\varepsilon_r \mu_r}} \frac{inches}{nsecs}$$ (7-5)

TIP In air, where the relative dielectric constant and relative permeability are both 1, the speed of light is about 12 inches/nsec. This is a really good rule of thumb to keep in mind.

For virtually all interconnect materials, the magnetic permeability of the dielectrics, μ_r, is 1. All polymers that do not contain a ferromagnetic material have a magnetic permeability of 1. Therefore, this term can be ignored.

In comparison, the relative dielectric constant of materials, ε_r, is never less than or equal to 1, except in the case of air. In all real interconnect materials, the dielectric constant is greater than 1. This means the speed of light in interconnects will always be less than 12 inches/nsec. The speed is:

$$v = \frac{12}{\sqrt{\varepsilon_r}} \frac{inches}{nsecs}$$ (7-6)

For brevity, we usually refer to the relative dielectric constant as just the "dielectric constant." This number characterizes some of the electrical properties of an insulator. It is an important electrical property. For most polymers, it is roughly 4. For glass, it is about 6, and for ceramics, it is about 10.

It is possible in some materials for the dielectric constant to vary with frequency. In other words, the speed of light in a material may be frequency dependent. We call this property *dispersion*, a frequency dependence to the speed of light in the material. In general, dielectric constant decreases with higher frequency. This makes the speed of light in the material increase as we go toward higher frequency. In most applications, dispersion is very small and can be ignored.

In most common materials, such as FR4, the dielectric constant varies very little from 500 MHz to 10 GHz. Depending on the ratio of epoxy resin to fiberglass, the dielectric constant of FR4 can be from 3.5 to 4.5. Most interconnect laminate materials have a dielectric constant of about 4. This suggests a simple, easy-to-remember generalization.

TIP   A good rule of thumb to remember is that the speed of light in most interconnects is about 12 inches/nsec/sqrt(4) = 6 inches/nsec. When evaluating the speed of signals in a board-level interconnect, we can assume that they travel at about 6 inches/nsec.

As pointed out in Chapter 5, "The Physical Basis of Capacitance," when the field lines see a combination of dielectric materials, as in a microstrip where there are some field lines in the bulk material and some in the air above, the effective dielectric constant that affects the signal speed is a combination of the different materials. The only way to predict the effective dielectric constant when the materials are inhomogeneous throughout the cross section is with a 2D field solver. In the case of stripline, for example, all the fields see the same material, and the effective dielectric constant is the bulk dielectric constant.

The time delay, TD, and length of an interconnect are related by:

$$TD = \frac{Len}{v} \qquad (7\text{-}7)$$

where:

TD = time delay, in nsec

Len = interconnect length, in inches

v = speed of the signal, in inches/nsec

This means that to travel down a 6-inch length of interconnect in FR4, for example, the time delay is about 6 inches/6 inches/nsec, or about 1 nsec. To travel 12 inches takes about 2 nsec.

The *wiring delay*, the number of psec of delay per inch of interconnect, is also a useful metric. It is just the inverse of the velocity: $1/v$. For FR4, the wiring delay is about 1/6 inches/nsec = 0.166 nsec/inch, or 170 psec/inch. This is the delay per inch of a signal propagating down a transmission line in FR4. Every inch of interconnect has a propagation delay of 170 psec. The total delay through the 0.5 inch of a BGA lead is 170 psec/inch × 0.5 inches = 85 psec.

7.6 Spatial Extent of the Leading Edge

Every signal has a rise time, RT, usually measured from the 10% to 90% voltage levels. As a signal moves down a transmission line, the leading-edge spreads out on the transmission line and has a spatial extent. If we could freeze time and look at the size of the voltage distribution as it moves out, we would find something like Figure 7-7.

The length of the rise time, Len, on the transmission line depends on the speed of the signal and the rise time:

$$Len = RT \times v \qquad\qquad (7\text{-}8)$$

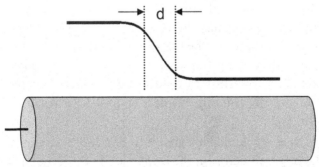

Figure 7-7 Spatial extent of the leading edge of the signal as it propagates down a transmission line.

where:

Len = spatial extent of the rise time, in inches

RT = rise time of the signal, in nsec

v = speed of the signal, in inches/nsec

For example, if the speed is 6 inches/nsec and the rise time is 1 nsec, the spatial extent of the leading edge is 1 nsec × 6 inches/nsec = 6 inches. As the leading edge moves down the circuit board, it is really a 6-inch section of rising voltage moving down the board. A rise time of 0.1 nsec has a spatial extent of 0.6 inch.

TIP   Many of the signal-integrity problems related to imperfections in the transmission line depend on the relative size of the discontinuity compared to the spatial extent of the leading edge. It is always a good idea to be aware of the spatial extent of the leading edge for all signals.

7.7 "Be the Signal"

All a signal cares about is how fast it moves down the line and what impedance it sees. As we saw previously, speed is based on the material properties of the dielectric and their distribution. To evaluate the impedance a signal sees in propagating down the line, we will take a microstrip transmission line and launch a signal into one end. A microstrip is a uniform but unbalanced transmission line. It has a narrow-width signal line and a wide return path.

The analysis we will do of this line is identical to that for any transmission line. We'll make it 10 feet long, so that we can actually walk down it and, in a Zen way, "be the signal" to observe what the signal would see. With each step along the way, we will ask *What impedance do we see?* We will answer this question by determining the ratio of the voltage applied, 1 v, and the current coming out of our foot to drive the signal down the transmission line.

In this case, we launch a signal into one end by connecting a 1-v battery between the two conductors at the front end. At the initial instant that we have launched the signal into the line, there has not been enough time for the signal to travel very far down the line.

Just to make it easier, let's assume that we have air between the signal and return paths, so the speed of propagation is 1 foot per nsec. After the first nsec, the voltage on the far end of the line is still zero, as the signal hasn't had enough time to get very far. The signal along the line would be about 1 volt for the first foot and zero for the remaining length of the line.

Let's freeze time after the first nsec and look at the charges on the line. What we see is illustrated in Figure 7-8. Between the signal- and return-path conductors in the first 12 inches, there will be a 1-volt difference. This is, after all, the signal. We know that because the signal and return paths are two separated conductors, there will be some capacitance between the conductors in this region. If there is a 1-v difference between them, there must also be some charge on the signal conductor and an equal and opposite amount on the return-path conductor.

In the next 1 nsec, we, being the signal, will move ahead another 12 inches. Let's stop time again. We now have the first 2 feet of line charged up. We see that having made this last step, we have brought the signal to the second foot-long section and created a voltage difference between the signal and return conductors. There is now a charge difference between the two conductors, at the point of each footstep, where there was none a nsec ago.

As we walk down the line, we are bringing a voltage difference to the two conductors and charging them up. In each nsec, we take another 1-foot step and charge up this new section of the line. Each step of the signal will leave another foot of charged transmission line in our wake.

Figure 7-8 Charge distribution on the transmission line after a 1-v signal has propagated for 1 nsec. There is no charge ahead of us (the signal) at this instant in time.

The charge that flowed to charge up each footstep came from the signal, as each foot came down, and ultimately from the battery. The fact that the signal is propagating down the line means that the capacitance between the signal and return paths is getting charged up. How much charge has to flow from our foot into the line in each footstep? In other words, what is the current that must flow to charge up successive regions of the transmission line as the signal propagates?

If the signal is moving down the line at a steady speed and the line is uniform—that is, it has the same capacitance per length—then we are injecting the same amount of charge into the line with each footstep. Each step charges up the same amount of capacitance to the same voltage. If we are always taking the same time per step, then we are injecting the same charge per unit of time for the signal to charge up the line. The same amount of charge per nsec flowing into the line means that there is a constant current flowing into the line from our foot.

> **TIP**   From the signal's perspective, as we walk down the line at our speed of 1 foot/nsec, we are charging up each foot of line in the same amount of time. Coming out of the bottom of our foot is the charge that is added to the line to charge it up. An equal charge out of our foot in an equal time interval means we are injecting a constant current into the line.

What affects the current coming out of our foot to charge up the line? If we are moving down the line at a constant speed, and if we increase the width of the signal path, the capacitance we need to charge up will increase, and the charge that must come out of our foot in the time we have till the next step will increase. Likewise, if anything is done to decrease the capacitance per length of the interconnect, the current coming out of our foot to charge up the reduced capacitance will decrease. For the same reason, if the capacitance per length stays the same, but our speed increases, we will be charging up more length per nsec, and the current needed will increase.

In this way, we can deduce that the current coming out of our foot will scale directly with both the capacitance per length and the speed of the signal. If either increases, the current out of our foot with each step will increase. If either decreases, the current from the signal to charge up the line will decrease. We have deduced a simple relationship between the current coming out of our foot and the properties of the line:

$$I \sim v \times C_L \qquad\qquad (7\text{-}9)$$

where:

I = current out of our foot

v = speed with which we move down the line, charging up regions

C_L = capacitance per length of the line

As we, the signal, move down the transmission line, we will constantly be asking *What is the impedance of the line?* The basic definition of *impedance* of any element is the ratio of voltage applied to current through it. So, as we move down the line, we will constantly be asking with each footstep *What is the ratio of the voltage applied to the current being injected into the line?*

The voltage of the signal is fixed at the signal voltage. The current into the line depends on the capacitance of each footstep and how long it takes to charge up each footstep. As long as the speed of the signal is constant and the capacitance of each footstep is constant, the current injected into the line by our foot will be constant, and the impedance of the line the signal sees will be constant.

We call the impedance the signal sees with each step the *instantaneous impedance*. If the interconnect features are uniform, the instantaneous impedance will be the same for each footstep. A uniform transmission line is called a controlled impedance transmission line because the instantaneous impedance is controlled to be the same everywhere on the line.

Suppose the line width were to suddenly widen. The capacitance of each footstep would be larger, and the current coming out of each footstep to charge up this capacitance would be larger. A higher current for the same voltage means we see the transmission line having a lower impedance in this region. The instantaneous impedance is lower in this part of the transmission line.

Likewise, if the line were to suddenly narrow, the capacitance of each footstep would be less, the current required to charge it up would be less, and the instantaneous impedance the signal would see for the line would be higher.

TIP   We call the impedance the signal sees with each step the *instantaneous impedance* of the transmission line. As the signal propagates down the line, it is constantly probing the instantaneous impedance of each footstep, as the ratio of the voltage applied to the current required to charge the line and propagate toward the next step.

The instantaneous impedance depends on the speed of the signal, which is a material property, and the capacitance per length. For a uniform transmission line, the cross-sectional geometry is constant down the line, and the instantaneous impedance the signal sees will be constant down the line. As we will see, an important behavior of a signal interacting with a transmission line is that whenever the signal sees a change in the instantaneous impedance, some of the signal will reflect and some will continue distorted. In addition, signal integrity may be affected. This is the primary reason controlling the instantaneous impedance the signal sees is so important.

> **TIP**   The chief way of minimizing reflection problems is to keep the instantaneous impedance the signal sees constant by keeping the geometry constant. This is what is meant by a *controlled-impedance interconnect*, a line with a constant instantaneous impedance down its length.

7.8 The Instantaneous Impedance of a Transmission Line

We can quantify our analysis so far in this chapter by building a simple physical model of the transmission line. We can model the line as an array of little capacitor buckets, each one equal to the capacitance in the transmission line that spans a footstep and separated by the distance we, the signal, move in each footstep. We call this model (the simplest model we can come up with that provides engineering insight) the *zeroth-order model* of a transmission line (see Figure 7-9). It is a physics model and *not* an equivalent circuit model. Circuit models do not have lengths in them.

In this model, the size of each footstep is Δx. The magnitude of each little capacitor bucket is the capacitance per length, C_L, of the line times the length of each footstep:

$$C = C_L \times \Delta x \qquad (7\text{-}10)$$

We can calculate the current out of our foot, I, using this model. The current is the charge that flows out of our foot to charge up each bucket in the time interval between each step. The charge we dump in each capacitor bucket, Q, is the capacitance of the bucket times the voltage applied, V. For each step we take, we are dumping the charge Q into the line, in a time interval of a step.

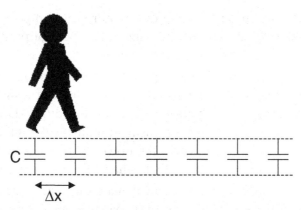

Figure 7-9 Zeroth-order model of a transmission line composed of an array of capacitors. With each footstep, another capacitor is charged up. The spacing is the size of our footstep.

The time between steps, Δt, is the length of our step, Δx, divided by our speed down the line, v. Of course, as the real signal propagates, each footstep is really small, but the time interval between steps gets really small as well. The ratio of the charge that flows to each time interval is a constant value, which is the current that flows into the line as the signal propagates:

$$I = \frac{Q}{\Delta t} = \frac{CV}{\left(\frac{\Delta x}{v}\right)} = \frac{C_L \Delta x v V}{\Delta x} = C_L v V \qquad (7\text{-}11)$$

where:

I = current from the signal

Q = charge in each footstep

C = capacitance of each footstep

Δt = time to step from capacitor to capacitor

C_L = capacitance per length of the transmission line

Δx = distance between the capacitors or each footstep

v = speed of walking down the line

V = voltage of the signal

This says the current coming out of our foot and going into the line is simply related to the capacitance per length, the speed of propagation, and the voltage of the signal—exactly as we reasoned earlier.

This is the defining relationship for the current-voltage (I-V) behavior of a transmission line. It says the instantaneous current of a signal anywhere on a transmission line is directly proportional to the voltage. Double the voltage applied, and the current into the transmission line will double. This is exactly how a resistor behaves. With each step down the transmission line, the signal sees an instantaneous impedance that behaves like a resistive load.

From this relationship, we can calculate the instantaneous impedance a signal would see at each step as it propagates down a transmission line. The instantaneous impedance is the ratio of the voltage applied to the current through the device:

$$Z = \frac{V}{I} = \frac{V}{C_L v V} = \frac{1}{C_L v} = \frac{83\,\Omega}{C_L} \sqrt{\varepsilon_r} \qquad (7\text{-}12)$$

where:

Z = instantaneous impedance of the transmission line, in Ohms

C_L = capacitance per length of the line, in pF/inch

v = speed of light in the material

ε_r = dielectric constant of the material

The instantaneous impedance a signal sees depends on only two terms, both of which are intrinsic to the line. It doesn't depend on the length of the line. The instantaneous impedance of the line depends on the cross section of the line and the material properties. As long as these two terms are constant as we move down the line, a signal would see the same constant, instantaneous impedance. And, of course, the units we use to measure the instantaneous impedance of the line are Ohms, as with any impedance.

Since the speed of the signal depends on a material property, we can relate the capacitance per length of the transmission line to the instantaneous impedance. For example, if the dielectric constant is 4 and the capacitance per length of the line is 3.3 pF/inch, the instantaneous impedance of the transmission line is:

$$Z = \frac{83}{C_L}\sqrt{\varepsilon_r} = \frac{83}{3.3}\sqrt{4} = 50\,\Omega \qquad (7\text{-}13)$$

What about the inductance of the line? Where does that come into play in this model? The answer is that this zeroth-order model is not an electrical model; it is a physical model. Rather than approximating the transmission line with Ls and Cs, we added the observation that the speed of the signal is the speed of light in the material.

In reality, the finite speed of the signal arises partly because of the series loop inductance per length of the signal and return paths. If we used a first-order, equivalent circuit model, including the inductance per length, it would derive for us the current into the transmission line and the finite propagation speed, but the model would be more complicated mathematically.

These two models are really equivalent, considering the connection between the propagation speed and the inductance per length. As we shall see, the propagation delay is directly related to the capacitance per length combined with the inductance per length. The speed of the signal has in it some assumptions about the inductance of the conductors.

7.9 Characteristic Impedance and Controlled Impedance

For a uniform line, anywhere we choose to look, we will see the same instantaneous impedance as we propagate down the line. There is one instantaneous impedance that characterizes the transmission line, and we give it the special name: *characteristic impedance.*

> **TIP**   There is one instantaneous impedance that is *characteristic* for a uniform transmission line. We refer to this constant, instantaneous impedance as the impedance that is characteristic of the line, or the *characteristic impedance* of the line.

To distinguish that we are referring of the special case of the *characteristic impedance* of the line, which is an intrinsic property of the line, we give it the special symbol Z_0 (Z with a subscript zero). Characteristic impedance is given in units of Ohms. Every uniform transmission line has a characteristic impedance, which is one of the most important terms describing its electrical properties and how signals will interact with it.

TIP   The characteristic impedance of the line will tell us the instantaneous impedance a signal will see as it propagates down the line. As we will see, this is the chief factor that influences signal integrity in transmission line circuits.

Characteristic impedance is numerically equal to the instantaneous impedance of the line and is intrinsic to the line. It depends only on the dielectric constant and the capacitance per length of the line. It does not depend on the length of the line.

For a uniform line, the characteristic impedance is:

$$Z_0 = \frac{83}{C_L} \sqrt{\varepsilon_r} \qquad\qquad (7\text{-}14)$$

If the line is uniform, it has only one instantaneous impedance, which we call the *characteristic impedance*. One measure of the uniformity of the line is how constant the instantaneous impedance is down the length. If the line width varies down the line, there is no one single value of instantaneous impedance for the whole line. By definition, a nonuniform line has no characteristic impedance. When the cross section is uniform, the impedance a signal sees as it propagates down the interconnect will be constant, and we say the impedance is controlled. For this reason, we call uniform cross-section transmission lines *controlled-impedance* lines.

TIP   We call a transmission line that has a constant instantaneous impedance down its length a controlled-impedance line. A board that is fabricated with all its interconnects as controlled-impedance lines, all with the same characteristic impedance, is called a controlled-impedance board. All high-speed digital products, with boards larger than about 6 inches and clock frequencies greater than 100 MHz, are built with controlled-impedance boards.

When the geometry and material properties are constant down the line, the instantaneous impedance of the line is uniform, and one number fully characterizes the impedance of the line.

A controlled-impedance line can be fabricated with virtually any uniform cross section. There are many standard cross-sectional shapes that can have controlled impedance, and many of these families of shapes have special names. For example, two round wires twisted together are called a *twisted pair*. A center conductor surrounded by an outer conductor is a *coaxial*, or *coax*, *transmission line*. A narrow strip of signal line over a wide plane is a *microstrip*. When the return path is two planes and the signal line is a narrow strip between them, we call this a *stripline*. The only requirement for a controlled-impedance interconnect is that the cross section be constant.

With this connection between the capacitance per length and the characteristic impedance, we can now relate our intuition about capacitance to our new intuition about characteristic impedance: What increases one will decrease the other.

We generally have a pretty good intuitive feel for capacitance and capacitance per length for the two conductors in a transmission line. If we make the two conductors wider, we increase the capacitance per length. This will decrease the characteristic impedance. If we move them farther apart, we make the capacitance per length lower and the characteristic impedance higher.

For a microstrip using FR4 dielectric, when the line width is twice the dielectric thickness, the characteristic impedance is about 50 Ohms. What happens to the characteristic impedance when we make the dielectric separation larger? It's not obvious initially. However, we now know that the characteristic impedance of a transmission line is inversely proportional to the capacitance per length between the conductors.

Therefore, if we move the conductors farther apart, the capacitance will decrease, and the characteristic impedance will increase. Making the microstrip signal trace wider will increase the capacitance per length and decrease the characteristic impedance. This is illustrated in Figure 7-10.

In general, a wide conductor with a thin dielectric will have a low characteristic impedance. For example, the characteristic impedance of the transmission line formed from the power and ground planes in a PCB will have a low characteristic impedance, generally less than 1 Ohm. Narrow conductors with a thick dielectric will have a high characteristic impedance. Signal traces with narrow lines will have high characteristic impedance, typically between 60 Ohms and 90 Ohms.

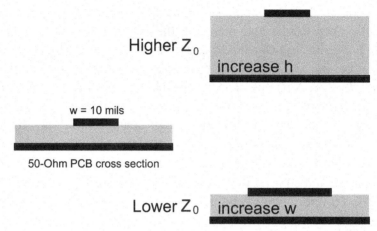

Higher Z_0 — increase h

w = 10 mils

50-Ohm PCB cross section

Lower Z_0 — increase w

Figure 7-10 If line width increases, capacitance per length increases, and characteristic impedance decreases. If dielectric spacing increases, capacitance per length decreases, and characteristic impedance increases.

7.10 Famous Characteristic Impedances

Over the years, various specs have been established for specialized controlled-impedance interconnects. A number of these are listed in Figure 7-11. One of the most common ones is RG58. Virtually all general-purpose coax cables used in the lab, with bayonet type BNC connectors, are made with RG58 cable. This spec defines an inner- and outer-conductor diameter and a dielectric constant. In addition, when the spec is followed, the characteristic impedance is about 52 Ohms. Look on the side of the cable, and you will see "RG58" stamped.

There are other cable specs as well. RG174 is useful to know about. It is a thinner cable than RG58 and is much more flexible. When trying to snake a cable around in tight spaces or when low stress is required, the flexibility of RG174 is useful. It is specified with a characteristic impedance of 50 Ohms.

The coax cable used in cable TV systems is specified at 75 Ohms. This cable will have a lower capacitance per length than a 50-Ohm cable and in general is thicker than a comparable 50-Ohm cable. For example, RG59 is thicker than RG58.

Twisted pairs, typically used in high-speed serial links, small computer system interface (SCSI) applications, and telecommunications applications are made with 18- to 26-gauge wire. With the typical insulation thickness commonly used, the characteristic impedance is about 100 Ohms to 130 Ohms. This is typically a higher impedance than used in circuit boards, but it matches the differential impedance of typical board traces. Differential impedance is introduced in Chapter 11, "Differential Pairs and Differential Impedance."

RG174	50Ω
RG58	52Ω
RG59	75Ω
RG62	93Ω
TV Antenna	300Ω
Cable TV	75Ω
Twisted pairs	100–130Ω

Figure 7-11   Some famous controlled-impedance interconnects, based on their specified characteristic impedance.

There is one characteristic impedance that has special, fundamental significance: free space. As described previously, a signal propagating in a transmission line is really light, the electric and magnetic fields trapped and guided by the signal- and return-path conductors. As a propagating field, it travels at the speed of light in the composite dielectric medium.

Without the conductors to guide the fields, light will propagate in free space as waves. These are waves of electric and magnetic fields. As the wave propagates through space, the electric and magnetic fields will see an impedance.

The impedance a wave sees is related to two fundamental constants—the permeability of free space and the permittivity of free space:

$$Z_0 = \sqrt{\frac{\mu_0}{\varepsilon_0}} = 120\pi = 376.99 \sim 377\Omega \qquad (7\text{-}15)$$

The combination of these two constants is the instantaneous impedance a propagating wave will see. We call this the *characteristic impedance of free space*, and it is approximately 377 Ohms. This is a fundamental number. The amount of radiated energy from an antenna is optimized when its impedance matches the 377 Ohms of free space. There is only one characteristic-impedance value that has fundamental significance, and it is 377 Ohms. All other impedances are arbitrary. The characteristic impedance of an interconnect can be almost any value, limited by manufacturability constraints.

But what about 50 Ohms? Why is it so commonly used? What's so special about 50 Ohms? Its use became popular in the early 1930s, when radio communications and radar systems became important and drove the first requirements for using high-performance transmission lines. The application was to transmit the radio signal from the not-very-efficient generator to the radio antenna with the minimum attenuation.

As we show in Chapter 9, "Lossy Lines, Rise-Time Degradation, and Material Properties," the attenuation of a coax cable is related to the series resistance of the inner conductor and outer conductor divided by the characteristic impedance. If the outer diameter of the cable is fixed, using the largest-diameter cable possible, there is an optimum inner radius that results in minimum attenuation.

With too large an inner radius, the resistance is lower, but the characteristic impedance is lower as well, and the attenuation is higher. Too small a diameter of the inner conductor, and the resistance and attenuation are both high. When you explore the optimum value of inner radius, you find that the value for the lowest attenuation is also the value that creates 50 Ohms.

The reason 50 Ohms was chosen almost 100 years ago was to minimize the attenuation in coax cables for a fixed outer diameter. It was adopted as a standard to improve radio and radar system efficiency, and it was easily manufactured. Once adopted, the more systems using this value of impedance, the better their compatibility. If all test and measurement systems matched to this standard 50 Ohms, then reflections between instruments were minimized, and signal quality was optimized.

In FR4, a 50-Ohm microstrip can be easily fabricated if the line is twice as wide as the dielectric is thick. A broad range of characteristic impedances around 50 Ohms can also be fabricated, so it is a soft optimum in printed circuit board technology.

In high-speed digital systems, a number of trade-offs determine the optimum characteristic impedance of the entire system. Some of these are illustrated in Figure 7-12. A good starting place is 50 Ohms. Using a higher characteristic impedance with the same pitch means more cross talk; however, higher characteristic impedance connectors or twisted-pair cables will cost less because they are easier to fabricate. Lower characteristic impedance means lower cross talk and lower sensitivity to delay adders caused by connectors, components, and vias, but it also means higher power dissipation when terminated. This is important in high-speed systems.

Property	Lower Z_0	Higher Z_0
Board costs	better	worse
Delay adders	better	worse
Cross talk	better	worse
Attenuation	better	worse
Connector costs	worse	better
Twisted pair/ cable costs	worse	better
Driver design	worse	better
Power dissipation	worse	better

Figure 7-12 Trade-offs in various system issues based on changing the characteristic impedance of the interconnects. Deciding on the optimum characteristic impedance, balancing performance and cost, is a difficult process. 50 Ohms is a good compromise in most systems.

Every system will have its own balance for the optimum characteristic impedance. In general, it is a very soft optimum, and the exact value chosen is not critically important as long as the same impedance is used throughout the system. Unless there is a strong driving force otherwise, 50 Ohms is usually used. In the case of Rambus memory, timing was critically important, and a low impedance of 28 Ohms was chosen to minimize the impact from delay adders. Manufacturing such a low impedance requires wide lines. But since the interconnect density in Rambus modules is low, the wider lines have only a small impact.

7.11 *The* Impedance of a Transmission Line

What impedance does a battery see when connected to the front of a transmission line? As soon as the battery is connected to the transmission line, the voltage signal sees the instantaneous impedance of the transmission line, and the signal starts propagating down the line. As long as the signal continues to propagate, the input impedance always looks like the instantaneous impedance, which is the characteristic impedance.

The impedance the battery sees when looking into the line is thus the same as the instantaneous impedance of the line—*as long as the signal is propagating down the line*. It looks like a resistor, and the current into the line is directly proportional to the voltage applied.

A circuit element that has a constant current through it for a constant voltage applied is an ideal resistor. From the battery's perspective, when the terminals are attached to the front end of the line and the signal propagates down the line, the transmission line draws a constant current and acts like a resistor to the battery. The impedance of the transmission line, as seen by the battery, is a constant resistance, as long as the signal is propagating down the line. There is no test the battery can do that to distinguish the transmission line from a resistor, at least while the signal is propagating down and back on the line.

> **TIP**   We've introduced the idea of the characteristic impedance of an interconnect. We often use the terms *characteristic impedance* and *impedance* of a line interchangeably. But they are not always the same, and it's important to emphasize this distinction.

What does it mean to refer to the *impedance* of a cable? An RG58 cable is often referred to as a 50-Ohm cable. What does this really mean? Suppose we were to take a 3-foot length of RG58 cable and measure the impedance at the front end, between the signal and return paths. What impedance will we measure? Of course, we can measure the impedance with an Ohmmeter. If we connect an Ohmmeter to the front end of the 3-foot transmission line, between the central signal line and the outer shield, as shown in Figure 7-13, what will it read? Will it be an open, a short, or 50 Ohms?

Let's be more specific. Suppose we were to measure the impedance using a digital multimeter (DMM) with a liquid crystal display (LCD) that updates in a second. What impedance will be measured?

Of course, if we wait long enough, the short length of cable will look like an open, and we would measure infinite as the input impedance. So, if the input impedance of this short cable is infinite, what does it mean to have a 50-Ohm cable? Where does the characteristic-impedance attribute come in?

To explore this further, consider a more extreme, very long length of RG58 cable. It's so long, in fact, that the line stretches from the earth to the moon. This is about 240,000 miles long. As we might recall from high school, the speed of light in a vacuum is about 186,000 miles per second, or close to 130,000 miles per second in the dielectric of the RG58 cable. It would take light about 2 seconds to go from one end of the cable to the far end and another 2 seconds to come back. If we attach our DMM to the front end of this long line, what will it see as the impedance? Remember, the DMM finds the resistance by connecting a 1-volt source to the device under test and measuring the ratio of the voltage applied to the current draw.

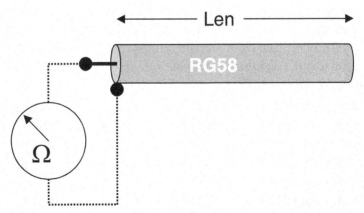

Figure 7-13   Measuring the input impedance of a length of RG58 cable with an Ohmmeter.

This is exactly the case of driving a transmission line, provided that we do the impedance measurement in less than the 4-second round-trip time of flight. During the first 4 seconds, while the signal is propagating out to the end of the interconnect and back, the current into the front of the line will be a constant amount equal to the current required by the signal to charge up successive sections of the cable as it propagates outward.

The impedance the source sees looking into the front of the line, the "input" impedance, will be the same as the instantaneous impedance the signal sees, which is the characteristic impedance of the line. In fact, the source won't know that there is an end to the transmission line until a round-trip time of flight, or 4 seconds, later. During the first 4 seconds of measurement, the Ohmmeter should read the characteristic impedance of the line, or 50 Ohms, in this case.

TIP   As long as the measurement time is less than the round-trip time of flight, the input impedance of the line, measured by the Ohmmeter, will be the characteristic impedance of the line.

But we know that if we wait a day with the Ohmmeter connected, we will eventually measure the input impedance of the cable to be an open. Here are the two extremes: Initially, we measure 50 Ohms, but after a long time, we measure an open. So, what is *the* input impedance of the line?

The answer is that there is no one value. It is time dependent; it changes. This example illustrates that the input impedance of a transmission line is time dependent; it depends on how long we are doing the measurement compared to the round-trip time of flight. This is illustrated in Figure 7-14. Within the round-trip time of flight, the impedance looking into the front end of the transmission line is the characteristic impedance of the line. After the round-trip time of flight, the input impedance can be anywhere from infinite to zero, depending on what is at the far end of the transmission line.

When we refer to the impedance of a cable or a line as being 50 Ohms, we are really saying the instantaneous impedance a signal would see propagating down the line is 50 Ohms. Or, the characteristic impedance of the line is 50 Ohms. Or, initially, if we look for a time, short compared to a round-trip time of flight, we will see 50 Ohms as the input impedance of the interconnect.

Even though these words sound similar, there is an important difference between *the* impedance, the *input* impedance, the *instantaneous* impedance, and the *characteristic* impedance of an interconnect. Just saying "the impedance" is ambiguous.

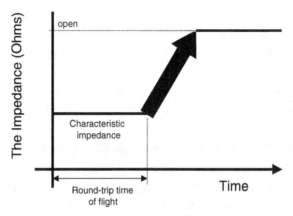

Figure 7-14   The input impedance, looking into a transmission line, is time dependent. During the round-trip time of flight, the measured input impedance will be the characteristic impedance. If we wait long enough, the input impedance will look like an open.

TIP   The input impedance of an interconnect is what a driver would see launching a signal into the front end of the interconnect. It is time dependent. It can be an open, it can be a short, or it can be anywhere in between, all for the same transmission line, depending on how the far end is terminated, how long the transmission line is, and how long we look.

The instantaneous impedance of the transmission line is the impedance the signal sees as it propagates down the line. If the cross section is uniform, the instantaneous impedance will be the same down the line. However, it may change where there are discontinuities, for example, at the end. If the end is open, the signal will see an infinite instantaneous impedance when it hits the end of the line. If there is a branch, it will see a drop in the instantaneous impedance at the branch point.

The characteristic impedance of the interconnect is a physical quality of a uniform transmission line that characterizes the transmission line due to its geometry and material properties. It is equal to the one value of instantaneous impedance the signal would see as it propagates down the uniform cross section. If the transmission line is not uniform, the instantaneous impedance changes, and there is not one impedance that characterizes the line. A characteristic impedance applies *only* to a uniform transmission line.

Everyone who works in the signal-integrity field gets lazy sometimes and just uses the term *impedance*. We therefore must ask the qualifying question of which impedance we mean or look at the context of its use to know which of these three impedances we are referring to. Knowing the distinction, we can all try to use the right one and be less ambiguous.

When the rise time is shorter than the round-trip time of flight of an interconnect, a driver will see the interconnect with a resistive input impedance equal to the characteristic impedance of the line during the rising edge. Even though the line may be open at the far end, during the transition time, the front of the line will behave like a resistor.

The round-trip time of flight is related to the dielectric constant of the material and the length of the line. With rise times for most drivers in the sub-nanosecond regime, any interconnect longer than a few inches will look long and behave like a resistive load to the driver during the transition. This is one of the important reasons that the transmission line behavior of all interconnects must be considered.

TIP   In the high-speed regime, an interconnect longer than a few inches does not behave like an open to the driver. It behaves like a resistor during the transition. When the length is long enough to show transmission line behavior, the input impedance the driver sees may be time dependent. This property will strongly affect the behavior of the signals that propagate on the interconnects.

Given this criterion, virtually all interconnects in high-speed digital systems will behave like transmission lines, and these properties will dominate the signal-integrity effects. For a transmission line on a board that is 3 inches long, the round-trip time of flight is about 1 nsec. If the integrated circuit (IC) driving the line has a rise time less than 1 nsec, the impedance it will see looking into the front of the line during the rising or falling edge will be the characteristic impedance of the line. The driver IC will see an impedance that acts resistive. If the rise time is very much longer than 1 nsec, it will see the impedance of the line as an open. If the rise time of the signal is somewhere between, the driver will see a very complicated, changing impedance as the edge bounces around between the low impedance driver and the open receiver. The received voltage can often be analyzed only using simulation tools. These tools are described in Chapter 8 "Transmission Lines and Reflections."

The round-trip time of flight is a very important parameter of a transmission line. To a driver, the line will seem resistive for this time. Figure 7-15 shows the round-trip time of flight for various-length transmission lines made with air ($\varepsilon_r = 1$), FR4 ($\varepsilon_r = 4$), and ceramic ($\varepsilon_r = 10$) dielectric materials. In most systems with clock frequencies higher than 200 MHz, the rise time is less than 0.5 nsec. For these systems, all transmission lines longer than about 1.5 inches will appear resistive during the rise time. This means for virtually all high-speed drivers, when they drive a transmission line during the transition, the input impedance they see will act like a resistor.

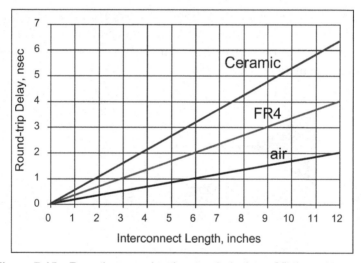

Figure 7-15 For a time equal to the round-trip time of flight, a driver will see the input impedance of the interconnect as a resistive load with a resistance equal to the characteristic impedance of the line.

7.12 Driving a Transmission Line

For a high-speed driver launching a signal into a transmission line, the input impedance of the transmission line during the transition time will behave like a resistance that is equivalent to the characteristic impedance of the line. Given this equivalent circuit model, we can build a circuit of the driver and transmission line and calculate the voltage launched into the transmission line. The equivalent circuit is shown in Figure 7-16.

The driver can be modeled as a voltage source element that switches on fast and as a source resistance. The voltage source has a voltage that is specified depending on the transistor technology. For CMOS devices, it ranges from 5 v to 1.5 v, depending on the transistor generation. Older CMOS devices use 5 v, while PCI and some memory buses use 3.3 v. The fastest processors use 2.4 v and lower for their output rails and 1.5 v and lower for their core. These voltages are the supply voltages and are very close to the output voltage when the device is driving an open circuit.

The value of the source resistance also depends on the device technology. It is typically in the 5-Ohm to 60-Ohm range. When the driver suddenly turns on, some current flows through the source impedance to the transmission line, and there is a voltage drop internal to the gate before the signal comes out the pin. This means that the full, open-circuit drive voltage does not appear across the output pins of the driver.

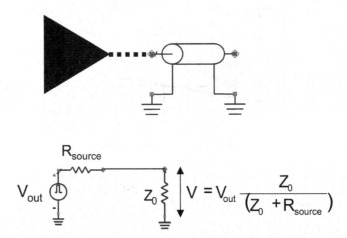

Figure 7-16   Top: Output gate driving a transmission line. Bottom: Equivalent circuit model showing the voltage source, which is the driver, the output-source impedance of the driver gate itself, and the transmission line modeled as a resistor, which is valid during the round-trip time of flight of the transmission line.

The actual voltage launched into the transmission line can be calculated by modeling this circuit as a resistive-voltage divider. The signal sees a voltage divider composed of the source resistance and the transmission line's impedance. The magnitude of the voltage initially launched into the line is the ratio of the impedance of the line to the series combination of the line and source resistance. It is given by:

$$V_{launched} = V_{ouput}\left(\frac{Z_0}{R_{source} + Z_0}\right)$$ (7-16)

where:

$V_{launched}$ = voltage launched into the transmission line

V_{output} = voltage from the driver when driving an open circuit

R_{source} = output-source impedance of the driver

Z_0 = characteristic impedance of the transmission line

When the source resistance is high, the voltage launched into the line will be low—usually not a good thing. In Figure 7-17, we plot the percentage of the source voltage that actually gets launched into the transmission line and propagates down it, for a characteristic impedance of 50 Ohms. When the output-source impedance is also 50 Ohms, we see that only half the open-circuit voltage is actually launched into the line. If the output is 3.3 volts, the signal launched into the line is only 1.65 volts. This is probably not enough to reliably trigger a gate that may be connected to the line. However, as the output resistance of the driver decreases, the signal voltage into the line increases.

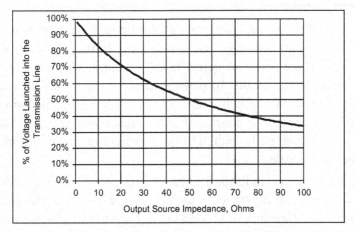

Figure 7-17 Amount of voltage launched into a 50-Ohm transmission line as the output source impedance of the driver varies.

> **TIP**   In order for the voltage, initially launched into the line, to be close to the source voltage, the driver's output source resistance must be small— significantly less than the characteristic impedance of the line.

We say that in order to "drive a transmission line"—in other words, launch a voltage into the line that is close to the open-circuit source voltage— we need an output impedance of the driver that is very small compared to the characteristic impedance of the line. If the line is 50 Ohms, we need a source impedance less than 10 Ohms, for example.

Output devices that have exceptionally low output impedances, 10 Ohms or less, are often called *line drivers* because they will be able to inject a large percentage of their voltage into the line. Older-technology CMOS devices were not able to drive a line since their output impedances were in the 90-Ohm to 130-Ohm range. Since most interconnects behave like transmission lines, current-generation, high-speed CMOS devices must all be able to drive a line and are designed with low-output impedance gates.

7.13 Return Paths

In the beginning of this chapter, we emphasized that the second trace is not the ground but the return path. We should always remember that *all* current, without exception, travels in loops.

> **TIP**   There is *always* a current loop, and if some current goes out to somewhere, it will always come back to the source.

Where is the current loop in a signal propagating on a transmission line? Suppose we have a microstrip that is very long. In this first case, we'll make it so long that the one-way time delay, TD, is 1 second. This is about the distance from the earth to the moon. To make it easier to think about for now, we will short the far end. We launch a signal into the line. This is shown in Figure 7-18. We have said in this chapter that this means we have a constant current going into the signal path, related to the voltage applied and the characteristic impedance of the line.

If current travels in loops and must return to the source, eventually we'd expect to see the current travel to the end of the line and flow back down the return path. But how long does this take? The current flow in a transmission line is very subtle. When do we see the current come out the return path?

Does it take 2 seconds—1 second to go down and 1 second to come back? What would happen then if the far end were really open? If there is insulating dielectric material between the signal and return conductors, how could the current possibly get from the signal to the return conductor, except at the far end?

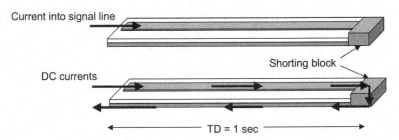

Figure 7-18 Current injected into the signal path of a transmission line and the current distribution after a long time. When does the current actually exit the return path?

The best way of thinking about it is by going back to the zeroth-order model, which describes the line as a bunch of tiny capacitors. This is shown in Figure 7-19. Consider the current flow initially. As the signal launches into the line, it sees the first capacitor. As we described in Chapter 5, "The Physical Basis of Capacitance," if the voltage across the initial capacitor is constant, there will be no current flow through this capacitor. The only way current flows through a capacitor is if the voltage across it changes. As the signal is launched into the transmission line, the voltage across the signal- and return-path conductors ramps up. It is during this transition time, as the edge passes by, that the voltage is changing and current flows through the initial capacitor. As current flows into the signal path to charge up the capacitor, exactly the same amount of current flows out of the return path, having gone through the capacitor.

In the first picosecond, the signal has not gotten very far down the line, and it has no idea how the rest of the line is configured, whether it is open, shorted, or whether it has some radically different impedance. The current flow back to the source, through the return path, depends only on the immediate environment and the region of the line where the voltage is changing—that is, where the signal edge is.

The current from the source flows into the signal conductor and, through displacement current, passes through the capacitance between the signal and return path, and back out the return path. This is the current loop. As the voltage transition edge propagates down the line, this current loop wavefront propagates down the transmission line, flowing between the signal and return path by displacement current.

We can extend the transmission line model to include the rest of the signal and return paths with all the various distributed capacitors between them. As the signal propagates down the line, there is current—the return current—flowing through the capacitance to the return-path conductor and looping back to the source. However, this displacement current loop from the signal path to the return path, flows between them only where the signal voltage is changing.

A few nsec after the signal launch, near the front end, the signal edge has passed by and the voltage is constant, and there is no current flow between the signal to the return path. There is just constant current flowing into the signal conductor and back out the return conductor. Likewise, in front of the signal edge, before the edge has gotten to that region of the line, the voltage is constant, and there is no current flow between the signal and return paths. It is only at the signal edge that current flows through the distributed capacitance.

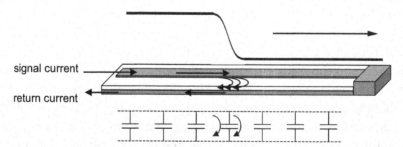

signal current

return current

Figure 7-19 Signal current gets to the return path through the distributed capacitance of the transmission line. Current is flowing only from the signal conductor to the return conductor where the signal voltage is changing—where there is a dV/dt.

Once the signal is launched into the line, it will propagate down the line as a wavefront, at the speed of light. Current will flow down the signal line, pass through the capacitance of the line, and travel back through the return path as a loop. The front of this current loop propagates outward coincident with the voltage edge. We see that the signal is not only the voltage wave front, but it is also the current loop wavefront, which is propagating down the line. The instantaneous impedance the signal sees is the ratio of the signal voltage to the signal current.

Anything that disturbs the current loop will disturb the signal and cause a distortion in the impedance, compromising signal integrity. To maintain good signal integrity, it is important to control both the current wave front and the voltage wave front. The most important way of doing this is to keep the impedance the signal sees constant.

TIP Anything that affects the signal current *or* the return-current path will affect the impedance the signal sees. This is why the return path should be designed just as carefully as the signal path, whether it is on a PCB, a connector, or an IC package.

When the return path is a plane, it is appropriate to ask where the return current flows? What is its distribution in the plane? The precise distribution is slightly frequency dependent and is not easy to calculate with pencil and paper. This is where a good 2D field solver comes in handy.

An example of the current distribution in a microstrip and a stripline for 10-MHz and 100-MHz sine waves of current is shown in Figure 7-20. We can see two important features. First, the signal current is only along the outer edge of the signal trace. This is due to skin depth. Second, the current distribution in the return path is concentrated in the vicinity of the signal line. The higher the sine-wave frequency, the closer to the surface this current distribution will be.

As the frequency increases, the current in the signal and return path will take the path of lowest impedance. This translates into the path of lowest loop inductance, which means the return current will move as close to the signal current as possible. The higher the frequency, the greater the tendency for the return current to flow directly under the signal current. Even at 10 MHz, the return current is highly localized.

In general, for frequencies above about 100 kHz, most of the return current flows directly under the signal trace. Even if the trace snakes around a curvy path or makes a right-angle bend, the return current in the plane follows it. By taking this path, the loop inductance of the signal and return will be kept to a minimum.

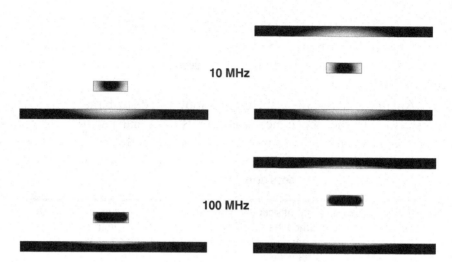

Figure 7-20 Current distribution in the signal and return paths for a microstrip and a stripline at 10 MHz and 100 MHz. In both cases, the line width is 5 mils, and it is 1-ounce copper. Lighter color indicates higher current density. Results calculated with Ansoft's 2D Extractor.

TIP   Anything that prevents the return current from closely following the signal current, such as a gap in the return path, will increase the loop inductance and increase the instantaneous impedance the signal sees, causing a distortion.

We see that the way to engineer the return path is to control the signal path. Routing the signal path around the board will also route the return current path around the board. This is a very important principle of circuit board routing.

7.14 When Return Paths Switch Reference Planes

Cables are specifically designed with a return path adjacent to the signal path. This is true for coax and twisted-pair cables. The return path is easy to follow. In planar interconnects in circuit boards, the return paths are usually designed as planes, as in multilayer boards. For a microstrip, there is one plane directly underneath the signal path, and the return current is easy to identify. But what if the plane adjacent to the signal path is not the plane that is being driven? What if the signal is introduced between the signal path and another plane, as shown in Figure 7-21? What will the return path do?

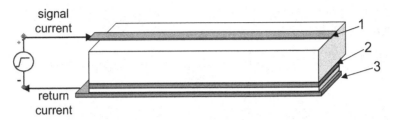

Figure 7-21   Driving a transmission line with the adjacent plane not being the driven return-path plane. What is the return-current distribution?

Current will always distribute so as to minimize the impedance of the signal-return loop. Right at the start of the line, the return path will couple between the bottom plane on layer 3 to the middle plane on layer 2 and then back to the signal path on layer 1.

One way to think about the currents is in the following way: The signal current in the signal path will induce eddy currents in the upper surface of the floating middle plane, and the return current in the bottom plane will induce eddy currents in the underside surface of the middle plane. These induced eddy currents will connect at the edge of the plane where the signal and return currents are being injected in the transmission line. The current flows are diagrammed in Figure 7-22.

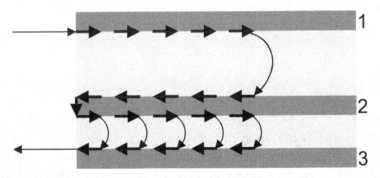

Figure 7-22   Side view of the current flow when driving a transmission line with the adjacent plane not being the driven return-path plane.

The precise current distribution in the planes will be frequency dependent, driven by skin-depth effects. In general, the currents will distribute in each plane to decrease the total loop inductance in the signal-return path. They can only be accurately calculated with a field solver. Figure 7-23 shows an example of the calculated current distribution, viewed from the end, for this sort of configuration for 2-mil-thick conductors at 20 MHz.

Figure 7-23 Current distribution, viewed end on, in the three conductors when the top trace and bottom plane are driven with the middle plane floating. Induced eddy currents are generated in the floating plane. Lighter colors are higher-current density. Calculated with Ansoft's 2D Extractor.

What input impedance will the driver see when looking into the transmission line between the signal line and the bottom plane? The driver, driving a signal into the signal and return paths with the floating plane between them, will see two transmission lines in series. There will be the transmission line composed of the signal and plane on layer 2 and the transmission line composed of the two planes on layers 2 and 3. This is illustrated in Figure 7-24. The input series impedance the driver will see is:

$$Z_{\text{input}} = Z_{1-2} + Z_{2-3} \qquad (7\text{-}17)$$

The smaller the impedance between the two planes, Z_{2-3}, the closer to Z_{1-2} the driver will see as the impedance looking into this transmission line.

This means that even though the driver is connected to the signal line and the bottom plane, the impedance the driver sees is really dominated by the impedance of the transmission line formed by the signal path and its most adjacent plane. *This is true no matter what the voltage of the adjacent plane is.* This is a startling result.

TIP   The impedance a driver will see for a transmission line in a multilayer board will be dominated by the impedance between the signal line and the most adjacent planes, regardless of which plane is actually connected to the driver's return.

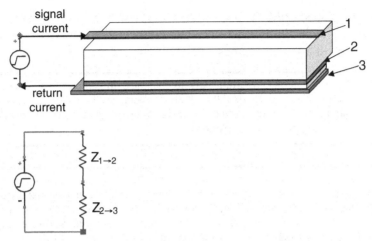

Figure 7-24 Top: Physical configuration of driver driving the transmission line with a floating plane between them. Bottom: Equivalent circuit model showing the impedance the driver sees as the sum of the impedance from the signal to the floating plane and between the two planes.

The smaller the impedance between the two planes compared to the impedance between the signal and its adjacent plane, the closer the driver will just see the impedance between the signal and floating plane.

The characteristic impedance between two long, wide planes, provided that h << w, can be approximated by:

$$Z_0 = \frac{377\Omega}{\sqrt{\varepsilon_r}} \frac{h}{w} \tag{7-18}$$

where:

Z_0 = characteristic impedance of the planes

h = dielectric thickness between the planes

w = width of the planes

ε_r = dielectric constant of the material between the planes

For example, in FR4, for planes that are 2 inches wide with a 10-mil separation, the characteristic impedance a signal will see between the planes is about 377 Ohms × 2 × 0.01/2 = 3.8 Ohms. If the separation were 2 mils, the impedance between the planes would be 377 Ohms × 2 × 0.002/2 = 0.75 Ohm.

When the plane-to-plane impedance is much less than 50 Ohms, it is less important which plane is actually DC-connected to the driver and the more likely the most adjacent planes will dominate the impedance.

TIP   The most important way to minimize the impedance between adjacent planes is to use the thinnest dielectric between the planes possible. This will keep the impedance between the planes low and provide tight coupling between the planes.

TIP   When planes are tightly coupled and there is a low impedance between them—as there should be for low rail collapse anyway—it doesn't matter to which plane the driver actually connects. The coupling between the planes will provide a low-impedance path for the return current to get as close as possible to the signal current.

What if the signal path changes layers in mid-trace? How will the return current follow? Figure 7-25 shows a four-layer board with the signal path starting on layer 1 and transitioning through a via to layer 4. In the first half of the board, the return current must be on the plane directly underneath the signal path, on layer 2. In addition, for sine-wave-frequency components of the current above 10 MHz, the current is actually flowing on the top surface of the layer 2 plane.

In the bottom half of the board, where the signal is on layer 4, where will the return current be? It has to be on the plane adjacent to the signal layer, or on plane layer 3. And, it will be on the undersurface of the plane. In the regions of uniform transmission line, the return currents are easy to follow. The via is clearly the path the signal current takes to go from layer 1 to layer 4. How does the return current make the transition from layer 2 to layer 3?

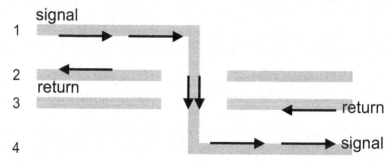

Figure 7-25   Cross section of a four-layer board with signal current transitioning from layer 1 to layer 4 through a via. How does the return current transition from layer 2 to layer 3?

If the two planes are at the same potential and have a via shorting them, the return current would take this path of low impedance. There might be a small jog for the return current, but it would be through a short length of plane, and a plane has low total inductance so would not have a large impedance discontinuity. This would be the preferred stack-up. If there were no other constraints such as cost, keeping the nearest reference planes at the same voltage and shorting them together close to the signal via is the optimum design rule. To reduce the voltage drop in the return path, always consider adding a return via adjacent to the signal via.

However, to minimize the total number of layers, it is sometimes necessary to use adjacent return layers with different voltages. If plane 2 is a 5-v plane and plane 3 is a 0-v plane, there would be no DC path between them. How will the return current flow from plane 3 to plane 2?

It can only go through the distributed impedance between them. At the inside of the clearance hole, the return current will snake around and change surfaces on the same plane. The return current will then spread out on the inner surfaces of the planes and couple through the plane-to-plane distributed impedance. The current will spread out between the planes at the speed of light in the dielectric. The current flows in the return path are shown in Figure 7-26. The two return-path planes create a transmission line, and the return current will see an impedance that is the instantaneous impedance of the two planes.

TIP   Whenever the return-path current switches planes and the planes are DC isolated, the return current will couple through the planes and see an impedance equal to the instantaneous impedance of the transmission line created by the planes.

The return current will have to go through this impedance, and there will be a voltage drop in the return path. We call this voltage drop in the return path *ground bounce*. The larger the impedance of the return path, the larger the voltage drop, and the larger the ground-bounce noise generated. All other signal lines that are also changing return planes will contribute to this ground-bounce voltage noise and encounter the ground-bounce noise created by the other signals.

TIP   The goal in designing the return path is to minimize the impedance of the return path to minimize the ground-bounce noise generated in the return path. As we will see, this is primarily implemented by keeping the impedance between the return planes as low as possible, usually by keeping them on adjacent layers with the thinnest possible dielectric between them.

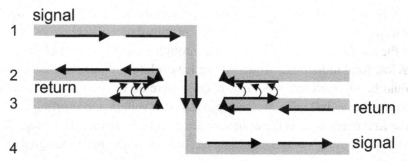

Figure 7-26 The return current transitions from layer 2 to layer 3 by capacitive coupling between the layers.

As the return current spreads out between the two return-path planes in an ever-expanding circle from the signal via, it will see an ever-decreasing instantaneous impedance. The capacitance per length that the signal sees gets larger as the radius gets larger. This makes the analysis complicated except for specific cases and, in general, requires a field solver.

However, we can build a simple model to estimate the instantaneous impedance between the two planes and to provide insight into how to optimize the design of a stack-up and minimize this form of ground bounce.

To calculate the instantaneous impedance the signal sees as it propagates radially outward between the two planes, we need to calculate the capacitance per length of the radial-transmission line and the speed of the signal. The capacitance per length the signal will see is the change in capacitance per incremental change in radius. The total capacitance the return current sees is:

$$C = \varepsilon_0 \varepsilon_r \frac{A}{h} \qquad (7\text{-}19)$$

and the area between the planes is:

$$A = \pi r^2 \qquad (7\text{-}20)$$

Combining these relationships shows the capacitance increasing with distance as:

$$C = \varepsilon_0 \varepsilon_r \frac{A}{h} = \varepsilon_0 \varepsilon_r \frac{\pi r^2}{h} \quad (7\text{-}21)$$

where:
C = coupling capacitance between the planes
ε_0 = permittivity of free space, 0.225 pF/in
ε_r = dielectric constant of the material between the planes
A = area of overlap of the return current in the planes
h = spacing between the planes
r = increasing radius of the coupling circle, expanding at the speed of light

The incremental increase in capacitance as the radius increases (i.e., the capacitance per length) is:

$$C_L = 2\pi \varepsilon_0 \varepsilon_r \frac{r}{h} \quad (7\text{-}22)$$

As expected, as the return current moves farther from the via, the capacitance per length increases. The instantaneous impedance this current will see is:

$$Z = \frac{1}{vC_L} = \frac{\sqrt{\varepsilon_r}}{c} \times \frac{h}{2\pi r \varepsilon_0 \varepsilon_r} = \frac{377\Omega}{2\pi} \frac{h}{r\sqrt{\varepsilon_r}} = 60\Omega \frac{h}{r\sqrt{\varepsilon_r}} \quad (7\text{-}23)$$

where:
Z = instantaneous impedance the return current sees between the planes
C_L = coupling capacitance per length between the planes
v = speed of light in the dielectric
ε_0 = permittivity of free space, 0.225 pF/in
ε_r = dielectric constant of the material between the planes

h = spacing between the planes

r = increasing radius of the coupling circle, expanding at the speed of light

c = speed of light in a vacuum

For example, if the dielectric thickness between the planes is 10 mils, the impedance the return current will see by the time it is 1 inch away from the via is $Z = 60 \times 0.01/(1 \times 2) = 0.3$ Ohm. And, this impedance will get smaller as the return current propagates outward. This says that the farther from the via, the lower the impedance the return current sees, and the lower the ground-bounce voltage will be across this impedance.

We can relate the impedance the return current sees (which is the impedance in series with the signal current) to time since the return current spreads out at the speed of light in the material and $r = v \times t$:

$$Z = 60\Omega \frac{h}{r\sqrt{\varepsilon_r}} = 60\Omega \frac{h}{vt\sqrt{\varepsilon_r}} = 60\Omega \frac{h\sqrt{\varepsilon_r}}{ct\sqrt{\varepsilon_r}} = 5\Omega \frac{h}{t} \qquad (7\text{-}24)$$

where:

Z = instantaneous impedance the return current sees between the planes, in Ohms

v = speed of light in the dielectric

ε_r = dielectric constant of the material between the planes

c = speed of light in a vacuum

h = spacing between the planes, in inches

t = time the return current is propagating, in nsec

For example, for a spacing of 0.01 inch, the impedance the return current sees after the first 0.1 nsec is $Z = 5 \times 0.01/0.1 = 0.5$ Ohm. The very first leading edge of the signal will see an initial impedance that can be as large as 0.5 Ohm. If the signal current were 20 mA in the first 100 psec, corresponding to a 1-v signal in a 50-Ohm impedance, the ground-bounce voltage drop across the switching planes that are in series with the signal voltage, for the first 0.1 nsec, would be 20 mA × 0.5 Ohm = 10 mV.

This may not seem like a lot compared to the 1-v signal, but if 10 signals transition simultaneously between the same reference planes, all within less than 0.6 inch of each other, they will each see the same common 0.5-Ohm impedance, and there will be 20 mA × 10 = 200 mA of current through the return-path impedance. This will generate a ground-bounce noise of 200 mA × 0.5 Ohm = 100 mV, which is now 10% of the signal voltage and can be significant. All signal paths that also transition through this path will see the 100 mV of ground-bounce noise, even if they are not switching.

The initial impedance the return current will see can be high, if there is substantial current flowing in a very short time initially. All current flowing during this initial short time will see a high impedance and generate ground-bounce voltages. Figure 7-27 shows the impedance the return current sees over time. From this plot, it is clear that the impedance of the return path is significant only for very fast rise times, substantially less than 0.5 nsec.

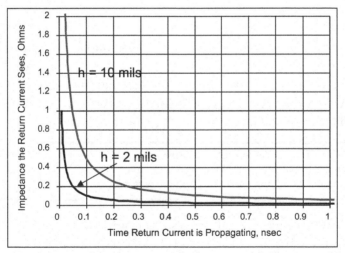

Figure 7-27   The impedance the return current sees between the planes for 2-mil spacing and 10-mil spacing, as the signal propagates outward from the via.

When the impedance is significant—about 5% of 50 Ohms—for one signal line switching, the impedance of the return path is important. When there are n signal paths switching through these planes, the maximum acceptable impedance of the return path is 2.5 Ohms/n.

> **TIP**   This analysis indicates that for fast edges and multiple signals switching simultaneously between reference planes, significant ground-bounce voltage can be generated across the return path. The only way to minimize this ground-bounce voltage is by decreasing the impedance of the return path.

This is primarily accomplished with the following steps:

1. Make sure that as the signal path transitions layers, it always has an adjacent plane with the same voltage level for its return and there is a shorting via between the switching planes in close proximity to any signal vias.
2. Keep the spacing between different DC-voltage-level return planes as thin as possible.
3. Space out adjacent switching vias so that the return currents do not overlap during the initial transients when the impedance of the return path can be high.

It is sometimes believed that adding a decoupling capacitor between two return planes for which there is a switching return current will help decrease the impedance of the return path. It is hoped that the added discrete capacitor will provide a low-impedance path for the return current to flow from one return plane to the other.

> **TIP**   To provide any effectiveness, the real capacitor must keep the impedance between the planes less than 5% × 50 Ohms, or 2.5 Ohms, at the bandwidth of the rise time's frequency components.

A real capacitor has some loop inductance and some equivalent series resistance associated with it. This will limit the usefulness of discrete decoupling capacitors for very short rise-time signals. The low impedance of the radial transmission line formed by the planes will provide the low impedance for the higher-frequency components.

It is not its capacitance that determines the real capacitor's impedance above its self-resonant frequency; it is its equivalent series inductance. Figure 7-28 shows the expected impedance of a real 1-nF capacitor with a loop inductance of 0.5 nH. This is an extremely optimistic loop inductance and can be achieved only by either a multiple via in-pad configuration or with the use of interdigitated capacitors (IDCs).

This real capacitor will provide a low impedance for the return path for signal bandwidths only up to 1 GHz. Higher capacitance than 1 nF will provide no added value since the high-frequency components will be sensitive only to the loop inductance.

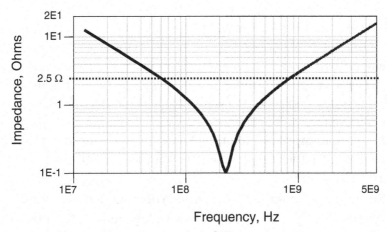

Figure 7-28   Impedance of a capacitor with 1-nF capacitance and only 0.5 nH of loop inductance.

TIP   When using a discrete capacitor to decrease the impedance of the return path, it is far more important to keep the series inductance low than to use more than 1 nF of capacitance.

Unfortunately, even the 0.5 nH of loop inductance of a well-designed capacitor will still have a large impedance at frequencies above 1 GHz, where the planes do not have low impedance and their ground bounce may be a problem.

TIP   A capacitor between different DC-voltage planes will not be very effective at managing the ground bounce from switching planes. It may, however, provide additional decoupling for lower-frequency noise but will not solve the ground-bounce problem as rise times continue to decrease.

TIP   In a multilayer board, when signal paths must change different voltage-level return layers, the only way of minimizing the ground-bounce voltage generated is by using a dielectric between the return planes that is as thin as possible.

One additional problem is created when a signal changes return planes and current is driven in the transmission line formed by the two adjacent planes: Where does this current end up? As the current propagates outward, eventually it will hit the edges of the board. The current injected between the planes when the signal current switches will rattle around between the planes, causing transient voltages between the two planes.

The impedance between the planes is very low, much less than 1 Ohm, so the transient voltages created will be low. However, with multiple signals switching between planes, each one injects some noise between the planes. This noise will get larger with more signals switching. The current injected into the planes is determined by the signal voltage and the impedance, roughly 50 Ohms. The voltage noise generated between the planes depends on the signal current and the impedance of the planes. To minimize this rattling voltage, the impedance of the planes should be kept to a minimum by keeping their dielectric spacing thin.

Signals transitioning return planes is a dominant source of high-frequency noise injected between the planes. This voltage noise will rattle around and contribute to the noise in the PDN. In low-noise systems, it can be a major source of cross talk to very sensitive lines, such as rf receivers or analog-to-digital converter inputs or voltage references. To minimize the noise in these systems, care should be taken to minimize the return current injected between the planes by the careful selection of return plane voltages, return vias, and low-inductance decoupling capacitors.

We sometimes call this voltage, which is rattling between the edges of the board between the adjacent plane layers, *resonances in the planes*. These resonances will eventually die out from conductor and dielectric attenuation. The resonances will have frequency components that match the round-trip time of flight between the board dimensions. For boards with a dimension of 10 to 20 inches on a side, the resonant frequencies will be in the 150-MHz to 300-MHz range. This is why capacitors between the different voltage planes can provide some value. Their low loop inductance provides the low impedance in this frequency regime and helps maintain a low impedance between the planes (in the board resonant-frequency range) and keep the voltages between the planes low. However, as we saw, these capacitors do not affect the transient ground-bounce voltage during the very fast transition.

TIP   To minimize the resonant rattling-around voltage, especially in small multilayer packages, it is important to avoid having any return currents switch between different voltage planes. The adjacent return layers should be at the same DC-voltage level, and a return via should connect the return path in proximity to the signal via. This will prevent injecting any currents between planes and avoid driving any plane resonances.

These problems will get more important as rise times decrease, especially below 100 psec.

7.15 A First-Order Model of a Transmission Line

An ideal transmission line is a new ideal circuit element that has the two special properties of a constant instantaneous impedance and a time delay associated with it. This ideal model is a "distributed" model in the sense that the properties of an ideal transmission line are distributed over its length rather than being concentrated in a single lumped point.

Physically, a controlled-impedance transmission line is composed of just two conductors with length and having a uniform cross section for the whole length. Earlier in this chapter, we introduced a zeroth-order model, which described a transmission line as a collection of capacitors spaced some distance apart. This was a physical model, not an equivalent electrical model.

We can further *approximate* the physical transmission line by describing the sections of signal- and return-path conductors as a loop inductance. The simplest equivalent circuit model for a transmission line would have small capacitors separated by small loop inductors, as shown in Figure 7-29. The C is the capacitance between the conductors, and the L is the loop inductance between the sections.

Each segment of the signal path has some partial self-inductance associated with it, and each segment of the return path has some partial self-inductance associated with it. There is some partial mutual inductance between each of the signal- and return-path segments between the discrete capacitors. For an unbalanced transmission line, such as a microstrip, the partial self-inductances of the signal- and return-path segments are different. In fact, the partial self-inductance of the signal path can be more than 10 times larger than the partial self-inductance of the return path.

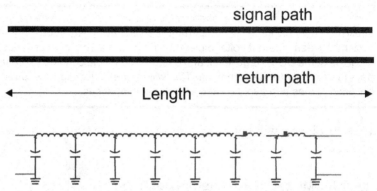

Figure 7-29 Top: Physical structure of a uniform transmission line. Bottom: First-order equivalent circuit model approximation for a transmission line based on combinations of Ls and Cs.

However, from a signal's perspective as it propagates down the line, it is a current loop, from the signal to the return path, that propagates. In this sense, all the signal current sees is the loop inductance down the signal-path segment and returning back through the return-path segment. For signal propagation in a transmission line and most cross-talk effects, the partial inductances of the signal and return paths do not play a role. It is only the loop inductance that is important. When approximating an ideal distributed-transmission line as a series of LC segments, the inductance the signal sees, as represented in the model, is really a loop inductance.

TIP   It is important to keep in mind that this lumped-circuit model is an approximation of an ideal transmission line. In the extreme case, as the size of the capacitors and inductors gets diminishingly small and the number of each increases, the approximation gets better and better.

In the extreme, when each capacitor and inductor is infinitesimal and there is an infinite number of elements, there is a uniform capacitance per length, C_L, and a uniform loop inductance per length, L_L. These are often called the *line parameters* of a transmission line. Given the total length of the line, Len, the total capacitance is given by:

$$C_{total} = C_L \times Len \qquad (7\text{-}25)$$

and the total inductance is given by:

$$L_{total} = L_L \times Len$$

(7-26)

where:

C_L = capacitance per length

L_L = loop inductance per length

Len = length of the transmission line

Just by looking at this LC circuit, it is difficult to get a feel for how a signal will interact with it. At first glance, we might think there will be a lot of oscillations and resonances. But what happens when the size of each element is infinitely small?

The only way to really know how a signal will interact with this circuit model is to apply network theory and solve the differential equations represented by this LC network. The results indicate that a signal traveling down the network will see a constant instantaneous impedance at each node. This constant instantaneous impedance is the same as the instantaneous impedance we discovered for an ideal distributed-transmission-line element. The instantaneous impedance is numerically the characteristic impedance of the line. Likewise, there will be a finite delay between the time the signal is introduced into the front of the LC network and the time it comes out.

Using network theory, we can calculate how the characteristic impedance and time delay depend on the line parameters and total length of the line:

$$Z_0 = \sqrt{\frac{L_L}{C_L}}$$

(7-27)

$$TD = \sqrt{C_{total}L_{total}} = Len \times \sqrt{C_L L_L} = \frac{Len}{v}$$

(7-28)

$$v = \frac{Len}{TD} = \frac{1}{\sqrt{C_L L_L}}$$

(7-29)

where:

Z_0 = characteristic impedance, in Ohms

L_L = loop inductance per length of the transmission line

C_L = capacitance per length of the transmission line

TD = time delay of the transmission line

L_{total} = total loop inductance in the transmission line

C_{total} = total capacitance in the transmission line

v = speed of the signal in the transmission line

Without having to invoke a finite speed for the propagation of a signal down a transmission line, the electrical properties of an LC network predict this behavior. Likewise, though it is hard to tell by looking at the circuit model, network theory predicts that there is a constant impedance the signal sees at each node along the circuit.

These two predicted properties—the characteristic impedance and the time delay—must match the same values we derived based on the zeroth-order physics model of finite speed and collection of capacitor buckets. By combining the results from these two models, a number of very important relationships can be derived.

Because the speed of a signal depends on the dielectric constant of the material and on the capacitance per length and inductance per length, we can relate the capacitance per length to the inductance per length:

$$v = \frac{c}{\sqrt{\varepsilon_r}} = \frac{1}{\sqrt{C_L L_L}} \qquad (7\text{-}30)$$

$$L_L = 7 \frac{\varepsilon_r}{C_L} \frac{nH}{inch} \qquad (7\text{-}31)$$

$$C_L = 7 \frac{\varepsilon_r}{L_L} \frac{pF}{inch} \qquad (7\text{-}32)$$

From the relationship between the characteristic impedance and velocity, the following relationships can be derived:

$$C_L = \frac{1}{vZ_0} = \frac{1}{cZ_0}\sqrt{\varepsilon_r} = \frac{83}{Z_0}\sqrt{\varepsilon_r}\frac{pF}{inch} \qquad (7\text{-}33)$$

$$L_L = \frac{Z_0}{v} = 0.083Z_0\sqrt{\varepsilon_r}\frac{nH}{inch} \qquad (7\text{-}34)$$

And, from the time delay and the characteristic impedance of the transmission line, the following can be derived:

$$C_{total} = \frac{TD}{Z_0} \qquad (7\text{-}35)$$

$$L_{total} = TD \times Z_0 \qquad (7\text{-}36)$$

where:

Z_0 = characteristic impedance, in Ohms

L_L = loop inductance per length of the transmission line, in nH/inch

C_L = capacitance per length of the transmission line, in pF/inch

TD = time delay of the transmission line, in nsec

L_{total} = total loop inductance in the transmission line, in nH

C_{total} = total capacitance in the transmission line, in pF

v = speed of the signal in the transmission line, in in/nsec

For example, a line that is 50 Ohms and has a dielectric constant of 4 has a capacitance per length of C_L = 83/50 × 2 = 3.3 pF/inch. This is a startling conclusion.

TIP   *All* 50-Ohm transmission lines with a dielectric constant of 4 have the same capacitance per length—about 3.3 pF/inch. This is a very useful rule of thumb to remember.

If the line width doubles, the dielectric spacing would have to double to maintain the same characteristic impedance, and the capacitance would stay the same. An interconnect in a BGA package that has been designed as a 50-Ohm controlled-impedance line, 0.5 inches long, has a capacitance of 3.3 pF/inch × 0.5 inch = 1.6 pF.

Likewise, the inductance per length of a 50-Ohm line made with FR4 is $L_L = 0.083 \times 50 \times 2 = 8.3$ nH/inch.

TIP   *All* 50-Ohm transmission lines with a dielectric constant of 4 have the same loop inductance per length of about 8.3 nH/inch. This is a very useful rule of thumb to remember.

If the time delay of a line is 1 nsec and it is 50 Ohms, the total capacitance in the line is C_{total} = 1 nsec/50 = 20 pF. If the line is 6 inches long, this 20 pF of capacitance is distributed over 6 inches and the capacitance per length is 20 pF/6 inches = 3.3 pF/inch. For this same line, the total loop inductance going down the signal line and looping back through the return path is L_{total} = 1 nsec × 50 Ohms = 50 nH. Distributed over the 6 inches of length, this is an inductance per length of 50 nH/6 inches = 8.3 nH/inch.

These relationships that relate the capacitance, inductance, characteristic impedance, and dielectric constant, which are associated with a transmission line, apply to any transmission line. They make no assumptions on the cross-section geometry. They are very powerful tools to help us estimate one or more of these terms using existing approximations or field solvers. If we know any two, we can always find the others.

It is tempting to consider this LC model of a transmission line as representing how a real transmission line behaves. If we ignore the inductance, for example, we might consider a real transmission line as just some capacitance. Won't we see an RC charging time for the transmission line when driven by the output impedance of a driver?

In fact, a real transmission line is not just a capacitor, or even just an inductor; it is a distributed LC interconnect. It is the adjacent L and C, microscopically small in size, which transforms this interconnect into a new behavior. A fast edge entering the interconnect maintains this rise time throughout the interconnect.

As the fast edge propagates, it does not see the capacitance and a need to charge it through an RC, nor does it see an L and a slower rise time as the L/R. Rather, it sees a brand-new emergent property of an instantaneous impedance that supports any rise-time signal. A transmission line does not look like an L or a C to the signal edge; it looks like a resistive element.

It is possible to calculate the total capacitance and total loop inductance in a transmission line interconnect, but it would be incorrect to think of the transmission line as just a C or just an L. They are forever connected together, and both elements are equally important.

Of course, this lumped-circuit LC ladder model is just an approximation. Instead of thinking of a real transmission line in terms of the LC ladder model, recalibrate your engineering intuition to see a transmission line as a brand-new ideal circuit element with the new, emergent behavior of supporting any rise-time signal once launched and seeing an instantaneous impedance.

7.16 Calculating Characteristic Impedance with Approximations

Engineering a particular target characteristic impedance is really a matter of adjusting the line widths, dielectric thickness, and dielectric constants. If we know the length of the transmission line and the dielectric constant of the material around the conductors and we can calculate the characteristic impedance, we can use the relationships above to calculate all other parameters.

Of course, every different type of cross-sectional geometry will have a different relationship between the geometrical features and the characteristic impedance. In general, there are three types of analysis we can use to calculate the characteristic impedance from the geometry:

1. Rules of thumb
2. Approximations
3. 2D field solvers

The two most important rules of thumb relate the characteristic impedance of a microstrip and stripline fabricated with FR4. The cross sections for 50-Ohm transmission lines are illustrated in Figure 7-30.

TIP   As a good rule of thumb, a 50-Ohm microstrip in FR4 has a line width twice the dielectric thickness. A 50-Ohm stripline has a total dielectric spacing between the planes equal to twice the line width.

There are only three cross-section geometries that have exact equations: coax, twin-round-wire, and round-wire-over-plane geometries (see Figure 7-31).

All others are approximations. The relationship between characteristic impedance and geometry for coax is:

$$Z_0 = \frac{377\Omega}{2\pi\sqrt{\varepsilon_r}}\ln\left(\frac{b}{a}\right) = \frac{60\Omega}{\sqrt{\varepsilon_r}}\ln\left(\frac{b}{a}\right) \tag{7-37}$$

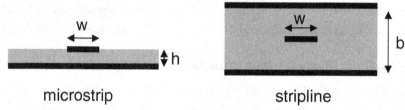

Figure 7-30   Scaled cross sections for 50-Ohm transmission lines in FR4. Left: 50-Ohm microstrip with w = 2 × h. Right: 50-Ohm stripline with b = 2 × w.

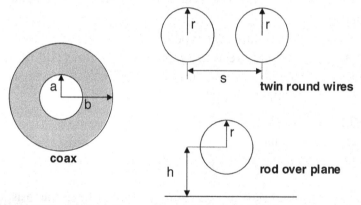

Figure 7-31   The only three cross sections for which there are exact equations for the characteristic impedance. All others are approximations.

For twin parallel round wires, the characteristic impedance is:

$$Z_0 = \frac{120\Omega}{\sqrt{\varepsilon}}\ln\left(\frac{s}{2r} + \sqrt{\left(\frac{s}{2r}\right)^2 - 1}\right) \tag{7-38}$$

For the case of a round rod over a plane, the characteristic impedance is:

$$Z_0 = \frac{60\Omega}{\sqrt{\varepsilon_r}} \ln\left(\frac{h}{r} + \sqrt{\left(\frac{h}{r}\right)^2 - 1} \right) \tag{7-39}$$

where:

Z_0 = characteristic impedance, in Ohms

a = inner radius of the coax, in inches

b = outer radius of the coax, in inches

r = radius of the round rod, in inches

s = center-to-center spacing of the round rod, in inches

h = height of the center of the rod over the plane, in inches

ε_r = dielectric constant of the materials

These relationships assume that the dielectric material completely and uniformly fills all space wherever there are electric fields. If this is not the case, the effective dielectric constant that affects the speed of the signal down the conductor will be some complicated combination of the mixture of dielectrics. This can often be calculated only with a field solver.

For homogeneous dielectric distributions, these are useful relationships to use to calibrate 2D field solvers since they are exact.

TIP With very few exceptions, virtually all other equations used to relate the characteristic impedance to the geometry are approximations. When being off by more than 5% may increase design time and costs excessively, approximations should not be used for final sign-off of the design of transmission lines. When accuracy counts, a verified 2D field solver should be used.

The value of approximations is that they show the relationship between the geometrical terms and can be used for sensitivity analysis in a spreadsheet. The most popular approximation for microstrip recommended by the IPC is:

$$Z_0 = \frac{87\Omega}{\sqrt{1.41+\varepsilon_r}} \ln\left(\frac{5.98h}{0.8w+t} \right) \tag{7-40}$$

For stripline, the IPC recommended approximation is:

$$Z_0 = \frac{60\Omega}{\sqrt{\varepsilon_r}} \ln\left(\frac{2b+t}{0.8w+t}\right) \tag{7-41}$$

where:

Z_0 = characteristic impedance, in Ohms

h = dielectric thickness below the signal trace to the plane, in mils

w = line width, in mils

b = plane-to-plane spacing, in mils

t = metal thickness, in mils

ε_r = dielectric constant

If we ignore the impact from the trace thickness, t, then the characteristic impedance for both structures depends only on the ratio of the dielectric thickness to the line width. This is a very important relationship.

TIP   To first order, the characteristic impedance of stripline and microstrip will scale with the ratio of the dielectric thickness to the line width. As long as this ratio is constant, the characteristic impedance will be constant.

For example, if the line width is doubled and the dielectric spacing is doubled, the characteristic impedance, as a first-order approximation, will be unchanged.

Though these equations look complicated, this is no measure of their accuracy. The only way to know the accuracy of an approximation is to compare it to the results from a verified field solver.

7.17 Calculating the Characteristic Impedance with a 2D Field Solver

If we need better than 10% accuracy or are worrying about second-order effects, such as the effect of trace thickness, solder-mask coverage, or side-wall shape, we should not be using one of these approximations. The most important tool that should be in every engineer's toolbox for doing impedance calculations is a 2D field solver.

The basic assumption made by all 2D field solvers is that the geometry is uniform down the length of the line. That is, as we move down the length of the transmission line, the cross-sectional geometry stays the same. This is the basic definition of a controlled-impedance line—that the cross section is uniform. In such a case, there is only one characteristic impedance that describes the line. This is why a 2D field solver is the right tool to provide an accurate prediction of the characteristic impedance of uniform transmission lines. Only the 2D cross-sectional information is important.

Whenever accuracy is important, a 2D field solver should be used. This means that before sign-off on a design and committing to hardware, the stack-up should be designed with a 2D field solver. Sometimes the argument is made that if the manufacturing tolerance is 10%, we should not care if the accuracy of the prediction is even 5%.

Accuracy is important because of yield. Any inaccuracy in the predicted impedance will shift the center position of the distribution of the manufactured impedances. The better we can center the distribution on the required target, the better we can increase yield. Even an inaccuracy of 1% will shift the distribution off center. Any product close to the spec limits could be shifted outside the limits and contribute to a yield hit. As we showed in Chapter 5, the accuracy of a field solver can be better than 0.5%.

By using a 2D field solver, the characteristic impedance of a microstrip can be calculated. These results are compared to the predictions based on the IPC approximations. Figure 7-32 shows the predicted characteristic impedance of a microstrip with dielectric constant of 4, with 1/2-ounce copper and dielectric thickness of 10 mils, as the line width is varied. Near and above 50 Ohms, the agreement is very good. However, for lower impedances, the IPC approximation can be off by as much as 25%.

The same comparison is made for stripline and is shown in Figure 7-33. The agreement close to 50 Ohms is very good, but for low impedances, the approximation can be off by more than 25%. Approximations should not be used when accuracy counts.

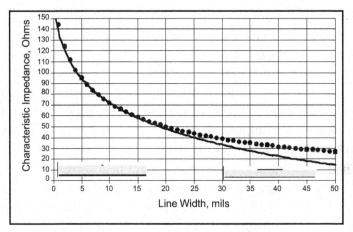

Figure 7-32   Comparison of field-solver results (circles) and the IPC approximation (line) for a microstrip with 10-mil-thick FR4 dielectric and 1/2-ounce copper. Field-solver results from the Ansoft 2D Extractor.

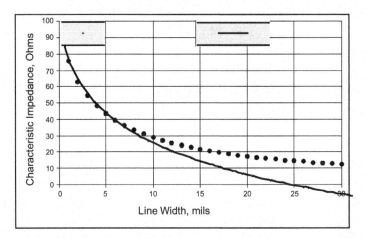

Figure 7-33   Comparison of field-solver results (circles) and the IPC approximation (line) for a stripline with 10-mil-thick FR4 dielectric and 1/2-ounce copper. Field-solver results from the Ansoft 2D Extractor.

In addition to providing an accurate estimate for the characteristic impedance, a 2D field solver can also provide insight into the impact from second-order effects, such as:

• The width of the return path
• The thickness of the signal-trace conductor

- The presence of solder mask over a surface trace
- The effective dielectric constant

How does the width of the return path influence the characteristic impedance? Applying the principles of Bogatin's rule 9, we would anticipate that if the width of the return path is narrow, the capacitance should be smaller and the characteristic impedance higher. Using a field solver, we can calculate how wide the return path has to be so that it doesn't matter anymore.

Figure 7-34 shows the calculated characteristic impedance of a microstrip with dielectric constant of 4 and trace thickness of 0.7 mil, corresponding to 1/2-ounce copper, dielectric thickness of 5 mils, and line width of 10 mils. This is nominally a 50-Ohm line. The width of the return path is varied and the characteristic impedance calculated. If the width of the extension on each side of the signal trace is greater than 15 mils, the characteristic impedance is less than 1% off from its infinite-width value. Because the fringe fields along the edge of the trace scale with the dielectric thickness, this is the important scale factor. The extent of the return path on either side of the signal path should be roughly three times the dielectric thickness, h.

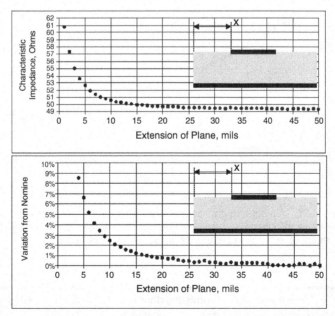

Figure 7-34   Top: Calculated characteristic impedance of a microstrip with 10-mil-wide trace and 5-mil-thick dielectric, as the return-path plane is widened. Bottom: Variation from the nominal value. Field-solver results from the Ansoft 2D Extractor.

TIP   As a rough rule of thumb, the return path should extend at least three times the dielectric thickness on either side of the signal trace for the characteristic impedance to not exceed more than 1% of the value when the return path is infinitely wide.

Again following rule 9, we would anticipate that as trace thickness increases, there will be more fringe fields coming off the sides, and the capacitance per length would increase. This would make the characteristic impedance lower as the conductor thickness increases.

Using a 2D field solver, we can calculate the characteristic impedance as the trace thickness changes from 0.1 mil thick to 3 mils thick. This is plotted in Figure 7-35. Each point on the curve is the calculated characteristic impedance for a different trace thickness. Sure enough, as the metal thickness increases, the fringe field capacitance increases, and the characteristic impedance decreases. Thicker metal means higher capacitance between the signal trace and the return path, which also means lower characteristic impedance. However, as we can see from the calculated results, this is not a large effect; it is second order.

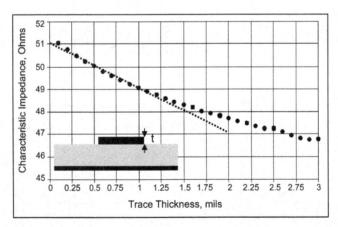

Figure 7-35   Calculated characteristic impedance of a nominal 50-Ohm microstrip as the trace thickness is changed. The circles are calculated results; the line is 2 Ohms/mil. Field-solver results from the Ansoft 2D Extractor.

TIP   As a rough rule of thumb, the decrease in characteristic impedance is about 2 Ohms per mil of signal-trace thickness.

If there is a thin solder-mask coating on top of the microstrip, the fringe field capacitance should increase, and the characteristic impedance should decrease. For the same microstrip as above, but using a conductor only 0.1 mil thick, the characteristic impedance is calculated for an increasing solder-mask coating, using a dielectric constant of 4. Figure 7-36 shows the decrease in characteristic impedance. It is about 2 Ohms/mil of coating for thin coatings. After about 10-mil-thick coating, the characteristic impedance is no longer affected, as all the external fringe fields are contained in the first 10 mils of coating. This is a measure of how far the fringe fields extend above the surface.

Of course, solder mask is more typically 0.5 mil to 2 mils. In this regime, we see that the presence of the solder mask can lower the characteristic impedance by as much as 2 Ohms—a significant amount. Reaching a target impedance with solder mask requires the line width to be narrower than nominal, so the solder mask would bring the impedance back down to the target value.

This last example illustrates that the characteristic impedance depends on the dielectric distribution above the trace in a microstrip. Of course, in a stripline, all the fields are contained by the dielectric, so solder mask on top of the upper plane has no impact on the characteristic impedance.

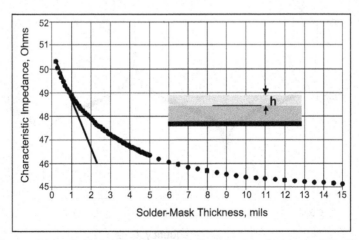

Figure 7-36   Calculated characteristic impedance of a nominal 50-Ohm microstrip as a solder mask with dielectric constant of 4 is added. Circles are the field-solver results; the line is 2 Ohms/mil. Field-solver results from the Ansoft 2D Extractor.

In addition to the characteristic impedance affected by the inhomogeneous dielectric distribution, the effective dielectric constant that the fields experience is also affected by the dielectric distribution. In microstrip, the effective dielectric constant, which determines the speed of the signal, depends on the specific geometry of the dielectric. Often, the effective dielectric constant can be accurately calculated only with a field solver. This was discussed in detail in Chapter 5.

7.18 An n-Section Lumped-Circuit Model

The ideal transmission-line circuit element is a distributed element that very accurately predicts the measured performance of real interconnects. Figure 7-37 shows the comparison of the measured and simulated impedance of a 1-inch-long transmission line in the frequency domain. We see the excellent agreement even up to 5 GHz, the bandwidth of the measurement.

TIP   Real interconnects match the behavior of an ideal transmission-line model to high bandwidth. An ideal transmission-line model is a very good model for real interconnects.

We can approximate this ideal model with a combination of LC lumped-circuit sections. How do we know how many LC sections to use for a given level of accuracy? What happens if we use too few sections?

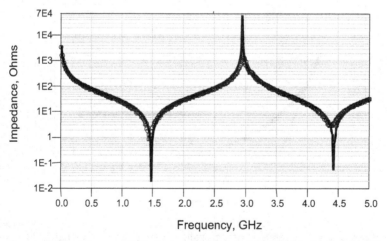

Figure 7-37   Measured (circles) and simulated (line) impedance of a 1-inch-long, 50-Ohm transmission line. The model is an ideal, lossless transmission line. The agreement is excellent up to the bandwidth of the measurement.

These questions can be explored using a simulation tool like SPICE. We will first work in the frequency domain to evaluate the impedance by looking into the front end of a transmission line and then interpret this result in the time domain.

In the frequency domain, we can ask *What is the impedance when looking into the front end of a transmission line, with the far end open?* In this example, we will use a 50-Ohm line that is 6 inches long with a dielectric constant of 4. Its time delay, TD, is 1 nsec.

The total capacitance is given by $C_{total} = TD/Z_0 = 1$ nsec/50 Ohms = 20 pF. The total loop inductance is given by $L_{total} = Z_0 \times TD = 50$ Ohms $\times 1$ nsec = 50 nH.

The simplest approximation for a transmission line is a single LC model, with the L and C values as the total values of the transmission line. This is the simplest lumped-circuit model for an ideal transmission line.

Figure 7-38 shows the predicted impedance for an ideal, distributed transmission line and the calculated impedance of a single-section LC lumped-circuit model using these values. In the low-frequency range, the LC model matches the performance really well. The bandwidth of this model is about 100 MHz. The limitation to the bandwidth occurs because, in fact, this ideal transmission line does not have all its capacitance in one place. Instead, it is distributed down the length, and between each capacitor is some loop inductance associated with the length of the sections. However, it is clear from this comparison that a transmission line, open at the far end, will look exactly like an ideal capacitor at low frequency.

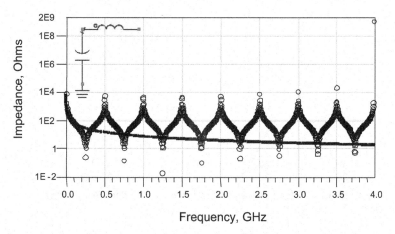

Figure 7-38 Simulated impedance of an ideal transmission line (circles) and simulated impedance of a single-section LC lumped-circuit model (line). The agreement is excellent up to bandwidth of about 100 MHz.

The impedance of the ideal transmission line shows the resonance peaks occurring when the frequency matches another half wavelength that can fit in the length of the transmission line. The peak resonant frequencies, f_{res}, are given by:

$$f_{res} = m \times \frac{f_0}{2} = m \times \frac{1}{2TD} \qquad (7\text{-}42)$$

where:

f_{res} = frequency for the peaks in the impedance

m = number of the peak, also the number of half waves that fit in the transmission line

TD = time delay of the transmission line

f_0 = frequency at which one complete wave fits in the transmission line

The first resonance for m = 1 is for 1×1 GHz/2 = 0.5 GHz. Here, just one half wave will fit evenly in the length of the transmission line, with a TD of 1 nsec. The second resonance for m = 2 is at 2×1 GHz/2 = 1 GHz. Here, exactly one wave will fit in the transmission line. These standing wave patterns are shown in Figure 7-39.

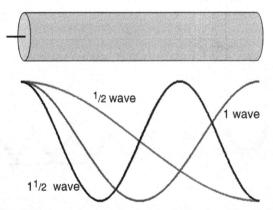

Figure 7-39 Voltage waves on the transmission line. Resonances occur when an additional half wavelength can fit in the line.

The bandwidth of the single-section LC model is about one-quarter the frequency of the first resonance, up to about 125 MHz. We can improve the bandwidth of the model by breaking the transmission line into more sections. If we break it up in two sections, each section can be modeled as an identical LC model, and the values of each L and C would be $L_{total}/2$ and $C_{total}/2$. The predicted impedance of this two-section LC model compared with the ideal T line is shown in Figure 7-40. The bandwidth of this model is roughly at half the first resonance peak. This is a frequency of about 250 MHz.

We can further increase the bandwidth of this lumped-circuit model by breaking up the length into more LC sections. Figure 7-41 shows the comparison with an ideal transmission line and using 16 different LC sections, with each L and C being $L_{total}/16$ and $C_{total}/16$. As we increase the number of sections, we are able to better approximate, to a higher bandwidth, the impedance behavior of an ideal transmission line. The bandwidth of this model is about up to the fourth resonance peak, at 2 GHz.

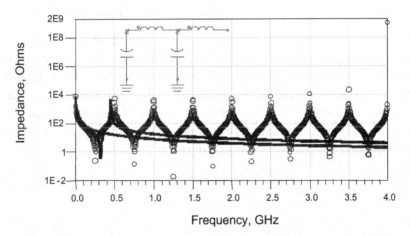

Figure 7-40 Simulated impedance of an ideal transmission line (circles) and simulated impedance of a one-section and two-section LC lumped-circuit model (lines).

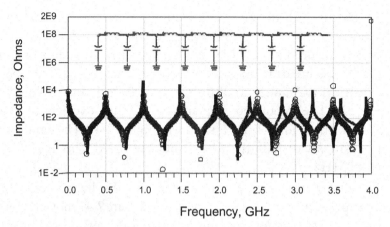

Figure 7-41 Simulated impedance of an ideal transmission line (circles) and simulated impedance of a 16-section LC lumped-circuit model (line).

TIP   An n-section LC circuit is an approximation of an ideal transmission line. The more sections, the higher the bandwidth of the approximation.

We can estimate the bandwidth of an n-section lumped-circuit model based on the time delay of this ideal transmission line. These last examples illustrate that the more segments in the LC model, the higher the bandwidth. One section has a bandwidth up to one-quarter of the first resonant peak; 2 sections up to half the first resonance; and 16 sections up to the second resonant frequency. We can generalize that the highest frequency at which we have good agreement—the bandwidth of the model—is:

$$BW_{model} = \frac{n}{4} \times \frac{f_0}{2} \sim n \times \frac{f_0}{10} \qquad (7\text{-}43)$$

or:

$$n = 10 \times \frac{BW_{model}}{f_0} = 10 \times BW_{model} \times TD \qquad (7\text{-}44)$$

where:

BW_{model} = bandwidth of the n-section lumped-circuit model

n = number of LC sections in the model

TD = time delay of the transmission line

f_0 = resonant frequency for one complete wavelength = 1/TD

We have approximated the relationship to be a little more conservative and a little easier to remember by using $n = 10 \times BW_{model} \times TD$ rather than $n = 8 \times BW_{model} \times TD$.

T I P   This is a very important rule of thumb to keep in mind. It says that to reach a bandwidth of the model equal to the 1/TD, 10 LC sections are required. It also says that because this frequency corresponds to having one wavelength in the transmission line, for a good approximation, there should be one LC section for every 1/10 wavelength of the signal.

For example, if the interconnect has a TD = 1 nsec, and we would like an n-section LC model with a bandwidth of 5 GHz, then we need at least $n = 10 \times$ 5 GHz × 1 nsec = 50 sections. At this highest frequency, there will be 5 GHz × 1 nsec = 5 wavelengths on the transmission line. For each wave, we need 10 sections; therefore, we need 5 × 10 = 50 LC sections for a good approximation.

If the TD of a line is 0.5 nsec, and we need a bandwidth of 2 GHz, the number of sections required is $n = 10 \times 2$ GHz × 0.5 nsec = 10 sections.

We can also evaluate the frequency to which we can use a single LC section to model a transmission line. In other words, up to what frequency does a transmission line look like a simple LC circuit? The bandwidth of one section is:

$$BW = n \times \frac{1}{10 \times TD} = 1 \times \frac{1}{10 \times TD} = 0.1 \times \frac{1}{TD}$$

(7-45)

For the case of a transmission line with TD = 1 nsec, the bandwidth of a single-section LC model for this line is 0.1 × 1/1 nsec = 100 MHz. If TD = 0.16 nsec (roughly 1 inch long), the bandwidth of a simple LC model for this line is 0.1 × 1/0.16 nsec = 600 MHz. The longer the time delay of a transmission line, the lower the frequency at which we can approximate it as a simple LC model.

In evaluating the number of sections we need to describe a transmission line for a required bandwidth, we have found that we needed about 10 LC sections per wavelength of the highest-frequency component of the signal and a total number of LC segments depending on the number of wavelengths of the highest frequency component of the signal that can fit in the transmission line.

If we have a signal with a rise time, RT, the bandwidth associated with the signal (the highest-sine-wave frequency component that is significant) is BW_{sig} = 0.35/RT. If we have a transmission line that has a time delay of TD, and we wish to approximate it with an n-section lumped-circuit model, we need to make sure the bandwidth of the model, BW_{model}, is at least > BW_{sig}:

$$BW_{model} > BW_{sig} \qquad (7\text{-}46)$$

$$n \times \frac{1}{10 \times TD} > \frac{0.35}{RT} \qquad (7\text{-}47)$$

$$n > 3.5 \frac{TD}{RT} \qquad (7\text{-}48)$$

where:
BW_{sig} = bandwidth of the signal
BW_{model} = bandwidth of the model
RT = rise time of the signal
TD = time delay of the transmission line
n = minimum number of LC sections needed for an accurate model

For example, with a rise time of 0.5 nsec and time delay of 1 nsec, we would need n > 3.5 × 1/0.5 = 7 sections for an accurate model.

When the rise time is equal to the TD of the line we want to model, we need at least 3.5 sections for an accurate model. In this case, the spatial extent of the rise time is the length of the transmission line. This suggests a very important rule of thumb, given in the following tip.

This rule of thumb is illustrated in Figure 7-42. In FR4, if the rise time is 1 nsec, the spatial extent of the leading edge is 6 inches. We need 3.5 LC sections for every 6 inches of length, or about 1.7 inches per section. We can generalize this: If the rise time is RT, and the speed of the signal is v, then the length for each LC section is $(RT \times v)/3.5$. In FR4, where the speed is about 6 inches/nsec, the length of each LC section required for a rise time, RT, is $1.7 \times RT$, with rise time in nsec.

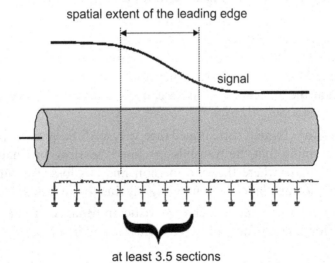

spatial extent of the leading edge

signal

at least 3.5 sections

Figure 7-42 As a general rule of thumb, there should be at least 3.5 LC sections per spatial extent of the rise time for an accurate model of the interconnect at the bandwidth of the signal.

If the rise time is 1 nsec, the length of each single LC should be less than 1.7 inches. If the rise time is 0.5 nsec, the length of each LC section should be no longer than $0.5 \times 1.7 = 0.85$ inches.

TIP   Of course, an ideal, distributed transmission-line model is always a good model for a uniform interconnect, at low frequency and at high frequency.

The analysis in this section evaluated the minimum number of many LC sections required to model a real transmission with adequate accuracy up to the bandwidth of the signal. But this is still an approximation with limited bandwidth. This is why the first choice when selecting a model for a real transmission line should always be an ideal transmission line, defined by a characteristic impedance and time delay. Only under very rare conditions, when the question is phrased in terms of the L or C values, should an n-section lumped model ever be used to model a real transmission line. *Always* start with an ideal transmission-line model.

7.19 Frequency Variation of the Characteristic Impedance

So far, we have been assuming that the characteristic impedance of a transmission line is constant with frequency. As we have seen, the input impedance, looking into the front of a transmission line, is strongly frequency dependent. After all, at low frequency, the input impedance of a transmission line open at the far end looks like a capacitor, and the impedance starts high and drops very low.

Does the characteristic impedance vary with frequency? In this section, we are assuming that the transmission line is lossless. In Chapter 9, we will look at the case where the transmission line has loss. We will see that the characteristic impedance does vary slightly due to the losses.

As we have seen, the characteristic impedance of an ideal lossless transmission line is related to the capacitance and inductance per length as:

$$Z_0 = \sqrt{\frac{L_L}{C_L}} \qquad\qquad (7\text{-}49)$$

Provided that the dielectric constant of the interconnect is constant with frequency, the capacitance per length will be constant. This is a reasonable assumption for most materials, though in some cases, the dielectric constant will vary slightly.

As we saw in Chapter 6, "The Physical Basis of Inductance," the loop inductance per length of a line will vary with frequency due to skin-depth effects. In fact, the loop inductance will start out higher at low frequency and decrease as all the currents distribute to the outer surface. This would suggest that characteristic impedance will start out higher at low frequency and decrease to a constant value at higher frequency.

At frequencies well above the skin depth, we would expect all the currents to be on the outer surface of all the conductors and not to vary with frequency beyond this point. The loop inductance should be constant, and the characteristic impedance should be constant. We can estimate this frequency for a 1-ounce copper conductor. The skin depth for copper is about 20 microns at 10 MHz. The thickness of 1-ounce copper is about 34 microns. We would expect to see the characteristic impedance start to decrease around 1 MHz to 10 MHz and stop decreasing at about 100 MHz, where the skin depth is only 6 microns.

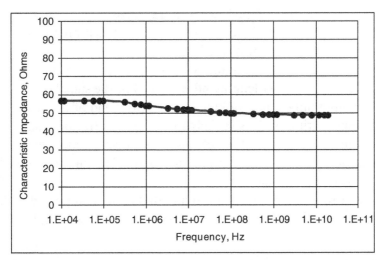

Figure 7-43 Calculated frequency variation of the characteristic impedance due to skin-depth effects, for a 50-Ohm line in FR4 with 1-ounce copper traces. Calculated with Ansoft's 2D Extractor.

We can calculate the frequency dependence of the characteristic impedance for a 50-Ohm microstrip with 1-ounce copper traces by using a 2D field solver. The result is shown in Figure 7-43. At low frequency, the characteristic impedance is high, it starts dropping at about 1 MHz, and it continues to drop until about 50 MHz. The total drop from DC to high frequency is about 7 Ohms, or less than 15%.

> **TIP**   Above about 50 MHz, the characteristic impedance of a transmission line is constant with frequency. This is the "high-frequency" characteristic impedance and is the value typically used for all evaluation of the behavior of high-speed signals.

7.20 The Bottom Line

1. A transmission line is a fundamentally new ideal-circuit element that accurately describes all the electrical properties of a uniform cross-sectional interconnect.
2. Forget the word *ground*. Think *return path*.
3. Signals propagate down a transmission line at the speed of light in the material surrounding the conductors. This primarily depends on the dielectric constant of the insulation.
4. The characteristic impedance of a transmission line describes the instantaneous impedance a signal would see as it propagates down the line. It is independent of the length of the line.
5. The characteristic impedance of a line primarily depends inversely on the capacitance per length and the speed of the signal.
6. The input impedance looking into the front end of a transmission line changes with time. It is initially the characteristic impedance of the line during the round-trip time of flight, but it can end up being anything, depending on the termination, the length of the line, and how long we measure the impedance.
7. A controlled-impedance board has all its traces fabricated with the same characteristic impedance. This is essential for good signal integrity.

8. A signal propagates through a transmission line as a current loop with the current going down the signal path and looping back through the return path. Anything that disturbs the return path will increase the impedance of the return path and create a ground-bounce voltage noise.

9. A real transmission line can be approximated with an n-section LC lumped-circuit model. The higher the bandwidth required, the more LC sections required. But it will always be an approximation with limited bandwidth.

10. For good accuracy, there should be at least 3.5 LC sections along the spatial extent of the leading edge.

11. An ideal transmission-line model is always a good starting model for a real interconnect, independent of the rise time and interconnect length. An ideal transmission-line model will always have the highest potential bandwidth of any model.

End-of-Chapter Review Questions

The answers to the following review questions can be found in Appendix D, "Review Questions and Answers."

7.1 What is a real transmission line?

7.2 How is an ideal transmission line model different from an ideal R, L, or C model?

7.3 What is ground? Why is it a confusing word for signal integrity applications?

7.4 What is the difference between chassis and earth ground?

7.5 What is the difference between the voltage on a line and the signal on a line?

7.6 What is a uniform transmission line, and why is this the preferred interconnect design?

7.7 How fast do electrons travel in a wire?

7.8 What is the difference between conduction current, polarization current, and displacement current?

7.9 What is a good rule of thumb for the speed of a signal on an interconnect?

7.10 What is a good rule of thumb for the aspect ratio of a 50-Ohm microstrip?

7.11 What is a good rule of thumb for the aspect ratio of a 50-Ohm stripline?

7.12 What is the effect called when the dielectric constant and the speed of a signal are frequency dependent?

7.13 What are two possible reasons the characteristic impedance of a transmission line would be frequency dependent?

7.14 What is the difference between the instantaneous impedance and the characteristic impedance and the input impedance of a transmission line?

7.15 What happens to the time delay of a transmission line if the length of the line increases by 3x?

7.16 What is the wiring delay of a transmission line in FR4?

7.17 If the line width of a transmission line increases, what happens to the instantaneous impedance?

7.18 If the length of a transmission line increases, what happens to the instantaneous impedance in the middle of the line?

7.19 Why is the characteristic impedance of a transmission line inversely proportional to the capacitance per length?

7.20 What is the capacitance per length of a 50-Ohm transmission line in FR4? What happens to this capacitance per length if the impedance doubles?

7.21 What is the inductance per length of a 50-Ohm transmission line in FR4? What if the impedance doubles?

7.22 What can you say about the capacitance per length of an RG59 cable compared to an RG58 cable?

7.23 What do we mean when we refer to "the impedance" of a transmission line?

7.24 A TDR can measure the input impedance of a transmission line in a fraction of a nanosecond. What would it measure for a 50-Ohm transmission line, open at the end, that is 2 nsec long? What would it measure after 5 seconds?

7.25 A driver has a 10-Ohm output resistance. If its open circuit output voltage is 1 V, what voltage is launched into a 65-Ohm transmission line?

7.26 What three design features could be engineered to reduce the impedance of the return path when a signal changes return path planes?

7.27 Which is a better starting model to use to describe an interconnect up to 100 MHz: an ideal transmission line, or a 2-section LC network?

7.28 An interconnect on a board is 18 inches long. What is an estimate of the time delay of this transmission line?

7.29 In a 50-Ohm microstrip, the line width is 5 mils. What is the approximate dielectric thickness?

7.30 In a 50-Ohm stripline, the line width is 5 mils. What is the length of the transmission line?

CHAPTER 8

Transmission Lines and Reflections

Reflections and distortions from impedance discontinuities can cause false triggering and bit errors. Reflections from impedance changes are the chief root cause of signal distortions and degradation of signal quality.

In some cases, this looks like ringing. The undershoot, when the signal level drops, can eat into the noise budget and contribute to false triggering. Or, on a falling signal, the peak can rise up above the low-bit-threshold and cause false triggering. One example of the reflection noise generated from impedance discontinuities at the ends of a short-length transmission line is shown in Figure 8-1.

Reflections occur whenever the instantaneous impedance the signal sees changes. This can be at the ends of lines or wherever the topology of the line changes, such as at corners, vias, branches, connectors, and packages. By understanding the origin of these reflections and arming ourselves with the tools to predict their magnitude, we can engineer a design with acceptable system performance.

TIP   For optimal signal quality, the goal in interconnect design is to keep the instantaneous impedance the signal sees as constant as possible.

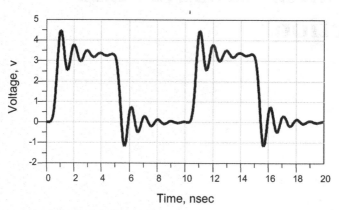

Figure 8-1   "Ringing" noise at the receiver end of a 1-inch-long controlled-impedance interconnect created because of impedance mismatches and multiple reflections at the ends of the line.

First, this means keeping the instantaneous impedance of the line constant—hence the growing importance in manufacturing controlled-impedance boards. All the various design guidelines, such as minimizing stub lengths, using daisy chains rather than branches, and using point-to-point topology, are methods to keep the instantaneous impedance constant.

Second, this means managing the impedance changes at the ends of the line with a termination strategy. No matter how efficient we may be at constructing a uniform transmission line, the impedance will always vary at the ends of the line. The rattling around from the reflections at the ends will result in ringing noise unless it is controlled. That's what a termination strategy handles.

Third, even with controlled impedance interconnects, well terminated, the specific routing topology can influence the reflections. When a signal line splits into two branches, there is an impedance discontinuity. Maintaining a linear routing topology with no branches or stubs is an important strategy to minimize impedance changes and reflection noise.

8.1 Reflections at Impedance Changes

As a signal propagates down a transmission line, it sees an instantaneous impedance for each step along the way. If the interconnect is a controlled impedance, then the instantaneous impedance will be constant and equal to the characteristic impedance of the line. If the instantaneous impedance ever changes, for whatever reason, some of the signal will reflect back in the opposite direction, and some of it will continue with a different amplitude. We call all locations where the instantaneous impedance changes *impedance discontinuities* or just *discontinuities*.

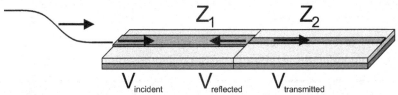

Figure 8-2   Whenever a signal sees a change in the instantaneous impedance, there will be some reflected signal, and the transmitted signal will be distorted.

The amount of signal that reflects depends on the magnitude of the change in the instantaneous impedance. This is illustrated in Figure 8-2. If the instantaneous impedance in the first region is Z_1 and the instantaneous impedance in the second region is Z_2, the magnitude of the reflected signal compared to the incident signal will be given by:

$$\frac{V_{reflected}}{V_{incident}} = \frac{Z_2 - Z_1}{Z_2 + Z_1} = \rho \qquad (8\text{-}1)$$

where:

$V_{reflected}$ = reflected voltage

$V_{incident}$ = incident voltage

Z_1 = instantaneous impedance of the region where the signal is initially

Z_2 = instantaneous impedance of the section where the signal just enters

ρ = reflection coefficient, in units of rho

The greater the difference in the impedances in the two regions, the greater the amount of reflected signal. For example, if a 1-v signal is moving on a 50-Ohm characteristic-impedance transmission line, it will see an instantaneous impedance of 50 Ohms. If it hits a region where the instantaneous impedance changes to 75 Ohms, the reflection coefficient will be $(75 - 50)/(75 + 50) = 20\%$, and the amount of reflected voltage will be $20\% \times 1$ v = 0.2 v.

For every part of the waveform that hits the interface, exactly 20% of it will reflect back. This is true no matter the shape of the waveform. In the time domain, it can be a sharp edge, a sloping edge, or even a Gaussian edge. Likewise, in the frequency domain, where all waveforms are sine waves, each sine wave will reflect, and the amplitude and phase of the reflected wave can be calculated from this relationship.

It is often the reflection coefficient, ρ (or rho), that is of interest. The reflection coefficient is the ratio of the reflected voltage to the incident voltage.

TIP   The most important thing to remember about the reflection coefficient is that it is equal to the ratio of the second impedance minus the first, divided by their sum. This distinction is particularly important in determining the sign of the reflection coefficient.

When considering signals on interconnects, keeping track of their direction of travel on the interconnect is critically important. If a signal is traveling down a transmission line and hits a discontinuity, a second wave will be generated at the discontinuity. This second wave will be superimposed on the first wave but will be traveling back toward the source. The amplitude of this second wave will be the incident voltage times rho.

8.2 Why Are There Reflections?

The reflection coefficient describes the fraction of the voltage that reflects back to the source. In addition, there is a transmission coefficient that describes the fraction of the incident voltage that is transmitted through the interface into the second region. This property of signals to reflect whenever the instantaneous impedance changes is the source of all signal-quality problems related to signal propagation on one net.

To minimize signal-integrity problems that arise because of this fundamental property, we must implement the following four important design features for all high-speed circuit boards:

1. Use controlled impedance interconnects.
2. Provide at least one termination at the ends of a transmission line.
3. Use a routing topology that minimizes the impact from multiple branches.
4. Minimize any geometry discontinuities.

But what causes the reflection? Why does a signal reflect when it encounters a change in the instantaneous impedance? The reflected signal is created to match two important boundary conditions.

Consider the interface between two regions, labeled as region 1 and region 2, each with a different instantaneous impedance. As the signal hits the interface, we must see only one voltage between the signal- and return-path conductors and one current loop flowing between the signal- and return-path conductors. Whether we look from the region 1 side or change our perspective and look from the region 2 side, we must see the same voltage and the same current on either side of the interface. We must not have a voltage discontinuity across the boundary because if we did, we would have an infinitely large electric field in the boundary. We must not have a current discontinuity, as this would mean we are building up a net charge at the interface.

A voltage difference across the infinitesimally short distance of the boundary would result in an infinitely large electric field. This could blow up the universe. A net current into the boundary would mean a charge buildup from nothing. If we waited long enough, we could have so much charge building up that the universe could explode. The reflected voltage is generated to prevent the destruction of the universe.

Without the creation of a reflected voltage heading back to the source and while maintaining the same voltage and current across the interface, we would have the condition of $V_1 = V_2$ and $I_1 = I_2$. But, $I_1 = V_1/Z_1$ and $I_2 = V_2/Z_2$. If the impedances of the two regions are not the same, there is no way all four conditions can be met.

To keep harmony in the universe—literally, to keep the universe from exploding—a new voltage is created in the first region that reflects back to its source. Its sole purpose is to take up the mismatched current and voltage between the incident and transmitted signals. Figure 8-3 illustrates what happens at the interface.

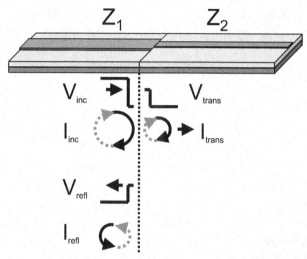

Figure 8-3 As the incident signal tries to pass through the interface, a reflected voltage and current are created to match the voltage and current loops on both sides of the interface.

The incident voltage, V_{inc}, moves toward the interface, while the transmitted voltage, V_{trans}, moves away from the interface. A new voltage is created as the incident signal tries to pass through the interface. This new wave is traveling only in region 1, back to the source. At any point in region 1, the total voltage between the signal and return conductors is the sum of the voltages traveling in the two directions: the incident signal plus the reflected signal.

The condition of the same voltages on both sides of the interface requires:

$$V_{inc} + V_{refl} = V_{trans} \tag{8-2}$$

The condition on the currents is a little more subtle. The total current at the interface, in region 1, is due to two current loops, which travel in opposite directions and circulate in opposite directions. At the interface, the direction of circulation of the incident current loop is clockwise. The direction of circulation of the reflected current loop is counterclockwise. If we define the positive direction of circulation as clockwise, then the net current at the interface in region 1 is $I_{inc} - I_{refl}$. In region 2, the current loop circulation is clockwise and is just I_{trans}. The condition of the same current viewed from either side of the interface is:

$$I_{inc} - I_{refl} = I_{trans} \tag{8-3}$$

The final condition is that the ratio of voltage to the current in each region is the impedance of each region:

$$\frac{V_{inc}}{I_{inc}} = Z_1 \tag{8-4}$$

$$\frac{V_{refl}}{I_{refl}} = Z_1 \tag{8-5}$$

$$\frac{V_{trans}}{I_{trans}} = Z_2 \tag{8-6}$$

Using these last relationships, we can rewrite the condition of the current as:

$$\frac{V_{inc}}{Z_1} - \frac{V_{refl}}{Z_1} = \frac{V_{trans}}{Z_2} \tag{8-7}$$

With a little bit of algebra, we get:

$$\frac{V_{inc}}{Z_1} - \frac{V_{refl}}{Z_1} = \frac{V_{inc} + V_{refl}}{Z_2} \tag{8-8}$$

and:

$$V_{inc}\left(\frac{Z_2 - Z_1}{Z_2 Z_1}\right) = V_{refl}\left(\frac{Z_2 + Z_1}{Z_2 Z_1}\right) \tag{8-9}$$

and finally:

$$\frac{V_{refl}}{V_{inc}} = \frac{Z_2 - Z_1}{Z_2 + Z_1} = \rho \qquad (8\text{-}10)$$

which is the definition of the reflection coefficient. Using the same approach, we can derive the transmission coefficient as:

$$t = \frac{V_{trans}}{V_{inc}} = \frac{2 \times Z_2}{Z_2 + Z_1} \qquad (8\text{-}11)$$

Dynamically, what actually creates the reflected voltage? No one knows. We only know that if it is created, we are able to match the same voltage on one side of the interface with that on the other side. The voltage is continuous across the interface. Likewise, the current loop is exactly the same on both sides of the interface. The current is continuous across the interface. The universe is in balance.

8.3 Reflections from Resistive Loads

There are three important special cases to consider for transmission line terminations. In each case, the transmission line characteristic impedance is 50 Ohms. The signal will be traveling in this transmission line from the source and hit the far end with a particular termination impedance.

> **TIP**   It is important to keep in mind that in the time domain, the signal is sensitive to the instantaneous impedance. It is not necessary that the second region be a transmission line. It might also be a discrete device that has some impedance associated with it, such as a resistor, a capacitor, an inductor, or some combination thereof.

When the second impedance is an open, as is the case when the signal hits the end of a transmission line with no termination, the instantaneous impedance at the end is infinite. The reflection coefficient is (infinite − 50)/(infinite + 50) = 1. This means a second wave of equal size to the incident wave will be generated at the open but will be traveling in the opposite direction— back to the source.

If we look at the total voltage appearing at the far end, where the open is, we will see the superposition of two waves. The incident wave, with an amplitude of 1 v, will be traveling toward the open end. The other signal, the reflected wave, also with an amplitude of 1 v, will be traveling in the opposite direction. When we measure the voltage at the far end, we measure the sum of these two propagating voltages, or 2 v. This is illustrated in Figure 8-4.

TIP   It is often said that when a signal hits the end of a transmission line, it doubles. While this is technically true, it is not what is really happening. The total voltage, the sum of the two propagating signals, is twice the incident voltage. However, if we think of it as a doubling, we mis-calibrate our intuition. It is better to think of the voltage at the far end as the sum of the incident and the reflected signals.

The second special case is when the far end of the line is shorted to the return path. The impedance at the end is 0 in this case. The reflection coefficient is $(0 - 50)/(0 + 50) = -1$. When a 1-v signal is incident to the far end, a -1-v signal is generated by the reflection. This second wave propagates back through the transmission line to the source.

The voltage that would be measured at the shorting discontinuity is the sum of the incident signal and the reflected signal, or $1 \text{ v} + -1 \text{ v} = 0$. This is reasonable because if we really have a short at the far end, by definition, we can't have a voltage across a short. We see now that the reason it is 0 v is that it is the sum of two waveforms: a positive one traveling in the direction from the source and a negative one traveling back toward it.

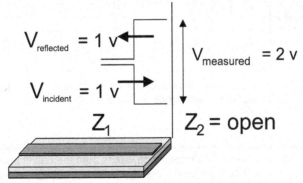

Figure 8-4   If the second impedance is an open, the reflection coefficient is 1. At the open, there will be two oppositely traveling waves superimposed.

The third important impedance at the far end to consider is when the impedance matches the characteristic impedance of the transmission line. In this example, it could be created by adding a 50-Ohm resistor to the end. The reflection coefficient would be (50 − 50)/(50 + 50) = 0. There would be no reflected voltage from the end. The voltage appearing across the 50-Ohm termination resistor would just be the incident-voltage wave.

If the instantaneous impedance the signal sees does not change, there will be no reflection. By placing the 50-Ohm resistor at the far end, we have matched the termination impedance to the characteristic impedance of the line and reduced the reflection to zero.

For any resistive load at the far end, the instantaneous impedance the signal will see will be between 0 and infinity. Likewise, the reflection coefficient will be between −1 and +1. Figure 8-5 shows the relationship between terminating resistance and reflection coefficient for a 50-Ohm transmission line.

When the second impedance is less than the first impedance, the reflection coefficient is negative. The reflected voltage from the termination will be a negative voltage. This negative-voltage wave propagates back to the source.

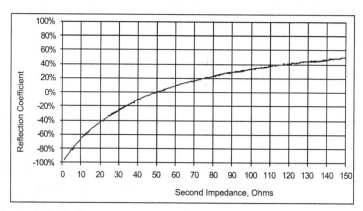

Figure 8-5 Reflection coefficient for the case of the first impedance being 50 Ohms and a variable impedance for the second region.

TIP   When the second impedance is less than the first impedance, the reflection coefficient is negative, and the reflected voltage subtracts from the incident voltage. This means the measured voltage appearing across the resistor will always be less than the incident voltage.

For example, if the transmission line characteristic impedance is 50 Ohms and the termination is 25 Ohms, the reflection coefficient is (25 − 50)/ (25 + 50) = −1/3. If 1 v were incident to the termination, a −0.33 v will reflect back to the source. The actual voltage that would appear at the termination is the sum of these two waves, or 1 v + −0.33 v = 0. 67 v.

Figure 8-6 shows the measured voltage that would appear across the termination for a 1-v incident voltage and a 50-Ohm transmission line. As the termination impedance increases from 0 Ohms, the actual voltage measured across the termination increases from 0 v up to 2 v when the termination is open.

8.4 Source Impedance

When a signal is launched into a transmission line, there is always some impedance of the source. For typical CMOS devices, this can be about 5 Ohms to 20 Ohms. For older-generation transistor-transistor logic (TTL) gates, this can be as high as 100 Ohms. The source impedance will have a dramatic impact on both the initial voltage launched into the transmission line and the multiple reflections.

When the reflected wave finally reaches the source, it will see the output-source resistance as the instantaneous impedance right at the driver. The value of this output-source impedance will determine how the reflected wave reflects again from the driver.

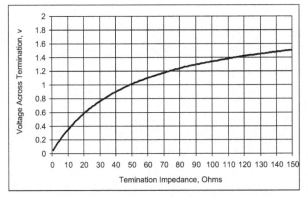

Figure 8-6 Voltage across the termination, for a 1-v incident signal. This voltage is the sum of the incident going and reflected wave.

If we have a model for the driver, either SPICE or IBIS based, a good estimate of the output impedance of the driver can be extracted with a few simple simulations. We assume that an equivalent circuit model for the driver is an ideal voltage source in series with a source resistor, illustrated in Figure 8-7. We can extract the output voltage of the ideal source when driving a high-output impedance. If we connect a low impedance like 10 Ohms to the output and measure the output voltage across this terminating resistor, we can back out the internal-source resistance from:

$$R_s = R_t \left(\frac{V_o}{V_t} - 1 \right) \tag{8-12}$$

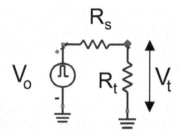

Figure 8-7   Simple, implicit model of an output driver with a terminating resistor connected.

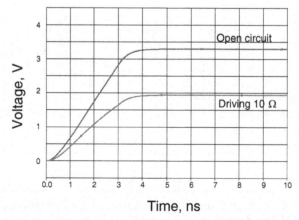

Figure 8-8   Simulated output voltage from a CMOS driver with a 10-kOhm resistor and then a 10-Ohm resistor connected. From the two voltage levels, the source impedance of the driver can be calculated. Simulation of a typical CMOS IBIS-driver model using the HyperLynx simulator.

where:

R_s = source resistance of the driver

R_t = terminating resistance connected to the output

V_o = open-circuit output voltage from the driver

V_t = voltage across the terminating resistor

To calculate the source resistance, we simulate the output voltage from the driver in two cases: when a very high resistance is attached (e.g., 10 kOhms) and when a low impedance is attached (e.g., 10 Ohms). An example of the simulated voltage using a behavioral model for a common CMOS driver is shown in Figure 8-8. The open-circuit voltage is 3.3 v, and the voltage with the 10-Ohm resistor attached is 1.9 v. From the equation above, the output-source impedance can be calculated as 10 Ohms × (3.3 v/1.9 v − 1) = 7.3 Ohms.

An alternative approach is to vary the load resistance until the loaded output voltage is exactly one-half the unloaded output voltage. The output source impedance is then equal to the load resistance.

8.5 Bounce Diagrams

As shown in Chapter 7, "The Physical Basis of Transmission Lines," the actual voltage launched into the transmission line, or the initial voltage propagating on the transmission line, is determined by the combination of the source voltage and the voltage divider made up of the source impedance and the transmission line input impedance.

Knowing the time delay of the transmission line, TD, and the impedance of each region where the signal will propagate, and knowing the initial voltage from the driver, we can calculate all the reflections at all the interfaces and predict the voltages that would be measured at any point in time.

For example, if the source voltage, driving an open termination, were 1 v and the source impedance were 10 Ohms, the actual voltage launched into a 1-nsec-long, 50-Ohm transmission line would be 1 v × 50/(10 + 50) = 0.84 v. This 0.84-v signal would be the initial voltage propagating down the transmission line.

Suppose the end of the transmission line is an open termination. The 0.84-v signal will hit the end of the line 1 nsec later, and a +0.84-v signal will be generated by the reflection, traveling back to the source. At the end of the line, the total voltage measured across the open will be the sum of the two waves, or 0.84 v + 0.84 v = 1.68 v.

When the 0.84-v reflected wave hits the source end another 1 nsec later, it will see an impedance discontinuity. The reflection coefficient at the source is (10 − 50)/(10 + 50) = −0.67. With a 0.84-v signal incident on the driver, a total of 0.84 v × −0.67 = −0.56 v will reflect back to the end of the line. Of course, this new wave will reflect again from the far end. A step voltage change of −0.56 v will reflect back. Measured at the far end, across the open, will be four simultaneous waves: 2 × 0.84 v or 1.68 v from the first wave, and 2 × −0.56 v or −1.12 v from the second reflection, for a total voltage of 0.56 v.

The backward-propagating −0.56-v wave will hit the source impedance and reflect yet again. The reflected voltage will be +0.37 v. At the far end, there will be the 0.56 v from the first two waves plus the new set of incident and reflected 0.37-v waves, for a total of 0.56 v + 0.37 v + 0.37 v = 1.3 v. Keeping track of these multiple reflections is straightforward but tedious. Before the days of simple, easy-to-use simulation tools, these reflections were diagrammed using *bounce*, or *lattice*, diagrams. An example is shown in Figure 8-9.

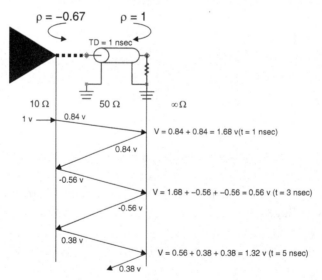

Figure 8-9   Lattice, or bounce, diagram used to keep track of all the multiple reflections and the time-varying voltages at the receiver on the far end.

When the source impedance is less than the characteristic impedance of the transmission line, we see that the reflection at the source end will be negative. This will contribute to the effect we normally call ringing.

Figure 8-10 shows the voltage waveform at the far end of the transmission line for the previous example, when the rise time of the signal is very short compared to the time delay of the transmission line. This analysis was done using a SPICE simulator to predict the waveforms at the far end, taking into account all the multiple reflections and impedance discontinuities.

Two important features should be apparent. First, the voltage at the far end eventually approaches the source voltage, or 1 v. This must be the case because we have an open, and we must ultimately see the source voltage across an open.

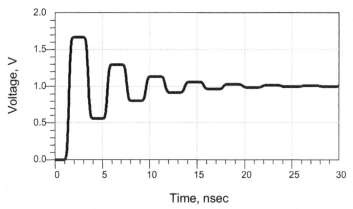

Figure 8-10   Simulated voltage at the far end of the transmission line described with the previous lattice diagram. Simulation performed with SPICE.

The second important effect is that the actual voltage across the open exceeds the voltage coming out of the source. The source has only 1 v, yet we would measure as much as 1.68 v at the far end. How was this higher voltage generated? The higher voltage is a feature of the resonance of the transmission line structure. Remember that there is no such thing as conservation of voltage, just conservation of energy.

8.6 Simulating Reflected Waveforms

Using the definition of the reflection coefficient above, the reflected signal from any arbitrary impedance can be calculated. When the terminating impedance is a resistive element, the impedance is constant, and the reflected voltages are easy to calculate. When the termination has a more complicated impedance behavior (such as a capacitive or an inductive termination, or some combination of the two), calculating the reflection coefficient and how it changes as the incident waveform changes is difficult and tedious if done by hand. Luckily, there are simple, easy-to-use circuit-simulation tools that make this calculation much easier.

The reflection coefficients and the resulting reflected waveforms from arbitrary impedances and for arbitrary waveforms can be calculated using SPICE or other circuit simulators. With such tools, sources are created, ideal transmission lines are added, and terminations are connected. The voltage appearing across the termination and any other nodes can be calculated as the incident waveform hits the end and reflects from all the various discontinuities.

There are many possible combinations of source impedance, transmission line characteristic impedance, time delay, and end termination. Each of these can be easily varied using a simulation tool. Figure 8-11 shows the simulated voltage across the termination as the rise time of the signal is increased from a 0.1 nsec to 1 nsec and, in a separate simulation, the signal waveform as the source terminating resistance is varied from 0 Ohms to 90 Ohms.

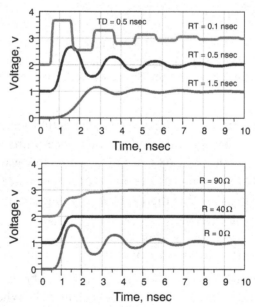

Figure 8-11   Examples of the variety of simulations possible with SPICE. Top: For a 10-Ohm driver and 50-Ohm characteristic-impedance line, showing the far-end voltage with different rise times for the signal. Bottom: Changing the series-source terminating resistor and displaying the voltage at the far end.

TIP   Using a SPICE or other circuit simulator, the performance of any arbitrary transmission-line circuit can be simulated, taking into account all specific features.

8.7 Measuring Reflections with a TDR

In addition to simulating the waveforms associated with transmission-line circuits, it is possible to measure the reflected waveforms from physical interconnects using a specialized instrument usually referred to as a *time-domain reflectometer* (*TDR*). This is the appropriate instrument to use when characterizing a passive interconnect that does not have its own voltage source. Of course, when measuring the actual voltages in an active circuit, a fast oscilloscope with a high-impedance, active probe is the best tool.

A TDR will generate a short rise-time step edge, typically between 35 psec and 150 psec, and measure the voltage at an internal point of the instrument. Figure 8-12 is a schematic of the workings inside a TDR. It is important to keep in mind that a TDR is nothing more than a fast step generator and a very fast sampling scope.

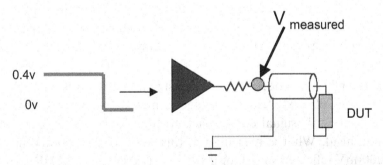

Figure 8-12   Schematic of the inside of a TDR. A very fast pulse generator creates a fast rising-voltage pulse. It travels through a precision 50-Ohm resistor in series with a short length of 50-Ohm coax cable to the front panel, where the DUT is connected. The total voltage at an internal point is measured with a very fast sampling oscilloscope and displayed on the front screen.

The voltage source is a very fast step generator, which outputs a step amplitude of about 400 mV. Right after the voltage source is a calibrated resistor of 50 Ohms. This assures that the source impedance of the TDR is a precision 50 Ohms. After the resistor is the actual detection point where the voltage is measured by a fast sampling amplifier. Connected to this point is a short-length coax cable that brings the signal to the front-panel SMA connector. This is where the DUT is connected. The signal from the source enters the DUT, and any reflected voltage is detected at the sampling point.

Before the step signal is generated, the voltage measured at the internal point will be 0 v. Where the voltage is actually measured, the signal encounters a voltage divider. The first resistor is the internal calibration resistor. The second resistor is the transmission line internal to the TDR. As the 400-mV step reaches the calibration resistor, the actual voltage measured at the detection point will be the result after going through the voltage divider.

The voltage detected is 400 mV × 50 Ohms/(50 Ohms + 50 Ohms) = 200 mV. This voltage is measured initially and is displayed by the fast sampling scope. The 200-mV signal continues moving down the internal coax cable to the DUT.

If the DUT is a 50-Ohm termination, there is no reflected signal, and the only voltage present at the sampling point is the forward-traveling wave of 200 mV, which is constant. If the DUT is an open, the reflected signal from the DUT is +200 mV. A short time after launch, this 200-mV reflected-wave signal comes back to the sampling point, and what is measured and displayed is the 200-mV incident voltage plus the 200-mV reflected wave. The total voltage displayed is 400 mV.

If the DUT is a short, the reflected signal from the DUT will be –200 mV. Initially, the 200-mV incident voltage is measured. After a short time, the reflected –200-mV signal comes back to the source and is measured by the sampling head. What is measured at this point is the 200-mV incident plus the –200-mV reflected signal, or 0 voltage. The measured-TDR plots of these cases are shown in Figure 8-13.

TIP   The TDR will measure the reflected voltage from any interconnect attached to the front SMA connector of the instrument and how this voltage changes with time as the signal propagates down the interconnect, reflecting from all discontinuities.

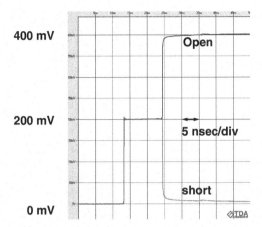

400 mV

200 mV

0 mV

Figure 8-13   Measured-TDR response when the DUT is an open and a short. Data measured with an Agilent 86100 DCA and displayed with TDA Systems IConnect software.

As the transmitted signal continues down the DUT, if there are other regions where the instantaneous impedance changes, a new reflected voltage will be generated and will travel back to the internal measurement point to be displayed. In this sense, the TDR is really indicating changes in the instantaneous impedance the signal encounters and when the signal encounters it.

Since the incident signal must travel down the interconnect and the reflected signal must travel back down the interconnect to the detection point, the time delay measured on the front screen is really a round-trip delay to any discontinuity.

For example, if a uniform 4-inch-long, 50-Ohm transmission line is the DUT, we will see an initial small reflected voltage at the entrance to the DUT because it is not exactly 50 Ohms, and then we will see a larger reflected signal when the incident signal reaches the open at the far end and reflects back to the detection point. This time delay is the round-trip time delay of the transmission line. If the transmission line is not 50 Ohms, multiple reflections will take place at both ends of the line. The TDR will display the superposition of all the voltage waves that make it back to the internal measurement point. An example of the TDR response from a 50-Ohm transmission line and a 15-Ohm transmission line, both open at the far end, is shown in Figure 8-14.

The combination of understanding the principles and leveraging simulation and measurement tools will allow us to evaluate many of the important impedance discontinuities a signal might encounter. We will see that many of them are important and must be carefully engineered or avoided, while some of them are not important and can be ignored in some cases.

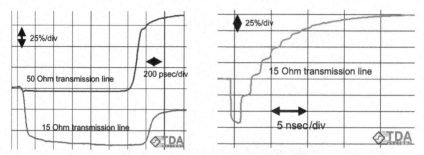

Figure 8-14 Measured TDR response of 4-inch-long transmission lines open at the far end: 50 Ohms and 15 Ohms. Left: Time base of 200 psec/div. Right: Reflections from the 15-Ohm line on expanded time base, 5 nsec/div. Measured with Agilent 86100 DCA and a GigaTest Labs Probe Station and displayed with TDA Systems IConnect.

TIP   It is only by applying engineering discipline and "putting in the numbers" that the important effects can be identified and managed while the unimportant ones are identified and ignored.

8.8 Transmission Lines and Unintentional Discontinuities

Wherever a signal sees an impedance change, there will be a reflection. Reflections can have a serious impact on signal quality. Predicting the impact on the signals from the discontinuities and engineering acceptable design alternatives is an important part of signal-integrity engineering.

Even if a circuit board is designed with controlled-impedance inter-connects, there is still the opportunity for a signal to see an impedance discontinuity from such features as:

- The ends of the line
- A package lead
- An input-gate capacitance
- A via between signal layers
- A corner
- A stub

- A branch
- A test pad
- A gap in the return path
- A neck down in a via field
- A crossover

When we model these effects, there are three common equivalent-circuit models we can use to electrically describe the unintentional discontinuity: an ideal capacitor, an ideal inductor, or a short-length ideal transmission line (either in series or in shunt). These possible equivalent circuit models are shown in Figure 8-15. These circuit elements can occur at the ends of the line or in the middle.

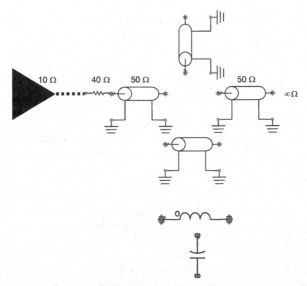

Figure 8-15 Transmission-line circuit used to illustrate the specific impedance from the three types of discontinuities: short transmission line in series and shunt, shunt capacitive, and series inductive.

The two most important parameters that influence the distortion of the signal from the discontinuity are the rise time of the signal and the size of the discontinuity. For an inductor and capacitor, their instantaneous impedance depends on the instantaneous rate at which either the current is changing or the voltage is changing and the value of the C or L.

As the signal passes across the element, the slope of the current and the voltage will change with time, and the impedance of the element will change in time. This means the reflection coefficient will change with time and with the specific features of the rise or fall time. The peak reflected voltage will scale with the rise time of the signal.

In general, the impact of a discontinuity is further complicated by the impedance of the driver and the characteristic impedance of the initial transmission line influencing the multiple bounces.

> **TIP**   These factors as well as the impact from the discontinuity itself can only be fully taken into account by converting the physical structure that creates the discontinuity into its equivalent electrical-circuit model and performing a simulation. Rule-of-thumb estimates can only provide engineering insight and offer rough guidelines for when a problem might arise.

Any impedance discontinuity will cause some reflection and distortion of the signal. It is not impossible to design an interconnect with absolutely no reflections. How much noise can we live with, and how much noise is too much? This depends very strongly on the noise budget and how much noise voltage has been allocated to each source of noise.

> **TIP**   Unless otherwise specified, as a rough rule of thumb, the reflection noise level should be kept to less than 10% of the voltage swing. For a 3.3-v signal, this is 330 mV of noise. Some noise budgets might be more conservative and allocate no more than 5% to reflection noise. The tighter the noise budget, typically, the more expensive the solution. Often, the noise allocated to one source may be tightened up because the fix to correct it is less expensive to implement, while another might be loosened because it is more expensive to fix. As a rough rule of thumb, we should definitely worry about the reflection noise if it approaches or exceeds 10% of the signal swing. In some designs, less than 5% may be too much.

By evaluating a few simple cases, we can see what physical factors influence the signal distortions and how to engineer them out of the design before they become problems. Ultimately, the final evaluation of whether a design is acceptable or not must come from a simulation. This is why it is so important that every practicing engineer with a concern about signal integrity have easy access to a simulator to be able to evaluate specific cases.

8.9 When to Terminate

The simplest transmission-line circuit has a driver at one end, a short length of controlled-impedance line, and a receiver at the far end. As we saw previously, the signal will bounce around between the high-impedance open at the far end and the low impedance of the driver at the near end. When the line is long, these multiple bounces will cause signal-quality problems, which we lump in the general category of ringing. But if the line is short enough, though the reflections will still happen, they may be smeared out with the rising or falling edge and may not pose a problem. Figure 8-16 illustrates how the received waveform changes when the time delay increases from 20% of the rise time to 30% of the rise time and then 40% of the rise time.

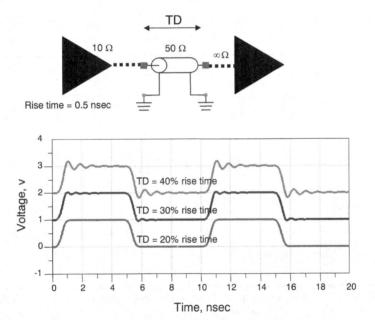

Figure 8-16 100-MHz clock signal at the far end of an unterminated transmission line as its length is changed from 20% of the rise time to 30% and 40% of the rise time. When the time delay of the transmission line is greater than 20% of the rise time, ringing noise may cause a problem.

When the TD of the interconnect is 0.1 nsec long, all the reflections take place, but they rattle around back and forth every 0.2 nsec, the round-trip time of flight. If this is short compared with the rise time, the multiple bounces will be smeared over the rise time and will be barely discernible, not posing a potential problem. In Figure 8-16, as a rough estimate, it appears that when the TD is less than 20%, the reflections are virtually invisible, but if the TD is greater than 20% of the rise time, the ringing begins to play a significant role.

> **TIP**   We use, as a rough rule of thumb, the threshold of TD > 20% of the rise time as the boundary of when to start worrying about ringing noise due to an unterminated line. If the TD of the transmission line is greater than 20% of the rise time, ringing will play a role and must be managed; otherwise, it will potentially cause a signal-integrity problem. If the TD < 20% of the rise time, the ringing noise may not be a problem, and the line may not require termination.

For example, if the rise time is 1 nsec, the maximum TD for a transmission line that might be used unterminated is ~ 20% × 1 nsec = 0.2 nsec. In FR4, the speed of a signal is about 6 inches/nsec, so the maximum physical length of an unterminated line is roughly 6 inches/nsec × 0.2 nsec = 1.2 inches.

This allows us to generalize a very useful rule of thumb that the maximum length for an unterminated line before signal-integrity problems arise is roughly:

$$\text{Len}_{max}\,[\text{inches}] < \text{RT}\,[\text{n sec}] \qquad (8\text{-}13)$$

where:

Len_{max} = maximum length for an unterminated line, in inches
RT = rise time, in nsec

> **TIP**   This is a very useful and easy-to-remember result: As a rough rule of thumb, the maximum length of an unterminated line (in inches) is the rise time (in nsec).

If the rise time is 1 nsec, the maximum unterminated length is about 1 inch. If the rise time is 0.1 nsec, the maximum unterminated length is 0.1 inch. As we will see, this is the most important general rule of thumb to identify when ringing noise will play a significant role. This is also why signal integrity is becoming a significant problem in recent years and might have been avoided in older-generation technologies.

When clock frequencies were 10 MHz, the clock periods were 100 nsec, and the rise times were about 10 nsec. The maximum unterminated line would be 10 inches. This is longer than virtually all traces on a typical motherboard. Back in the days of 10-MHz clocks, though the interconnects always behaved like transmission lines, the reflection noise never caused a problem, and the interconnects were "transparent" to the signals. We never had to worry about impedance matching, terminations, or transmission line effects.

However, the form factor of products is staying the same, and lengths of interconnects are staying fixed, but rise times are decreasing. Therefore, it is inevitable that we will reach a high enough clock frequency with a short enough rise time that virtually *all* the interconnects on a board will be longer than the maximum possible unterminated length, and termination will be important.

These days, with rise times of signals as short as 0.1 nsec, the maximum unterminated length of a transmission line before ringing noise becomes important is about 0.1 inch. Virtually 100% of all interconnects are longer than this. A termination strategy is a must in all of today's and future-generations' products.

8.10 The Most Common Termination Strategy for Point-to-Point Topology

We have identified the origin of the ringing as the impedance discontinuities at the source and the far end and the multiple reflections back and forth. If we eliminate the reflections from at least one end, we can minimize the ringing.

> **TIP** Engineering the impedance at one or both ends of a transmission line to minimize reflections is called *terminating the line*. Typically, one or more resistors are added at strategic locations.

When one driver drives one receiver, we call this a *point-to-point topology*. Figure 8-17 illustrates four techniques to terminate a point-to-point topology. The most common method is to use a resistor in series at the driver. This is called *source-series termination*. The sum of the terminating resistor and the source impedance of the driver should add up to the characteristic impedance of the line.

If the driver-source impedance is 10 Ohms and the characteristic impedance of the line is 50 Ohms, then the terminating resistor should be about 40 Ohms. With this terminating resistor in place, the 1-v signal coming out of the driver will now encounter a voltage divider composed of the 50-Ohm total resistance and the 50-Ohm transmission line. In this case, 0.5 v will be launched into the transmission line.

At first glance, it might seem that half the source voltage will not be enough to affect any triggering. However, when this 0.5-v signal hits the open, far end of the transmission line, it will again see an impedance discontinuity. The reflection coefficient at the open is 1, and the 0.5-v incident signal will reflect back to the source with an amplitude of 0.5 v. At the far end, the total voltage across the open termination will be the 0.5-v incident voltage and the 0.5-v reflected voltage, or a total of 1 v.

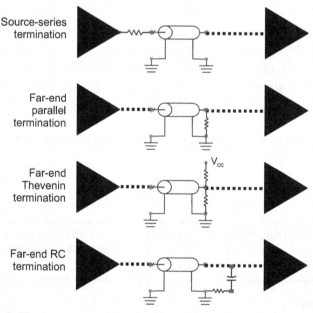

Figure 8-17   Four common termination schemes for point-to-point topologies. The top one, source-series termination, is the most commonly used approach.

The 0.5-v reflected signal travels back to the source. When it reaches the series terminating resistance, the impedance it sees looking into the source is the 40 Ohms of the added series resistor plus the 10 Ohms of the source, or 50 Ohms. It is already in a 50-Ohm transmission line; therefore, the signal will encounter no impedance change, and there will be no reflection. The signal will merely be absorbed by the terminating resistor and the source resistor.

At the far end, all that is seen is a 1-v signal and no ringing. Figure 8-18 shows the waveform at the far end for the case of no terminating resistor and the 40-Ohm source-series terminating resistor.

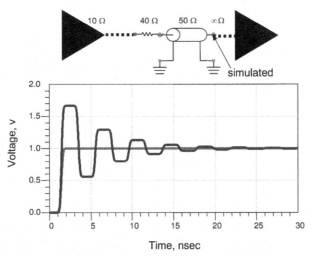

Figure 8-18 Voltage signal of a fast edge, at the far end of the transmission line with and without the source-series terminating resistor.

TIP Understanding the origin of the reflections allows us to engineer a solution to eliminate reflections at one end and prevent ringing. The resulting waveform is very clean and free of signal-quality problems.

At the near end, coming out of the source, right after the source-series terminating resistor, the initial voltage that would be measured is just the incident voltage launched into the transmission line. This is about half the signal voltage. Looking at the source end, we would have to wait for the reflected wave to come by to bring the total voltage up to the full voltage swing.

For a time equal to the round-trip time of flight, the voltage at the source end, after the series resistor, will encounter a shelf. The longer the round-trip time delay of the transmission line compared to the rise time, the longer the shelf will last. This is a fundamental characteristic of source-series terminated lines. An example of the measured voltage at the source end is shown in Figure 8-19.

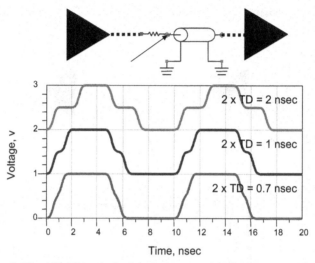

Figure 8-19   100-MHz clock signal, measured at the source end of the transmission line with a source-series resistor as the length of the line is increased. Rise time of the signal is 0.5 nsec.

As long as there are no other receivers near the source to see this shelf, it will not cause a problem. It is when other devices are connected near the source that the shelf may cause a problem, and other topologies and termination schemes might be required.

In the following examples, we are always assuming that the source impedance has been matched to the 50-Ohm characteristic impedance of the initial transmission line.

8.11 Reflections from Short Series Transmission Lines

Many times the line width of a trace on a board must neck down, as it might in going through a via field or routing around a congested region of the board. If the line width changes for a short length of the line, the characteristic impedance will change, typically increasing. How much change in impedance and over what length might start to cause a problem?

Three features determine the impact from a short transmission line segment: the time delay (TD) of the discontinuity, the characteristic impedance of the discontinuity (Z_0), and the rise time of the signal (RT). When the time delay is long compared to the rise time—in other words, the discontinuity is electrically long—the reflected voltage will saturate. The maximum value of the reflection coefficient will be related to the reflection from the front of the discontinuity:

$$\rho = \frac{Z_2 - Z_1}{Z_2 + Z_1} \qquad\qquad (8\text{-}14)$$

For example, if the neck down causes an impedance change from 50 Ohms to 75 Ohms, the reflection coefficient is 0.2. Some examples of the reflected and transmitted signals from electrically long transmission-line discontinuities are shown in Figure 8-20.

These impedance discontinuities cause the signal to rattle around, contributing to reflection noise. This is why it is so important to design interconnects with one uniform characteristic impedance. Keeping the reflection noise to less than 5% of the voltage swing requires keeping the characteristic-impedance change to less than 10%. This is why the typical spec for the control of the impedance in a board is ±10%.

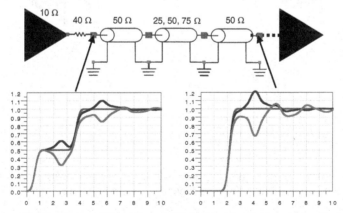

Figure 8-20   Reflected and transmitted signal in a transmission-line circuit with an electrically long but uniform discontinuity, as the impedance of the discontinuity is changed.

Note that whatever the reflection is from the first interface, it will be equal but of opposite sign from the second interface, since Z_1 and Z_2 would be reversed. If the discontinuity can be kept short, the reflections from the two ends might cancel, and the impact on signal quality can be kept negligible. Figure 8-21 shows the reflected and transmitted signal with a short-length discontinuity that is 25 Ohms. If the TD of the discontinuity is shorter than 20% of the rise time, the discontinuity may not cause problems. This gives rise to the same general rule of thumb as before; that is, the maximum acceptable length for an impedance discontinuity is:

$$\text{Len}_{\text{max}}\,[\text{inches}] < \text{RT}\,[\text{n sec}] \qquad (8\text{-}15)$$

where:

Len_{max} = maximum length for a discontinuity, in inches

RT = rise time, in nsec

For example, if the rise time is 0.5 nsec, neck-downs shorter than 0.5 inch may not cause a problem.

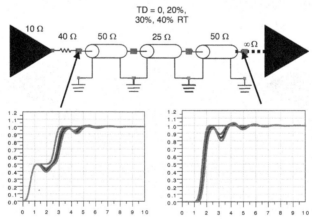

Figure 8-21   Reflected and transmitted signal in a transmission-line circuit with an electrically short but uniform discontinuity, as the time delay of the discontinuity is increased from 0% to 40% of the rise time.

If the TD of the discontinuity can be kept shorter than 20% of the rise time, the impact from the discontinuity may be negligible. This is the same as the rule of thumb which says the length of the discontinuity, in inches, should be less than the rise time of the signal, in nsec.

8.12 Reflections from Short-Stub Transmission Lines

Often, a branch is added to a uniform transmission line to allow the signal to reach multiple fanouts. When the branch is short, it is called a *stub*. A stub is commonly found on BGA packages to allow busing all the pins together so the bonding pads can be gold-plated. The buses are broken off during manufacturing, leaving small, short-length stubs attached to each signal line.

The impact from a stub is complicated to analyze because of all the many reflections that must be taken into account. As the signal leaves the driver, it will first encounter the branch point. Here it will see a low impedance from the parallel combination of the two transmission-line segments, and a negative reflection will head back to the source. A fraction of the signal will continue down both branches. When the signal in the stub hits the end of the stub, it will reflect back to the branch point, again reflecting back to the end and rattling around in the stub. At the same time, at each interaction with the branch point, some fraction of the signal in the stub will head back to the source and to the far end. Each interface acts as a point of reflection.

The only practical way of evaluating the impact of a stub on signal quality is by using a SPICE or a behavioral simulator. The two important factors that determine the impact of the stub on signal quality are the rise time of the signal and the length of the stub. In this example, we assume that the stub is located in the middle of the transmission line and has the same characteristic impedance as the main line. Figure 8-22 shows the simulated reflected signal and transmitted signal as the stub length is increased from 20% of the rise time to 60% of the rise time.

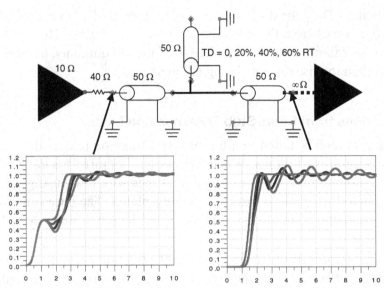

Figure 8-22 Reflected and transmitted signal in a transmission-line circuit with a short stub in the middle, while the time delay of the stub is increased from 20% to 60% of the rise time.

TIP   As a rough rule of thumb, if the stub length is kept shorter than 20% of the spatial extent of the rise time, the impact from the stub may not be important. Likewise, if the stub is longer than 20% of the rise time, it may have an important impact on the signal and must be simulated to evaluate whether it will be acceptable.

For example, if the rise time of the driver is 1 nsec, a stub with a time delay shorter than 0.2 nsec might be acceptable. The length of the stub would be about 1 inch. Once again, the rule of thumb is:

$$\text{Len}_{\text{stub-max}}\,[\text{inches}] < RT\,[\text{nsec}] \qquad (8\text{-}16)$$

where:

$L_{\text{stub-max}}$ = maximum acceptable stub length, in inches
RT = signal rise time, in nsec

This is a simple, easy-to-remember rule of thumb. For example, for a 1-nsec rise time, keep stubs shorter than 1 inch. If the rise time is 0.5 nsec, keep stubs shorter than 0.5 inch. It is clear that as rise times decrease, it gets harder and harder to engineer stubs short enough not to impact signal integrity.

In BGA packages, it is often not possible to avoid plating stubs used in the manufacture of the packages. These stubs are typically less than 0.25-inch long. When the rise time of the signals is longer than 0.25 nsec, these plating stubs may not cause a problem, but as rise times drop below 0.25 nsec, they definitely will cause problems, and it may be necessary to pay extra for packages that are manufactured without plating stubs.

8.13 Reflections from Capacitive End Terminations

All real receivers have some input-gate capacitance. This is typically on the order of 2 pF. In addition, the receiver's package-signal lead might have a capacitance to the return path of about 1 pF. If there is a bank of three memory devices at the end of a transmission line, there might be as much as a 10-pF load at the end of the transmission line.

When a signal travels down a transmission line and hits an ideal capacitor at the end, the actual instantaneous impedance the signal sees, which determines the reflection coefficient, will change with time. After all, the impedance of a capacitor, in the time domain, is related to:

$$Z = \frac{V}{C \frac{dV}{dt}} \tag{8-17}$$

where:

Z = instantaneous impedance of the capacitor

C = capacitance of the capacitor

V = instantaneous voltage in the signal

If the rise time is short compared with the charging time of the capacitor, then initially, the voltage will rise very quickly, and the impedance will be low. But as the capacitor charges, the voltage across it gets smaller, and the dV/dt slows down. As the capacitor charges up, the rate at which the voltage across it changes will slow down. This will cause the impedance of the capacitor to increase dramatically. If we wait long enough, the impedance of a capacitor, after it has charged fully, is open.

This means the reflection coefficient will change with time. The reflected signal should suffer a dip and then move up to look like an open. The exact behavior will depend on the characteristic impedance of the line (Z_0), the capacitance of the capacitor, and the rise time of the signal. The simulated reflected and transmitted voltage behavior for 2-pF, 5-pF, and 10-pF capacitances is shown in Figure 8-23.

The long-term response of the transmitted voltage pattern looks like the charging of a capacitor by a resistor. The presence of the capacitor filters the rise time and acts as a "delay adder" for the signal at the receiver. We can estimate the new rise time and the increase in time delay for the signal to transition through the midpoint (i.e., the delay adder) because it is very similar to the charging of an RC circuit, where the voltage increases with an exponential time constant:

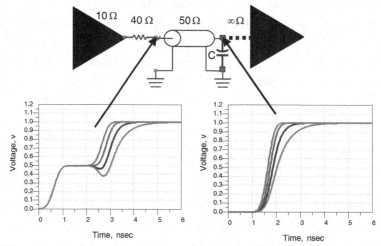

Figure 8-23   Reflected and transmitted signal in a transmission-line circuit with a capacitive load at the far end, for a 0.5-nsec rise time and capacitances of 0, 2 pF, 5 pF, and 10 pF.

$$\tau_e = R \times C \tag{8-18}$$

This time constant is how long it takes for the voltage to rise up to 1/e, or 37% of the final voltage. The 10%–90% rise time is related to the RC time constant by:

$$\tau_{10-90} = 2.2 \times \tau_e = 2.2 \times R \times C \tag{8-19}$$

At the end of a transmission line with a capacitive load, it looks like the voltage is charging up with an RC behavior. The C is the capacitance of the load. The R is the characteristic impedance of the transmission line, Z_0. The 10–90 rise time for the transmitted signal, if dominated by the RC charging, is roughly:

$$\tau_{10-90} = 2.2 \times Z_0 \times C \tag{8-20}$$

For example, if the transmission line has a characteristic impedance of 50 Ohms and the capacitance is 10 pF, the 10–90 charging time will be 2.2 × 50 Ohms × 10 pF = 1.1 nsec. If the initial-signal rise time is short compared with this 1.1-nsec charging time, the presence of the capacitive load at the end of the line will dominate and will now determine the rise time at the receiver. If the initial rise time of the signal is long compared to the 10–90 charging rise time, the capacitor at the end will add a delay to the rise time, and it will be roughly equal to the 10–90 rise time.

> **TIP**   Always be aware of the 10–90 RC rise time, which is based on the characteristic impedance of the line and the typical capacitive load of the input receiver. When the 10–90 rise time is comparable to the initial-signal rise time, the capacitive load at the far end will affect the timing.

A typical case is with a capacitance of 2 pF and characteristic impedance of 50 Ohms. The 10–90 rise time is about 2.2 × 50 × 2 = 0.2 nsec. When rise times are 1 nsec, this additional 0.2-nsec delay adder is barely discernible and may not be important. But when rise times are 0.1 nsec, the 0.2-nsec RC delay can be a significant delay adder. When driving multiple loads grouped at the far end, it is important to include the RC delay adder in all timing analysis.

In an IBIS model of a receiver, the input gate capacitance term is the C_comp term. This will automatically include the impact of the gate capacitance in any circuit simulation. If there are ESD protection diodes in the receiver, this could be as high as 5–8 pF. Generally, it can be kept as low as 2–3 pF.

8.14 Reflections from Capacitive Loads in the Middle of a Trace

A test pad, a via, a package lead, or even a small stub attached to the middle of a trace will act as a lumped capacitor. Figure 8-24 shows the reflected voltage and the transmitted voltage when a capacitor is added to the middle of a trace. Since the capacitor has a low impedance initially, the signal reflected back to the source will have a slight negative dip. If there were a receiver connected near the front end of the trace, this dip might cause problems as it would look like a non-monotonic edge.

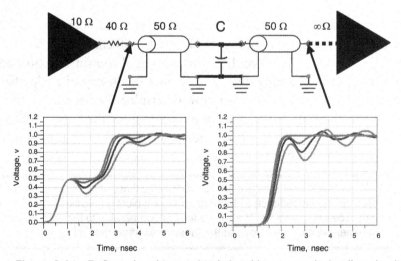

Figure 8-24 Reflected and transmitted signal in a transmission-line circuit with a small capacitive discontinuity in the middle of the trace for a 0.5-nsec rise time and capacitances of 0, 2 pF, 5 pF, and 10 pF.

The transmitted signal initially might not experience a large impact on the first pass of the signal, but after the signal reflects from the end of the line, it will head back to the source. It will then hit the capacitor again, and some of the signal, now with a negative sign, will reflect back to the far end. This reflection back to the receiver will be a negative voltage and will pull the received signal down, causing undershoot.

The impact of an ideal capacitor in the middle of a transmission line will depend on the rise time of the signal and the size of the capacitance. The larger the capacitor, the lower the impedance it will have and the larger the negative reflected voltage, which will contribute to a larger undershoot at the receiver. Likewise, the shorter the rise time, the lower the impedance of the capacitor and the greater the undershoot. If a certain capacitance, C_{max}, is barely acceptable for a certain rise time, RT, and if the rise time were to decrease, the maximum allowable capacitance would have to decrease as well. It is as though the ratio of RT/C_{max} must be greater than some value to be acceptable.

This ratio of the rise time to the capacitance has units of Ohms. But what is it the impedance of? The impedance of a capacitor, in the time domain, is:

$$Z_{cap} = \frac{V}{C\dfrac{dV}{dt}} \tag{8-21}$$

If the signal were a linear ramp, with a rise time, RT, then the dV/dt would be V/RT, and the impedance of a capacitor would be:

$$Z_{cap} = \frac{V}{C\dfrac{dV}{dt}} = \frac{V}{C\dfrac{V}{RT}} = \frac{RT}{C} \tag{8-22}$$

where:

Z_{cap} = impedance of the capacitor, in Ohms

C = capacitance of the discontinuity, in nF

RT = rise time of the signal, in nsec

It is as though the capacitor between the signal and return paths is a shunting impedance, of Z_{cap}, during the time interval of the rise time. This shunting impedance across the transmission line causes the reflections (see Figure 8-25). In order for this impedance to not cause serious problems, we would want it to be large compared to the impedance of the transmission line. In other words, we would want $Z_{cap} \gg Z_0$. As a starting place, this can be translated as $Z_{cap} > 5 \times Z_0$. The limit on the capacitance and rise time is translated as:

$$Z_{cap} > 5 \times Z_0 \tag{8-23}$$

$$\frac{RT}{C_{max}} > 5 \times Z_0 \tag{8-24}$$

$$C_{max} < \frac{RT}{5 \times Z_0} \tag{8-25}$$

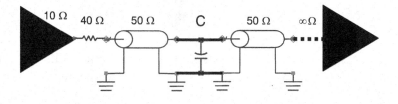

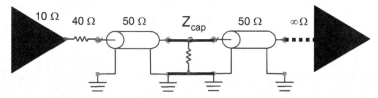

Figure 8-25 Describing the capacitive discontinuity shunting a transmission line as a shunt impedance, during the time interval when the edge passes by.

where:

Z_{cap} = impedance of the capacitor during the rise time

Z_0 = characteristic impedance of the transmission line, in Ohms

RT = rise time, in nsec

C_{max} = maximum acceptable capacitance, in nF, before reflection noise may be a problem

For example, if the characteristic impedance is 50 Ohms, the maximum allowable capacitance is:

$$C_{max}[pF] < \frac{RT[psec]}{5 \times 50} = 4 \times RT[nsec]$$

(8-26)

where:

RT = rise time

C_{max} = maximum acceptable capacitance, in pF, before reflection noise may be a problem

---

TIP   Equation 8-26 is the origin of a very simple rule of thumb: To keep capacitive discontinuities from causing excessive undershoot noise, keep the capacitance, in pF, less than four times the rise time, with the rise time in nsec.

If the rise time is 1 nsec, the maximum allowable capacitance would be 4 pF. If the rise time is 0.25 nsec, the maximum allowable capacitance discontinuity before undershoot problems might arise would be $0.25 \times 4 = 1$ pF. Likewise, if the capacitive discontinuity is 2 pF, the shortest rise time we might be able to get away with would be 2 pF/4 = 0.5 nsec.

This rough limit suggests that if the rise time of a system is 1 nsec, it may be possible to get away with capacitive discontinuities on the order of 4 pF. Likewise, if the capacitance of an empty connector is 2 pF, for example, then this might be acceptable if rise times are longer than 0.5 nsec. However, if the rise time is 0.2 nsec, there may be a problem, and it is critical to simulate the performance before committing to hardware. It may be worth the effort to look for alternative connectors or designs.

8.15 Capacitive Delay Adders

A capacitive load will cause the first-order problem of undershoot noise at the receiver. There is a second, more subtle impact from a capacitive discontinuity: The received time of the signal at the far end will be delayed. The capacitor combined with the transmission line acts as an RC filter. The 10–90 rise time of the transmitted signal will be increased, and the time for the signal to pass the 50% voltage threshold will be increased. The 10−90 rise time of the transmitted signal is roughly:

$$RT_{10-90} = 2.2 \times RC = 2.2 \times \frac{1}{2} Z_0 C = Z_0 C \qquad (8\text{-}27)$$

The increase in delay time for the 50% point is referred to as the *delay adder*, and it is roughly:

$$\Delta TD = RC = \frac{1}{2} Z_0 C \qquad (8\text{-}28)$$

where:

RT_{10-90} = 10% to 90% rise time, in nsec

ΔTD = increase in time delay, in nsec for the 50% threshold

Z_0 = characteristic impedance of the line, in Ohms

C = capacitive discontinuity, in nF

The factor of 1/2 is there because the first half of the line charges the capacitor while the back half is discharging the capacitor, so the effective impedance charging the capacitor is really half the characteristic impedance of the line.

For example, in a 50-Ohm line, the increase in the 10−90 rise time of the transmitted signal for a 2-pF discontinuity will be about 50×2 pF = 100 psec. The 50% threshold delay adder will be about $0.5 \times 50 \times 2$ pF = 50 psec. Figure 8-26 shows the simulated rise time and delay in the received signal reaching the 50% threshold for three different capacitive discontinuities. The capacitor values of 2 pF, 5 pF, and 10 pF have expected delay adders of 50 psec, 125 psec, and 250 psec. This estimate is very close to the actual simulated values.

It is very difficult to keep some capacitive discontinuities that are created by test pads, connector pads, and via holes to less than 1 pF. Every 1-pF pad will add about $0.5 \times 50 \times 1$ pF = 25 psec of extra delay and increase the rise time of the signal. In very high-speed serial links, such as with OC-48 data rates and above, where the rise time of the signal may be 50 psec, every via pad or connector has the potential of adding 25 psec of delay and increasing the rise time of the signal by 50 psec. One via can easily double the rise time, causing significant timing problems.

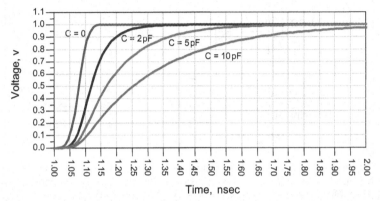

Figure 8-26   The resulting increase in delay time at the receiver for different values of a capacitive discontinuity in the middle of a 50-Ohm trace with a rise time of the signal of 50 psec. The estimates of the delay adders based on the simple rule of thumb are 50 psec, 125 psec, and 250 psec.

One way of minimizing the impact of delay adders is to use a low characteristic impedance. The lower the characteristic impedance, the less the delay adder for the same capacitive discontinuity.

8.16 Effects of Corners and Vias

As a signal passes down a uniform interconnect, there is no reflection and no distortion of the transmitted signal. If the uniform interconnect has a 90-degree bend, there will be an impedance change and some reflection and distortion of the signal. It is absolutely true that a 90-degree corner in an otherwise uniform trace will be an impedance discontinuity and will affect the signal quality. Figure 8-27 shows the measured TDR response of a 50-psec rise-time signal reflecting off the impedance discontinuity of two 90-degree bends in close proximity. This is an easily measured effect.

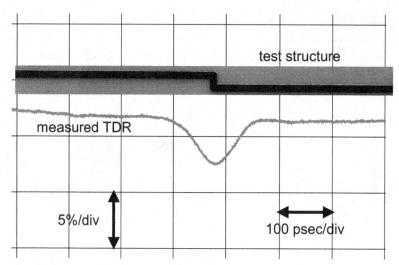

Figure 8-27   Measured TDR response of a uniform 50-Ohm line, 65 mils wide, with two 90-degree corners in close proximity. The rise time of the source is about 50 psec. Measured with an Agilent DCA 86100 and GigaTest Labs Probe Station.

Converting the 90-degree turn into two 45-degree bends will reduce this effect, and using a rounded bend of constant width will reduce the impact from a corner even more. But is the distortion from a corner a problem? Is the magnitude of the discontinuity large enough to worry about, and in what cases might a corner pose a problem? The only way to know the answers is to put in the numbers, and the only way to do this is to understand the root cause of why a corner affects signal integrity.

It is sometimes thought that a 90-degree bend causes the electrons to accelerate around the bend, resulting in excess radiation and distortion. As we saw in Chapter 7, the electrons in a wire are actually moving at the slow speed of about 1 cm/sec. In fact, the presence of the corner has no impact at all on the speed of the electrons. It is true that there will be high electric fields at the sharp point of the corner, but this is a DC effect and is due to the sharp radius of the outside edge. These high DC fields might cause enhanced filament growth and lead to a long-term reliability problem, but they will not affect the signal quality.

TIP   The only impact a corner has on the signal transmission is due to the extra width of the line at the bend. This extra width acts like a capacitive discontinuity. It is this capacitive discontinuity that causes a reflection and a delay adder for the transmitted signal.

If the trace were to make the turn with a constant width, the width of the line would not change, and the signal would encounter the same instantaneous impedance at each point around the turn, and there would be no reflection. We can roughly estimate the extra metal a corner represents; Figure 8-28 illustrates that a corner will represent a small fraction of a square of extra metal. It is definitely less than one square, and as a very rough approximation, it might be on the order of half a square of metal.

The capacitance of a corner can be estimated from the capacitance of a square and the capacitance per length of the trace:

$$C_{corner} = 0.5 \times C_{sq} = 0.5 \times C_L \times w \tag{8-29}$$

The capacitance per length of the line is related to the characteristic impedance of the line:

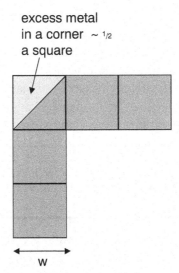

excess metal
in a corner ~ 1/2
a square

w

Figure 8-28 A simple estimate of the extra metal associated with a corner is about half of a square.

$$C_L = \frac{83}{Z_0} \sqrt{\varepsilon_r} \tag{8-30}$$

An estimate of the capacitance of a corner is roughly:

$$C_{corner} = 0.5 \times C_L \times w = 0.5 \times w \times \frac{83}{Z_0}\sqrt{\varepsilon_r} \sim \frac{40}{Z_0} \times \sqrt{\varepsilon_r} \times w \qquad (8\text{-}31)$$

where:

C_{corner} = capacitance per corner in pF

C_L = capacitance per length, in pF/inch

w = line width of the line, in inches

Z_0 = characteristic impedance of the line, in Ohms

ε_r = dielectric constant of the dielectric

For example, in the 0.065-inch-wide trace measured above, the estimated capacitance in each of the two 90-degree bends is about $40/50 \times 2 \times 0.065 = 0.1$ pF = 100 fF. In the above example, there are two corners in close proximity, so the total capacitance of the discontinuity is estimated as about 200 fF. Using the measured TDR response, we can estimate the excess capacitance caused by the discontinuity. Figure 8-29 shows the comparison of the measured response and the simulated response of a uniform line with a lumped capacitance of 200 fF in the middle. The excellent agreement suggests that the discontinuity caused by the two corners can be modeled as a 200-fF capacitor. This is very close to the 200-fF capacitance suggested by the simple model.

> **TIP**   We can generalize the estimate of the capacitance of a corner into a simple, easy-to-remember rule of thumb: In a 50-Ohm transmission line, the capacitance associated with a corner, in fF, is equal to 2 times the line width in mils.

As the line width gets narrower, while keeping the 50-Ohm impedance, the capacitance of a corner will decrease, and its impact will be less significant. For a typical signal line in a high-density board, 5 mils wide, the capacitance of a corner is about 10 fF. The reflection noise from a 10-fF capacitor will be important for rise times on the order of 0.010 pF/4 ~ 3 psec. The delay adder from a 10-fF capacitance will be about $0.5 \times 50 \times 0.01$ pF = 0.25 psec. It is unlikely that the capacitance of a corner will play a significant signal-integrity role for 5-mil-wide lines.

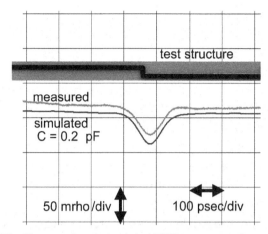

Figure 8-29 Measured and simulated TDR response of a uniform 50-Ohm line, 65 mils wide, with two 90-degree corners in close proximity. The rise time of the source is about 50 psec. The simulated response, based on a capacitance of 0.2 pF, is shifted down slightly for clarity. Measured with an Agilent DCA 86100 and GigaTest Labs Probe Station and simulated with TDA Systems IConnect software.

If a via connects a signal line to a test point or connects signal lines on adjacent layers but continues through the entire board, the barrel of the via may have excess capacitance to the various planes in the board. The residual via stub will cause a via often to look like a lumped capacitive load to a signal. The capacitance of a via stub will depend very strongly on the barrel size, the clearance holes, the size of the pads on the top and bottom of the board, and, of course, the length of the stub. The residual capacitance can range from 0.1 pF to more than 1 pF. Any vias touching the signal line will probably look like capacitive discontinuities, and in uniform, high-speed serial connections, such discontinuities are a chief limitation to the signal quality of the line.

The residual capacitance of a via can be estimated from a simple approximation. Unless special care is taken, the effective characteristic impedance of a via, including the return path through the various planes, is lower than 50 Ohms, on the order of 35 Ohms. The capacitance per length of a 50-Ohm line is 3.3 pF/inch. The capacitance per length of a via stub would be about 5 pF/inch, or about 5 fF/mil. This rough rule of thumb can be used to estimate the capacitive load of a via stub.

For example, a via stub 20 mils long would have a capacitance of 20 mils $\times$ 5 fF/mil = 100 fF. A via stub in a thick board 100 mils long would have a capacitance of 100 mils $\times$ 5 fF/mil = 500 fF or 0.5 pF.

Figure 8-30 shows the measured TDR response of a uniform line with and without a single through-hole via in the middle of a 15-inch-long line in a 10-layer board. The trace impedance is about 58 Ohms, and the line width is nominally 8 mils. The rise time of the signal is about 50 psec. In this trace, the capacitance associated with both the SMA connector's via and the through-hole via in the middle of the line is about 0.4 pF. The difference in reflected voltage between the two vias is due to the rise-time degradation of the signal from dielectric losses as the signal propagates to the middle of the board and back. The changing reflected voltage along the line is a measure of the impedance variation due to manufacturing process variations.

This particular via can be approximated as a capacitance of about 0.4 pF. We would expect a delay adder from this single via to be about $0.5 \times 50 \times 0.4$ pF = 10 psec. The transmitted signal in Figure 8-31 shows a 9-psec increase in delay time compared to a signal traveling through an identical line with no via. This is very close to the estimate based on the simple rule of thumb.

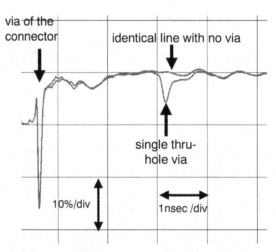

Figure 8-30 Measured TDR response of a uniform transmission line with and without a through-hole via in the middle, creating a capacitive discontinuity. The connector via at the front of both lines is also a capacitive discontinuity. Sample provided courtesy of Doug Brooks, UltraCAD. Measured with an Agilent DCA 86100, GigaTest Labs Probe Station, and TDA Systems IConnect software.

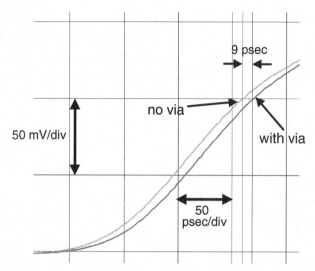

Figure 8-31 Measured transmitted signal after traveling 15 inches in a uniform transmission line and an identical line with a single through-hole via, showing a delay adder of 9 psec. Sample courtesy of Doug Brooks, UltraCAD. Measured with an Agilent DCA 86100, GigaTest Labs Probe Station, and TDA Systems IConnect software.

8.17 Loaded Lines

When there is one small capacitive load on a transmission line, the signal will be distorted and the rise time degraded. Each discrete capacitance acts to lower the impedance the signal sees in its proximity. If there are multiple capacitive loads distributed on the line (for example, 2-pF connector stubs spaced every 1.2 inches for a bus, or 3-pF package and input-gate capacitances distributed every 0.8 inch for a memory bus) and if the spacing is short compared to the spatial extent of the rise time, the reflections from each capacitive discontinuity may smear out. In this case, it appears as though the characteristic impedance of the line has been reduced. The transmission line with a distribution of uniformly spaced capacitive loads is called a *loaded line*.

Each discontinuity will look like a region of lower impedance. When the rise time is short compared to the time delay between the capacitances, each discontinuity acts like a discrete discontinuity to the signal. When the rise time is long compared to the time delay between them, the lower-impedance regions overlap, and the average impedance of the line looks lower.

An example of the reflected signal from a loaded line with three different rise times is shown in Figure 8-32. In this example, there are five 3-pF capacitors distributed every 1 inch on a nominal 50-Ohm line. The last 10 inches of the line are unloaded. Each capacitor has an intrinsic 10–90 rise time of about $2.2 \times 0.5 \times 50$ Ohms $\times 3$ pF = 150 psec. Even though the initial rise time in the first example is 50 psec, after the first capacitor, the rise time is increased to 150 psec and longer after each capacitor.

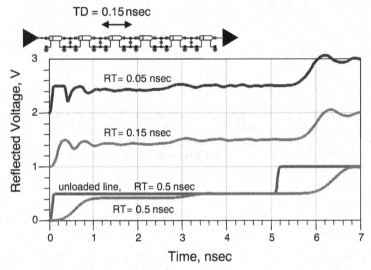

Figure 8-32   Reflected signal from a loaded line with time delay between the 3-pF capacitors of 0.15 nsec. As the rise time increases, the reflection from each capacitor smears out.

The first few capacitors are visible as discrete discontinuities, but the later ones are smeared out by the long rise time of the transmitted signal. When the rise time of the signal is long compared to the time delay between the capacitive discontinuities, the uniformly distributed capacitive loads act to lower the apparent characteristic impedance of the line. In such a loaded line, it is as though the capacitance per length of the line has been increased by the added board features. The higher capacitance per length means a lower characteristic impedance and a longer time delay.

In a uniform, unloaded transmission line, the characteristic impedance and time delay are related to the capacitance per length and inductance per length as:

$$Z_0 = \sqrt{\frac{L_L}{C_{0L}}} \qquad (8\text{-}32)$$

$$TD_0 = Len\sqrt{L_L C_{0L}} \qquad (8\text{-}33)$$

where:

Z_0 = characteristic impedance of the unloaded line, in Ohms

L_L = inductance per length of the line, in pH/inch

C_{0L} = capacitance per length of the unloaded line, in pF/inch

Len = length of the line, in inches

TD_0 = time delay of the unloaded line, in psec

When there are uniformly distributed capacitive loads, each of C_1 and spaced a distance of d_1, the distributed capacitance per length of the line is increased from C_{0L} to $(C_{0L} + C_1/d_1)$. This changes the characteristic impedance of the line and its time delay to:

$$Z_{Load0} = \sqrt{\frac{L_L}{C_{0L} + \dfrac{C_1}{d_1}}} = Z_0 \sqrt{\frac{C_{0L}}{C_{0L} + \dfrac{C_1}{d_1}}} = Z_0 \sqrt{\frac{1}{1 + \dfrac{C_1}{C_{0L} d_1}}} \qquad (8\text{-}34)$$

$$TD_{Load} = Len\sqrt{L_L\left(C_{0L} + \frac{C_1}{d_1}\right)} = TD_0 \sqrt{\left(1 + \frac{C_1}{C_{0L} d_1}\right)} \qquad (8\text{-}35)$$

where:

Z_0 = characteristic impedance of the unloaded line, in Ohms

Z_{Load0} = characteristic impedance of the loaded line, in Ohms

L_L = inductance per length of the line, in pH/inch

C_{0L} = capacitance per length of the unloaded line, in pF/inch

C_1 = capacitance of each discrete capacitor, in pF

d_1 = distance between each discrete capacitor, in inches

Len = length of the line, in inches

TD_0 = time delay of the unloaded line, in psec

TD_{Load} = time delay of the loaded-line region

In a 50-Ohm line, the capacitance per length is about 3.4 pF/inch. When the added distributed capacitive load is comparable to this, the characteristic impedance and time delay can be changed significantly. For example, if 3-pF loads from the input-gate capacitance of a memory bank are spaced every 1 inch in a multidrop bus, the additional loaded capacitance per length is 3 pF/inch. The loaded characteristic impedance is decreased to $0.73 \times Z_0$, and the time delay is increased to $1.37 \times TD_0$.

Since the characteristic impedance of the line is reduced, the terminating resistance should be reduced as well. Alternatively, in the region where the distributed capacitive loads are positioned, the line width could be reduced so the unloaded impedance is higher. When the loads are attached, the loaded line impedance would be closer to the target value of the impedance.

This effect of increased discrete capacitance added to a line acting to lower the characteristic impedance and increase the time delay is exactly what happens inside a via.

Any nonfunctional pads on each layer in the pad stack, or just the extra capacitance of the via barrel passing through the clearance holes of the planes, will look like added discrete capacitance. This contributes to the lower impedance of the via and to a time delay that is longer than would be expected based on the length of the via and the dielectric constant, Dk, of the laminate material.

It is as though there is a higher effective Dk, which can be as high as 8–15, compared to the bulk Dk of 4. This is all due to the higher discrete loaded capacitance of the via barrel to the planes.

8.18 Reflections from Inductive Discontinuities

Just about every series connection added to a transmission line will have some series loop inductance associated with it. Every via used to change signal layers, every series-terminating resistor, every connector, and every engineering change wire will have some extra loop inductance. The signal will see this loop inductance as an additional discontinuity above and beyond what is in the transmission line.

If the discontinuity is in the signal path, the loop inductance will be dominated by the partial self-inductance of the signal path, though there will still be some partial mutual inductance with the return path. If the discontinuity is in the return path, the partial self-inductance of the return path will dominate the loop inductance. In either case, it is the loop inductance to which the signal is sensitive because the signal is a current loop propagating down the transmission line between the signal and return paths.

A higher series loop inductance will initially look like a higher impedance to an incident, fast rise-time signal. This will cause a positive reflection back to the source. Figure 8-33 shows the measured reflected signal from a uniform transmission line that travels over a small gap in the return path.

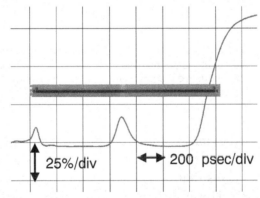

Figure 8-33   TDR-reflected signal from a uniform transmission line with an inductive discontinuity in the middle caused by a gap in the return path. The rise time is about 50 psec and was measured with an Agilent DCA 86100 and a GigaTest Labs Probe Station, analyzed with TDA Systems IConnect software.

Figure 8-34 shows the signal at the receiver and source for different values of an inductive discontinuity. The shape of the signal at the near end, going up and then back down, is called *non-monotonicity*. The signal is not increasing at a steady upward pace, monotonically. This feature, by itself, may not cause a signal-integrity problem. However, if there were a receiver located at the near end, and it received a signal increasing past the 50% point and then dropping down below 50%, it might be falsely triggered.

The initial, distorted rising or falling edge of the signal may not cause a bit error at the receiver if the distortion happens outside the setup and hold time. However, if the edge of a clock signal is distorted, it can result in a timing error, which can cause a bit error.

Non-monotonic behavior is to be avoided wherever possible. At the far end, the transmitted signal will show overshoot and a delay adder.

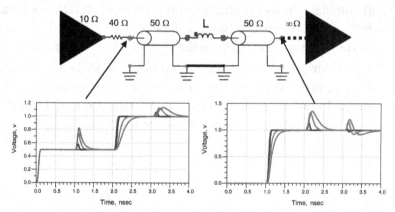

Figure 8-34 Signal at the source and the receiver with a 50-psec rise time interacting with an inductive discontinuity. Inductance values are L = 0, 1 nH, 5 nH, and 10 nH.

In general, the maximum amount of inductance that might be acceptable in a circuit depends on the noise margin and other features of the circuit. This means that each case must be simulated to evaluate what might be acceptable. However, as a rough measure of how much inductance might be too much, we can use the limit of when the series-impedance discontinuity of the discrete inductor increases to more than 20% of the characteristic impedance of the line. At this point, the reflected signal will be about 10% of the signal swing, usually the maximum acceptable noise allocated to reflection noise.

We can approximate the impedance of the inductor if it is small compared to the characteristic impedance and if the rise time is a linear ramp, while the rise time is passing through it, by:

$$Z_{inductor} = \frac{V}{I} = \frac{L\frac{dI}{dt}}{I} = \frac{L}{RT} \tag{8-36}$$

where:

$Z_{inductor}$ = impedance of the inductor, in Ohms

L = inductance, in nH

RT = rise time of the signal, in nsec

The estimate of the maximum acceptable inductive discontinuity is set by keeping the impedance of the inductor less than 20% of the impedance of the line:

$$Z_{inductor} < 0.2 \times Z_0 \tag{8-37}$$

$$\frac{L_{max}}{RT} < 0.2 \times Z_0 \tag{8-38}$$

$$L_{max} < 0.2 \times Z_0 \times RT \tag{8-39}$$

where:

L_{max} = maximum allowable series inductance, in nH

Z_0 = characteristic impedance of the line, in Ohms

RT = rise time of the signal, in nsec

For example, if the characteristic impedance of the line is 50 Ohms and the rise time is 1 nsec, the maximum acceptable series inductance would be about $L_{max} = 0.2 \times 50 \times 1$ nsec = 10 nH.

TIP Consider this simple rule of thumb: As a rough approximation, in a 50-Ohm line, the maximum allowable excess loop inductance (in nH) is 10 times the rise time (in nsec). Likewise, if there is some loop inductance from a discontinuity, the shortest rise time that might be acceptable before reflection noise exceeds the noise budget (in nsec) is L/10 with the inductance (in nH).

If there is a residual loop inductance from a connector of 5 nH, the shortest usable rise time for that connector might be on the order of 5 nH/10 = 0.5 nsec. If the rise time of the signal is 0.1 nsec, all inductive discontinuities should be kept less than $10 \times RT = 10 \times 0.1 = 1$ nH.

Based on this estimate, we can evaluate the useful rise time for an axial lead and an SMT terminating resistor. The series loop inductance in an axial lead resistor is about 10 nH, while in an SMT resistor, it is about 2 nH.

> **TIP**   The shortest rise time at which an axial-lead resistor might be used before reflection noise causes a problem is about 10 nH/10 ~ 1 nsec, while for an SMT resistor, it is about 2 nH/10 ~ 0.2 nsec.

When the rise times are in the sub-nsec regime, axial-lead resistors are not suitable components and should be avoided. As rise times approach 100 psec, it will be increasingly important to engineer the use of SMT resistors with the lowest loop inductance possible. The two most important design features of a high-performance SMT resistor are a short length and a return plane as close to the surface as possible. Alternatively, resistors that are integrated into the board or package can have considerably lower loop inductance than 2 nH and might be required.

An inductive discontinuity contributes to reflection noise and also to a delay adder. When the rise time is very short and the transmitted rise time is dominated by the series inductor, the 10–90 rise time of the transmitted signal is approximately:

$$TD_{10-90} = 2.2 \times \frac{L}{2Z_0} = \frac{L}{Z_0} \tag{8-40}$$

$$TD_{adder} = 0.5 \times \frac{L}{Z_0} \tag{8-41}$$

where:

TD_{10-90} = 10–90 rise time of the transmitted signal, in nsec

L = series loop inductance of the discontinuity, in nH

Z_0 = characteristic impedance of the line, in Ohms

TD_{adder} = delay adder at the 50% point, in nsec

For example, a 10-nH discontinuity will increase the 10–90 rise time to about 10/50 = 0.2 nsec. The time delay added to the midpoint is roughly half of this, or 0.1 nsec. Figure 8-35 shows the simulated time delay of the received signal with discontinuities of 1 nH, 5 nH, and 10 nH.

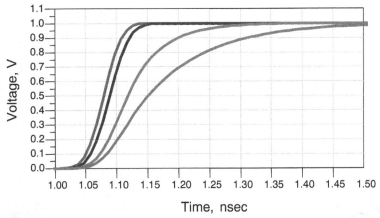

Figure 8-35 Delay adders in the received signal for a 50-psec rise time and inductive discontinuities of 0, 1 nH, 5 nH, and 10 nH. The estimated delay adders are 0, 10 psec, 50 psec, and 100 psec.

8.19 Compensation

Sometimes, it is unavoidable to have a series loop inductance in a circuit, as when a specific connector has already been designed in. Left to itself, it may cause excessive reflection noise. A technique called *compensation* is used to cancel out some of this noise.

The idea is to try to trick the signal into not seeing a large inductive discontinuity and have it instead see a section of transmission line that matches the characteristic impedance of the line. After all, an ideal transmission can be approximated to first order, as a single section of an LC network. In this case, the characteristic impedance of any section of the line is given by:

$$Z_0 = \sqrt{\frac{L_L}{C_L}} = \sqrt{\frac{L}{C}}$$

(8-42)

where:

Z_0 = characteristic impedance of the line, in Ohms

L_L = inductance per length of the line, in nH/inch

L = total inductance of any section of the line, in nH

C_L = capacitance per length of the line, in nF/inch

C = total capacitance of any section of the line, in nF

We can turn an inductive discontinuity into a transmission-line segment by adding a small capacitor to either side, as illustrated in Figure 8-36. In such a case, the apparent characteristic impedance of the inductor is:

$$Z_1 = \sqrt{\frac{L_1}{C_1}} \qquad (8\text{-}43)$$

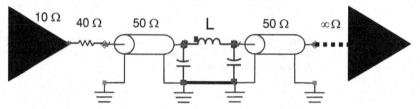

Figure 8-36   Compensation circuit for an inductive discontinuity. Add enough capacitance on either side to make the inductive discontinuity look like part of a 50-Ohm transmission line.

To minimize the reflected noise, we need to find values of the capacitors so that the apparent characteristic impedance of the connector, Z_1, is equal to the characteristic impedance of the rest of the circuit, Z_0. Using the relationship above, the capacitance to add is:

$$C_1 = \frac{L_1}{Z_0^2} \qquad (8\text{-}44)$$

where:

C_1 = total compensation capacitance to add, in nF

L_1 = inductance of the discontinuity, in nH

Z_0 = characteristic impedance of the line, in Ohms

For example, if the inductance of the connector is 10 nH and the characteristic impedance of the line is 50 Ohms, the total compensation capacitance to add is $10/50^2 = 0.004$ nF $= 4$ pF. For optimum compensation, the 4-pF capacitance should be distributed with 2 pF on each side of the inductor.

Figure 8-37 shows the reflected and transmitted signals for the three cases of no connector, an uncompensated connector, and the same connector compensated. Depending on the rise time of the system, the reflected noise can sometimes be reduced by more than 75%.

This technique applies to all inductive discontinuities such as vias and resistors as well. Depending on the relative amount of capacitance in the pads and the series inductance, a real discontinuity may look capacitive or inductive.

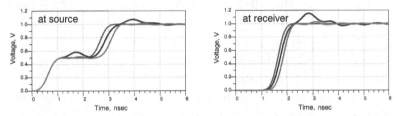

Figure 8-37 Source and transmitted signal for a 10-nH inductive discontinuity and 0.5-nsec rise time. The simulations are for no connector, for a connector but no compensation, and for 2-pF capacitance added to each side of the inductor to compensate for the inductance.

TIP The goal in designing interconnects is to engineer the pads and other features to make the structure look like a section of uniform transmission line. In this way, some inductive discontinuities, such as connectors, can be made to nearly disappear.

8.20 The Bottom Line

1. Wherever the instantaneous impedance a signal sees changes, there will be a reflection, and the transmitted signal will be distorted. This is the primary source of signal-quality problems associated with a single net.

2. As a rough rule of thumb, any transmission line longer, in inches, than the rise time, in nsec, will need a termination to prevent excessive ringing noise.

3. Source-series termination is the most common termination scheme for point-to-point connections. A series resistor should be added so that the sum of the resistor and the source impedance match the characteristic impedance of the line.

4. A SPICE or other circuit simulator is essential for every engineer involved with signal integrity. Many are low cost and easy to use. These tools allow simulation of all the multiple bounces due to impedance discontinuities.

5. As a rough rule of thumb, variations in the characteristic impedance of a line should be kept to less than 10% to keep reflection noise below 5%.

6. As a rough rule of thumb, if the length of a short transmission line discontinuity can be kept shorter, in inches, than the rise time of the signal, in nsec, the reflections from the discontinuity might not cause a problem.

7. As a rough rule of thumb, if the length of a short stub can be kept shorter, in inches, than the rise time of the signal, in nsec, the reflections from the stub might not cause a problem.

8. A capacitive load at the far end of a line will cause a delay adder but no signal-quality noise.

9. As a rough rule of thumb, a capacitive discontinuity in the middle of a line will cause excessive reflection noise if the capacitance, in pF, is greater than four times the rise time, in nsec.

10. The delay adder, in psec, for a capacitive discontinuity in the middle of a line will be about 25 times the capacitance, in pF.

11. A corner contributes a capacitance, in fF, of about two times the line width, in mils.

12. A uniformly spaced distribution of capacitive loads will lower the effective characteristic impedance of the line.

13. The magnitude of an acceptable inductive discontinuity, in nH, is about 10 times the rise time of the signal, in nsec.

14. The impact from an inductive discontinuity can be greatly minimized by adding capacitance on either side of the inductor to fool the signal into seeing a section of a uniform transmission line. In this way, vias can be engineered to be nearly invisible to high-speed signals.

End-of-Chapter Review Questions

8.1 What is the only thing that causes a reflection?

8.2 What two features influence the magnitude of the reflection coefficient?

8.3 What influences the sign of the reflection coefficient?

8.4 What two boundary conditions must be met on either side of any interface?

8.5 When viewed at the transmitter, how long does a reflection from a discontinuity last?

8.6 What is the difference in reflection coefficient when a signal reflection from a 50-Ohm line hits a 75-Ohm transmission line or a 75-Ohm resistor?

8.7 A signal travels to the end of a 50-Ohm line and sees a series 30-Ohm resistor at the high impedance of the receiver. What is the impact of having the 30-Ohm series resistor at the receiver?

8.8 How would you terminate a bidirectional bus by using source-series resistance? Where would you put the source resistor?

8.9 What is the raw measurement actually displayed on the screen of a TDR?

8.10 How do you convert this raw measurement into the instantaneous impedance the signal must have encountered?

8.11 The unloaded voltage from a driver is 1 V. What is the output impedance if the output voltage is 0.8 V when a 50-Ohm resistor shorts the output pin?

8.12 What is the shape of the TDR response from a capacitive discontinuity? Why is it this shape?

8.13 What is the shape of the TDR response from an inductive discontinuity? Why is it this shape?

8.14 If the output impedance of a driver is 35 Ohms, what value of source-series resistor should be used in the driver when connected to a 50-Ohm line? To a 65-Ohm line?

8.15 How can you tell the difference between a short low-impedance transmission line and a small capacitor in the middle of a uniform transmission line by using a TDR?

8.16 By looking at the TDR response, how would you tell the difference between a really long 75-Ohm line and a short 75-Ohm line connected to a 75-Ohm resistor?

8.17 What are the reflection and transmission coefficients when a signal is coming from a 40-Ohm environment and encounters an 80-Ohm environment?

8.18 What are the reflection and transmission coefficients when a signal is coming from an 80-Ohm environment and encounters a 40-Ohm environment?

8.19 Consider a driver with a 1-V unloaded output voltage and 10-Ohm output resistance. The transmission line it is connected to is 50 Ohms, and the line is terminated at the far end. What value resistor should be used to terminate the line? What are the high and low voltages at the receiver if the far end resistor is tied to Vss?

8.20 Redo Question 8.19 with the far-end resistor tied to Vcc.

8.21 Redo Question 8.19 with the far-end resistor tied to 1/2 × Vcc, sometimes call the termination voltage, or VTT.

8.22 If the rise time of a signal is 3 nsec, how long a line would you expect to be able to get away without having to terminate?

8.23 What happens to the size of the reflected voltage from a short transmission line discontinuity when its length gets shorter and shorter? At what length is it "transparent"?

8.24 Due to the capacitive loading of the nonfunctional capture pads on each layer of a via, will the via look electrically longer or shorter?

8.25 Suppose a 1-V signal from a 1-Ohm output impedance source, with a 0.1-nsec rise time and 50% duty cycle is launched on a 12-inch-long 50-Ohm transmission line. What is the average power consumption with a single 50-Ohm terminating resistor to the return path?

8.26 Suppose the far-end resistor were terminated not to ground, but to a voltage 1/2 the Vcc voltage. What would be the average power consumption?

8.27 In Question 8.26, suppose a 49-Ohm source series resistor were used to terminate the line. Would the power consumption in the source resistor be larger, smaller, or the same as with a far-end termination?

CHAPTER 9

Lossy Lines, Rise-Time Degradation, and Material Properties

A signal with a very fast edge going into a real transmission line will come out with a longer rise time. For example, the measured response of a 50-psec signal transported through a 36-inch-long, 50-Ohm backplane line in FR4 is shown in Figure 9-1. The rise time has been increased to almost 1 nsec. This rise-time degradation is due to the losses in the transmission line and is the dominant root cause of *intersymbol interference* (ISI) and the collapse of the eye diagram.

Loss in transmission lines is the principal signal-integrity problem for all signals with clocks higher than 1 GHz, transported for distances longer than 10 inches, such as in high-speed serial links and Gigabit Ethernet.

TIP   The reason the rise time is increased in propagating down a real transmission line is specifically because the higher-frequency components of the signal are preferentially attenuated more than the lower frequencies.

The simplest way to analyze the frequency-dependent losses is in the frequency domain. On the other hand, the problems created by lossy lines are time-domain dependent, and ultimately the final response must be analyzed in the time domain. In this chapter, we look first in the frequency domain to understand the loss mechanisms and then switch to the time domain to evaluate the impact on the signal-integrity response.

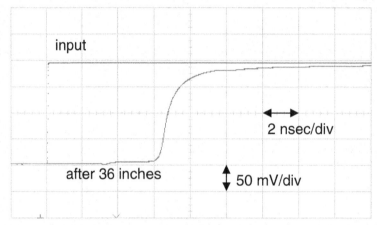

Figure 9-1 Measured input and output signals through a 50-Ohm line, 36 inches long. A 50-psec rise-time signal goes in, while a 1-nsec rise-time signal comes out.

9.1 Why Worry About Lossy Lines?

If the losses were independent of frequency and if low-frequency components were attenuated the same as high-frequency components, the entire signal waveform would uniformly decrease in amplitude, but the rise time would stay the same. This is illustrated in Figure 9-2. The effect of the constant attenuation could be compensated with some gain at the receiver. The rise time, the timing, and the jitter would remain unaffected by the constant attenuation.

It is not the loss that causes rise-time degradation, ISI, collapse of the eye, and deterministic jitter but the frequency dependence of the loss.

As a signal propagates down a real lossy transmission line, the amplitudes of the higher-frequency components are reduced, and the low-frequency components stay about the same. The bandwidth of the signal is decreased because of this selective attenuation. As the bandwidth of the signal decreases, the rise time of the signal increases. It is the frequency dependence of the losses that specifically drives the rise-time degradation.

If the rise-time degradation were small compared to the period for one bit, or the unit interval, the bit pattern would be very constant and independent of what came previously. By the time one bit cycle ended, the signal would have stabilized and reached the final value. The voltage waveform for one bit in the stream would be independent of what the previous bit was, whether it was a high or a low and for how long it was high or low. In this case, there would be no ISI.

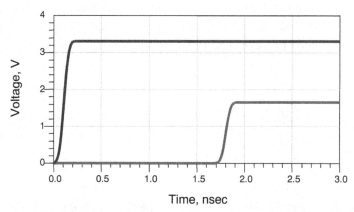

Figure 9-2 Simulated signal propagation of a 100-psec rise-time signal with losses that are independent of frequency. The only impact is a scaling of the signal.

However, if the rise-time degradation increases the received rise time significantly, comparable to the unit interval, the actual voltage level for a bit will depend on how long the signal was in a previous high or low state. If the bit pattern was high for a long time previously and the signal drops down for one bit and then up, the low voltage will not have had time to fall all the way to the lowest voltage. The precise voltage levels achieved by a single bit will depend on the previous bit pattern. This is called *intersymbol interference* (*ISI*) and is illustrated in Figure 9-3.

The important consequence of frequency-dependent loss and rise-time degradation is ISI: The precise waveform of the bit pattern will depend on the previous bits that have passed by. This will significantly affect the ability to tell the difference between a low and high level signal by the receiver, increasing the bit error rate.

TIP   In addition, the time for the signal to reach the switching threshold will change, depending on the previous data pattern. ISI is a significant contributor to jitter. If the rise time were short compared to the bit period, there would be no ISI.

One of the common metrics for describing the signal quality of a high-speed serial link, at the receiver, is an eye diagram. A pseudorandom bit stream, whose pattern represents all possible bit-stream patterns, is simulated or measured, using the clock reference as the trigger point. Each received cycle is taken out of the bit stream and overlaid on the previous cycle, synchronous with the clock, and hundreds of cycles are superimposed. This set of superimposed waveforms is called an eye diagram because it looks like an open eye.

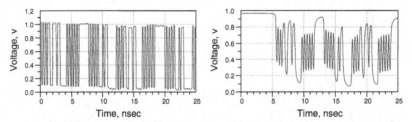

Figure 9-3   5-Gbps pseudorandom bit stream. Left: Bit pattern when the rise time is much shorter than the bit period. Right: Bit pattern when the rise time is comparable to the bit pattern, causing pattern-dependent voltage levels or intersymbol interference.

The closing of the eye diagram is a measure of the bit error rate. A valid 1 or 0 bit means the received voltage level is above the minimum input for a high or below a maximum input for a low and measured within the setup and hold time. These two conditions define a valid signal vertically and horizontally. We call these levels an *acceptance mask*. As long as the voltage for each bit is output the mask, the data will be read correctly.

If any voltage at the receiver drops inside the mask value, there is a good chance it will not be read correctly and will result in a bit error. A large eye opening means low bit error rate. A collapsing eye means a higher potential bit error rate, especially if the eye encroaches on the mask.

The horizontal width of the crossover region that separates eye openings is a measure of the jitter. A direct consequence of frequency-dependent lossy lines and an indirect measure of the ISI is the collapse of the opening of the eye.

Figure 9-4 shows the collapse of the eye diagram for the same 5-Gbps waveform, with and without losses.

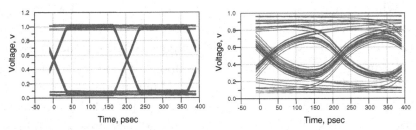

Figure 9-4   Eye diagrams of a 5-Gbps pseudorandom bit stream. Left: Little loss. Right: Same bit pattern when there is a lot of loss, showing the collapse of the eye diagram, and increased jitter, indicated by the widening of the crossover regions.

9.2 Losses in Transmission Lines

The first-order approximate model for a transmission line is an n-section LC model. This approximation is often referred to as the *lossless model*. It accounts for the two important features of a transmission line—characteristic impedance and time delay—but offers no mechanism to account for the loss of the voltage as the signal propagates.

The losses need to be added to this model so that the received waveform can be accurately predicted. There are five ways energy can be lost to the receiver while the signal is propagating down a transmission line:

1. Radiative loss
2. Coupling to adjacent traces
3. Impedance mismatches
4. Conductor loss
5. Dielectric loss

Each of these mechanisms will reduce or affect the received signal. It is dangerous to lump all of these processes in the general category "attenuation," as they have different root causes. As we emphasized in Chapter 1, "Signal Integrity Is in Your Future," the fastest way to solve a problem is to first identify its root cause. If we lump these five processes under attenuation, we lose the significance of their radically different root causes and design solutions. Instead, we include only the conductor loss and dielectric loss in the attenuation category. These are the mechanisms by which the signal loses energy into the materials of the transmission line, converting signal energy into heating up the transmission line.

While radiative loss is important when it comes to EMI, the amount of energy typically lost to radiation is very small compared to the other loss processes, and this loss mechanism will have no impact on the received signal.

Coupling to adjacent traces is important and can cause rise-time degradation. This effect can be very accurately modeled and the resulting waveforms on both the active and the quiet lines predicted. In tightly coupled transmission lines, the signal on one trace will be affected by the energy coupled to the adjacent one and must be included in a critical net simulation to accurately predict performance and the transmitted signal. This topic is covered in Chapter 10, "Cross Talk in Transmission Lines."

TIP   Impedance discontinuities can have a dramatic impact in distorting the transmitted signals. This will have the direct consequence of degrading the rise time of the received signal. Even a line with no loss will show rise-time degradation from impedance discontinuities. This is why it is so important to have accurate models for the transmission lines, board vias, and connectors—to accurately predict the signal quality from simulations. And this is why it is so important to minimize discontinuities in the design of high-speed interconnects.

If the rise time is degraded due to the removal of high-frequency components, where did the high-frequency components go? After all, capacitive or inductive discontinuities will not in themselves absorb energy. The high-frequency components are reflected back to the source and ultimately get absorbed and dissipated in any terminating resistors or the source impedance of the driver.

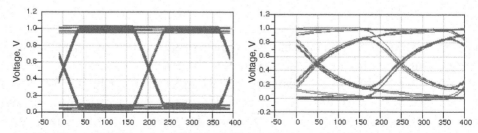

Figure 9-5   Eye diagrams of a 5-Gbps pseudorandom bit stream. Left: Little loss. Right: Same bit pattern still with no loss but a 4-pF capacitive discontinuity from four through-hole vias.

Figure 9-5 is an example of the 5-Gbps signal from above, as transmitted through a short, ideal, lossless transmission line with four via pads in series, each of 1-pF load, contributing a total of 4-pF capacitive load. The resulting 50% point rise-time degradation expected is about $1/2 \times 50 \times 4$ pF = 100 psec, equal to half the bit period. Impedance discontinuities and their impact on rise-time degradation are covered in Chapter 8, "Transmission Lines and Reflections."

The last two loss mechanisms represent the primary causes of attenuation in transmission lines that are not taken into account by other models. Conductor loss refers to energy lost in the conductors in both the signal and return paths. This is ultimately due to the series resistance of the conductors. *Dielectric loss* refers to the energy lost in the dielectric due to a specific material property—the dissipation factor of the material.

TIP   In general, with typical 8-mil-wide traces and 50-Ohm transmission lines in FR4, the dielectric losses are greater than the conductor losses at frequencies above about 1 GHz. For high-speed serial links with clock frequencies at or above 2.5 Gbps, the dielectric losses dominate. This is why the dissipation factor of laminate materials is such an important property.

TIP   When considering the attenuation in a transmission line, do not lump energy loss from coupling or energy loss from reflections. These processes are included when analyzing the cross talk to adjacent channels and signal quality from impedance discontinuities in the transmission line. Attenuation is a new and separate mechanism to include.

9.3 Sources of Loss: Conductor Resistance and Skin Depth

The series resistance a signal sees in propagating down the signal and return paths is related to the conductors' bulk resistivity and the cross section through which the current propagates. At DC, the current distribution in the signal conductor is uniform and the resistance is:

$$R = \rho \frac{Len}{w \times t} \tag{9-1}$$

where:

R = resistance of the line, in Ohms

ρ = bulk resistivity of the conductor, in Ohm-inches

Len = length of the line, in inches

w = line width, in inches

t = thickness of the conductor, in inches

The DC current distribution in the return path, if it is a plane, is spread out through the cross section, and the resistance is much smaller than the signal path's resistance and can be ignored.

For a typical 5-mil-wide trace, with 1.4-mil-thick copper (1-ounce copper), 1 inch in length, the resistance in the signal path at DC is about $R = 0.72 \times 10^{-6}$ Ohms-inches $\times$ 1 inch/(0.005 $\times$ 0.0014) = 0.1 Ohm.

The bulk resistivity of copper, and virtually all other metals, is absolutely constant with frequency until frequencies near 100 GHz. At first glance, we might expect the resistance of a trace to be constant with frequency. After all, this is how an ideal resistor behaves. However, as described in Chapter 6, the current will redistribute itself at higher frequencies due to skin-depth effects.

At a high frequency, the cross section through which current will be flowing in a copper conductor is in a thickness approximately equal to the skin depth, δ:

$$\delta = 2.1\sqrt{\frac{1}{f}}$$ (9-2)

where:

δ = skin depth, in microns

f = sine-wave frequency, in GHz

In copper, at 1 GHz, the current in the signal path of a microstrip, for example, is concentrated in a layer about 2.1 microns thick, on either side of the trace. At 10 MHz, or 0.01 GHz, it would be concentrated in a layer about 21 microns thick. This is a rough approximation. The actual current distribution in both the signal and return paths can be calculated with a 2D field solver. An example of the current distribution in a microstrip and a stripline for sine waves at 10 MHz is shown in Figure 9-6.

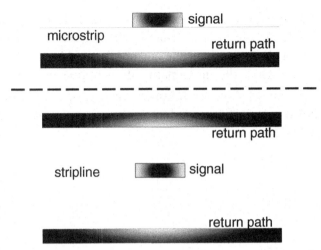

Figure 9-6   Current distribution in 1-ounce copper, for near 50-Ohm lines, at 10 MHz, showing onset of current redistribution due to skin-depth effects. Top: Microstrip. Bottom: Stripline. The lighter the color, the higher the current density. Simulated with Ansoft's 2D Extractor.

For 1-ounce copper, with a geometrical thickness of 35 microns, the skin depth is thinner for all frequencies above about 10 MHz. The current distribution is driven by the requirement of the current to find the path of least impedance or, at higher frequencies, the path of lowest loop inductance. This translates into two trends: The current in each conductor wants to spread out as far apart as possible to minimize the partial self-inductance of each conductor, and, simultaneously, the oppositely directed current in each conductor will move as close together as possible to maximize the partial mutual inductance between the two currents.

> **TIP**   This means that for all important signal-frequency components, the current distribution in most PC-board interconnects is always going to be skin-depth limited, and the resistance will always be frequency dependent for frequency components above 10 MHz.

The resistance of a signal will depend on the actual cross section of the conductor transporting current. At higher frequencies, the current will be using a thinner section of the conductor, and the resistance will increase with frequency. The frequency dependence of skin depth will cause a frequency dependence to resistance. It is important to note that the resistivity of copper, and most conductors, is very constant with frequency. What is changing is the cross section through which the current flows. Above about 10 MHz, the resistance per length of the signal path will be frequency dependent.

In this regime of skin-depth-limited current, if the current were flowing in just the bottom half of the conductor, the resistance of a conductor would be approximated by:

$$R = \rho \frac{Len}{w \times \delta} \tag{9-3}$$

where:

R = resistance of the line, in Ohms

ρ = bulk resistivity of the conductor, in Ohm-inches

Len = length of the line, in inches

w = line width, in inches

δ = skin depth of the conductor, in inches

As shown in Figure 9-6, the current flows in more than just the bottom half of the conductor, even for microstrip. There is substantial current also in the top of the conductor. These two regions are in parallel. To first order, the resistance of the signal path can be approximated as 0.5 × R, to take into account the two parallel paths in the signal path. The current distribution in the signal path of the microstrip and stripline are very similar.

Skin depth is driven by the need for the currents to take the path of lowest impedance, which is dominated by the loop inductance at higher frequencies. This mechanism also drives the current in the return path to redistribute and change with frequency. At DC, the return current will be distributed all throughout the return plane. When in the skin-depth-limited regime, the current distribution in the return path will concentrate close to the signal path, near the surface, to minimize the loop inductance.

To first order, in a microstrip, the width of the current distribution in the return path is roughly three times the width of the signal path, as observed in Figure 9-6. This resistance in the return path is in series with the resistance of the signal path. At frequencies above about 10 MHz, we would expect the total series resistance of the transmission line to be 0.5R + 0.3R = 0.8R. The total resistance of the signal path in a microstrip is expected to be about:

$$R = 0.8 \times \rho \frac{\text{Len}}{w \times \delta} \qquad (9-4)$$

where:

R = resistance of the line, in Ohms

ρ = bulk resistivity of the conductor, in Ohm-inches

Len = length of the line, in inches

w = line width, in inches

δ = skin depth of the conductor, in inches

0.8 = factor due to the specific current distribution in the signal and return paths

Figure 9-7 compares this simple first-order model with the results from a 2D field solver, which calculates the precise current distribution at each frequency. The agreement at both low-frequency and skin-depth-limited frequencies is excellent for so simple a model. The total resistance per length of a stripline should be slightly lower. This is also compared in the figure.

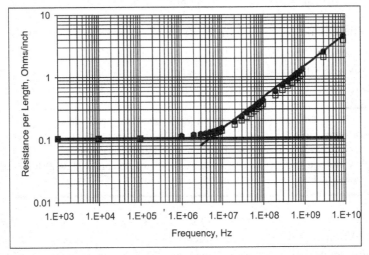

Figure 9-7 Calculated resistance versus frequency of a 5-mil-wide, nearly 50-Ohm microstrip and stripline based on a field solver and the simple approximation of DC resistance and skin-depth-limited current. The circles are the microstrip, and the squares are the stripline, calculated with Ansoft's 2D Extractor field solver. The line is the simple first-order model of DC resistance and skin-depth-limited resistance.

So far, all we have pointed out is that the series resistance of the conductors in a transmission line will increase with increasing frequency. The question of how this frequency-dependent resistance affects loss is addressed later in this chapter.

9.4 Sources of Loss: The Dielectric

An ideal capacitor with air as the dielectric has an infinite DC resistance. Apply a DC voltage, and no current will flow through it. However, if a sine wave of voltage $V = V_0 \sin(wt)$ is applied, a cosine wave of current will flow through the capacitor, determined by the capacitance and the frequency. The current through an ideal capacitor is given by:

$$I = C_0 \frac{dV}{dt} = C_0 \omega V_0 \cos(\omega t) \tag{9-5}$$

where:

I = current through the capacitor

C_0 = capacitance of the capacitor

ω = angular frequency, in radians/sec

V_0 = amplitude of the voltage sine wave applied across the capacitor

There is no loss mechanism in an ideal capacitor. The current that flows through the capacitor is exactly 90 degrees out of phase with the voltage sine wave. If the ideal capacitor were filled with an insulator with a dielectric constant of ε_r, the capacitance would increase to $C = \varepsilon_r \times C_0$.

> **TIP** The current through an ideal capacitor, when filled with an ideal lossless dielectric, will be increased by a factor equal to the dielectric constant. All of the current will be exactly 90 degrees out of phase with the voltage, no power will be dissipated in the material, and there is no dielectric loss.

However, real dielectric materials have some resistivity associated with them. When a real material is placed between the plates of a capacitor with a DC voltage across it, there will be some DC current that flows. This is usually referred to as *leakage current*. It can be modeled as an ideal resistor. The leakage resistance between the signal conductor and the return conductor, which is due to the material between these two conductors, can be approximated with the parallel-plate approximation as:

$$R_{leakage} = \rho \frac{h}{Len \times w} = \frac{1}{\sigma} \frac{h}{Len \times w} \tag{9-6}$$

The amount of leakage current that flows through this resistance is:

$$I_{leakage} = \frac{V}{R_{leakage}} = V \frac{1}{\rho} \frac{Len \times w}{h} = V\sigma \frac{Len \times w}{h} \tag{9-7}$$

where:

$I_{leakage}$ = leakage current through the dielectric

V = applied DC voltage

$R_{leakage}$ = leakage resistance associated with the dielectric

ρ = bulk-leakage resistivity of the dielectric

σ = bulk-leakage conductivity of the dielectric ($\rho = 1/\sigma$)

Len = length of the transmission line

w = line width of the signal path

h = dielectric thickness between the signal and return paths

The leakage current, since it is going through a resistor, has current that is precisely in phase with the voltage. This current will dissipate power in the material and contribute to loss. The power dissipated by a resistor, with a constant voltage across it, is:

$$P = \frac{V^2}{R} \sim \frac{1}{\rho} = \sigma \tag{9-8}$$

where:

P = power dissipated, in watts

V = voltage across the resistor, in Volts

R = resistance, in Ohms

σ = conductivity of the material

For most dielectrics, the bulk resistivity is very high, typically 10^{12} Ohm-cm, so the leakage resistance through a typical 10-inch-long, 50-Ohm transmission line, with $w \sim 2 \times h$, is very high (on the order for 10^{11} Ohms). The resulting DC power loss through this resistance will be insignificant, or less than 1 nWatt.

However, for most materials, the bulk-leakage resistivity of the material is frequency dependent, getting smaller at higher frequencies. This is due to the origin of the leakage current.

There are two ways of getting leakage current through a dielectric. The first is ionic motion. This is the dominant mechanism for DC currents. The reason the DC current is low in most insulators is that the density of mobile charge carriers (i.e., ions in the case of most insulators) is very low, and their mobility is low. In comparison, the free electrons in a metal have a high density, and a very high mobility.

The second mechanism for current flow in a dielectric is by the reorientation of permanent electric dipoles in the material. When a voltage is applied across a capacitor, an electric field is generated. This field will cause some randomly oriented dipoles in the dielectric to align with the field. The initial motion of the negative end of the dipole toward the positive electrode and the positive end of the dipole to the negative electrode looks like a transient current through the material. This is illustrated in Figure 9-8.

Of course, the charges move and the dipoles rotate only a very short distance and for only a very short time. If the voltage applied is a sine wave, then the dipoles will be sinusoidally rotated back and forth. This motion gives rise to an AC current. The higher the sine-wave frequency, the faster the charges will rotate back and forth and the higher the current. The higher the current, the lower the bulk resistivity at that frequency. The resistivity of the material decreases with increasing frequency.

The conductivity of the material is exactly the inverse of the resistivity, $\sigma = 1/\rho$. Just as the bulk resistivity relates to the ability of the material to resist the flow of current, the bulk conductivity relates to the ability of the material to conduct current. A higher conductivity means the material conducts better.

The bulk resistivity of a dielectric decreases with frequency, and the bulk conductivity increases with frequency. If the movements of the dipoles are able to follow the externally applied field and move the same distance for the same applied field, the current they create and the bulk conductivity of the material will increase linearly with frequency.

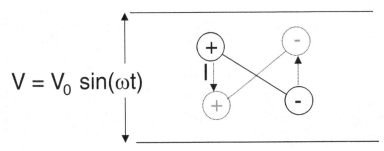

Figure 9-8   The reorientation of the permanent dipoles in the dielectric when the external field changes is really an AC current through the dielectric.

Most dielectrics behave in this way: Their conductivity is constant from DC until some frequency is reached, and then it begins to increase and continues increasing, in proportion to the frequency. Figure 9-9 shows the bulk conductivity of FR4 material, with the transition frequency of about 10 Hz.

At frequencies above this transition frequency, where the motion of dipoles plays a significant role, there can be very high leakage current through a real capacitor as frequency is increased. This current will be in phase with the voltage and will look like a resistor. At higher frequencies, the leakage resistance will go down, and the power dissipated will go up, causing the dielectric to heat up.

TIP   Effectively, the rotation of the dipoles translates electrical energy into mechanical energy. The friction of the motion of the dipoles with their neighbors and the rest of the polymer backbone causes the material to heat up ever so slightly.

The actual heat energy absorbed in normal cases is so small as to contribute to a negligible temperature rise. For the case of the previous 10-inch-long, 50-Ohm microstrip, even at 1 GHz, the leakage resistance of the dielectric is less than 1 kOhm, and the power dissipated is less than 10 mWatts. However, this is not always the case for dielectric loss. The notable exception is a microwave oven. The 2.45-GHz radiation is transferred from electrical energy in the microwaves to mechanical motion and heat by the rotation of water molecules, which strongly absorb the radiation.

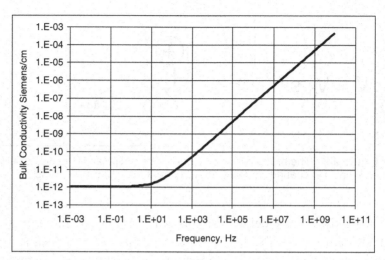

Figure 9-9 Simulated bulk conductivity of FR4 material from DC-leakage current and the AC-dipole motions.

In transmission lines, the dielectric's dipoles will suck energy out of the signal and cause less signal at the far end. It is not enough energy to heat the substrate very much, but it is enough to cause rise-time degradation. The higher the frequency, the higher the AC-leakage conductivity, and the higher the power dissipation in the dielectric.

9.5 Dissipation Factor

At low frequency, the leakage resistance of a dielectric material is constant, and a bulk conductivity is used to describe the electrical properties of the material. This bulk conductivity is related to the density and mobility of ions in the material.

At high frequency, the conductivity increases with frequency due to the increasing motion of the dipoles. The more dipoles in the material that can rotate, the higher the bulk conductivity of the material. The farther the dipoles can move with an applied field, the higher the conductivity. To describe this new material property that measures the dipoles in a material, a new material electrical property must be introduced. This new material property, relating to dipole motion, is called the *dissipation factor* of the material:

$$\sigma = 2\pi f \times \varepsilon_0 \varepsilon_r \times \tan(\delta) \qquad (9\text{-}9)$$

where:

σ = bulk AC conductivity of the dielectric

f = sine-wave frequency, in Hz

ε_0 = permittivity of free space, 8.89×10^{-14} F/cm

ε_r = relative dielectric constant, dimensionless

$\tan(\delta)$ = dissipation of the material, dimensionless

The dissipation factor, usually written as the tangent of the loss angle, $\tan(\delta)$, and also abbreviated sometimes as Df, is a measure of the number of dipoles in the material and how far each of them can rotate in the applied field:

$$\tan(\delta) \sim n \times p \times \theta_{max} \qquad (9\text{-}10)$$

where:

$\tan(\delta)$ = dissipation factor, Df

n = number density of dipoles in the dielectric

p = dipole moment, a measure of the charge and separation of each dipole

θ_{max} = how far the dipoles rotate in the applied field

As the frequency increases, the dipoles move the same distance but faster, so the current increases and the conductivity increases. This is taken into account by the frequency term.

It is important to note that in the definition of the dissipation factor, as the tangent of an angle δ, the angle δ is completely unrelated and separate from the skin-depth term, which, unfortunately, also is described using the Greek letter δ. It is an unfortunate coincidence that both terms relate to two different, unrelated loss processes in transmission lines. Don't confuse them.

In real materials, the dipoles can't really move the same for all applied frequencies. This slight, second-order variation of the dipole motions with frequency is taken into account by the frequency variation of θ_{max}, which causes the dissipation factor to be slightly frequency dependent. The ability of the dipoles to move in the applied field is strongly dependent on how they are attached to the polymer-backbone chain and on the mechanical resonances of the other molecules nearby. At high enough frequency, the dipoles will not be able to respond as fast as at lower frequency, and we would, therefore, expect the dissipation factor to decrease.

There is a whole field of study, called *dielectric spectroscopy*, focused on the frequency dependence of using both the dissipation factor and the dielectric constant to analyze the mechanical properties of polymer chains. Monitoring the dissipation factor and its frequency dependence is sometimes a measure of the degree of cure of a polymer. The more highly cross-linked the polymer, the tighter the dipoles are held and the lower the dissipation factor.

Material	ε	tan(δ)	Relative Cost
FR4	4.0–4.7	0.02	1
DriClad(IBM)	4.1	0.011	1.2
GETek	3.6–4.2	0.013	1.4
BT	4.1	0.013	1.5
Polyimide/glass	4.3	0.014	2.5
CyanateEster	3.8	0.009	3.5
NelcoN6000SI	3.36	0.003	3.5
RogersRF35	3.5	0.0018	5

Figure 9-10 Dissipation factor and dielectric constants of some common interconnect dielectrics.

As a rough rule of thumb, the more tightly the dipoles are held by the polymer, the lower the dielectric constant and the lower the dissipation factor. The polymers with lowest dielectric constants (i.e., Teflon, silicone rubber, and polyethylene) all have very low dissipation factors. Figure 9-10 lists some commonly used interconnect dielectrics and their dissipation factors and dielectric constants.

In most interconnect materials, the dissipation factor of the material is mostly constant with frequency. Often this small variation with frequency can be neglected and a constant value used to accurately predict the loss behavior. However, there may be variations in the dissipation factor from lot to lot, from board to board, and even across the same board, due to variations in the processing of the laminate material. If the material absorbs moisture from humidity, the higher density of absorbed water molecules will increase the dissipation factor. In Polyimide or Kapton flex films, humidity can more than double the dissipation factor.

TIP   To completely describe the electrical properties of a dielectric material, two material terms are required. The *dielectric constant* of the material describes how the material increases the capacitance and decreases the speed of light in the material. The *dissipation factor* describes the number of dipoles and their motion and is a measure of how much the conductivity increases in proportion to the frequency. Both terms may be weakly frequency dependent and will vary from lot to lot and board to board.

Since both terms affect electrical performance, to accurately predict performance, it is important to know how these material properties vary with frequency and how they vary from board to board. If there is uncertainty in the material properties, there will be uncertainty in the performance of the circuits. Some techniques to measure these material properties at high frequency are presented later in this chapter.

9.6 The Real Meaning of Dissipation Factor

Describing a term like dissipation factor as tan(δ) is a bit confusing. Why describe it as the tangent of an angle? What is it the angle of? In practice, it is irrelevant what the angle refers to, and it's really okay to not worry about it. If that is enough for you right now, skip this section. But if you are curious and want to understand the intrinsic behavior of materials, read on.

TIP   To use the term tan(δ) and to describe lossy lines, neither the origin of the term nor the angle to which it refers is important. It is simply a material property that relates to the number of freely moving dipoles in a material and how far they can move with frequency.

To investigate the underlying mechanism for the origin of loss and to look at how to design materials to control their dissipation factor, it is important to look more deeply, at the real origin of dissipation factor.

A dielectric material has two important electrical material properties. One property, the relative dielectric constant, which we introduced in Chapter 5, describes how dipoles re-align in an electric field to increase capacitance. This term describes how much the material will increase the capacitance between two electrodes and the speed of light in the material. However, it does not tell us anything about the losses in the material.

A second, new material-property term, the dissipation factor, is needed to describe how the dipoles slosh back and forth and contribute to a resistance with current that is in phase with the applied voltage sine wave.

However, both of these terms relate to the number of dipoles, how large they are, and how they are able to move. When viewed in the frequency domain with applied sine-wave voltages, one term relates to the motion of the dipoles that are out of phase with the applied field and contributes to increasing capacitance, while the other term relates to the motion of the dipoles that are in phase with the applied voltage and contribute to the losses.

The current through a real capacitor, with an applied sine-wave voltage, can be described by two components. One component of the current is exactly out of phase with the voltage and contributes to the current we think of passing through an ideal lossless capacitor. The other current component is exactly in phase with the applied-voltage wave and looks like the current passing through an ideal resistor, contributing to loss.

To describe these two currents, the out-of-phase and the in-phase components, a formalism has been established based on complex numbers. It is intrinsically a frequency-domain concept because it involves the use of sine-wave voltages and currents. To take advantage of the complex number formalism, we change the dielectric constant and make it a complex number. Using the complex number formalism, the applied voltage can be written as:

$$V = V_0 \exp(i\omega t) \tag{9-11}$$

The current through a capacitor is related to the capacitance by:

$$I = C \frac{dV}{dt} \tag{9-12}$$

Using the complex number formalism, the current through the capacitor, in the frequency domain, can be written as:

$$I = C \frac{dV}{dt} = i\omega C V \tag{9-13}$$

This relationship describes the current through an ideal, lossless capacitor as out of phase with the applied voltage. The "i" term, as used by physics convention, and "j," as used by electrical engineering convention, describes the phase between the current and the voltage as 90 degrees out of phase.

If the capacitor, with an empty capacitance of C_0, is filled with a material that has a dielectric constant, ε_r, then the current through the ideal capacitor is:

$$I = C\frac{dV}{dt} \, i\omega\varepsilon_r C_0 V \tag{9-14}$$

Now, here's the first trick. To describe these two material properties (i.e., the dielectric constant that influences currents out of phase and the dissipation factor that influences currents in phase with the voltage), we change the definition of the dielectric constant. If the dielectric constant is a single, real number, the only current generated is i, out of phase with the voltage. If we make the dielectric constant a complex number, the real part of it will still relate to the current i out of phase, but the imaginary term will convert some of the voltage into a current in phase with the voltage and relate to the losses.

Here is the second trick. If we just describe the new complex dielectric constant as a real and imaginary term—for example, a + ib—when it is multiplied by the factor of i from the voltage, the i from the b term will convert the i term from the voltage into −1. This would make the real part of the current negative, or 180 degrees out of phase from the current. To bring the real part of the current exactly in phase with the voltage, we define the complex dielectric constant with the negative of its imaginary term. The complex dielectric constant is defined in the form:

$$\varepsilon_r = \varepsilon_r{}' - i\varepsilon_r{}'' \tag{9-15}$$

where:

ε_r = complex dielectric constant

$\varepsilon_{r'}$ = real part of the complex dielectric constant

$\varepsilon_{r''}$ = imaginary part of the complex dielectric constant

We introduce the minus sign in the definition of the complex dielectric constant so that in this formalism, the real part of the current comes out as positive and exactly in phase with the voltage. The real part of the new, complex dielectric constant is actually what we have been calling simply the dielectric constant.

TIP   We see now that what we have traditionally called the dielectric constant is actually the real part of the complex dielectric constant. The imaginary part of the complex dielectric constant is the term that creates the current in phase with the voltage and relates to the losses.

Using this definition, the current through an ideal, lossy capacitor is given by:

$$I = i\omega\varepsilon_r C_0 V = i\omega(\varepsilon_r' - i\varepsilon_r'')C_0 V = i\omega\varepsilon_r' C_0 V + \omega\varepsilon_r'' C_0 V \qquad (9\text{-}16)$$

where:

I = current though an ideal, lossy capacitor, in the frequency domain

ω = angular frequency, $= 2\pi \times f$

C_0 = empty space capacitance of the capacitor

V = applied sine-wave voltage, $V = V_0 \exp(i\omega t)$

ε_r = complex dielectric constant

$\varepsilon_{r'}$ = real part of the complex dielectric constant

$\varepsilon_{r''}$ = imaginary part of the complex dielectric constant

By turning the dielectric constant into a complex number, the relationship between the in-phase and out-of-phase currents is compact. Using complex notation, we can generalize the current through a real capacitor. What makes it a little confusing is that the imaginary component of the current, which is 90 degrees out of phase with the voltage and contributes to the capacitive current with which we have been familiar, is actually related to the real part of the complex dielectric constant. The real part of the current—which is in phase with the voltage, behaves like a resistor, and contributes to loss—is actually related to the imaginary part of the dielectric constant.

As a complex number, the dielectric constant has a real part and an imaginary part. We can describe this number as a vector in the complex plane, as shown in Figure 9-11. The angle of the vector with the real axis is called the loss angle, δ. As previously noted, the use of the Greek letter δ to label the loss angle is an unfortunate coincidence with the use of the same Greek letter to label the skin depth. These two terms are completely unrelated; the loss angle relates to the dielectric material, and the skin depth relates to the conductor properties.

The tangent of the loss angle is the ratio of the imaginary to the real component of the dielectric constant:

$$\tan(\delta) = \frac{\varepsilon_r{''}}{\varepsilon_r{'}} \tag{9-17}$$

and:

$$\varepsilon_r'' = \varepsilon_r' \times \tan(\delta) = \varepsilon_r' \times Df \tag{9-18}$$

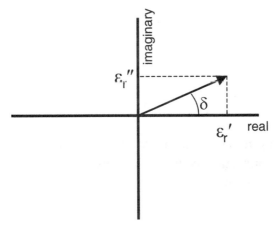

Figure 9-11   The complex dielectric constant plotted in the complex plane. The angle the dielectric-constant vector makes with the real axis is called the loss angle, δ.

It is conventional that rather than use the imaginary part of the dielectric constant directly, we use the tangent of the loss angle, tan(δ), also referred to as Df. In this way, the real part of the dielectric constant and tan(δ) and their possible frequency dependence completely describe the important electrical properties of an insulating material. We also, by habit, omit the distinction that it is the real part of the dielectric constant and simply refer to it as the dielectric constant.

From the relationship above, we can relate the AC-leakage resistance of the transmission line to the imaginary part of the dielectric constant and the dissipation factor as:

$$R_{leakage} = \frac{V}{Real(I)} = \frac{V}{\omega \varepsilon_r'' C_0 V} = \frac{1}{\omega \varepsilon_r'' C_0} = \frac{1}{\omega \varepsilon_r' \tan(\delta) C_0} = \frac{1}{\omega \tan(\delta) C} \qquad (9\text{-}19)$$

In any geometrical configuration of conductors, the same geometrical features that affect the capacitance between the conductors also affect the resistance—but inversely. This is most easily seen for a parallel-plate configuration. The resistance and capacitance are given by:

$$C = \varepsilon_0 \varepsilon_r' \frac{A}{h} \qquad (9\text{-}20)$$

$$\frac{A}{h} = \frac{C}{\varepsilon_0 \varepsilon_r'} \qquad (9\text{-}21)$$

$$R = \frac{1}{\sigma} \frac{h}{A} = \frac{1}{\sigma} \frac{\varepsilon_0 \varepsilon_r'}{C} \qquad (9\text{-}22)$$

Combining these two forms for the resistance between the conductors results in the connection between the bulk-AC conductivity of the material and the dissipation factor:

$$\sigma = \varepsilon_0 \varepsilon_r' \omega \tan(\delta) \qquad (9\text{-}23)$$

where:

σ = bulk-AC conductivity of the dielectric material

ε_0 = permittivity of free space = 8.89×10^{-14} F/cm

$\varepsilon_{r'}$ = real part of the dielectric constant

$\varepsilon_{r''}$ = imaginary part of the dielectric constant

$\tan(\delta)$ = dissipation factor of the dielectric

δ = loss angle of the dielectric

R = AC-leakage resistance between the conductors

C = capacitance between the conductors

h = dielectric thickness between the conductors

A = area of the conductors

ω = angular frequency = $2 \times \pi \times f$, with f = sine-wave frequency

---

TIP   Even though the dissipation factor is only weakly frequency dependent, once again, we see that the bulk-AC conductivity of a dielectric will increase linearly with frequency due to the ω term. Likewise, since the power dissipated by the leakage resistance is proportional to the bulk-AC conductivity, the power dissipation will also increase linearly with frequency. This is the fundamental origin of the main problem that lossy lines create for signal integrity.

9.7 Modeling Lossy Transmission Lines

The two loss processes for attenuating signals in a transmission line are the series resistance through the signal- and return-path conductors and the shunt resistance through the lossy dielectric material. Both of these resistors have resistances that are frequency dependent.

It is important to note that an ideal resistor has a resistance that is constant with frequency. We have shown that in an ideal lossy transmission line, the two resistances used to describe the losses are more complicated than simple ideal resistors. The series resistance increases with the square root of frequency due to skin-depth effects. The shunt resistance decreases with frequency due to the dissipation factor of the material and the rotation of dipole molecules.

In Chapter 7, we introduced a new, ideal circuit element, the ideal distributed transmission line. It was described by a characteristic impedance and a time delay. This model distributes the properties of the transmission line throughout its length. An ideal lossy distributed transmission-line model will add to this lossless model two loss processes: series resistance increasing with the square root of frequency and shunt resistance decreasing inversely with frequency. This is the basis of the new ideal lossy transmission line that is implemented in many simulators. The two factors that are specified, in addition to the characteristic impedance and time delay, are the dissipation factor and resistance per length, R_L, of the form:

$$R_L = R_{DC} + R_{AC}\sqrt{f} \tag{9-24}$$

where:

R_L = resistance per length of the conductors

R_{DC} = resistance per length at DC

R_{AC} = coefficient of the resistance per length that is proportional to $f^{0.5}$

To gain insight into the behavior of ideal lossy lines, we can start with the approximation of a transmission line as an n-section LC circuit and add the loss terms and evaluate the behavior of the circuit model.

In Chapter 7, we showed that an ideal, distributed lossless transmission line can be approximated with an equivalent circuit model consisting of lumped-circuit-model sections with a shunt C and series L. This model is sometimes called the *first-order model* of a transmission line, an *n-section lumped-circuit model*, or a *lossless model* of a transmission line. A piece of it is shown in Figure 9-12.

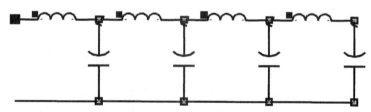

Figure 9-12 Four sections of an n-section LC model as an approximation of an ideal lossless, distributed transmission line.

This model is an approximation. However, it can be a very accurate approximation, to very high bandwidth, by using enough sections. We showed that the minimum number of sections required to achieve a bandwidth, BW, for a time delay, TD, is given by:

$$n = 10 \times BW \times TD \tag{9-25}$$

where:

n = number of sections for an accurate LC model

BW = bandwidth of the model, in GHz

TD = time delay of the transmission line being approximated, in nsec

For example, if the bandwidth required for the model is 2 GHz, and the time delay of the line is 1 nsec, which is about 6 inches long physically, the minimum number of sections required for an accurate model is n ~ 10 × 2 × 1 = 20.

However, a significant limitation to this ideal lossless model is that it would still always be a lossless model. Using this first-order equivalent circuit model as a starting place, we can modify it to account for the losses. In each section, we can add the effect of the series resistance and the shunt resistance. A small section of the n-section lumped-circuit approximation for an ideal lossy transmission would have four terms that describe it:

C = capacitance

L = loop self-inductance

R_{series} = series resistance of the conductors

R_{shunt} = dielectric-loss shunt resistance

If we double the length of the transmission line, the total C doubles, the total L doubles, and the total R_{series} doubles. However, the total R_{shunt} is cut in half. If we double the length of the line, there is more area through which the AC-leakage current can flow, so the shunt resistance decreases.

For this reason, it is conventional to use the conductance of the dielectric-leakage resistance rather than the resistance to describe it. The conductance, denoted by the letter G, is defined as G = 1/R. Based on the resistance, the conductance is:

$$R_{leakage} = \frac{1}{\omega \tan(\delta) C} \qquad (9\text{-}26)$$

$$G = \frac{1}{R_{leakage}} = \omega \tan(\delta) C \qquad (9\text{-}27)$$

If the length of the transmission line doubles, the shunt resistance is cut in half, but the conductance doubles. When we use G to describe the losses, we still model the losses as a resistor whose resistance decreases with frequency; we just use the G parameter to describe it. Using the conductance instead of the shunt resistance, the four terms that describe a lossy transmission line all scale with length. It is conventional to refer to their values per unit length. These four terms are referred to as the *line parameters* of a transmission line:

R_L = series resistance per length of the conductors

C_L = capacitance per length

L_L = series loop inductance per length

G_L = shunt conductance per length from the dielectric

We will use this ideal, second-order n-section lumped-circuit model to approximate an ideal lossy transmission line, which in turn is an approximation to a real transmission line. An example of the equivalent n-section RLGC transmission-line model is shown in Figure 9-13.

The number of sections we use will depends on the length of the line and the bandwidth of the model. The minimum number of sections required is still roughly $10 \times BW \times TD$.

This is an equivalent circuit model. We can apply circuit theory to this circuit and predict the electrical properties. The mathematics is complicated because coupled, second-order differential equations are involved and also because the equations are complex, and the resistances have values that vary with frequency. The easiest domain in which to solve the equations is the frequency domain. We assume that signals are sine waves of voltage and, from the impedance, that the sine waves of current can be calculated. The results only will be discussed.

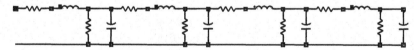

Figure 9-13 Four sections of an n-section RLGC model for an ideal lossy transmission line; an approximation of an ideal distributed lossy transmission line.

In a lossless line, the resistance and conductance are equal to zero. This lossless circuit model would predict an interconnect that propagates a signal undistorted. The instantaneous impedance the signal sees at each step along the way is equal to the characteristic impedance of the line:

$$Z_0 = \sqrt{\frac{L_L}{C_L}} \qquad (9\text{-}28)$$

The speed of a signal is given by:

$$v = \frac{1}{\sqrt{C_L \times L_L}} \qquad (9\text{-}29)$$

where:
Z_0 = characteristic impedance
v = speed of the signal
C_L = capacitance per length
L_L = inductance per length

In this model, the ideal L, C, and Z_0 terms as well as the time delay are all constant with frequency. These are the only terms that define the ideal, lossless transmission line. A signal entering at one end of the line will exit the other end, with no change in amplitude. The only impact on a sine wave other than a reflection due to a possible impedance change would be a phase shift in transmission.

However, when the R and G terms are added to the model, the behavior of the ideal lossy transmission line is slightly different from the ideal lossless transmission line. When the differential equations are solved, the results are rather complicated. The solution, in the frequency domain, makes no assumption on how C_L, L_L, R_L, or G_L vary with frequency. At each frequency, they may change, or they may be constant.

The final solution ends up having three important features:

1. The characteristic impedance is frequency dependent and complex.
2. The velocity of a sine-wave signal is frequency dependent.
3. A new term is introduced that describes the attenuation of the sine-wave amplitude as it propagates down the line. This attenuation is also frequency dependent.

When the dust clears, the exact values for the characteristic impedance, velocity, and attenuation per length are given by:

$$Z_0 = \sqrt{\frac{R_L + i\omega L_L}{G_L + i\omega C_L}} \qquad (9\text{-}30)$$

$$v = \frac{\omega}{\sqrt{\frac{1}{2}\left[\sqrt{(R_L^2 + \omega^2 L_L^2)(G_L^2 + \omega^2 C_L^2)} + \omega^2 L_L C_L - R_L G_L\right]}} \qquad (9\text{-}31)$$

$$\alpha_n = \sqrt{\frac{1}{2}\left[\sqrt{(R_L^2 + \omega^2 L_L^2)(G_L^2 + \omega^2 C_L^2)} - \omega^2 L_L C_L + R_L G_L\right]} \qquad (9\text{-}32)$$

where:

Z_0 = characteristic impedance

v = speed of a signal

α_n = attenuation per length of the amplitude, in nepers/length

ω = angular frequency of the sine wave, in radians/sec

R_L = series resistance per length of the conductors

C_L = capacitance per length

L_L = series loop inductance per length

G_L = shunt conductance per length from the dielectric

These equations are formidable, and though they can be implemented in a spreadsheet, are difficult to use to gain useful engineering insight. To simplify the algebra, one of the approximations that is commonly made is that the lines are lossy but not too lossy. This is called the *low-loss approximation*. The approximation is that the series resistance, R_L, is $R_L \ll \omega L_L$ and the shunt conductance, G_L, is $G_L \ll \omega C_L$.

This approximation assumes that the impedance associated with the series resistance of the conductors is small compared with the series impedance associated with the loop inductance. Likewise, the shunt current through the leakage resistance of the dielectric is small compared to the shunt current through the capacitance between the signal and return paths.

For 1-ounce copper traces, the series resistance above about 10 MHz will increase with the square root of frequency, and ωL_L will increase linearly with frequency. At some frequency, this approximation will be good, and it will get better at higher and higher frequencies.

The DC resistance per length of a 1-ounce copper trace is:

$$R_L = \frac{0.5}{w} \tag{9-33}$$

where:

R_L = resistance per length, in Ohms/inch

w = line width, in mils

Above 10 MHz, the current will flow in a thinner cross section, and rather than the geometrical thickness of 34 microns for 1-ounce copper, the skin depth will determine the thickness of the current distribution. The skin depth for copper is:

$$\delta = 66\sqrt{\frac{1}{f}} \tag{9-34}$$

where:

δ = skin depth, in microns

f = sine-wave-frequency component, in MHz

The AC resistance per length of a 1-ounce copper trace, above about 10 MHz, is about:

$$R_L = \frac{0.5t}{w\delta} = \frac{0.5 \times 34}{w \times 66}\sqrt{f} = \frac{0.25}{w}\sqrt{\frac{\omega}{2\pi \times 10^6}} = \frac{1 \times 10^{-4}}{w}\sqrt{\omega} \tag{9-35}$$

where:

R_L = resistance per length, in Ohms/inch

δ = skin depth, in microns

t = geometrical thickness, in microns for 1-oz copper = 34 microns

w = line width, in mils

f = sine-wave-frequency component, in MHz

ω = sine-wave-frequency component, in radians/sec

The inductance per length is roughly 9 nH/inch for a 50-Ohm line. The low-loss regime happens when $\omega L_L \gg R_L$ or:

$$\omega \times 9 \times 10^{-9} \gg \frac{1 \times 10^{-4}}{w} \sqrt{\omega} \tag{9-36}$$

$$\omega \gg \left(\frac{1}{w}\right)^2 \left(\frac{1 \times 10^{-4}}{9 \times 10^{-9}}\right)^2 = \frac{1 \times 10^8}{w^2} \tag{9-37}$$

$$f = \frac{\omega}{2\pi} \gg \frac{1 \times 10^8}{2\pi w^2} \sim \frac{2 \times 10^7}{w^2} \tag{9-38}$$

where:

R_L = resistance per length, in Ohms/inch

ω = sine-wave-frequency component, in radians/sec for the low-loss regime

f = sine-wave-frequency component, in Hz for the low-loss regime

w = line width, in mils

TIP   This is a startling result. The conclusion is that for line widths wider than 3 mils, the low-loss regime is for sine-wave-frequency components above 2 MHz. In this regime, the impedance of the series resistance is much less than the reactance of the series inductance. For lines wider than 3 mils, the low-loss regime begins at even lower frequencies. The very lossy regime is actually in the low frequency, below the frequency where skin depth plays a role.

The conductance will roughly increase linearly with frequency, and the capacitance will be roughly constant with frequency. The low-loss regime happens when $G_L \ll \omega C_L$. This is when $\tan(\delta) \ll 1$. For virtually all interconnect materials, the dissipation factor is less than 0.02, and the interconnect is always in the low-loss regime.

TIP   The low-loss regime for circuit-board interconnects with 3-mil-wide traces and wider is for frequencies above 2 MHz, which is where most important frequency components are.

The conclusion is that the low-loss approximation is a very good approximation for all important frequency ranges of interest in high-speed digital applications.

9.8 Characteristic Impedance of a Lossy Transmission Line

In an ideal lossy transmission line, the characteristic impedance becomes frequency dependent and is complex. The characteristic impedance is given by:

$$Z_0 = \sqrt{\frac{R_L + \omega L_L}{G_L + \omega C_L}} \qquad (9\text{-}39)$$

With a bit of algebra, after the dust clears, the real part and imaginary part of the characteristic impedance are given by:

$$\text{Re}(z_0) = \frac{1}{\sqrt{G_L^2 + \omega^2 C_L^2}} \sqrt{\frac{1}{2}\left[\sqrt{\left(R_L^2 + \omega^2 L_L^2\right)\left(G_L^2 + \omega^2 C_L^2\right)} + \omega^2 L_L C_L + R_L G_L\right]} \qquad (9\text{-}40)$$

$$\text{Imag}(z_0) = \frac{1}{\sqrt{G_L^2 + \omega^2 C_L^2}} \sqrt{\frac{1}{2}\left[\sqrt{\left(R_L^2 + \omega^2 L_L^2\right)\left(G_L^2 + \omega^2 C_L^2\right)} - \omega^2 L_L C_L + R_L G_L\right]} \qquad (9\text{-}41)$$

where:

$\text{Re}(Z_0)$ = real part of the characteristic impedance

$\text{Imag}(Z_0)$ = imaginary part of the characteristic impedance

R_L = series resistance per length of the conductors

C_L = capacitance per length

L_L = series loop inductance per length

G_L = shunt conductance per length from the dielectric

ω = angular frequency

In the low-loss regime, the characteristic impedance reduces to:

$$\text{Re}(Z_0) = \sqrt{\frac{L_L}{C_L}} \qquad (9\text{-}42)$$

$$\text{Imag}(Z_0) = 0 \qquad (9\text{-}43)$$

The low-loss approximation for the characteristic impedance is exactly the same as for the lossless characteristic impedance. The terms that affect the characteristic impedance vary as R^2 and G^2 compared to $\omega^2 L^2$ or $\omega^2 C^2$. Our assumption of the low-loss regime is seen to introduce less than 1% errors, when we are at frequencies above 10 times the boundary of roughly 2 MHz for a 3-mil-wide line.

We can use the magnitude of the characteristic impedance as a rough measure of the impact from the losses. The magnitude is given by:

$$\text{Mag}(Z_0) = \sqrt{\text{Re}(Z_0)^2 + \text{Imag}(Z_0)^2} \qquad (9\text{-}44)$$

Figure 9-14 plots the magnitude of the complex characteristic impedance of a 3-mil-wide, 50-Ohm microstrip in FR4 using the exact relationship above. This includes the conductor loss and the dielectric loss. We see that for frequencies above about 10 MHz, the complex characteristic impedance is very close to the lossless value. This transition frequency will move toward lower frequency for wider and less lossy lines.

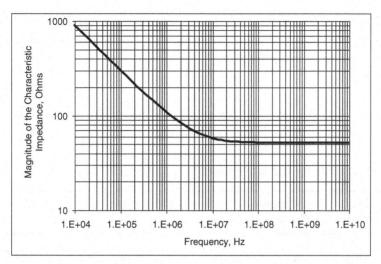

Figure 9-14   The magnitude of the complex characteristic impedance of a 50-Ohm lossy microstrip in FR4 shows that above about 10 MHz, the lossy characteristic impedance is very close to the lossless impedance. The low-loss regime is above 10 MHz.

T I P   In the low-loss regime, there is no impact on the characteristic impedance from the losses.

As described earlier, there may be some frequency dependence to the inductance from skin-depth effects. Above about 100 MHz, the skin depth is much thinner than the geometrical thickness, and the inductance is constant with frequency above this point. There may be some frequency dependence of the capacitance due to the real part of the dielectric constant changing with frequency. These terms may contribute to a slight frequency dependence of the characteristic impedance. In real interconnects, the impact from these effects is usually not noticeable.

9.9 Signal Velocity in a Lossy Transmission Line

A consequence of the solution to the circuit model of a lossy transmission line is that the velocity of a sine wave is complicated. The velocity is given by:

$$v = \frac{\omega}{\sqrt{\frac{1}{2}\left[\sqrt{\left(R_L^2 + \omega^2 L_L^2\right)\left(G_L^2 + \omega^2 C_L^2\right)} + \omega^2 L_L C_L - R_L G_L\right]}} \qquad (9\text{-}45)$$

In the low-loss regime, where the impedance of the resistance is much less than the reactance of the inductance and the dissipation factor is $\ll 0.1$, the velocity can be approximated by:

$$v = \frac{1}{\sqrt{L_L C_L}}$$

(9-46)

This is exactly the same result as for a lossless line.

TIP   The conclusion is that in the low-loss regime, the velocity of a signal is not affected by the losses.

Using the exact form for the velocity, we can evaluate just how constant the velocity is and at what point the speed will vary with frequency. This effect of a velocity that is frequency dependent is called *dispersion,* and in this case the dispersion is caused by loss. Figure 9-15 shows the frequency dependence to the speed of a signal for a worst case of a 3-mil-wide line in an FR4 50-Ohm micro-strip, including both the dielectric and conductor losses.

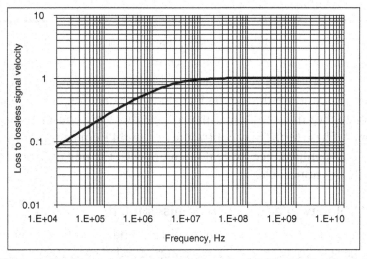

Figure 9-15   Dispersion due to losses for a 3-mil-wide, 50-Ohm line in FR4. The ratio of the lossy velocity to the lossless velocity is plotted.

The impact from the losses is to slow down the lower frequencies more than the higher frequencies. At lower frequencies, the series resistive impedance dominates over the series reactive impedance from the loop inductance. In addition, the line looks more lossy and the signal speed is reduced. When speed varies with frequency, we call this dispersion. It arises from two material properties of related mechanisms: the frequency-dependent dielectric constant and losses.

Dispersion will cause the higher-frequency components to travel faster than the lower-frequency components. In the time domain, the fast edge will arrive first, followed by a slowly rising tail, effectively increasing the rise time. However, if the losses are ever large enough to have a noticeable impact on rise-time degradation, the impact from the attenuation will usually be far larger than the impact from dispersion.

TIP   For the worst-case line, 3 mils wide in FR4, the low-loss regime is above about 10 MHz. In this regime, the speed is independent of frequency, and the dispersion from the losses is negligible.

Dispersion can also be caused by reflections. When the signals bounce back and forth due to reflections, there can be overlap of some of the signal with a delayed portion of the signal. This will cause an anomalous phase shift at some frequencies, which will appear as a small variation of the time delay and the velocity of the signal. We sometimes refer to this as *anomalous phase shift* or *anomalous dispersion*.

9.10 Attenuation and dB

The dominant effect on a signal from the losses in a line is a decrease in amplitude as the signal propagates down the length of the line. If a sine-wave-voltage signal with amplitude V_{in} is introduced in a transmission line, its amplitude will drop as it moves down the line. Figure 9-16 shows what the sine wave might look like at different positions, if we could freeze time and look at the sine wave as it exists on the line. This is for the case of a 1-GHz sine wave on a 40-inch-long, 50-Ohm microstrip in FR4 with a 10-mil-wide trace.

The amplitude drops off not linearly but exponentially with distance. This can be described with an exponent either to base e or to base 10. Using base e, the output signal is given by:

$$V(d) = V_{in} \exp(-A_n) = V_{in} \exp(-d \times \alpha_n) \qquad (9\text{-}47)$$

where:

$V(d)$ = voltage on the line at position d

d = position along the line, in inches

V_{in} = amplitude of the input voltage wave

A_n = total attenuation, in nepers

α_n = attenuation per length, in nepers/inch

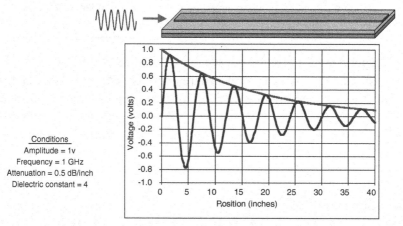

Figure 9-16   Amplitude of a 1-GHz sine-wave signal as it would appear on a 10-mil-wide, 50-Ohm transmission line in FR4.

When using the base e, the units of attenuation are dimensionless but are still labeled as nepers, after John Napier, the Scotsman who is credited with the introduction of the base e exponent published in 1614. One confusing aspect of Napiers is the spelling. Napiers, napers, and nepers all refer to the same unit and are commonly used alternative spellings of his name. Though dimensionless, we use the label to remind us it is the attenuation using base e.

For example, if the attenuation were 1 napier, the final amplitude would be $\exp(-1) = 37\%$ of the input amplitude. If the attenuation were 2 napiers, the output amplitude would be $\exp(-2) = 13\%$ of the input voltage.

Likewise, given the input and output amplitude, the attenuation can be found from:

$$A_n = -\ln\left(\frac{V(d)}{V_{in}}\right) \tag{9-48}$$

There is some ambiguity about the sign. In all passive interconnects, there will never be any gain. The output voltage is always smaller than the input voltage. An exponent of 0 will have exactly the same output amplitude as input amplitude. The only way to get a reduced amplitude is by having a negative exponent. Should the minus sign be placed explicitly in the exponent, or should it be part of the attenuation? Both ways are conventionally done. The attenuation is sometimes referred to as -2 napiers or 2 napiers. Since it is always referred to as an attenuation, there is no ambiguity.

Usually, the attenuation is considered to be a value greater than zero and a positive number. In this view, the negative sign is not part of the attenuation term but part of the exponent.

It is more common to describe attenuation using a base 10 than a base e. The output amplitude is of the form:

$$V(d) = V_{in} 10^{-\frac{A_{dB}}{20}} = V_{in} 10^{\left(-d \times \frac{\alpha_{dB}}{20}\right)} \tag{9-49}$$

where:
V(d) = voltage on the line at position d
d = position along the line, in inches
V_{in} = amplitude of the input voltage wave
A_{dB} = total attenuation in dB
α_{dB} = attenuation per length, in dB/inch
20 = factor to convert dB into amplitude, described below

TIP   The units used to describe attenuation are in decibels, or dB. These units are used throughout engineering and wherever they appear, they leave confusion in their wake. Understanding the origin of this unit will help remove the confusion.

The decibel was created over 100 years ago by Alexander Graham Bell. He started his career as a physician, studying and treating children with hearing problems. To quantify the degree of hearing loss, he developed a standardized set of sound intensities and quantified individuals' ability to hear them. He found that the sensation of loudness depended not on the power intensity of the sound but on the log of the power intensity. He developed a scale of loudness that started with the quietest whisper that could be heard as a unit of 0 and the sound that produced pain as 10.

All other sounds were distributed on this scale, in a log ratio of the actual measured power level. If the loudness increased from 1 to 2, for example, the loudness was perceived to have doubled. However, the actual power level in the sound was measured as having increased by $10^2/10^1 = 10$. What Bell established is that the quality of perceived loudness change depends not on the power-level change but on a unit that is proportional to the log of the power-level change.

The units of the Bell scale of loudness were called Bells, with 0 Bells as the quietest whisper that can be heard. The actual power density at the ear has since been quantified for each sound level. The quietest sound level that can be perceived at our peak sensitivity at about 2 kHz is the start of the scale, 0 Bell. It corresponds to a power density of 10^{-12} watts/m². For the loudest sound just at the pain threshold, 10 Bells, the power level is 10^{-2} watts/m².

As the Bell scale of perceived loudness gained wide acceptance, the last L was dropped, and it became the Bel scale of perceived loudness. Over time, it was found that the range of the scale, 0 Bels to 10 Bels, was too small, given the incredible range of perceived loudness. Instead of loudness measured in Bels, the scale was changed to deciBels, where the preface "deci" means 1/10. The perceived loudness scale now starts with 0 decibels (the *b* having been lowercased) at the low end for the whisper and has 100 decibels at the onset of pain. The decibel is usually abbreviated dB.

TIP   Over the years, the decibel scale has been adopted for other applications in addition to loudness, but in every case, the decibel has retained its definition as the log of the ratio of two powers. The most important property of the decibel scale is that it always refers to the log of the ratio of two powers.

In virtually all engineering applications, the log of the ratio of two powers, P_1 and P_0, is also measured in Bels: the number of Bels $= \log(P_1/P_0)$. Since 1 Bel = 10 decibels, the ratio, in dB, is:

$$\text{ratio(dB)} = 10 \times \log \frac{P_1}{P_0} \qquad (9\text{-}50)$$

For example, if the power increases by a factor of 1000, the increase, in Bels, is $\log(1000) = 3$ Bels. In dB, this is 10×3 Bels $= 30$ dB. A decrease in the power level, where the output power is only 1% the input power, is $\log(10^{-2}) = -2$ Bels, or 10×-2 Bels $= -20$ dB.

When the power level changes by any arbitrary factor, the change can be described in dB, but calculating the log value requires a calculator. If the power doubles, the change in dB is $10 \times \log(2) = 10 \times 0.3 = 3$ dB. We often use the expression "a 3 dB change" to refer to a doubling in the power level. If the power change were to drop by 50%, the change in dB would be $10 \times \log(0.5) = -3$ dB.

The ratio of the actual power levels can be extracted from the ratio in dB by:

$$\text{ratio} = \frac{P_1}{P_0} = 10^{\frac{\text{ratio(dB)}}{10}} \qquad (9\text{-}51)$$

The first step is to convert the dB into Bels. This is the exponent to base 10. For example, if the ratio in dB is 60, the power level ratio is $10^{60/10} = 10^6 = 1,000,000$. If the dB value is -3 dB, the power level ratio is $10^{-3/10} = 10^{-0.3} = 0.5$ or 50%.

It is important to always keep in mind three rules about the dB scale:

1. The dB scale *always* refers to the log of the ratio of two powers or energies.
2. When measured in dB, the exponent to base 10 of the ratio of two powers is just dB/10.
3. When converting from dB to the actual ratio of the powers, a factor of 10 is used.

The distinction about power is important whenever the ratio of two other quantities is measured. When the ratio, r, of two voltages is measured—for example, V_0 and V_1—the units are dimensionless: $r = \log(V_1/V_0)$. But we cannot use dB to measure the ratio since the dB unit refers to the ratio of two powers or energies. A voltage is not an energy; it is an amplitude.

We can refer to the ratio of the powers associated with the two voltage levels, or $r_{dB} = 10 \times \log(P_1/P_0)$. How are the power levels related to the voltage levels? The energy in a voltage wave is proportional to the square of the voltage amplitude, or $P \sim V^2$.

The ratio of the voltages, in dB, is really the ratio of the associated powers, as:

$$r_{dB} = 10 \times \log\left(\frac{P_1}{P_0}\right) = 10 \times \log\left(\frac{V_1^2}{V_0^2}\right) = 10 \times 2 \log\left(\frac{V_1}{V_0}\right) = 20 \log\left(\frac{V_1}{V_0}\right) \qquad (9\text{-}52)$$

TIP   Whenever the ratio of two amplitudes is measured in units of dB, it is calculated by taking the log of the ratio of their associated powers. This is equivalent to multiplying the log of the ratio of the voltages by 20.

For example, a change in the voltage from 1 v to 10 v, measured in dB, is $20 \times \log(10/1) = 20$ dB. The voltage increased by only a factor of 10. However, the underlying powers corresponding to the 1 v and 10 v increased by a factor of 100. This is reflected by the 20-dB change in the power level.

dB is *always* a measure of the change in power. When the amplitude decreases to half its original value, or is reduced by 50%, the ratio of the final to the initial value, in dB, is $20 \times \log(0.5) = 20 \times -0.3 = -6$ dB. If the voltage is reduced by 50%, the power in the signal must have decreased by $(50\%)^2$, or 25%. If the ratio of two powers is 25%, in dB, this is $10 \times \log(0.25) = 10 \times -0.6 = -6$ dB.

TIP   When referring to an energy or power, the factor of 10 is used in calculating the dB value. When referring to an amplitude, a factor of 20 is used. An amplitude is a quantity such as voltage, a current, or an impedance.

From the ratio in dB, the actual ratio of the voltages can be calculated as:

$$\text{ratio} = \frac{V_1}{V_0} = 10^{\frac{\text{ratio}_{dB}}{20}} \qquad (9\text{-}53)$$

For example, if the ratio in dB is 20 dB, the ratio of the amplitudes is $10^{20/20} = 10^1 = 10$. If the ratio in dB is -40 dB, the ratio of the voltages is $10^{-40/20} = 10^{-2} = 0.01$. If the ratio in dB is negative, this means the final value is always less than the initial value. Figure 9-17 lists a few examples of the ratio of the voltages, their associated powers, and the ratio in dB.

Voltage Ratio	Power Ratio	dB
100	10,000	40
10	100	20
2	4	6
1.4	2	3
1	1	0
0.7	0.5	−3
0.5	0.25	−6
0.1	0.01	−20
0.01	0.0001	−40

Figure 9-17   Ratio of the voltages, their corresponding powers, and their ratio in dB.

9.11 Attenuation in Lossy Lines

When a sine-wave signal propagates down a transmission line, the amplitude of the voltage decreases exponentially. The total attenuation, measured in dB, increases linearly with length. In FR4, a typical attenuation of a 1-GHz signal might be 0.1 dB/inch. In propagating 1 inch, the attenuation is 0.1 dB, and the signal amplitude has dropped to $V_{out}/V_{in} = 10^{-0.1/20} = 99\%$. In propagating 10 inches, the attenuation is 0.1 dB/inch $\times$ 10 inches = 1 dB, and the amplitude has dropped to $V_{out}/V_{in} = 10^{-1/20} = 89\%$.

Attenuation is a new term that describes a special property of lossy transmission lines. It is a direct result of the solution of the second-order, linear differential equation for the lossy RLCG circuit model. The attenuation per length, usually denoted by alpha, α_n, in nepers/length, is given by:

$$\alpha_n = \sqrt{\frac{1}{2}\left[\sqrt{\left(R_L^2 + \omega^2 L_L^2\right)\left(G_L^2 + \omega^2 C_L^2\right)} - \omega^2 L_L C_L + R_L G_L\right]} \tag{9-54}$$

In the low-loss approximation, it is approximated by:

$$\alpha_n = \frac{1}{2}\left(\frac{R_L}{Z_0} + G_L Z_0\right) \tag{9-55}$$

There is a simple conversion from a ratio of two voltages in nepers to the same ratio in dB. If r_n is the ratio of the two voltages in nepers and r_{dB} is the ratio of the same voltages in dB, then, since they equal the same ratio of voltages:

$$10^{\frac{r_{dB}}{20}} = e^{r_n} \tag{9-56}$$

$$r_{dB} = r_n \times 20 \log e = 8.68 \times r_n \tag{9-57}$$

Using this conversion, the attenuation per length of a transmission line, in dB/length, is:

$$\alpha_{dB} = 8.68 \alpha_n = 8.68 \times \frac{1}{2}\left(\frac{R_L}{Z_0} + G_L Z_0\right) = 4.34\left(\frac{R_L}{Z_0} + G_L Z_0\right) \tag{9-58}$$

where:

α_n = attenuation of the amplitude, in nepers/length

α_{dB} = attenuation, in dB/length

R_L = series resistance per length of the conductors

C_L = capacitance per length

L_L = series loop inductance per length

G_L = shunt conductance per length from the dielectric

Z_0 = characteristic impedance of the line, in Ohms

Surprisingly, though this is the attenuation in the frequency domain, there is no intrinsic frequency dependence to the attenuation.

TIP   If the series resistance per length of the conductors were constant with frequency and the shunt dielectric conductance per length were constant with frequency, the attenuation of the transmission line would be constant with frequency. Every frequency would see the same amount of loss.

Every frequency would be treated exactly the same in propagating through the transmission line. Though the amplitude of the signal would decrease as it propagates through the transmission line, every frequency would have the same attenuation, and the shape of the signal's spectrum would be preserved. The rise time would be unchanged. The result would be the same rise time coming out of the line as going in.

However, as we saw earlier, this is not how real lossy transmission lines on typical laminate substrates behave. In the real world, to a very good approximation, the resistance per length will increase with the square root of frequency due to skin depth, and the shunt conductance per length will increase linearly with frequency due to the dissipation factor of the dielectric. This means the attenuation will increase with frequency. Higher-frequency sine waves will be attenuated more than lower-frequency sine waves. This is the primary mechanism that will decrease the bandwidth of signals when propagating down a lossy line.

There are two parts to the attenuation per length. The first part relates the attenuation from the conductor. The attenuation from just the conductor loss is:

$$\alpha_{cond} = 4.34 \left(\frac{R_L}{Z_0} \right) \tag{9-59}$$

The second part of the attenuation relates to the losses from just the dielectric materials:

$$\alpha_{diel} = 4.34 (G_L Z_0)$$

(9-60)

The total attenuation is:

$$\alpha_{dB} = \alpha_{cond} + \alpha_{diel}$$

(9-61)

where:

α_{cond} = attenuation per length from just the conductor loss, in dB/length

α_{diel} = attenuation per length from just the dielectric loss, in dB/length

α_{dB} = total attenuation, in dB/length

R_L = series resistance per length of the conductors

C_L = capacitance per length

L_L = series loop inductance per length

G_L = shunt conductance per length from the dielectric

Z_0 = characteristic impedance of the line, in Ohms

In the skin-depth-limited regime, the resistance per length of a stripline, ignoring the slight resistance in the return path, was approximated in Equation 9-35, earlier in this chapter, with frequency in MHz, as:

$$R_L = \frac{0.5t}{w\delta} = \frac{0.5 \times 34}{w \times 66} \sqrt{f}$$

(9-62)

or for frequency in GHz:

$$R_L = \frac{0.5t}{w\delta} = \frac{0.5 \times 34}{w \times 66} \sqrt{1000} \sqrt{f} = \frac{8.14}{w} \sqrt{f}$$

(9-63)

where:

R_L = resistance per length, in Ohms/inch of a stripline

δ = skin depth, in microns

t = geometrical thickness, in microns (for 1-ounce copper) = 34 microns

w = line width, in mils

f = sine-wave-frequency component, in GHz

0.5 factor from assuming equal current flows in both surfaces of the trace

Combining these results, the attenuation from just the conductor in a stripline is approximately:

$$\alpha_{cond} = 4.34 \left(\frac{R_L}{Z_0} \right) = 4.34 \times \frac{1}{Z_0} \times \frac{8.14}{w} \sqrt{f} = \frac{36}{wZ_0} \sqrt{f} \tag{9-64}$$

The total attenuation from the conductor for the entire length of stripline transmission line is:

$$A_{cond} = Len \times \alpha_{cond} = Len \frac{36}{wZ_0} \sqrt{f} \tag{9-65}$$

where:

R_L = resistance per length, in Ohms/inch of the stripline

w = line width, in mils

f = sine-wave-frequency component, in GHz

Z_0 = characteristic impedance of the line, in Ohms

A_{cond} = total attenuation from just the conductor loss, in dB

Len = length of the transmission line, in inches

For example, at 1 GHz, a 50-Ohm stripline line with a line width of 10 mils will have an attenuation per length from just the conductor of $\alpha_{cond} = 36/(10 \times 50) \times 1 = 0.07$ dB/inch. If the line is 36 inches long, typical of a backplane application, the total attenuation from one end to the other would be 0.07 dB/inch $\times$ 36 inches = 2.5 dB. The ratio of the output voltage to the input voltage is $V_{out}/V_{in} = 10^{-2.5/20} = 75\%$. This means there will only be 75% of the amplitude left at the end of the line for 1-GHz frequency components as a result of *just* the conductor losses. Higher-frequency components will be attenuated even more. Of course, this is an approximation. A more accurate value can be obtained using a 2D field solver that allows calculation of the precise current distribution and how it changes with frequency.

Figure 9-18 compares this estimate for the attenuation from just the conductor loss to the calculated attenuation for the case of a 10-mil-wide, 50-Ohm micro-strip trace using a 2D field solver. The approximation is a reasonable estimate.

It should be noted that these estimates assume that the surfaces of the copper trace are smooth. When the surface roughness is comparable to the skin depth, the series resistance of that surface will increase and can be more than double that of smooth copper. For typical surface roughness on the order of 2 microns, the series resistance of a surface will double at frequencies above about 5 GHz. Since most copper foils are rough on one side and smooth on the other, the series resistance will double for only one side of the conductor. This means the impact from surface roughness could increase the series resistance by 35% over the estimate above for smooth copper.

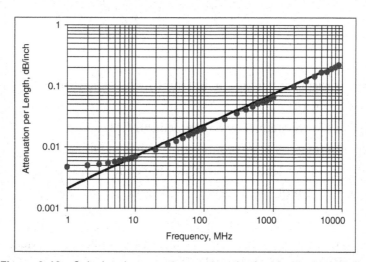

Figure 9-18 Calculated attenuation per length of a 10-mil-wide, 50-Ohm microstrip, assuming only conductor loss and no dielectric loss, comparing the simple model above (line) and the simulation using Ansoft's SI2D field solver (circles).

As shown previously, for all geometries, the conductance per length is related to the capacitance per length as:

$$G_L = \omega \tan(\delta) C_L \qquad (9\text{-}66)$$

Likewise, for all geometries, the characteristic impedance is related to the capacitance as:

$$Z_0 = \frac{\sqrt{\varepsilon_r}}{c C_L} \qquad (9\text{-}67)$$

From these two relationships, the attenuation per length from just the dielectric material can be rewritten as:

$$\alpha_{diel} = 4.34(G_L Z_0) = 4.34(\omega \tan(\delta) C_L)\left(\frac{\sqrt{\varepsilon_r}}{c C_L}\right) = \frac{4.34}{c}\omega \tan(\delta)\sqrt{\varepsilon_r} \qquad (9\text{-}68)$$

where:
α_{diel} = attenuation per length from just the dielectric loss, in dB/length
G_L = conductance per length
ω = angular frequency, in radians/sec
$\tan(\delta)$ = dissipation factor
C_L = capacitance per length
Z_0 = characteristic impedance
ε_r = real part of the dielectric constant
c = speed of light in a vacuum

If we use units of inches/nsec for the speed of light and GHz for the frequency, the attenuation per length from just the dielectric becomes:

$$\alpha_{diel} = 2.3 f \tan(\delta)\sqrt{\varepsilon_r} \qquad (9\text{-}69)$$

where:

α_{diel} = attenuation per inch from just the dielectric loss, in dB/inch

f = sine-wave frequency, in GHz

$\tan(\delta)$ = dissipation factor

ε_r = real part of the dielectric constant

It is interesting that the attenuation is independent of the geometry. If the line width is increased, for example, the capacitance will increase, so the conductance will increase, but the characteristic impedance will decrease. The product stays the same.

TIP   The attenuation due to the dielectric is only determined by the dissipation factor of the material. The attenuation due to the dielectric cannot be changed from the geometry; it is completely based on a material property.

This is not an approximation, but due to the fact that all the geometrical terms that affect the shunt conductance inversely affect the characteristic impedance, the product is always independent of geometry.

FR4 has a dielectric constant of about 4.3 and a dissipation factor of roughly 0.02. At 1 GHz, the attenuation per length of a transmission line using FR4 would be about $2.3 \times 1 \times 0.02 \times 2.1 = 0.1$ dB/inch. This should be compared with the result above of 0.07 dB/inch for the attenuation per inch from the conductor for a 50-Ohm line and with a 10-mil width.

This is the origin of a very valuable rule of thumb: The dielectric loss in FR4-type laminates is about 0.1 dB/inch/GHz. This is independent of the impedance of the line or any geometrical feature and is only related to material properties. This simple rule of thumb allows quick evaluation of the expected losses in a channel. However, this only includes the dielectric loss. In the case of narrow lines, the conductor loss can contribute an equal amount of loss. Typical channels will have an attenuation of roughly 0.1 to 0.2 dB/inch/GHz.

At 1 GHz, the attenuation from the dielectric is slightly greater than the attenuation from the conductor. At even higher frequency, the attenuation from the dielectric will only get larger faster than the attenuation from the conductor. This means that if the dielectric loss dominates at 1 GHz, it will become more important at higher frequency, and the conductor loss will become less important at higher frequency.

The attenuation from the dielectric will increase faster with frequency than the attenuation from the conductor. There is some frequency above which the attenuation will be dominated by the dielectric. Figure 9-19 shows the attenuation per length for a 50-Ohm line with an 8-mil-wide trace in FR4, comparing the attenuation from the conductor, from the dielectric, and for the total combined attenuation per length. For 50-Ohm traces wider than 8 mils, the transition frequency where dielectric and conductor losses are equal is less than 1 GHz. Above 1 GHz, the dielectric loss dominates. If the line width is narrower than 8 mils, the transition frequency is above 1 GHz.

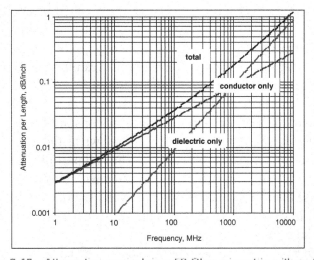

Figure 9-19 Attenuation per inch in a 50-Ohm microstrip with an 8-mil-wide trace, separating the attenuation from the conductor, the dielectric, and the total. In this geometry, with FR4, at frequencies above 1 GHz, the dielectric losses dominate the total losses.

When referring to the losses in a transmission line, many terms are used in the industry, and they are unfortunately used interchangeably, even though they all refer to different quantities.

The following are some of the terms used and their real definitions:

- *Loss*: This is a generic term referring to all aspects of lossy lines.
- *Attenuation*: This is a specific measure of the total attenuation of a line, which is a measure of the decrease in power of the transmitted signal (when measured in dB) or the decrease in amplitude (when described

as a percentage transmitted signal). When measured in dB, the total attenuation of a signal will increase linearly with the length of the line. When measured in percent voltage at the output, it will decrease exponentially with increasing line length.

- *Attenuation per length*: This is the total attenuation of the power, measured in dB, normalized to the length of the line, which is constant as long as the line parameters of the transmission line are constant. The attenuation per length is intrinsic and does not depend on the length of the interconnect.

- *Dissipation factor*: This is the specific, intrinsic material property of all dielectrics, which is a measure of the number of dipoles and how far they can move in an AC field. This is the material property that contributes to dielectric loss and may be slightly frequency dependent.

- *Loss angle*: This is the angle, in the complex plane, between the complex dielectric constant vector and the real axis.

- *tan(δ)*: This is the tangent of the loss angle, which is also the ratio of the imaginary part of the complex dielectric constant to the real part of the complex dielectric constant; it is also known as the *dissipation factor*.

- *Real part of the dielectric constant*: The real part of the complex dielectric constant is the term associated with how a dielectric will increase the capacitance between two conductors, as well as how much it will slow down the speed of light in the material. It is an intrinsic material property.

- *Imaginary part of the dielectric constant*: The imaginary part of the complex dielectric constant is the term associated with how a dielectric will absorb energy from electric fields due to dipole motion. It is an intrinsic material property related to the number of dipoles and how they move.

- *Dielectric constant*: Normally associated with just the real part of the complex dielectric constant, the dielectric constant relates how a dielectric will increase the capacitance between two conductors.

- *Complex dielectric constant*: This is the fundamental intrinsic material property that describes how electric fields will interact with the material. The real part describes how the material will affect the capacitance; the imaginary part describes how the material will affect the shunt leakage resistance.

When referring to the loss in a transmission line, it is important to distinguish the term to which we are referring. The actual attenuation is frequency dependent. However, the dissipation factor of the material or other properties of the material are generally only slowly varying with frequency. Of course, the only way to know this is by measuring real materials.

9.12 Measured Properties of a Lossy Line in the Frequency Domain

The ideal model of a lossy transmission line introduced here has three properties:

1. A characteristic impedance that is constant with frequency
2. A velocity that is constant with frequency
3. An attenuation that has a term proportional to the square root of frequency and another term proportional to the frequency

The assumption here is that the dielectric constant and dissipation factor are constant with frequency. This does not have to be the case; it just happens that it is a pretty good approximation for most materials in most cases. In real materials that have a frequency-dependent material property, it usually varies so slowly with frequency that it can be considered constant over wide-frequency ranges. The only way to know how it varies is to measure it.

Unfortunately, at frequencies in the GHz range, where knowing the material properties is important, there is no instrument that is a dissipation-factor meter. We cannot put a sample of the material in a fixture and read out the final dissipation factor at different frequencies. Instead, a slightly more complicated method must be used to extract the intrinsic material properties from laminate samples.

The first step is to build a transmission line with the laminate, preferably a stripline, so there is uniform dielectric everywhere around the signal path. In order to probe the transmission line, there will be vias at both ends. Using microprobes, the behavior of sine-wave voltages can be measured with minimal impact from the probes.

A vector-network analyzer (VNA) can be used to send in sine waves and measure how they are reflected and transmitted by the transmission line. The ratio of the reflected to the incident sine wave is called the *return loss*, or S_{11}, and the ratio of the transmitted to the incident sine wave is called the *insertion loss*, or S_{21}. These S-parameters terms are reviewed in Chapter 12, "S-Parameters for Signal-Integrity Applications."

These two terms, at every frequency, completely describe how sine waves interact with the transmission line. At each frequency, S_{11} or S_{21}, called S-parameters, or scattering parameters, describe the reflected or transmitted sine-wave amplitude and phase compared with the incident amplitude and phase. The further restriction in this definition is that reflected and transmitted sine waves are measured when a 50-Ohm source and load are connected to the ends of the transmission line.

If the characteristic impedance of the transmission line is different from 50 Ohms, there will be significant reflections, and depending on the length of the line, there will be periodic patterns in the S-parameters, as the sine waves find resonances due to the length of the line and impedance discontinuities. However, we can account for all of these effects if we know the characteristic impedance of the line and the model for the vias or connectors on the ends.

The actual measured insertion loss in a 4-inch length of 50-Ohm transmission line is shown in Figure 9-20. The transmitted signal, S_{21}, in dB, drops off roughly as the frequency increases, as expected. In this example, the transmission line is roughly 50 Ohms, and the vias do not contribute a very noticeable effect until above 8 GHz. The measured insertion loss is a rough approximation of the attenuation. When S_{21} is displayed in dB, the slope of the attenuation is seen to be pretty constant, as expected by the simple model.

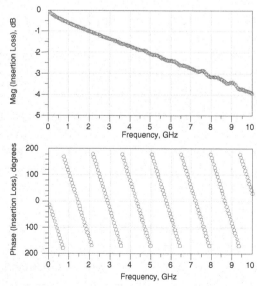

Figure 9-20 Measured insertion loss of a 4-inch-long, nearly 50-Ohm stripline made from FR4 measured with a GigaTest Labs Probe Station.

For a perfectly matched transmission line with no vias, S_{21} is exactly a measure of the attenuation as a function of frequency. However, in practice, this is almost impossible to engineer, so we must interpret S_{21} based on a model for the transmission-line test structure, which includes the vias.

To account for the electrical effects of the vias at the ends, we can model them as a simple pi circuit with a C-L-C topology. The real transmission-line test structure can be modeled with an ideal circuit model as shown in Figure 9-21. This model is fully defined in terms of just eight parameter values:

C_{via1} = capacitance of the first part of the via

L_{via} = loop inductance associated with the via

C_{via2} = capacitance of the second part of the via

Z_0 = lossless characteristic impedance

ε_r = real part of the dielectric constant, assumed to be constant over frequency

len = length of the transmission line, in this case, measured with a ruler as 4 inches

$\tan(\delta)$ = dissipation factor of the laminate, assumed to be constant over frequency

$\alpha_{cond}/f^{0.5}$ = term related to conductor loss and normalized to the square root of the frequency

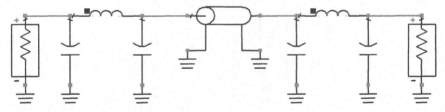

Figure 9-21 Circuit topology to model this 4-inch-long transmission line, including the VNA ports and vias at the two ends of the ideal lossy transmission line.

If we knew the values for each parameter of this model, we could simulate its insertion loss. If we had the right topology and we knew the values of each parameter, we should get very good agreement between the measured insertion loss and the simulated insertion loss.

After optimization, the best set of parameter values for each of the eight terms in the model, for this particular measured transmission line, is:

$C_{via1} = 0.025$ pF

$L_{via} = 0.211$ nH

$C_{via2} = 0.125$ pF

$Z_0 = 51.2$ Ohms

$\varepsilon_r = 4.05$

Len = 4 inches

$\tan(\delta) = 0.015$

$\alpha_{cond}/f^{0.5} = 3$ dB/m/sqr(f)

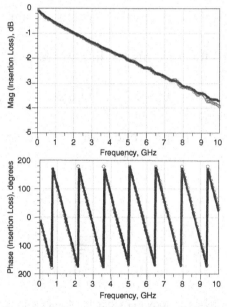

Figure 9-22 Comparing the measured and simulated insertion loss of a 4-inch-long, nearly 50-Ohm transmission line made from FR4.

Figure 9-22 shows the comparison of the simulated insertion loss and the actual measured insertion loss, using the ideal circuit model and these parameter values, which are all constant across the entire measurement bandwidth.

The agreement is excellent between the measured insertion loss and the predicted insertion loss, based on the simple ideal circuit model. This gives us confidence we have a good topology and parameter values and can conclude that the dielectric constant is 4.05 and the dissipation factor is 0.015. The assumption that they are both constant across the entire measurement bandwidth of 10 GHz fits the measured data.

In Figure 9-23, the measurement bandwidth is extended up to 20 GHz for this same sample. There is excellent agreement between the measured response and the predicted response, assuming a constant dielectric constant and dissipation factor, up to about 14 GHz. Above this frequency, it appears as though the actual dissipation factor of the laminate is increasing, and the dielectric constant is decreasing very slightly. From this measurement, we can identify the limit to the assumption about constant material properties. For this FR4 sample, the assumption of constant material properties is good to about 14 GHz.

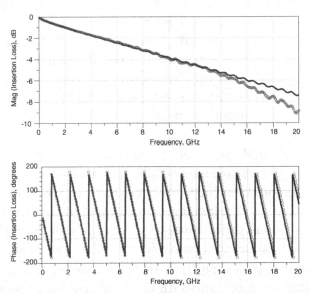

Figure 9-23 Measured and simulated insertion loss extended up to 20 GHz, showing 14 GHz as the bandwidth of the model. Above this frequency, it appears that the actual dissipation factor may be increasing with frequency.

TIP   The fact that we get such excellent agreement between the measured insertion loss and the simple ideal lossy transmission-line model supports the use of this simple model to describe the high-frequency properties of real lossy transmission lines. The only caveat is that for different material systems, it is important to measure the specific material properties.

9.13 The Bandwidth of an Interconnect

If we start with the spectrum of an ideal square wave and preferentially attenuate the higher-frequency components more than the low-frequency components, the bandwidth of the transmitted signal—the highest sine-wave frequency that is significant—will decrease. The longer we let the wave propagate, the more the attenuation of the higher-frequency components and the lower the bandwidth.

The entire concept of bandwidth as the highest sine-wave-frequency component that is significant is inherently only a rough approximation. As we have stated previously, if a problem is so sensitive to bandwidth that knowing its value to within 20% is important, the bandwidth term should not be used. Rather, the entire spectrum of the signal and actual insertion or return loss behavior of the interconnect across the whole frequency range should be used. However, the concept of bandwidth is very powerful in helping to feed our intuition and provide insight into the general behavior of interconnects.

There is a simple and fundamental connection between the bandwidth of an interconnect and the losses in the transmission line. The longer the line, the greater the high-frequency losses and the lower the bandwidth of the line. Being able to estimate the loss-limited bandwidth of an interconnect will allow us to establish some performance requirements on how much attenuation is too much and what material properties might be acceptable.

As we showed in Chapter 2, the bandwidth of a signal is the highest frequency that has less than −3 dB amplitude compared to an ideal square-wave amplitude. At each distance, Len, down the transmission line, we can calculate the frequency that has a −3 dB attenuation and is the bandwidth of the signal at that point. This frequency will be the intrinsic 3–dB bandwidth of the transmission line, BW_{TL}.

If we assume we are in a frequency regime where dielectric loss dominates over conductor loss, so we can ignore the conductor loss, the total attenuation at a frequency, f, and an interconnect length, Len, is:

$$A_{dB} = \alpha_{diel} \times Len = 2.3f \times \tan(\delta)\sqrt{\varepsilon_r} \times Len \qquad (9\text{-}70)$$

where:

A_{dB} = total attenuation, in dB

α_{diel} = attenuation per length from just the dielectric, in dB/inch

ε_r = real part of the complex dielectric constant

Len = length of the transmission line, in inches

f = frequency of the sine wave, in GHz

$\tan(\delta)$ = material's dissipation factor

The intrinsic 3-dB bandwidth of the transmission line, BW_{TL}, corresponds to the frequency that has an attenuation of just 3 dB. By substituting BW_{TL} for the frequency, f, and the attenuation as 3 dB, the connection between the 3-dB bandwidth and interconnect length can be found as:

$$BW_{TL} = \frac{3dB}{2.3 \times \tan(\delta) \times \sqrt{\varepsilon_r}} \times \frac{1}{Len} = \frac{1.3}{\tan(\delta) \times \sqrt{\varepsilon_r}} \times \frac{1}{Len} \qquad (9\text{-}71)$$

where:

BW_{TL} = intrinsic 3-dB bandwidth of the interconnect, in GHz, for a length, d, in inches

ε_r = real part of the complex dielectric constant

Len = length of the transmission line, in inches

$\tan(\delta)$ = material's dissipation factor

This says that the longer the interconnect length, the lower the bandwidth and the lower the frequency at which the attenuation has increased to 3 dB. Likewise, the higher the value of the dissipation factor, the lower the bandwidth of the interconnect.

The rise time of an ideal square wave is 0, and the bandwidth of its spectrum is infinite. If we do something to the spectrum to decrease the bandwidth, the rise time will increase. The resulting rise time, RT, is:

$$RT = \frac{0.35}{BW} \qquad \qquad (9\text{-}72)$$

where:

RT = rise time, in nsec

BW = bandwidth, in GHz

In a lossy interconnect, given its intrinsic bandwidth as limited by the dissipation factor of the material, we can calculate the resulting rise time of the waveform after propagating down the transmission line:

$$RT_{TL} = 0.35 \times \frac{\tan(\delta) \times \sqrt{\varepsilon_r}}{1.3} \times Len = 0.27 \times tab(\delta) \times \sqrt{\varepsilon_r} \times Len \qquad (9\text{-}73)$$

where:

RT_{TL} = intrinsic rise time of the transmission line, in nsec

ε_r = real part of the complex dielectric constant

Len = length of the transmission line, in inches

f = frequency of the sine wave, in GHz

$\tan(\delta)$ = material's dissipation factor

For example, a transmission line using FR4, and having a dissipation factor of 0.02 would have an intrinsic interconnect rise time for a 1-inch length of line of about $0.27 \times 0.02 \times 2 \times 1 = 10$ psec. For a length of 10 inches, the intrinsic interconnect rise time would be about 100 psec. If a 1-psec rise-time signal were sent into such a transmission line, after traveling 10 inches, its rise time would have been increased to about 100 psec due to all the high-frequency components being absorbed by the dielectric and converted into heat.

TIP   As a rough rule of thumb, the rise time of a signal propagating down an FR4 transmission line will increase its rise time by about 10 psec/ inch of travel.

The actual rise time of the signal will get longer and longer as the signal propagates down the line. This intrinsic interconnect rise time is dominated by the length of the line and the dissipation factor of the laminate, and it is the shortest rise time the interconnect will support. Figure 9-24 shows the intrinsic interconnect rise times for a variety of laminate materials. As shown, the rise time can range from 10 psec/inch in FR4 to less than 1 psec/inch in some Teflon-based laminates.

Material	ε	tan(δ)	Intrinsic Rise Time in psec/in
FR4	4.0-4.7	0.02	10
DriClad (IBM)	4.1	0.011	5.4
GETek	3.6-4.2	0.013	7
BT	4.1	0.013	7
Polyimide/glass	4.3	0.014	8
CyanateEster	3.8	0.009	4.7
NelcoN6000SI	3.36	0.003	1.5
RogersRF35	3.5	0.0018	0.9

Figure 9-24   Intrinsic interconnect rise times for various laminate materials, assuming that the bandwidth is limited only by dielectric loss.

In these examples, we are assuming that the line is wide enough that the chief limitation to attenuation is the dielectric. Of course, if the line is very narrow, especially for low-loss materials, the intrinsic rise time of the interconnect will be larger than the estimate, based only on dielectric loss.

When the rise time that enters is not 1 psec but some longer value, RT_{in}, even comparable to the intrinsic rise time, the resulting output rise time, RT_{out}, is related to the intrinsic interconnect rise time, by:

$$RT_{out} = \sqrt{RT_{in}^2 + RT_{TL}^2}$$ (9-74)

where:

RT_{out} = rise time coming out of the interconnect

RT_{in} = rise time of the signal going into the interconnect

RT_{TL} = intrinsic interconnect rise time

This is only a rough approximation, assuming that the shape of the rising edge is Gaussian. For example, Figure 9-25 shows a roughly 41-psec input-rise-time signal entering an 18-inch-long FR4 transmission line. The intrinsic interconnect rise time is about RT_{TL} = 10 psec/inch × 18 inches = 180 psec. We would expect the output rise time to be about:

$$RT_{out} = \sqrt{41^2 + 180^2} = 185 \text{ psec}$$ (9-75)

In fact, what is measured is a rise time of about 150 psec, close to this estimate.

If the intrinsic interconnect rise time is much smaller than the input rise time, the output rise time is roughly the same and is unchanged. The relative change in the output rise time to the input rise time is:

$$\frac{RT_{out}}{RT_{in}} = \sqrt{1 + \left(\frac{RT_{TL}}{RT_{in}}\right)^2}$$ (9-76)

To increase the output rise time by 25%, the intrinsic rise time has to be at least about 50% of the input rise time.

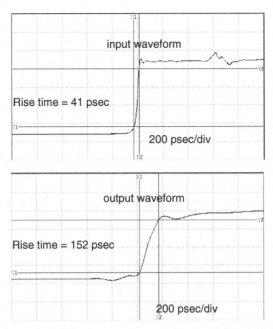

Figure 9-25 Measured rise-time degradation through an 18-inch length of FR4 trace with roughly a 50-Ohm characteristic impedance. Measured with an Agilent 54120 TDR.

TIP This suggests that for a lossy transmission line to not appreciably degrade the rise time of the signal by more than about 25%, the intrinsic interconnect rise time must be less than 50% of the input rise time. If the initial rise time of the signal is 100 psec, the intrinsic interconnect rise time should be less than 50 psec. If it is longer, we will end up with an output rise time grossly increased.

In FR4, with an intrinsic-rise-time degradation of about 10 psec/inch, or 0.01 nsec/inch, there is a simple rule of thumb relating the rise time and the interconnect length at which the lossy effects will be important:

$$RT_{TL} > 0.5 \times RT_{signal} \tag{9-77}$$

$$0.01 \times Len > 0.5 \times RT_{signal} \tag{9-78}$$

$$Len > 50 \times RT_{signal} \qquad\qquad (9\text{-}79)$$

where:

RT_{TL} = intrinsic interconnect rise time, in nsec/inch

RT_{signal} = rise time of the signal, in nsec

Len = interconnect length, in inches, where lossy effects are important

For example, if the rise time is 1 nsec, for transmission lines longer than 50 inches, the lossy effects will degrade the rise time and potentially cause ISI problems. If the lengths are shorter than 50 inches, the lossy effects in FR4 may not be a problem. However, if the rise time were 0.1 nsec, lossy effects might be a problem for lengths longer than only 5 inches.

This is why most motherboard applications with dimensions on the order of 12 inches and typical rise times of 1 nsec do not experience problems with lossy effects. However, for backplanes, where lengths are > 36 inches and rise times are less than 0.1 nsec, lossy effects can often dominate performance.

> **TIP**   This suggests a simple rule of thumb for estimating when to worry about lossy lines: In FR4, when the length of the line (in inches) is greater than 50 multiplied by the rise time (in nsec), the lossy effects may play a significant role.

Of course, this analysis is only a rough approximation. We have been assuming that we can actually use the 10−90 rise time to describe the output signal. In fact, the actual waveform gets distorted in a complex way because the higher-frequency components are decreased gradually, and the actual spectrum of the transmitted signal will change.

This simple rule of thumb describing the rise-time degradation through a lossy line is meant only as a way of estimating the point at which lossy-line properties will start to affect the signal quality. At this point, to accurately predict the actual waveforms and signal-quality effects, a lossy-line transient simulator should be used.

9.14 Time-Domain Behavior of Lossy Lines

If high-frequency components are attenuated more than low-frequency components, the rise time will increase as the signal propagates. Rise time is usually defined as the time for the edge to transition between 10% and 90% of the final values. This assumes that the edge profile of the signal looks somewhat Gaussian, with the middle as the fastest slope region. For this waveform, the 10%–90% rise time makes sense and has value.

However, due to the nature of the attenuation in lossy lines, the rise time is distorted, and the waveform is not a simple Gaussian edge. The initial part of the waveform is faster, and there is a long tail to the rising edge. If we use just one number, like the 10–90 rise time, to describe the rise time, we will have a distorted sense of when the signal has reached a level associated with a trigger threshold. In the lossy regime, rise time has less meaning and is more of a rule of thumb only.

Figure 9-26 shows an example of the measured input waveform and output waveform through a 15-inch-long lossy transmission line in FR4 having a dissipation factor of about 0.01. The resulting rising-edge waveform is not very Gaussian.

In comparing the actual measured S-parameters of real lossy transmission lines in the frequency domain with the predictions of ideal lossy transmission-line models, it is clear that the simple, ideal models can work very well at frequencies at least above 10 GHz, provided that we have the correct material properties.

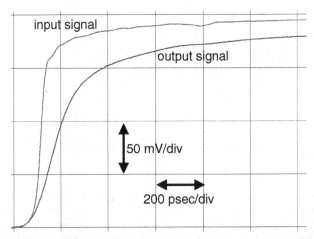

Figure 9-26   Measured rise-time degradation through a 15-inch length of FR4 trace with roughly a 50-Ohm characteristic impedance. Sample supplied by Doug Brooks and measured with an Agilent 86100, a GigaTest Labs Probe Station, and analyzed with TDA Systems IConnect.

This ideal lossy transmission-line model should be a good model to predict the time-domain behavior of real transmission lines as well. The basis of this ideal lossy transmission-line model is that the series resistance is proportional to the square root of the frequency, and the shunt conductance is proportional to the frequency. This is how most real lossy lines behave.

However, this is not how an ideal resistor behaves. An ideal resistor element is constant with frequency. If we use a time-domain simulator that simply has ideal resistor elements for both the series resistance and the shunt conductance, it will not be able to accurately simulate the effects of a lossy line. If the resistance is constant with frequency, the attenuation will be constant with frequency, and there will be no rise-time degradation. The output rise time will look identical to the input rise time, just with slightly less amplitude.

> **TIP**   A simulator with only a model of a resistor element with resistance constant with frequency is worthless as a lossy-line simulator. It will miss the most important impact on performance.

By using a lossy-line simulator with the ideal lossy-line model having a frequency-dependent resistance and conductance, the time-dependent waveform can be evaluated. An example of a transient simulation using a lossy-line simulator is shown in Figure 9-27.

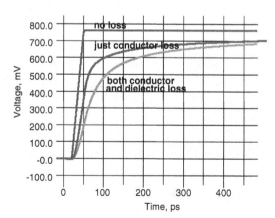

Figure 9-27 Simulated transmitted signals at the output of a 30-inch-long transmission line for a roughly 50-psec rise-time input signal with no loss, with conductor loss for an 8-mil-wide trace, and with combined conductor and dielectric loss for a dissipation factor of 0.02, showing the increasing rise-time degradation. Simulation performed with Mentor Graphics HyperLynx.

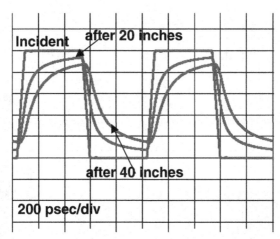

Figure 9-28 Simulated transmitted signals at the output of a 20-inch-and a 40-inch-long transmission line for a 1-Gbps signal, assuming 50-Ohm line, 8-mil-wide trace, and FR4 dielectric. These signals are compared to the received signal if there were no losses. Simulation performed with Mentor Graphics HyperLynx.

If a similar interconnect were used with a 1-GHz clock, the resulting signal at the far end would look similar to that shown in Figure 9-28, comparing the lossless simulation with the lossy simulation for the cases of 20-inches long and 40-inches long.

The most effective way of evaluating the impact of a lossy transmission line is by displaying the transmitted signal in an eye diagram. This shows the degree to which each bit pattern can be discerned for all combinations of bits. A pseudorandom bit pattern is synthesized, and the transmitted signal through the interconnect is simulated. Each bit is overlaid with the previous one, synchronized to the clock. If there is no ISI, the eye pattern will be perfectly open. In other words, each bit, no matter what the previous pattern was, would look the same and would be identical to the previous bit. Its eye diagram would look like just one cycle.

The degree of ISI from losses and other effects, such as capacitive discontinuities of vias, will collapse the eye diagram. If the eye collapses more than the noise margin of the receiver, the bit error rate will increase and may cause faults.

Figure 9-29 shows the simulated eye diagram for the case of a 50-Ohm, 36-inch-long backplane trace in FR4 with no losses or discontinuities and then successively turning on the resistive loss, the dielectric loss, and one 0.5 pF via at either end of the line. The line width in this example is 4 mils. The stimulating source has a bit period of 200 psec, corresponding to a bit rate of 5 Gbps.

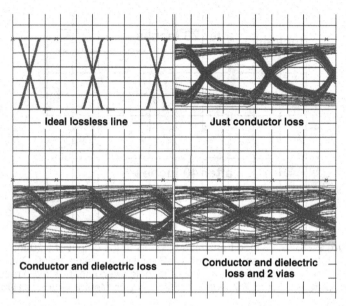

Figure 9-29   Simulated transmitted signals at the output of a 36-inch-long, 50-Ohm FR4 backplane trace, successively adding the effects of conductor loss, dielectric loss, and two vias. Simulation performed with Mentor Graphics HyperLynx.

In the final simulation, which includes the losses and capacitive load from the vias, the eye is substantially closed and would not be usable at this bit rate. For acceptable performance, the transmission line would have to be improved or signal-processing techniques used to enhance the opening of the eye diagram.

9.15 Improving the Eye Diagram of a Transmission Line

Three board-design factors influence the quality of the eye diagram:

1. Discontinuities from the via stubs
2. Conductor loss
3. Dielectric loss

If the rise-time degradation is a problem, these are the only board-level features that will affect the performance.

The first step is to design the sensitive lines with vias having a minimum stub length by restricting layer transitions, using blind and buried vias or backdrilling long stubs. The next step is to reduce the size of capture pads and increase the size of antipad clearance holes to match the via impedance closer to 50 Ohms. This will minimize the rise-time degradation.

TIP   In general, the biggest impact a via will have is from its stub. Reduce the stub length to less than 10 mils, and the via may look pretty transparent up to 10 GHz. Then worry about matching the via to 50 Ohms.

The line width of the signal trace will be the dominant term affecting the attenuation from the conductor losses, provided that the dielectric thickness is changed to keep the impedance of the line at a fixed target. A wider line width will decrease the conductor loss. If the line width is increased, the dielectric thickness will have to increase as well. This is not always practical and thus sets a limit on how wide a line can be used.

Depending on the bandwidth of interest, there is little impact from making the line too wide because the dielectric loss may dominate. Figure 9-30 shows the impact on the attenuation per length of a 50-Ohm line, with FR4 dielectric. If decreasing attenuation is important, the first goal is to use wide lines and avoid line widths less than 5 mils if possible. However, making the line width wider than 10 mils does not significantly decrease the attenuation because of the dielectric losses in the FR4.

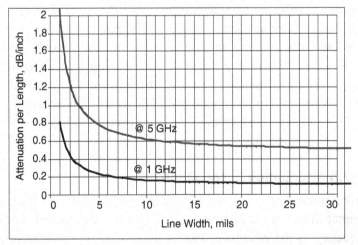

Figure 9-30   Total attenuation per length for a 50-Ohm line in FR4, as the line width is increased, assuming that the dielectric thickness is also increased to keep the impedance constant.

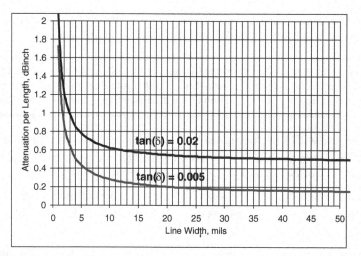

Figure 9-31   Total attenuation per length at 5 GHz for a 50-Ohm line as the line width is increased, for two different dissipation-factor materials.

> **TIP**   This suggests that with FR4-laminate lines, the optimum line width for minimal attenuation is around 5 mils to 10 mils.

If vias are optimized and the line width is kept to 10 mils, the only other term to adjust is the dissipation factor of the laminate. Figure 9-31 shows a similar attenuation curve for two different dissipation-factor materials, both at 5 GHz. A lower-dissipation factor will always contribute to a lower attenuation. Again, we see that there is a point of diminishing returns in increasing the line width even for a low-loss laminate. Much beyond 20 mils, the attenuation is dominated by the laminate material. This is why an important element to predicting high-speed performance of interconnects is having accurate values for the material properties.

9.16 How Much Attenuation Is Too Much?

There are a number of ways of estimating how much attenuation is too much in a channel. Because the attenuation is frequency dependent, we have to pick one frequency as the reference. This is usually the Nyquist frequency, which is the underlying clock frequency of the data pattern. The Nyquist is 1/2 the data rate. For example, a 2 Gbps data rate has a Nyquist of 1 GHz. The most important frequency component is the first harmonic of the Nyquist. In a lossy channel, the Nyquist is the highest-sine-wave-frequency component of the signal. After all, whatever attenuation the Nyquist has, the next frequency component, the third harmonic, will have three times the attenuation. If the first harmonic amplitude is marginal, the third harmonic will be irrelevant.

When the total attenuation at the Nyquist is about 10 dB, the eye will be sufficiently closed that it will fail most mask specs. This is a good rule of thumb for the highest acceptable attenuation. For example, if the interconnect is 20 inches long, and it is a very lossy channel with 0.2 dB/inch/GHz attenuation, the highest Nyquist that could be transmitted through the channel is 10 dB/ (20 inches × 0.2 dB/inch/GHz) = 2.5 GHz. This is a data rate of 5 Gbps. This is a significant limitation of lossy channels and why loss is so important for data rates above 1 Gbps.

After everything is done to reduce the attenuation through physical design and materials selection, there is one other approach to increase the data rate the channel will support. If we send a short rise-time step through the interconnect, it will be distorted by the time it exits the interconnect. The losses will cause the rise time to increase, and when it becomes comparable to the unit interval, ISI results, and they eye closes.

If we can predict the degree of distortion, we can pre-distort the signal so that after propagating down the interconnect, the signal will approximate the sharp voltage step. There are three ways of distorting the waveform. We lump all three of these approaches under the heading *equalization*. The high-frequency signal components will be attenuated more than the low-frequency signal components. This distorts the spectrum from the 1/f drop-off of a short rise-time signal. If we add a high-pass filter that cuts out the low frequencies and lets through the high frequencies, the product of the attenuation from the interconnect and the filter will be equal across the bandwidth, and there will not be frequency-dependent loss.

When we filter out the low-frequency components to match them to the attenuated high-frequency components, we call this *equalizing the channel with a continuous-time linear equalizer (CTLE)*. If we add gain to the filter and increase the strength of the high-frequency components, we call this *active CTLE*. A CTLE filter can recover an open eye if the attenuation at the Nyquist is as high as 15 dB.

In the second approach, we add extra high-frequency components to the initial signal from the transmitter so that by the time the edge reaches the far end, these higher-frequency components have attenuated back in line with the low-frequency components. This is called *feed-forward equalization (FFE)*. When it is accomplished by affecting just one bit adjacent to the initial bit, we sometimes refer to it as *pre-emphasis* or *de-emphasis*, as a special case of FFE.

The third technique accomplishes effectively the same thing but at the receiver. This is called *decision-feedback equalization (DFE)*.

The combination of CTLE, FFE, and DFE techniques can recover a closed eye with a total attenuation as much as 25 to 35 dB at the Nyquist. The challenge is not that the attenuation is so large and the received signal is so small. The challenge is that this signal is frequency dependent. If the attention at the Nyquist is 35 dB, this means that some of the low-frequency components, corresponding to data patterns with many consecutive 1s and 0s, will have a signal strength 30 times the signal at the Nyquist. This results in a huge amount of signal distortion.

Using any equalization technique requires that the distortions of the interconnect be predictable and reproducible. This can happen only if the material properties of the laminate are known. Equalization methods are powerful techniques to compensate for lossy interconnects and are used in all high-end, high-speed serial links.

9.17 The Bottom Line

1. The fundamental problem caused by lossy lines is rise-time degradation, which results in pattern-dependent noise, also called intersymbol interference (ISI).

2. The frequency-dependent losses in circuit-board interconnects arise from the conductor losses and dielectric losses.

3. The rise time of a signal is increased when propagating down a lossy line because the higher frequencies are attenuated more than the low frequencies. This results in a decrease in the bandwidth of the propagating signal.

4. At about 1 GHz, for line widths of about 8 mils, the contributions from the two losses are comparable. At a higher frequency, the dielectric losses increase in a manner that is proportional to the frequency, while the conductor losses increase in a manner that is proportional to the square root of the frequency.

8.1

9.

10.
th
m
15
N/y

to see

d?

m in a

iciently

with

e dissipation

e voltage
ne current

age across it, what
hrough it? What is

what is the ratio of
pacitor? What is the
ne energy lost per cycle?
mission line model for a
o a 10-GHz bandwidth, how
ue to just conductor loss for a
e at 50 Ohms? At 1 GHz and

9.20 What is the attenuation per inch in an FR4 channel for a 5-mil-wide trace, at the Nyquist of a 5-Gbps signal, from both the conductor and dielectric loss? Which is larger?

9.21 What is the skin depth of copper at 1 GHz? At 10 GHz?

9.22 The data rate of a signal is 5 Gbps. What is the UI? If the rise time of the signal is 25 psec, would you expect to see any ISI?

9.23 A signal line is 5 mils wide in ½-ounce copper. What is its resistance per length at DC and at 1 GHz? Assume that current is on both the top and bottom of the signal trace.

9.24 For the worst-case lossy interconnect with 3-mil-wide lines in FR4, above what frequency does a transmission line behave like a low loss versus lossy line?

9.25 What impact do the losses have on the characteristic impedance? In the high loss and low loss regimes?

9.26 What impact do the losses have on the speed of a signal? In the high loss and low loss regimes?

9.27 What is the most significant impact on the properties of a transmission line from the losses?

9.28 What is the ratio of the amplitudes of two signals if they have a ratio of −20 dB? −30 dB? −40 dB?

9.29 What is the ratio of two amplitudes in dB, if their ratio is 50%, 5%, and 1%?

9.30 If the line width of a transmission line stays constant, but the impedance decreases, what happens to the conductor loss? How do you decrease the impedance while keeping the line width constant? In what interconnect structures might this be the situation?

9.31 The attenuation per length per GHz from just dielectric loss depends on material properties. This makes it a useful figure of merit for a material. What is this figure of merit for FR4 and for Megtron 6? Select another laminate material and calculate this figure of merit from the data sheet.

9.32 Some lossy line simulators use an ideal series resistor and an ideal shunt resistor. What is the problem with this sort of model?

9.33 Roughly, how much attenuation at the Nyquist might be too much and result at an eye at the receiver that is too closed? For a 5-Gbps signal, how long an FR4 interconnect might be the maximum unequalized length? What about for a 10-Gbps interconnect?

CHAPTER 10

Cross Talk in Transmission Lines

Cross talk is one of the six families of signal-integrity problems. It is the transfer of an unwanted signal from one net to an adjacent net, and it occurs between every pair of nets. A net includes both the signal and the return path, and it connects one or more nodes in a system. We typically call the net with the source of the noise the *active net* or the *aggressor net*. The net on which the noise is generated is called the *quiet net* or the *victim net*.

> **TIP**   Cross talk is an effect that happens between the signal and return paths of one net and the signal and return paths of a second net. The entire signal–return path loop is important—not just the signal path.

In single-ended digital signaling systems, the noise margin is typically about 15% of the total signal-voltage swing, but it varies among device families. Of this 15%, about one-third, or 5%, of the signal swing is typically allocated to cross talk. If the signal swing were 3.3 v, the maximum allocated cross talk might be about 160 mV. This is a good starting place for the maximum allowable cross-talk noise. Unfortunately, the magnitude of the noise generated in typical traces on a board can often be larger than 5%. This is why it is important to be able to predict the magnitude of cross talk, identify the origin of excessive noise, and actively work to minimize the cross talk in the design of packages, connectors, and board-level interconnects. Understanding the origin of the problem and how to design interconnects with reduced cross talk is increasingly important as rise time decreases.

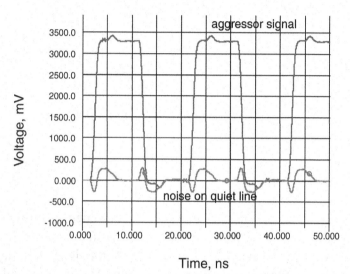

Figure 10-1   Simulated cross talk on a quiet line with aggressor lines on either side. Each line is a source-series terminated 50-Ohm microstrip in FR4 with a 10-mil line and space. Simulated with Mentor Graphics HyperLynx.

Figure 10-1 shows the noise on the receiver of a quiet line when aggressor lines on either side have 3.3-v signals. In this example, the noise at the receiver is more than 300 mV.

In mixed-signal systems, such as with analog or rf elements, the acceptable noise on sensitive lines can be much less than 5% the signal swing. It can be as low as –100 dB below the signal swing, which is 0.001% of the signal. When evaluating design rules for reduced cross talk, the first step is to establish an acceptable specification, keeping in mind that, generally, the lower the acceptable cross talk, the lower the achievable interconnect density, and the higher the potential cost of the system. Be aware of requirements that recommend substantially below 5% maximum allowable coupled noise. Always verify if they really need such low cross talk, as it generally will not be free.

10.1 Superposition

Superposition is an important principle in signal integrity and is critical when dealing with cross talk. Superposition is a property of all linear, passive systems, of which interconnects are a subset. It basically says multiple signals on the same net do not interact and are completely independent of each other. The amount of voltage that might couple onto a quiet net from an active net is completely independent of the voltage that might already be present on the quiet net.

TIP   The noise coupled to a quite line is independent of any signal that might also be present.

Suppose the noise generated on a quiet line were 150 mV from a 3.3-v driver when the voltage level on the quiet line is 0 v. There would also be 150 mV of noise generated on the quiet line when the quiet line is driven directly by a driver to a level of 3.3 v. The total voltage appearing on the quiet line would be the direct sum of the signals that may be present and the coupled noise. If there are two active nets coupling noise to the same quiet line, the amount of noise appearing on the quiet line would be the sum of the two noise sources. Of course, they may have a different time dependency based on the voltage pattern on the two active lines.

Based on superposition, if we know the coupled noise when the quiet line has no additional signal on it, we can determine the total voltage on the quiet line by adding the coupled noise and any signal that might also be present.

Once the noise is on the quiet line, it is subject to the same behavior as the signal: Once generated at some location on the quiet line, it will immediately propagate and see the same impedance, and it will suffer reflections and distortions from any impedance discontinuities that may be present in the quiet line.

TIP   Noise voltages on a quiet line behave exactly as signal voltages. Once generated on the quiet line, they will propagate and be subject to reflections from discontinuities.

If a quiet line has an active line on either side of it, and each active line couples an equal amount of noise to the quiet line, the maximum allowable noise between one pair of lines would be $\frac{1}{2} \times 5\% = 2.5\%$. In a bused topology, it is important to be able to calculate the worst-case total number of adjacent traces that might couple to determine the worst-case coupled noise. This will put a limit on how much noise might be allowable between just two traces.

10.2 Origin of Coupling: Capacitance and Inductance

When a signal propagates down a transmission line, there are electric-field lines between the signal and return paths and rings of magnetic-field lines around the signal- and return-path conductors. These fields are not confined to the immediate space between the signal and return paths. Rather, they spread out into the surrounding volume. We call these fields that spread out *fringe fields*.

TIP   As a rough rule of thumb, the capacitance contributed by the fringe fields in a 50-Ohm microstrip in FR4 is about equal to the capacitance from the field lines that are directly beneath the signal line.

Of course, the fringe fields drop off very quickly as we move farther away from the conductors. Figure 10-2 shows the fringe fields between a signal path and a return path and how they might interact with a second net when it is far away and then when it is close.

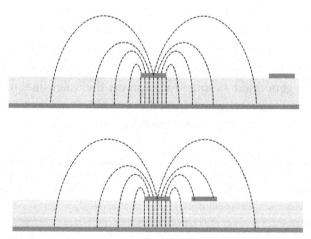

Figure 10-2   Fringe fields near a signal line. When a second trace is far away, there is little fringe-field coupling and little cross talk. When the second net is in the vicinity of the fringe fields, there can be excessive coupling and cross talk.

If we are unfortunate enough to route another signal and its return path in a region where there are still large fringe fields from another net, the second trace may pick up noise from these fringe fields. The only way noise will be picked up in the quiet line is when the signal voltage and current in the active line change. This will cause current to flow through the changing electrics as displacement current and as induced currents from the changing magnetic fields.

Engineering interconnects to reduce cross talk is about reducing the overlap of the electric and magnetic fringe fields between the two signal- and return-path pairs. This is usually accomplished in two ways. First, the spacing between the two signal lines can be increased. Second, the return planes can be brought closer to the signal lines. This will couple the fringe field lines closer to the plane, and less will leak out to the adjacent signal line.

> **TIP** Ultimately, it is the fringe fields that cause cross talk. An important way to minimize cross talk is to space nets far enough apart so their fringe fields are reduced to acceptable levels. Another design feature is to bring the return plane closer to the signal lines to confine the fringe fields more in the vicinity of the signal line.

While the actual coupling mechanism is by electric and magnetic fields, we can approximation this coupling by using capacitor and mutual inductor circuit elements.

Between every two nets in a system, there will always be some combination of capacitive coupling and inductive coupling arising from these fringe fields. We refer to the coupling capacitance and the coupling inductance as the *mutual capacitance* and the *mutual inductance*. Obviously, if we were to move the two adjacent signal- and return-path traces farther apart, the mutual-capacitance and mutual-inductance parameter values would decrease.

Being able to predict the cross talk based on the geometry is an important step in evaluating how well a design will meet the performance specification. This means being able to translate the geometry of the interconnects into the equivalent mutual capacitance and inductance and relating how these two terms contribute to the coupled noise.

Though both mutual capacitance and mutual inductance play a role in cross talk, there are two regimes to consider. When the return path is a wide, uniform plane, as is the case for most coupled transmission lines in a circuit board, the capacitively coupled current and inductively coupled current are of the same order of magnitude, and both must be considered to accurately predict the amount of cross talk. This is the regime of cross talk in transmission lines on circuit boards as part of a bus, and the noise will have a special signature.

When the return path is not a wide uniform plane but is a single lead in a package or a single pin in a connector, there is still capacitive and inductive coupling, but in this case, the inductively coupled currents are much larger than the capacitively coupled currents. In this regime, the noise behavior is dominated by the inductively coupled currents. The noise on the quiet line is driven by a dI/dt in the active net, which usually happens at the rising and falling edges of the signal when the driver switches. This is why this type of noise is usually referred to as *switching noise*.

These two extremes are considered separately.

10.3 Cross Talk in Transmission Lines: NEXT and FEXT

The noise between two adjacent transmission lines can be measured in the configuration shown in Figure 10-3. A signal is injected into one end of the line, with the far end terminated to eliminate the reflection at the end of the line. The voltage noise is measured on the two ends of the adjacent quiet line. Connecting the ends of the quiet line to the input channels of the fast scope will effectively terminate the quiet line. Figure 10-4 shows the measured voltage noise in a quiet line adjacent to an active signal line that is driven by a fast-rising edge. In this case, the two 50-Ohm microstrip transmission lines are about 4 inches long, with a spacing about equal to their line width. The ends of each line are terminated in 50 Ohms, so the reflections are negligible.

The measured noise voltage has a very different pattern on each end. To distinguish the two ends, we label the end nearest the source *the near end* and the end farthest from the source *the far end*. The ends are also defined in terms of the direction the signal is traveling. The far end is in the *forward* direction to the signal-propagation direction. The near end is in the *backward* direction to the signal-propagation direction.

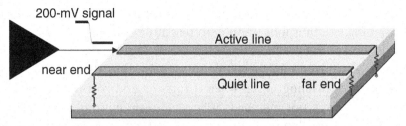

Figure 10-3 Configuration to measure the cross talk between an active net and a quiet net, looking on the near end and far end of the quiet line.

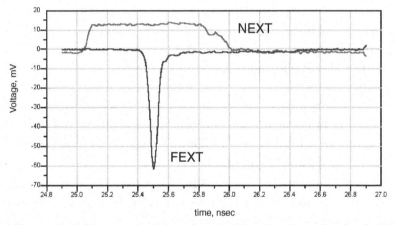

Figure 10-4 Measured noise on the quiet line when the active line is driven with a 200-mV, 50-psec rise-time signal. Measured with an Agilent DCA TDR and GigaTest Labs Probe Station.

When the ends of the lines are terminated so multiple reflections do not play a role, the patterns of noise appearing at the near and far ends have a special shape. The near-end noise rises up quickly to a constant value. It stays up at this level for a time equal to twice the time delay of the coupling length and then drops down. We label the constant, saturated amount of near-end noise the *near-end cross talk* (or *NEXT*) coefficient. This is the ratio of the near-end noise voltage to the signal voltage. In the example shown here, the incident signal voltage is 200 mV, and the measured noise voltage is 13 mV. This makes the NEXT coefficient = 13 mV/200 mV = 6.5%.

The NEXT value is special in that it is defined as the near-end noise when the coupling length is long enough to reach the constant, flat value and in the special case of matched terminations. Changing the terminations at the end of the lines will not change the coupled noise into the quiet line. However, the backward-traveling noise, when it hits the end of the line, may reflect if the termination is not matched. When the terminations are matched to the characteristic impedance of the line, the NEXT is a measure of the noise generated in the line. Once this is known, the impact on this voltage when it encounters a different termination can easily be estimated.

Obviously, the value of the NEXT will depend on the separation of the traces. Unfortunately, the only way of decreasing the NEXT is to move the traces farther apart or bring the return plane closer to the signal traces.

The far end has a signature very different from the near end. There is no far-end noise until one time of flight after the signal enters the active line. Then it comes out very rapidly and lasts for a short time. The width of the pulse is the rise time of the signal. The peak voltage value is labeled the *far-end cross talk*. In the example above, the far-end cross talk voltage is about 60 mV. The FEXT coefficient is the ratio of the peak far-end voltage to the signal voltage. In this example, with a signal of 200 mV, the FEXT coefficient is 60 mV/ 200 mV = 30%. This is a huge amount of noise.

If the terminations are not matched, and reflections affect the magnitude of the noise appearing at the ends, we still refer to the far-end cross talk, but the magnitude is no longer related to FEXT. This coefficient is the special case when the terminations are matched.

> **TIP**   Four factors decrease the FEXT: bringing the return plane closer, decreasing the coupling length, increasing the rise time, and moving the traces farther apart.

10.4 Describing Cross Talk

One way of describing the coupling that contributes to cross talk is in terms of the equivalent circuit model of the coupled lines. This model allows simulations that take into account the specific geometry and the terminations when predicting voltage waveforms. Two different models are generally used to model the coupling in transmission lines.

An ideal, distributed coupled transmission-line model for two lines describes a differential pair. The terms that describe the coupling are the odd- and even-mode impedances and the odd- and even-mode time delays. These four terms describe all the transmission-line and coupling effects. Many simulation engines, such as SPICE engines, especially those that have an integrated 2D field solver, use this type of model. The bandwidth of this model is as high as the bandwidth of an ideal lossless transmission line. This is the same model as a differential pair and is reviewed in detail in Chapter 11, "Differential Pairs and Differential Impedance."

An alternative, widely used model to describe coupling uses the n-section lumped-circuit-model approximation. In this model, each of the two transmission lines is described with an n-section lumped-circuit model, and the coupling between them is described with mutual-capacitor and mutual-inductor elements. The equivalent circuit model of just one section is shown in Figure 10-5.

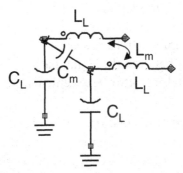

Figure 10-5 Equivalent circuit model of one section of an n-section coupled transmission-line model.

The actual capacitance and loop inductance between the signal and return paths and their mutual values are distributed uniformly down the length of the transmission lines. For uniform, coupled transmission lines, the per-length values describe the transmission lines and the coupling. As shown below, these values can be displayed in a matrix, and this matrix formalism can be scaled and expanded to represent any number of coupled transmission lines. In some simulators, this matrix representation is the basis of describing the coupling, even though the actual simulation engine uses a true, distributed-transmission-line model.

We can approximate this distributed behavior by small, discrete lumped elements placed periodically down the length. The approximation gets better and better as we make the discrete lumped elements smaller. The number of sections needed, as shown in a Chapter 7, "The Physical Basis of Transmission Lines," depends on the required bandwidth and the time delay, with the minimum number being:

$$n > 10 \times BW \times TD \qquad (10\text{-}1)$$

where:

n = minimum number of lumped sections for an accurate model

BW = required bandwidth of the model, in GHz

TD = time delay of each transmission line, in nsec

Two coupled transmission lines can be described with two independent n-section lumped-circuit models. If the lines are symmetrical, the L and C values in each segment will be the same for each line. To this uncoupled model, we need to add the coupling. In each section, the coupling capacitance can be modeled as a capacitor between the signal paths. The coupling inductance can be modeled as a mutual inductor between each of the loop inductors in the n-section model.

Each single-ended transmission line is described by a capacitance per length, C_L, and a loop self-inductance per length, L_L. The coupling is described by a mutual capacitance per length, C_{ML}, and a loop mutual inductance per length, L_{ML}. For a pair of uniform transmission lines, the mutual capacitance and mutual inductance are distributed uniformly down the two lines.

TIP   Everything about the two coupled transmission lines can be described by these four line parameters. When there are more than two transmission lines, the model can be scaled directly, but it looks more complicated. Between every pair of sections of the transmission lines, there is a mutual capacitor. Between every pair of signal- and return-loop sections, there is a mutual inductor.

Each of the mutual capacitors and loop mutual inductors scales with length, and we refer to their mutual capacitance and mutual inductance per length. To keep track of each of these additional mutual capacitors and inductors, we can take advantage of a simple formalism based on matrices.

10.5 The SPICE Capacitance Matrix

For a collection of multiple transmission lines, we can label each signal path with an index number. If there are five lines, for example, we would label each one from 1 to 5. The return path, by convention, we label as conductor 0. An example of the cross section of five conductors and a common return plane is shown in Figure 10-6. We first look at the capacitor elements. Later in this chapter, we look at the inductor elements.

Every pair of conductors in the collection has a capacitance between them. For every signal line, there is a capacitor to the return path. Between every pair of signal lines, there is a coupling capacitor. To keep track of all the pairs, we can label the capacitors based on the index numbers. The capacitance between conductors 1 and 2 is labeled C_{12}, and the capacitance between capacitors 2 and 4 is labeled C_{24}. The capacitance between the signal line and the return we might label as C_{10} or C_{30}.

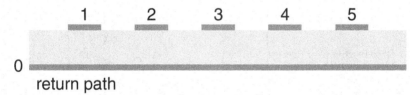

Figure 10-6 Five coupled transmission lines, in cross section, with each conductor labeled using the index convention.

To take advantage of the powerful formalism of matrix notation, we rename the capacitor labels that describe the capacitance between the signal paths and their return paths. Instead of C_{10}, we reserve the diagonal element for the capacitance between the signal and its return path and label it C_{11}. Likewise, the other capacitors between the signals and their returns become C_{22}, C_{33}, C_{44}, and C_{55}. In this way, we end up with a 5×5 matrix of capacitors that labels the capacitance between every pair of conductors. The equivalent circuit and corresponding matrix of parameter values are shown in Figure 10-7.

Of course, even though there is a matrix entry for C_{14} and C_{41}, it is the same capacitor value, and there is only one instance of this capacitor in the model.

TIP   In the capacitor matrix, the diagonal elements are the capacitance between the signal and the return paths. The off-diagonal elements are the coupling or mutual capacitance. For uniform transmission lines, each matrix element is the capacitance per length, usually in units of pF/inch.

The matrix is a handy, convenient, compact way of keeping track of all the capacitor values. This matrix is often called the SPICE capacitance matrix to distinguish it from some of the other matrices. It is a place to store the parameter values for the SPICE equivalent circuit model, shown earlier in this chapter. Each matrix element represents the value of a capacitor that would be present in the complete circuit model for the coupled transmission lines.

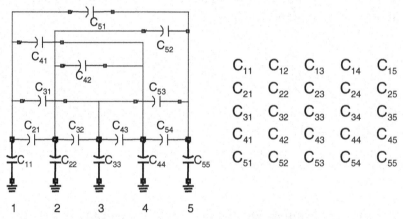

Figure 10-7   Equivalent capacitance model of five coupled transmission lines and the corresponding matrix of capacitance parameter values.

Each element is the capacitance per length. To construct the actual transmission lines' approximate model, we would first identify how many sections are needed in the lumped-circuit model, from $n > 10 \times BW \times TD$. From the length of the transmission lines and the number of sections required, the length of each section can be calculated as Length per section $= Len/n$. The value of the capacitor for each section is the matrix element of the capacitance per length times the length of each section. For example, the coupling capacitance of each section would be $C_{21} \times Len/n$.

The actual values for each capacitor-matrix element can be found by either calculation or by measurement. Few approximations are very good. Rather, a few simple rules of thumb can be used, and when a more accurate value of the coupling capacitance is required, a 2D field solver should be used. Many field solver tools are commercially available; they are easy to use and generally very accurate. An example of the SPICE capacitance matrix for a collection of five microstrip conductors, as calculated with a 2D field solver, is shown in Figure 10-8.

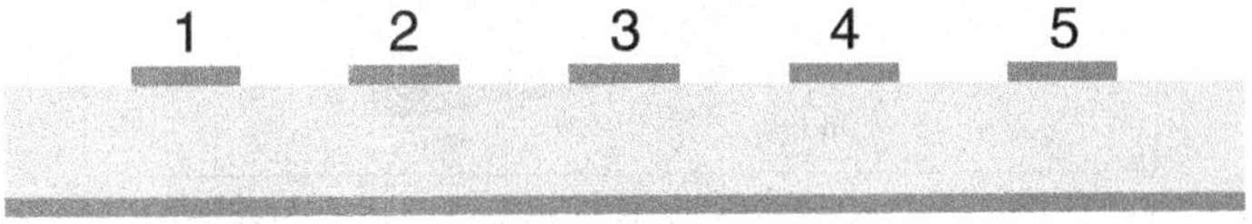

1	2	3	4	5
2.812	0.151	0.016	0.008	0.005
0.151	2.682	0.149	0.016	0.008
0.016	0.149	2.675	0.149	0.017
0.008	0.016	0.149	2.684	0.151
0.005	0.008	0.017	0.151	2.813

Figure 10-8   Five coupled transmission lines, each of 5-mil line width and 5-mil space and the SPICE capacitance matrix, in pF/inch, calculated with the Ansoft SI2D field-solver tool.

When it is difficult to get a good physical feel for the values of the capacitance-matrix elements and how quickly they drop off by just looking at the numbers, the matrix can be plotted in 3D. The vertical axis is the magnitude of the capacitance. These same matrix elements are shown in Figure 10-9. At a glance, it is apparent that all the diagonal elements have about the same values, and the off-diagonal elements drop off very fast.

In this particular example, the conductors are 50-Ohm microstrips, with 5-mil line width and 5-mil space, placed as close together as possible. We see that the coupling between conductors 1 and 3 is negligible compared to that between conductors 1 and 2. The farther apart the traces, the more rapidly the off-diagonal elements drop off. This is a direct indication of how quickly the fringe electric fields drop off with spacing.

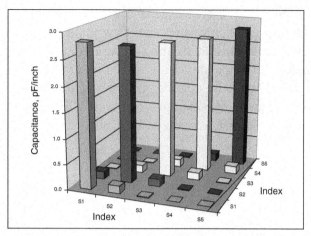

Figure 10-9   Plotting the SPICE capacitance-matrix elements, showing how quickly off-diagonal elements drop off.

In the SPICE capacitance matrix, it is important to keep in mind that each element is the parameter value of a circuit element that appears in an equivalent circuit model. The value of each element is a direct measure of the amount of capacitive coupling between the two conductors. This will directly determine, for example, the capacitively coupled current that would flow between each pair of conductors for a given dV/dt. The larger the matrix element, the larger the capacitive coupling, and the more fringe fields between the conductors.

T I P   In coupled transmission lines, the size of the off-diagonal element should always be compared with the diagonal element. In this geometry example of five coupled 50-Ohm lines, with a spacing equal to the line width (the tightest spacing manufacturable), the relative coupling between adjacent traces is about 5%. The coupling between one trace and the trace two traces away is down to less than 0.6%. These are good values to remember as a rough rule of thumb.

The physical configuration of the traces will affect the parameter values, but for a given configuration of traces, the circuit model itself will not change as the geometry of the traces changes. Obviously, if we move the traces farther apart, the parameter values will decrease.

If we change the line width of one line, to first order, it will affect the diagonal element of that line and the coupling between that line and the adjacent traces on either side. However, it may also affect, to second or third order, the coupling between the lines on either side of it. The only way to know for sure is to put in the numbers with a 2D field solver.

10.6 The Maxwell Capacitance Matrix and 2D Field Solvers

Unfortunately, there is more than one capacitance matrix, and this creates confusion. Earlier in this chapter, we introduced the SPICE capacitance matrix, whose elements were the parameter values of the equivalent circuit model for the coupled lines. There is also a capacitance matrix that is the result of a field-solver calculation; it is referred to as the *Maxwell capacitance matrix*. Even though they are both called capacitance matrices, their definitions are different.

A field solver is basically a tool that solves one or more of Maxwell's Equations for a specific set of boundary conditions. A circuit topology for the collection of conductors is assumed, and from the fields, all the parameter values are calculated. The equation that is solved to calculate the capacitances of an array of conductors is LaPlace's Equation. In its simplest differential form, it is:

$$\nabla^2 V = 0 \tag{10-2}$$

This differential equation is solved under the specific geometry boundary conditions of the conductors and dielectric materials. Solving this equation allows for the calculation of the electric fields at every point in space.

For example, suppose there is a collection of five conductors, as shown in Figure 10-10. Conductor 0 is defined as the ground reference and is always at 0-v potential. To calculate the capacitance between the conductors, there are six steps:

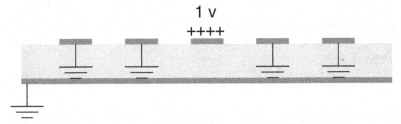

Figure 10-10 Conductor setup to calculate the capacitance matrix for this collection of transmission lines using a 2D field solver that solves LaPlace's Equation.

1. A 1-v potential is set on conductor k, and the potential of every other conductor is set to 0 v.
2. Given this boundary condition, LaPlace's Equation is solved to find the potential everywhere in space.
3. Once the potential is solved, the electric field is calculated at the surface of each conductor, from:

$$E = -\nabla V \tag{10-3}$$

4. The total charge is calculated on each conductor by integrating the electric field on the surface of each conductor:

$$Q_j = \oint_j E \bullet da_j \tag{10-4}$$

5. From the charge on each conductor, the capacitance is calculated from the definition of the Maxwell Capacitance matrix:

$$C_{jk} = \frac{Q_j}{V_k} \tag{10-5}$$

6. This process is repeated with a 1-v potential sequentially placed on each of the conductors.

The definition of the Maxwell capacitance-matrix elements is different from the SPICE capacitance-matrix elements. The SPICE matrix elements are the parameter values for the corresponding equivalent circuit model. The value of each element is a direct measure of the amount of capacitively coupled current that would flow between each pair of conductors for a given dV/dt between them.

The Maxwell capacitance-matrix elements are really defined based on:

$$C_{jk} = \frac{Q_j}{V_k} \tag{10-6}$$

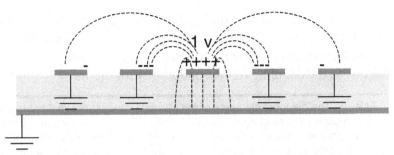

Figure 10-11 Charge distribution for the five conductors with conductor 3 set to 1 v and all others at ground potential.

Each capacitance-matrix element between two conductors is a measure of how much excess charge will be on one conductor when the other is at a 1-v potential and *all other conductors* are grounded. This is a very specialized condition and gives rise to much confusion.

Suppose a 1-v potential is placed on conductor 3, and all the other conductors are at 0-v potential. To do this will require placing some extra plus charge on conductor 3 to raise it to a 1-v potential with respect to ground. This plus charge will attract some negative charge to all the other nearby conductors. The negative charge will come out of the ground reservoir to which each other conductor is connected. How much charge is attracted to each of the other conductors is a measure of how much capacitive coupling there is to the conductor with the 1 v applied. The charge distribution is illustrated in Figure 10-11.

From the definition of the Maxwell capacitance matrix, the capacitance between conductor 3 and the reference ground, which is the diagonal element, is the ratio of the charge on conductor 3, Q_3, and the voltage on it, $V_3 = 1$ v. This is the charge on conductor 3 when all the other conductors are also connected to ground. This capacitance is often called the loaded capacitance of conductor 3:

$$C_{33} = \frac{Q_3}{V_3} = C_{loaded} \qquad (10\text{-}7)$$

The loaded capacitance will always be larger than the SPICE diagonal capacitance.

When there is plus charge on conductor 3 to raise it up to the 1-v potential, the charge induced on all the other conductors is negative. Even though all the other conductors are at ground potential, they will have some net negative charge due to the coupling to the conductor with the 1 v.

> **TIP**   The diagonal elements of the Maxwell capacitance matrix are the loaded capacitances of each conductor. It is not just the capacitance to the return path, the ground reference; it is the capacitance to the return path *and* to all the other conductors that are also tied to ground. This is not the same as the diagonal element of the SPICE capacitance matrix, which just includes the coupling of the conductor to the return path and not to any other signal path.

By the definition of the Maxwell capacitance matrix, the off-diagonal-matrix element between conductor 3 and 2 will be:

$$C_{23} = \frac{Q_2}{V_3} \qquad (10\text{-}8)$$

> **TIP**   Since the charge induced on conductor 2 is negative and the induced charge on every other conductor is also negative, every off-diagonal capacitor-matrix element must be negative.

The negative sign really means that the induced charge on conductor 2 will be negative when a 1-v potential is placed on conductor 3.

An example of the Maxwell capacitance matrix for a collection of five microstrip traces is shown in Figure 10-12. At first glance, it is bizarre to see capacitance values that are negative. What could it possibly mean to have negative capacitances? Is this an inductance? In fact, they are negative because they are not SPICE capacitance values but Maxwell capacitance values, and the definition of the Maxwell capacitor-matrix elements is different from the SPICE elements.

The output from most commercially available field solvers is typically in the form of Maxwell capacitance values. This is usually because the software developer who wrote the code did not really understand the end user's applications and did not realize that most signal-integrity engineers want to see the SPICE capacitance-matrix elements. The Maxwell capacitance matrix is not wrong; it is just not an engineer's first choice of what to see.

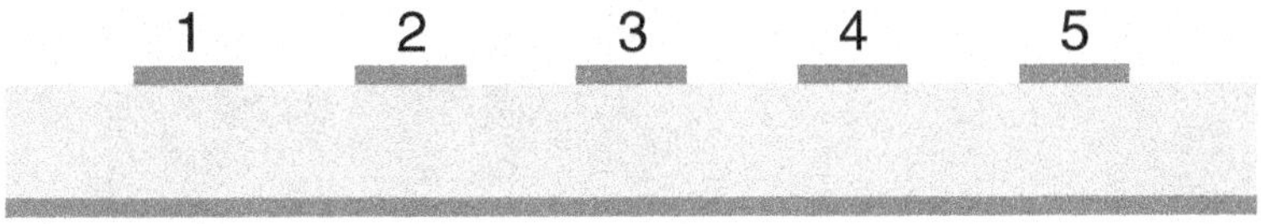

1	2	3	4	5
2.992	-0.151	-0.016	-0.008	-0.005
-0.151	3.005	-0.149	-0.016	-0.008
-0.016	-0.149	3.006	-0.149	-0.017
-0.008	-0.016	-0.149	3.007	-0.151
-0.005	-0.008	-0.017	-0.151	2.994

Figure 10-12 Maxwell capacitance matrix, in pF/inch, for the collection of five closely spaced, 50-Ohm transmission lines calculated with Ansoft's SI2D field solver.

It is very easy to convert from one matrix to the other. The off-diagonal elements are very similar, with just the sign difference:

$$C_{ij}(\text{SPICE}) = -C_{ij}(\text{Maxwell}) \tag{10-9}$$

The off-diagonals relate to the number of field lines that couple the two conductors and directly relate to the capacitively coupled current that might flow between the conductors for a given dV/dt between them.

However, the diagonals are a little more complicated. The diagonal elements of the Maxwell matrix are the loaded capacitance of each conductor. The diagonal elements of the SPICE matrix are the capacitance between just the diagonal conductor and the return path. The SPICE diagonal element is counting only the field lines coupling between the signal line to the return path. Based on this comparison, diagonal elements of the SPICE and Maxwell matrices can be converted by:

$$C_{ij}(\text{Maxwell}) = \sum_i C_{ij}(\text{SPICE}) \tag{10-10}$$

$$C_{ij}(\text{SPICE}) = \sum_i C_{ij}(\text{Maxwell}) \qquad (10\text{-}11)$$

The easiest way to determine which matrix is reported by the field solver is to look for negative signs. If there are negative signs, it's usually not numerical accuracy; it is the Maxwell capacitance matrix.

In either matrix, the off-diagonal elements are a direct measure of the coupling between signal lines and the strength of the fringe fields that couple the conductors. The greater the spacing, the fewer the fringe-field lines between the traces and the lower the coupling. In both matrices, the physical presence of any conductor between two traces will affect the field lines between them and will be taken into account by the matrix-element values.

Each matrix element will depend on the presence of the other conductors. For example, for two conductors and their return path, the diagonal SPICE capacitance of one line, C_{11}, will depend on the position of the adjacent conductor. C_{11} is the capacitance of line 1 to the return path. If we bring the adjacent trace in proximity, it will begin to steal some of the fringe-field lines between line 1 and the return and decrease C_{11}. This is illustrated in Figure 10-13.

When the spacing is more than about two line widths or four dielectric thicknesses, the presence of the adjacent trace has very little impact on the diagonal element of the SPICE capacitance matrix.

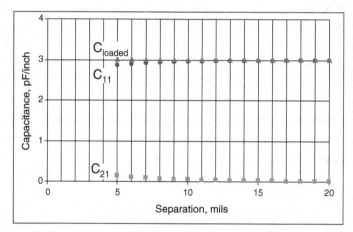

Figure 10-13 Variation of the diagonal and off-diagonal elements of the SPICE capacitance matrix and the loaded capacitance of conductor 1 as the spacing between the two 5-mil-wide, 50-Ohm lines increases. Simulated with Ansoft's SI2D.

Since the loaded capacitance of line 1 is a measure of all the fringe-field lines between the signal line and all the other conductors, it doesn't change much as the adjacent trace is brought closer. Field lines not going from 1 to the return, stolen by trace 2, are accounted for by the new field lines between 1 and 2.

The off-diagonal elements will also depend on the geometry and the presence of other conductors. If the spacing is increased, the off-diagonal element will decrease. Also, if another conductor is added between two traces, this conductor will steal some of the field lines between the two conductors and decrease the off-diagonal SPICE capacitance element.

Figure 10-14 shows three geometry configurations and the resulting SPICE capacitance-matrix element that is calculated. In each case, the signal line is 5 mils wide and roughly 50 Ohms.

TIP   When the spacing is also 5 mils wide, the coupling capacitance is 0.155 pF/in. This is about 5% of the on-diagonal element of 2.8 pF/inch. When the spacing is increased to 15 mils, so the spacing is three times the width, the capacitive coupling is 0.024 pF/inch, or 0.9% of the diagonal element. If another 5-mil-wide trace is added in this space, the coupling capacitance between the two outer conductors is reduced to 0.016 pF/inch, or 0.6% of the on-diagonal element.

By adding a conductor between the two signal lines, the coupling capacitance between them is reduced. This is the basis of the use of guard traces, discussed in detail later in this chapter. Of course, we have only considered one type of coupling; there is also inductive coupling to consider.

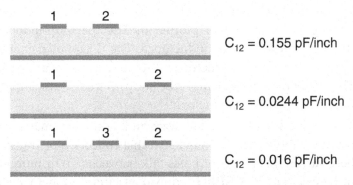

$C_{12} = 0.155$ pF/inch

$C_{12} = 0.0244$ pF/inch

$C_{12} = 0.016$ pF/inch

Figure 10-14   Three geometries and the corresponding capacitance matrix elements between the two signal lines. The presence of the metal between the conductors decreases the capacitive coupling about 35%.

10.7 The Inductance Matrix

Just as a matrix is used to store all the capacitor values in a collection of signal- and return-path conductors, a matrix is used to store the values of the loop self-inductance and the loop mutual inductances associated with a collection of conductors. It is important to keep in mind that the inductance elements are loop inductances. As the signal propagates down a transmission line, the current loop travels down the signal path and immediately returns through the return path. This current loop is probing the loop inductance in the immediate vicinity of the edge of the signal transition. Of course, the loop self-inductance is related to the partial self-inductances of the signal and return paths, and their mutual inductance, by:

$$L_{loop} = L_{self\text{-}signal} + L_{self\text{-}return} - 2 \times L_{mutual} \qquad (10\text{-}12)$$

where:

L_{loop} = loop inductance per length of the transmission line

$L_{self\text{-}signal}$ = partial self-inductance per length of the signal path

$L_{self\text{-}return}$ = partial self-inductance per length of the return path

L_{mutual} = partial mutual inductance per length between the signal and return paths

In the inductance matrix, the diagonal elements are the loop self-inductances of each signal and return path. The off-diagonal elements are the loop mutual inductance between every pair of signal and return paths. The units are in inductance per length, typically nH/inch.

For example, the five microstrip traces above have an inductance matrix shown in Figure 10-15. When plotted in 3D, the loop-inductance matrix reveals the basic properties of inductance. The diagonal elements—the loop self-inductances of each conductor and its return path—are all basically the same. The off-diagonal elements—the loop mutual inductances—drop off very rapidly the farther apart the pair.

The combination of the capacitance and inductance matrices contains all the information about the coupling between a collection of transmission lines. From these values, all aspects of cross talk between two or more transmission lines can be calculated. A SPICE equivalent circuit model can be built that could be used to simulate the behavior of a collection of coupled traces.

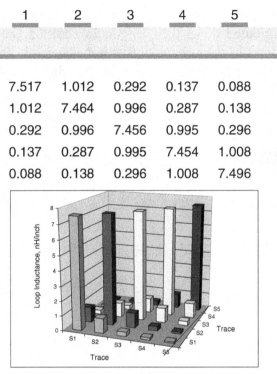

Figure 10-15 Five transmission lines, each 50 Ohms and with a 5-mil width and 5-mil spacing, their inductance matrix, and the values of each matrix element as extracted using the Ansoft SI2D field solver.

TIP These two matrices contain all the fundamental information about the coupling between multiple transmission lines.

10.8 Cross Talk in Uniform Transmission Lines and Saturation Length

For the case of two coupled transmission lines, the C and L matrices are simple, two-by-two matrices. The off-diagonal elements describe the amount of mutual capacitance and mutual inductance. The easiest way to understand the generation of the noise in the quiet line and the particular near-end and far-end signature is to walk down the line and observe the noise coupling-over at each step.

Consider two 50-Ohm microstrip transmission lines that have some coupling distributed down their length. In addition, we will terminate the ends of the lines in their characteristic impedance of 50 Ohms to eliminate any effects from reflections. This equivalent circuit model is illustrated in Figure 10-16.

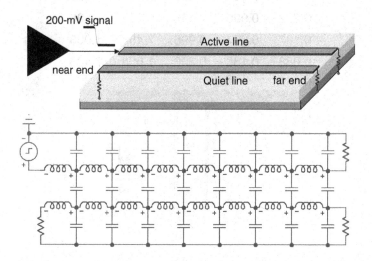

Figure 10-16   A pair of tightly coupled transmission lines and an equivalent circuit model using an n-section lumped-circuit approximation.

As a signal propagates down the active line, it will see the mutual capacitors and mutual inductors connecting it to the quiet line. The only way noise current will flow from the active line to the quiet line is through these elements, and the only way current flows through a capacitor, or is induced in a mutual inductor, is if either the voltage or current changes. This is illustrated in Figure 10-17. If the leading edge is approximated by a linear ramp with a rise time of RT, the noise is approximately proportional to V/RT and I/RT.

TIP   As the signal propagates down the active line, the only place there is coupled-noise current to the quiet line is in the specific region where the edge of the signal is, where there is a dV/dt or a dI/dt. Everywhere else along the line, the voltage and current are constant, and there is no coupled-noise current.

The edge of the signal acts as a current source moving down the line. At any instant, the total current that flows through the mutual capacitors is:

$$I_C = C_m \frac{V}{RT}$$ (10-13)

where:

I_C = capacitively coupled-noise current from the active line to the quiet line

V = signal voltage

RT = 10–90 rise time of the signal

C_m = mutual capacitance that couples over the length of the signal rise time

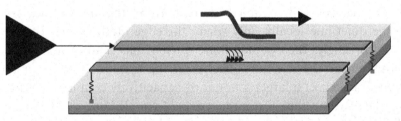

Figure 10-17 The only region in which coupled noise flows from the active line to the quiet line is at the signal wavefront where the voltage and current change.

The total capacitance that is coupling is the capacitance along the spatial extent of the rise time, where there is a changing voltage:

$$C_m = C_{mL} \times \Delta x = C_{mL} \times v \times RT$$ (10-14)

where:

C_m = mutual capacitance that couples over the length of the signal rise time

C_{mL} = mutual capacitance per length (C_{12})

Δx = spatial extent of the leading edge as it propagates over the active line

v = signal-propagation speed

RT = signal rise time

The total, instantaneous, capacitively coupled current injected into the quiet line is:

$$I_C = C_{mL} \times v \times RT \times \frac{V}{RT} = C_{mL} \times v \times V \qquad (10\text{-}15)$$

where:

I_C = capacitively coupled-noise current from the active line to the quiet line

C_{mL} = mutual capacitance per length (C_{12})

v = signal-propagation speed

RT = signal rise time

V = signal voltage

The capacitively coupled current from the active line is injected into the quiet line locally only where the edge of the signal is in the active line. Surprisingly, the coupled-noise current has a total magnitude that is independent of the rise time. The faster the rise time, the larger the dV/dt, so we would expect a larger amount of capacitively coupled current. But, the faster the rise time, the shorter the region of coupled line where there is a dV/dt and the less capacitance is available to do the coupling. The capacitively coupled current depends only on the mutual capacitance per length.

By the same analysis, the instantaneous voltage induced in the mutual inductor in the quiet line is:

$$V_L = L_m \frac{dI}{dt} = L_{mL} \times v \times RT \times \frac{I}{RT} = L_{mL} \times v \times I \qquad (10\text{-}16)$$

where:

V_L = inductively coupled-noise voltage from the active line to the quiet line

I = signal current in the active line

L_{mL} = mutual inductance per length (L_{12})

v = signal-propagation speed

RT = signal rise time

Again, we see that the inductively coupled noise crosses to the quiet line only where the signal voltage is changing on the active line. Also, the amount of noise voltage generated in the quiet line does not depend on the rise time of the signal but only on the mutual inductance per length.

Four important properties emerge about the coupled-noise to the quiet line:

1. The amount of instantaneous coupled voltage and current noise depends on the signal strength. The larger the signal voltage and current, the larger the amount of instantaneously coupled noise.
2. The amount of instantaneously coupled voltage and current noise depends on the amount of coupling per length, as measured by the mutual capacitance and mutual inductance per length. If the coupling per length increases as the conductors are brought closer together, the instantaneously coupled noise will increase.
3. It appears that the higher the velocity, the higher the instantaneously coupled total current. This is due to the fact that the higher the speed, the longer the spatial extent of the rise time and the longer the region that we will see coupling at any one instant. If the velocity of the signal increases, the coupled length over which current flows will increase, and the total coupling capacitance or inductance will increase.
4. Surprisingly, the rise time of the signal does not affect the total instantaneously coupled-noise current or voltage. While it is true that the shorter rise time will increase the coupled noise through a single mutual C or L element, with a shorter rise time, the spatial extent of the edge is shorter, and there is less total mutual C and mutual L that couple at any one instant.

This last property is based on an assumption that the length of the coupled region is longer than half the spatial extent of the rise time. This is the most confusing and subtle aspect of near-end noise.

Consider a pair of coupled lines with a TD of the coupling region very long compared to the rise time of the signal. As the signal starts out from the driver and enters the coupling region, the amount of coupled noise flowing between the aggressor and victim lines will begin to increase and appear as increasing near-end noise. The near-end noise will continue to increase as long as more rising edge enters the coupling region. The near-end noise increases for a time equal to the rise time. After this period of time, the near-end noise has reached its maximum value and has "saturated."

When the beginning of the leading edge of the signal finally leaves the coupled region, the coupled current flowing from the aggressor line to the victim line will begin to decrease. Of course, it will take one TD of the coupled region for the beginning of the rising edge to traverse the coupled region. Once the coupled current begins to decrease, at the far end of the line, it will take another TD for this reduced current to make it back to the near end and be recorded as reduced near-end noise. The near-end noise will begin to decrease in a time equal to 2 × TD from the very beginning of the near-end noise starting.

As the coupling TD between the two transmission lines is decreased, there will be a point where the near-end noise reaches its peak value, a rise time after it begins, just as it begins to decrease, 2 × TD after it enters the coupled region. This condition is that the rise time = 2 × TD. This is the condition for the coupling length being just long enough to saturate the near-end noise.

When the rise time of the signal is 2 × TD, the coupled lines are saturated. This condition is when the TD is half the rise time, or when the coupling length is half the spatial extent of the rising edge. We give this length the special name *saturation length*:

$$\text{Len}_{\text{sat}} = \frac{1}{2} \times \text{RT} \times v \sim \text{RT} \times 3 \frac{\text{inch}}{\text{nano second}} \tag{10-17}$$

where:

Len_{sat} = saturation length for near-end cross talk, in inches

RT = rise time of the signal in nsec

v = speed of the signal down the active line in inches/nsec

If the rise time is 1 nsec, in a transmission line composed of FR4, with a velocity of roughly 6 inches/nsec, the saturation length is ½ nsec × 6 inches/nsec = 3 inches. If the rise time were 100 psec, the saturation length would be only 0.3 inch. For short rise times, the saturation length is usually shorter than a typical interconnect length, and near-end noise is independent of coupled length. The saturation length is illustrated in Figure 10-18.

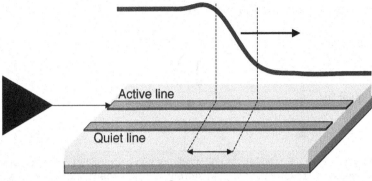

Figure 10-18 The saturation length is half the spatial extent of the leading edge. If the length of the coupled region is longer than the saturation length, the amount of near-end noise on the quiet line is independent of the rise time and independent of the coupling length.

Once the noise current transfers from the active line to the quiet line, it will propagate in the quiet line and give rise to the effects we see as near-end and far-end noise. Even though a constant current is transferring to the quiet line, the features of the propagation in the quiet line will shape this distributed-current source into very different patterns at the near and far ends. To understand the details of the origin of the near- and far-end signatures, we will first look at how the capacitively coupled currents behave at the two ends and then at the inductively coupled currents and add them up.

10.9 Capacitively Coupled Currents

Figure 10-19 shows the redrawn equivalent circuit model with just the mutual-capacitance elements. In this example, we assume that the coupled length is longer than the saturation length. The rising edge will act as a current source moving down the active line. Because current flows through the mutual capacitors only when there is a dV/dt, it is only at the rising edge that there is capacitively coupled current flowing into the quiet line.

Once this current appears in the quiet line, which way will it flow? The primary factor that will determine the direction of the current flow is the impedance the noise current sees. When the noise current looks up and down the quiet line, it sees exactly the same impedance in both directions: 50 Ohms. An equal amount of noise current will flow in both the forward and backward directions.

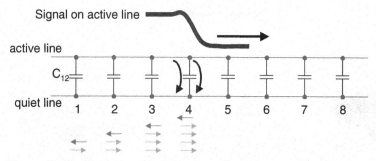

Figure 10-19 Equivalent circuit model of two coupled lines just showing the coupling capacitors, the coupled current, and the spatial extent of the signal edge.

TIP The direction of the capacitively coupled-current loop in the quiet line is from the signal path to the return path. It is a positive voltage between the signal and return paths of the quiet line that will propagate in both directions.

As the signal initially emerges from the driver, there will be some capacitively coupled current to the quiet line. Half of this will travel backward to the near end. The other half will travel in the forward direction. The current, flowing through the terminating resistor on the near end of the quiet trace, will flow in the positive direction, from the signal path to the return path. It will start out at 0 v, and as the rising edge emerges from the driver, it will rise up. As the signal edge moves down the line, the backward-flowing capacitively coupled-noise current will continue back to the near end at a steady rate. It is as though the active signal is leaving a constant, steady amount of current flowing back toward the near end in its wake.

After a time equal to the rise time, the current appearing at the near end will reach its peak value. After the beginning of the rising edge in the active line has left the coupled region and reached the far-end terminating resistor, the coupled-noise current will begin to decrease, taking a time equal to the rise time. There is still the backward-moving current in the quiet line that has yet to reach the near end of the quiet line. It will continue flowing back to the near end of the quiet line, taking additional time equal to the time delay, TD, of the coupled region.

The signature of the near-end, capacitively coupled current is a rise up to a constant value in a time equal to the signal's rise time and staying at a constant value lasting for a time equal to $2 \times TD$ – rise time, and then falling to zero in a rise time. This is illustrated in Figure 10-20.

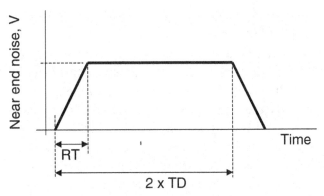

Figure 10-20 Typical signature of the capacitively coupled voltage at the near end of the quiet line, through the terminating resistor.

The magnitude of the saturated, capacitively coupled current at the near end will be:

$$I_C = \frac{1}{2} \times \frac{1}{2} \times C_{mL} \times v \times V = \frac{1}{4} \times C_{mL} \times v \times V \tag{10-18}$$

where:

I_C = capacitively coupled, saturated noise current at the near end of the quiet line

C_{mL} = mutual capacitance per length (C'_{12})

v = signal-propagation speed

V = signal voltage

½ factor = comes from half the current going to the near end and the other half to the far end

½ factor = accounts for the backward-flowing noise current spread out over $2 \times TD$

The second factor of ½ comes from the fact that the current source is moving in the forward direction, while the near end's portion of the induced current is moving in the backward direction. In every short interval of time, a total amount of charge is transferred into the quiet line, which is moving in the backward direction—but over a spatial extent that is expanding in both directions. The total current, which is the charge that flows past a point per unit time, is spread over two units of time.

While half the capacitively coupled-noise current is flowing backward to the near end, the other half of the capacitively coupled-noise current is moving in the forward direction. The forward current in the quiet line is moving to the far end at exactly the same speed as the signal edge is moving to the far end in the active line. At each step along its path, half of it is added to the already present noise moving in the forward direction. It is as though the forward-moving capacitively coupled current were growing like a snowball down a hill, building up more and more each step along its way.

At the far end, no current is present until the signal edge reaches the far end. Coincident with the signal hitting the far end, the forward-moving, capacitively coupled current reaches the far end. This current is flowing from the signal path to the return path. Through the terminating resistor across the quiet line, the voltage drop will be in the positive direction.

Since the capacitively coupled current to the quiet line scales with dV/dt, the actual noise profile in the quiet line, moving to the far end, will be the derivative of the signal edge. If the signal edge is a linear ramp, the capacitively coupled-noise current will be a short rectangular pulse, lasting for a time equal to the rise time. The capacitively induced noise signature at the far end of the quiet line is illustrated in Figure 10-21.

The total amount of current that couples over from the active to the quiet line will be concentrated in this narrow pulse. The magnitude of the current pulse, translated into a voltage by the terminating resistor, will be:

$$I_C = \frac{1}{2} \times C_{mL} \times Len \times \frac{V}{RT} \tag{10-19}$$

where:

I_C = total capacitively coupled-noise current from the active line to the quiet line

½ factor = fraction of capacitively coupled current moving to the far end

C_{mL} = mutual capacitance per length (C_{12})

RT = signal rise time

V = signal voltage

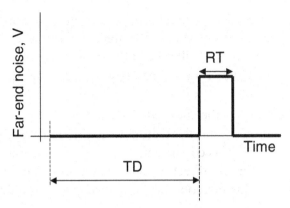

Figure 10-21 Typical signature of the capacitively coupled voltage at the far end of the quiet line, through the terminating resistor.

T I P   The magnitude of the capacitively coupled current at the far end scales directly with the mutual capacitance per length and with the coupled length of the pair of lines, and it scales inversely with the rise time. A shorter rise time will increase the far-end noise.

Unlike the backward-propagating noise, the far-end propagating noise voltage will scale with the length of the coupled region and will scale inversely with the rise time of the signal. The current direction for the forward-propagating capacitively coupled current will be in the positive direction, from the signal line to the return path, therefore generating a positive voltage across the terminating resistor.

10.10 Inductively Coupled Currents

Inductively coupled currents behave in a similar way to capacitively coupled currents. These currents are driven by a dI/dt in the active line through the mutual inductor, which creates a voltage in the quiet line. The noise voltage induced in the quiet line will see an impedance and will drive an associated current.

The changing current in the active line is moving from the signal to the return path, propagating down the line. If the direction of propagation is from left to right, the direction of the current loop is clockwise, as a signal-return path loop. This is also the direction of the increasing current, the dI/dt in the active line, a clockwise circulating loop. This changing current loop will ultimately induce a current loop in the quiet line. But in what direction will be the induced current loop? Will it be in the same direction as the signal current loop, the clockwise direction, or the opposite direction, the counterclockwise direction?

The direction of the induced current loop is based on the results of Maxwell's Equations. While it is tedious to go through the analysis to determine the direction of the induced current, it is easy to remember based on Lentz's Law, which states that the direction of the induced current loop in the quiet line will be in the opposite direction of the inducing current loop in the active line.

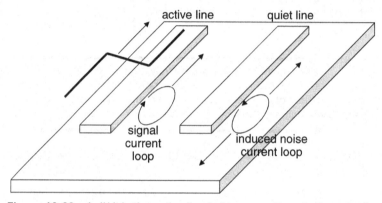

Figure 10-22   A dI/dt in the active line induces a voltage in the quiet line, which in turn creates a dI/dt in the quiet line. Half of the current loop will propagate in each direction in the quiet line.

The direction of the induced current loop in the quiet line will be circulating in the counterclockwise direction, located in the quiet line right where the signal edge is in the active line. This is illustrated in Figure 10-22.

Once this counterclockwise current loop is generated in the quiet line, which direction will it propagate? Looking up and down the line, it will see the same impedance, so it will propagate equal amounts of current in the two directions. This is a very subtle and confusing point. Half of the current in the induced current loop in the quiet line will propagate back to the near end. The other half of the current in the induced current loop in the quiet line will propagate in the forward direction.

T I P   Moving in the backward direction, the counterclockwise current loop will be flowing from the signal path to the return path. This is the same direction the capacitively coupled currents flow. At the near end, the capacitively and inductively coupled-noise currents will add together.

T I P   Moving in the forward direction, the counterclockwise current loop in the quiet line is flowing from the return path up to the signal path. The capacitively coupled and inductively coupled currents circulate in the opposite directions when they are moving in the forward direction. When the coupled currents reach the far-end terminating resistor on the quiet line, the net current through the terminating resistor will be the difference between the capacitively coupled current and the inductively coupled current.

The backward-moving inductively coupled-noise current will have exactly the same signature as the capacitively generated noise current. It will start out at zero and rise up as the signal emerges from the driver. After a time equal to the rise time, the backward-flowing current will reach a constant value and stay at that level. The signal edge will act as a current source for the inductively coupled currents and couple a constant amount of current as it propagates down the whole coupled length.

After the rising edge of the signal has just reached the terminating resistor at the far end of the active line, there will still be backward-flowing inductively coupled-noise current in the quiet line. It will take another TD for all this current to finally flow back to the near end of the quiet line. The current flows for the backward- and forward-moving noise currents are shown in Figure 10-23.

Moving in the forward direction, the inductively coupled noise will travel at the same speed as the signal edge in the active line. Each step along the way, there will be more and more inductively coupled noise current coupled over. The far-end noise will grow larger with coupled length. The shape of the inductively coupled current at the far end will be the derivative of the rise time, as it is directly proportional to the dI/dt of the signal.

The direction of the inductively coupled current at the far end is counterclockwise, from the return path up to the signal path. This is the opposite direction of the capacitively coupled current. At the far end, the capacitively coupled noise and inductively coupled noise will be in opposite directions. The net far-end noise will actually be the difference between the two.

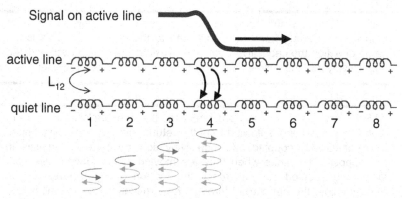

Figure 10-23   Induced-current loops propagating in the forward and backward direction as the signal in the active trace moves down the line.

10.11 Near-End Cross Talk

The near-end noise voltage is related to the net coupled current through the terminating resistor on the near end. The general signature of the waveform is displayed in Figure 10-24. There are four important features of the near-end noise:

1. If the coupling length is longer than the saturation length, the noise voltage will reach a constant value. The magnitude of this maximum voltage level is defined as the near-end cross-talk (NEXT) value. It is usually reported as a ratio of the near-end noise voltage in the quiet line to the signal in the active line. If the voltage on the active line is V_a and the maximum backward voltage on the quiet line is V_b, the NEXT is NEXT = V_b/V_a. In addition, this ratio is also defined as the near-end cross-talk coefficient, $k_b = V_b/V_a$.

2. If the coupling length is shorter than the saturation length, the voltage will peak at a value less than the NEXT. The actual noise-voltage level will be the peak value, scaled by the actual coupling length to the saturation length. For example, if the saturation length is 6 inches—that is, the signal has a rise time of 2 nsec in FR4—and the coupled length is 4 inches, the near-end noise is V_b/V_a = NEXT × 4 inches/6 inches = NEXT × 0.66. Figure 10-25 shows examples of the near-end noise for coupling lengths ranging from 20% of the saturation length to two times the saturation length.

3. The total time the near-end noise lasts is $2 \times TD$. If the time delay for the coupled region is 1 nsec, the near-end noise will last 2 nsec.

4. The turn on for the near-end noise is the rise time of the signal.

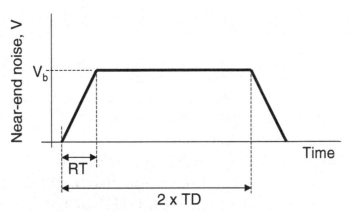

Figure 10-24   Near-end cross-talk voltage signature when the signal is a linear ramp.

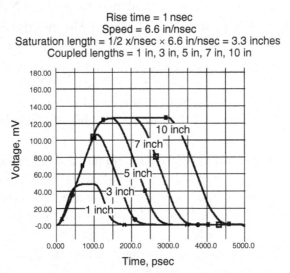

Figure 10-25   Near-end cross-talk voltage as the coupling length increases from 20% of the saturation length to two times the saturation length. Rise time is 1 nsec, speed is 6.6 inch/nsec, and saturation length is 0.5 nsec × 6.6 in/nsec = 3.3 inches. Simulated with Mentor Graphics HyperLynx.

The magnitude of the NEXT will depend on the mutual capacitance and the mutual inductance. It is given by:

$$NEXT = \frac{V_b}{V_a} = k_b = \frac{1}{4}\left(\frac{C_{mL}}{C_L} + \frac{L_{mL}}{L_L}\right) \qquad (10\text{-}20)$$

where:

NEXT = near-end cross-talk coefficient

V_b = voltage noise on the quiet line in the backward direction

V_a = voltage of the signal on the active line

k_b = backward coefficient

C_{mL} = mutual capacitance per length, in pF/inch (C_{12})

C_L = capacitance per length of the signal trace, in pF/inch (C_{11})

L_{mL} = mutual inductance per length, in nH/inch (L_{12})

L_L = inductance per length of the signal trace, in nH/inch (L_{11})

As the two transmission lines are brought together, the mutual capacitance and mutual inductance will increase, and the NEXT will increase.

The only practical way of calculating the matrix elements and the backward cross-talk coefficient is with a 2D field solver. Figure 10-26 shows the calculated near-end cross-talk coefficient, k_b, for two geometries, a microstrip pair and a stripline pair. In each case, each line is 50 Ohms, and the line width is 5 mils. The spacing is varied from 4 mils up to 50 mils. It is apparent that when the spacing is greater than about 10 mils, the stripline geometry has lower near-end cross talk.

As a rough rule of thumb, the maximum acceptable cross talk allocated in a noise budget is about 5% of the signal swing. If the quiet line is part of a bus, there could be as much as two times the near-end noise appearing on a quiet line. This is due to the sum of the noise from the two adjacent traces on either side and any that are farther away. To estimate a design rule for near-end noise, the spacing should be large enough so that the near-end noise between just two adjacent traces is less than 5% ÷ 2 ~ 2%.

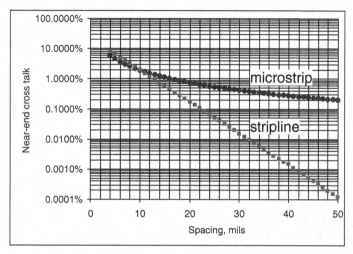

Figure 10-26   Calculated near-end cross-talk coefficient for microstrip and stripline traces, each 50 Ohms and 5 mils wide, in FR4, as the spacing is increased. Results from the Ansoft SI2D field solver.

T I P   In the case of 5-mil-wide lines in microstrip and stripline, the minimum spacing that creates less than 2% near-end noise is about 10 mils. This is a good rule of thumb for acceptable noise: The edge-to-edge spacing of signal traces should be at least two times the line width.

If the spacing between adjacent signal lines is greater than two times the line width, the maximum near-end noise will be less than 2%. In the worst-case coupling between one victim line and many aggressor lines on both sides, the maximum possible near-end noise coupled to the victim line will be less than 5%, which is within most typical noise budgets.

From this plot, two other rules of thumb can be generated. These apply for the special case of 50-Ohm lines in FR4 with dielectric constant of 4. The near-end cross talk will scale with the ratio of the line width to the spacing. Of course, it is the dielectric thickness that is important, but a specific line width and a 50-Ohm characteristic impedance also defines a dielectric thickness. While it is the dielectric thickness that really determines the fringe-field extent from the edge, it is the line width that we see when looking at the traces on the board, and we can use it to estimate the spacing.

Figure 10-27 summarizes the coupling in microstrip and stripline for spacings of $1 \times w$, $2 \times w$, and $3 \times w$. These are handy values to remember.

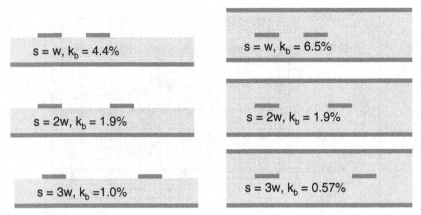

Figure 10-27 Near-end cross-talk coefficients for microstrip and stripline for a few specific spacings. These are handy rules of thumb to remember.

10.12 Far-End Cross Talk

The far-end noise voltage is related to the net coupled current through the terminating resistor on the far end. This, after all, is the voltage that is propagating down the quiet line in the forward direction. The general signature of the waveform is displayed in Figure 10-28. There are four important features of the far-end noise:

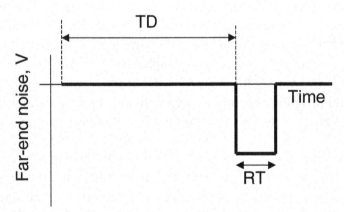

Figure 10-28 General signature of far-end cross-talk voltage noise when the signal is a linear ramp.

1. No noise appears until a TD after the signal is launched. The noise has to travel down the end of the quiet line at the same speed as the signal.
2. The far-end noise appears as a pulse that is the derivative of the signal edge. The coupled current is generated by a dV/dt and a dI/dt, and this generated noise pulse travels down the quiet line in the forward direction, coincident with the active signal traveling down the aggressor line. The width of the pulse is the rise time of the signal. Figure 10-29 shows the far-end noise with different rise times. As the rise time decreases, the width of the far-end noise decreases, and its peak value increases.
3. The peak value of the far-end noise scales with the coupling length. Increase the coupling length, and the peak value increases.
4. The FEXT coefficient is a direct measure of the peak voltage of the far-end noise, V_f, usually expressed relative to the active signal voltage, V_a. FEXT $= V_f/V_a$. This noise value scales with the two extrinsic terms (coupling length and rise time) in addition to the intrinsic terms that are based on the cross section of the coupled lines. The FEXT is related to:

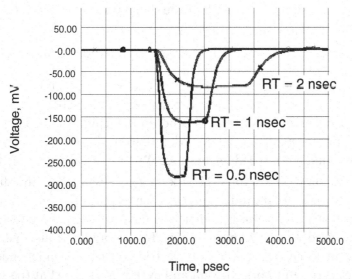

Figure 10-29 Far-end noise between two 50-Ohm microstrips in FR4 with 5-mil line and space, for the case of three different signal rise times but the same 10-inch-long coupled length. Simulated with Mentor Graphics HyperLynx.

$$\text{FEXT} = \frac{V_f}{V_a} = \frac{\text{Len}}{\text{RT}} \times k_f = \frac{\text{Len}}{\text{RT}} \times \frac{1}{2v} \times \left(\frac{C_{mL}}{C_L} - \frac{L_{mL}}{L_L} \right) \tag{10-21}$$

and

$$k_f = \frac{1}{2v} \times \left(\frac{C_{mL}}{C_L} - \frac{L_{mL}}{L_L} \right) \tag{10-22}$$

where:

FEXT = far-end cross-talk coefficient

V_f = voltage at the far end of the quiet line

V_a = voltage on the signal line

Len = length of the coupled region between the two lines

k_f = far-end coupling coefficient that depends only on intrinsic terms

v = speed of the signal on the line

C_{mL} = mutual capacitance per length, in pF/inch (C_{12})

C_L = capacitance per length of the signal trace, in pF/inch (C_{11})

L_{mL} = mutual inductance per length, in nH/inch (L_{12})

L_L = inductance per length of the signal trace, in nH/inch (L_{11})

The k_f term, or the far-end coupling coefficient, depends only on intrinsic qualities of the line: the relative capacitive and inductive coupling and the speed of the signal. It does not depend on the length of the coupling region nor on the rise time of the signal. What does this term mean? The inverse of k_f, $1/k_f$, has units of a speed, inches/nsec. What speed does it refer to?

As we show in Chapter 11, $1/k_f$ is really related to the difference in speed between an odd-mode signal and an even-mode signal. Another way of looking at far-end noise is that it is created when the odd mode has a different speed than the even mode. In a homogeneous distribution of dielectric material, the effective dielectric constant is independent of any voltage pattern, and both the odd and even modes travel at the same speed. There is no far-end cross talk.

TIP   If the dielectric material distribution surrounding all the conductors is uniform and homogeneous, as with two coupled, fully embedded microstrip traces or with two coupled striplines, the relative capacitive coupling is exactly the same as the relative inductive coupling. There will be no far-end cross talk in this configuration.

If there is any inhomogeneity in the distribution of dielectric materials, the fields will see a different effective dielectric constant, depending on the specific voltage pattern between the signal lines and the return path, and there will be a difference in the relative capacitive and inductive coupling. This will result in far-end noise.

If all the space surrounding the conductors in a pair of coupled lines is filled with air, and there is no other dielectric nearby, the relative capacitive coupling and inductive coupling are both equal, and the far-end coupling coefficient, k_f, is 0.

If all the space surrounding the conductors is filled with a material with dielectric constant ε_r, the relative inductive coupling will not change, as magnetic fields do not interact with dielectric materials at all.

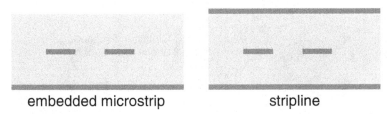

embedded microstrip   stripline

Figure 10-30   Two structures with homogeneous dielectric and no far-end cross talk: fully embedded microstrip and stripline.

The capacitive coupling will increase proportionally to the dielectric constant. The capacitance to the return path will also increase proportionally to the dielectric constant. However, the ratio will stay the same. There will be no far-end cross talk. An example of a fully embedded microstrip with no far-end cross talk is shown in Figure 10-30.

If the dielectric is removed above the embedded microstrip traces, the relative inductive coupling will not change at all since inductance is completely independent of any dielectric materials. However, the capacitance terms will be affected by the dielectric distribution. Figure 10-31 shows the change in the two capacitance terms as the dielectric thickness above the traces is decreased. Though the capacitance to the return path decreases as the dielectric thickness above decreases, it does so only by a relatively small amount. The coupling capacitance decreases much more. The coupling capacitance C_{mL}, is strongly dependent on the dielectric constant of the material where the coupling fields are strongest—between the signal traces. As the dielectric on top is removed, it dramatically reduces the coupling capacitance.

In the fully embedded case, the relative coupling capacitance is as large as the relative coupling inductance. In the pure microstrip case with no dielectric above the traces, the relative coupling capacitance has actually decreased from the fully embedded case.

TIP　　Here is a situation where we are actually decreasing the coupling capacitance, yet the far-end noise increases.

What is often reported is not k_f but $v \times k_f$, which is dimensionless:

$$v \times k_f = \frac{1}{2} \times \left(\frac{C_{mL}}{C_L} - \frac{L_{mL}}{L_L} \right)$$

(10-23)

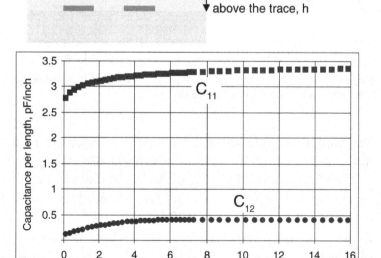

Figure 10-31　Calculated C_{11} and C_{12} terms as the dielectric thickness above the traces is increased. The diagonal element increases slightly, but the off-diagonal term increases much more. Results calculated with Ansoft's SI2D field solver.

Using this term, the FEXT can be written as:

$$\text{FEXT} = \frac{V_f}{V_a} = \frac{\text{Len}}{\text{RT} \times v} \times v \times k_f = \frac{\text{TD}}{\text{RT}} \times v \times k_f \qquad (10\text{-}24)$$

The term $v \times k_f$ is also an intrinsic term and depends only on the cross-sectional properties of the coupled lines. This is a measure of how much far-end noise there might be when the time delay of the coupled region is equal to the spatial extend of the rise time, or when TD = RT. When $v \times k_f = 5\%$, there will be 5% far-end cross talk on the quiet line when TD = RT. If the coupled length doubles, the far-end noise will double to 10%.

Figure 10-32 shows how $v \times k_f$ varies with separation for the case of two 50-Ohm FR4 microstrip traces with line widths of 5 mils. From this curve, we can develop a simple rule of thumb to estimate the far-end noise. If the spacing is equal to the line width, $v \times k_f$ will be about 4%. For example, if the rise time is 1 nsec, the coupled length is 6 inches, and the TD = 1 nsec, the far-end noise from just one adjacent aggressor line will be $v \times k_f \times \text{TD/RT} = 4\% \times 1 = 4\%$.

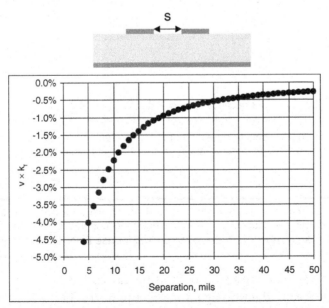

Figure 10-32 Calculated value of $v \times k_f$ for the case of two 50-Ohm coupled microstrip traces in FR4 with 5-mil-wide traces, as the spacing increases. Simulated with Ansoft's SI2D.

TIP   As a good rule of thumb, for two 50-Ohm microstrip lines with FR4, and with the tightest spacing manufacturable, a spacing equal to the line width, the far-end cross-talk noise will be –4% × TD/RT.

If the coupled length increases, the far-end noise voltage will increase. If the rise time decreases, the far-end noise will increase. If there are aggressors on either side of the signal line, there will be an equal amount of far-end noise from each active line. With a line width of 5 mils and spacing of 5 mils, the far-end noise on the quiet line would be 8% when TD = RT.

The length of surface traces on a board usually does not shrink much from one product generation to the next. The coupled time delay will also be roughly the same. However, with each product generation, rise times generally decrease. This is why far-end noise will become an increasing problem.

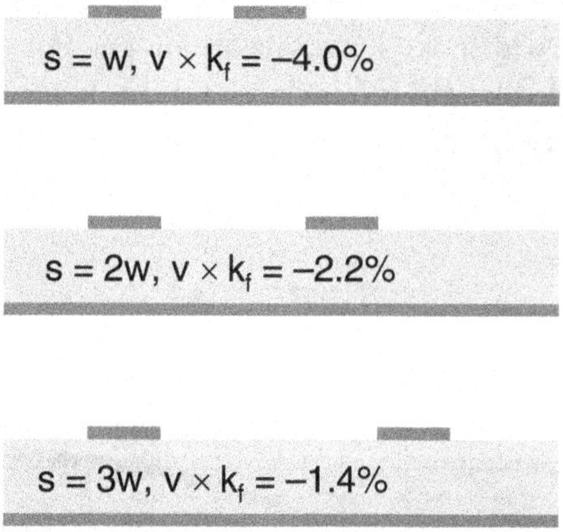

$$s = w, v \times k_f = -4.0\%$$

$$s = 2w, v \times k_f = -2.2\%$$

$$s = 3w, v \times k_f = -1.4\%$$

Figure 10-33 Simple rules of thumb for estimating the far-end cross talk for a pair of coupled 50-Ohm microstrips in FR4 for different spacings.

TIP   As rise times shrink, far-end noise will increase. For a circuit board with the closest pitch and 6-inch-long coupled traces, the far-end noise can easily exceed the noise budget for rise times of 1 nsec or less.

One important way of decreasing far-end noise is by increasing the spacing between adjacent signal paths. Figure 10-33 lists the value of $v \times k_f$

for three different spacings of two coupled, 50-Ohm microstrip transmission lines in FR4. Since the capacitance and inductance matrix elements scale with the ratio of line width to dielectric thickness, this table offers a handy way of estimating the amount of far-end cross talk for any line width, as long as each line is 50 Ohms.

10.13 Decreasing Far-End Cross Talk

There are four general guidelines for decreasing far-end noise:

1. Increase the spacing between the signal traces. Increasing the spacing from $1 \times w$ to $3 \times w$ will decrease the far-end noise by 65%. Of course, the interconnect density will also decrease and may make the board more expensive.

2. Decrease the coupling length. The amount of far-end noise will scale with the coupling length. While $v \times k_f$ can be as large as 4% for the tightest spacing (i.e., equal to the line width), if the coupled length can be kept very short, the magnitude of the far-end noise can be kept small. For example, if the rise time is 0.5 nsec, and the coupling length time delay, TD, is less than 0.1 nsec, the far-end noise can be kept below $4\% \times 0.1/0.5 = 0.8\%$. A TD of 0.1 nsec is a length of about 0.6 inch. A tightly coupled region under a BGA or connector field, for example, may be acceptable if it is kept short. The maximum co-parallel run length is a term that can typically be set up in the constraint file of a layout tool.

3. Add dielectric material to the top of the surface traces. When surface traces are required and the coupling length cannot be decreased, it is possible to decrease the far-end noise by adding a dielectric coating on top of the traces. This could be with a thicker solder mask, for example. Figure 10-34 shows how $v \times k_f$ varies with the thickness of a top coat, assuming that a dielectric constant the same as the FR4, or 4, and assuming that the trace-to-trace separation is equal to the line width.

   Adding dielectric above the trace will also increase the near-end noise and decrease the characteristic impedance of the traces. These have to be taken into account when adding a top coating.

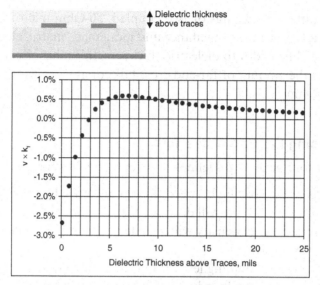

Figure 10-34 Variation of $v \times k_f$ as the coating thickness above the signal lines increases. The bus uses tightly coupled 50-Ohm microstrip traces with 5-mil-line and 5-mil spacing in FR4, assuming a dielectric coating with the same dielectric constant. Simulated with Ansoft's SI2D.

As the coating thickness is increased, the far-end noise initially decreases and actually passes through zero. Then it goes positive and finally drops back and approaches zero. In a fully embedded microstrip, the dielectric is homogeneous, and there is no far-end noise. This is when the dielectric thickness is roughly five times the line width. This complex behavior is due to the precise shape of the fringe fields between the traces and also between the traces and the return plane, as well as to how they penetrate into the dielectric material as the coating thickness is increased.

It is possible to find a value of the coating thickness so the far-end noise for surface traces is exactly zero. In this particular case, it is with a thickness equal to the dielectric thickness between the traces and the return path, or about 3 mils.

TIP   In general, the optimum coating thickness will depend on all the geometry features and dielectric constants. Even a thin solder-mask coating will provide some benefit by decreasing the far-end noise a small amount.

---

4. Route the sensitive lines in stripline. Coupled lines in buried layers, as stripline cross sections, will have minimal far-end noise. If far end is a problem, the surest way of minimizing it is to route the sensitive lines in stripline.

In practice, it is usually not possible to use a perfectly homogenous dielectric material even in a stripline. There will always be some variations in the dielectric constant due to combinations of core and prepreg materials. Usually, the prepreg is resin rich and has a lower dielectric constant than the core laminate. This will give rise to an inhomogeneous dielectric distribution and some far-end noise.

Far-end noise can be the dominant source of noise in microstrip lines. These are the surface traces. When the spacing equals the line width, and the rise time is 1 nsec, the far-end noise will be more than 8% on a victim line in the middle of a bus with a coupling length longer than 6 inches. As the rise time decreases or coupling length increases, the far-end cross talk will increase. This is why far-end noise can often be the dominant problem in low-cost boards where many of the signal lines are routed as microstrip.

Whenever you use microstrip traces, a small warning should go off that there is the potential for too much far-end noise, and an estimate should be performed of the expected far-end noise to have confidence it won't be a problem. If it is likely to be problematic, increase the routing pitch, decrease the coupling length, add some more solder mask, or route the long lines in stripline.

10.14 Simulating Cross Talk

In the special case of two uniform transmission lines with perfect terminations on each end, the expected voltage noise on the near and far ends can be calculated given the cross-section geometry and material properties. A 2D field solver will allow the calculation of k_b and NEXT. The 2D field solver will also calculate k_f and from the TD of the coupled length and the rise time, we can get the FEXT.

But what if the termination changes, and what if the coupled length is just part of a larger circuit? This requires a circuit simulator that includes a model for a coupled transmission line. If the mutual capacitance per length and mutual inductance per length are known, an n-section coupled-transmission-line model can be created to allow the simulation for any termination strategy, provided that enough sections are used. The difficulty with an n-section lumped-circuit model is its complexity and computation time.

For example, if the time delay is 1 nsec, as in a 6-inch-long interconnect, and the required model bandwidth is 1 GHz, a total of $10 \times 1 \text{ GHz} \times 1 \text{ nsec} = 10$ lumped-circuit sections would be required in the coupled model. For longer lengths, even more sections are required.

There is a class of simulators that allow the creation of coupled transmission lines using an ideal, distributed, coupled transmission-line model. These tools usually have an integrated 2D field solver. The geometry of the cross section is input, and the distributed coupled model is automatically created. The details of an ideal, distributed, coupled transmission line, often called an *ideal differential pair*, are reviewed in Chapter 11.

> **TIP**   Coupled transmission-line circuits with arbitrary drivers, loads, and terminations can be simulated with a tool that has an integrated 2D field solver and allows automatic generation of a distributed, coupled transmission-line model from the cross-sectional information. These tools are incredibly powerful in predicting the performance of coupled noise in real systems.

Figure 10-35 shows an example of the near- and far-end noise predicted in an ideal case, with a low-impedance driver on an active line and terminating resistors on all the ends. The geometry in this case is closely coupled, 50-Ohm microstrips, with 5-mil-wide traces and 5-mil-wide spacing.

When source-series termination is used, the results are slightly more complicated. Figure 10-36 shows the circuit for source-series termination. In this case, the noise at the far end of the quiet line is important as this is the location of the receiver that would be sensitive to any noise. The actual noise appearing at the receiver end of the quiet line is subtle.

Initially, the noise at the receiver end of the quiet line is due to the far-end noise from the active signal. This is a large negative spike. However, as the signal on the active line reflects off the open end of the receiver on the active line, it travels back to the source of the active line. As it travels back to the source, the receiver on the quiet line is now the backward end for this reflected signal wave. There will be backward noise on the quiet line at the receiver.

Even though the receiver on the quiet line is at the far end of the quiet line, it will see both far-end and near-end noise because of the reflections of the signal in the active line.

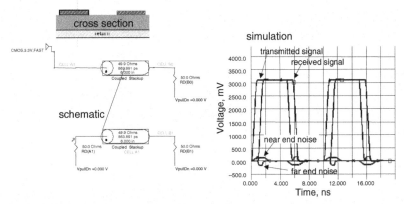

Figure 10-35 Simulating coupled transmission-line circuits with Mentor Graphics HyperLynx, showing the cross section of a tightly coupled pair of microstrip lines, the resulting circuit with the coupled lines with drivers and terminating resistors, and the simulated voltages on the active and quiet lines.

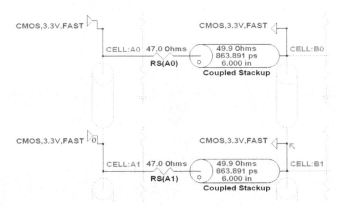

Figure 10-36 Circuit schematic for a source-series pair of coupled lines using Mentor Graphics HyperLynx.

The near-end noise on the quiet-line receiver is about the same as what is expected based on the ideal NEXT. Figure 10-37 shows the comparison of the ideal case of NEXT and FEXT in the quiet line, with the noise at the receiver on the quiet line for this source-series-terminated topology. This illustrates that the NEXT and FEXT terms are actually good figures of merit for the expected noise on the quiet line. Of course, the parasitics associated with the package models and the discrete termination components will complicate the received noise compared with the ideal NEXT and FEXT. A simulator tool will automatically take these details into account.

However, this is the case of having just one aggressor line coupling to the victim line. In a bus, there will be multiple traces coupling to a victim line. Each of the aggressor lines can add additional noise to the victim line. How many aggressors on either side should be included? The only way to know is to put in the numbers from a calculation.

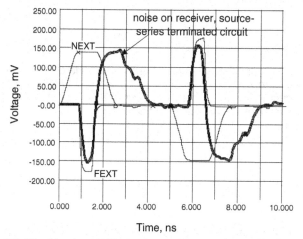

Figure 10-37 Simulated NEXT and FEXT in a far-end-terminated circuit compared with the receiver noise on the quiet line of a similar pair of lines but with source-series termination. Simulated with Mentor Graphics HyperLynx.

In Figure 10-38, a bus circuit is set up, with one victim line and five aggressor lines on either side. Each line is configured with a source-series termination and receivers on the far ends. In this example, aggressive design rules are used for 50-Ohm striplines with 5-mil-wide traces and 5-mil spaces. The speed of a signal in this transmission line is about 6 inches/nsec. The signal has a rise time of about 1 nsec and a clock frequency of 100 MHz. The saturation length is ½ nsec × 6 inches/nsec = 3 inches. We use a length of 10 inches so the near-end noise will be saturated and thus a worst case.

There will be no far-end noise in this stripline. However, the signal reflecting off the far end of the aggressor lines will change direction, and its backward noise will appear at the victim's far end. The signal in each aggressor is 3.3 volts.

Figure 10-39 shows the simulated noise at the receiver of the victim line. With just one of the adjacent lines switching, the noise on the quiet line is 195 mV. This is about 6%, probably too large all by itself for any reasonable noise budget. This geometry really is a worst case.

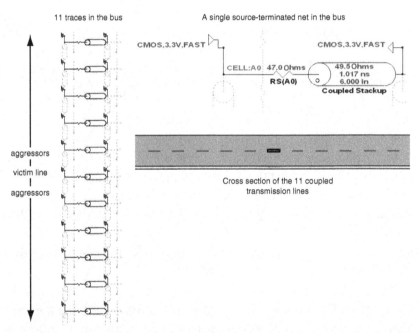

Figure 10-38 Circuit diagram used to simulate the noise on a victim line when it is surrounded by five aggressor lines on each side. Each line has source-series termination, 5-mil-wide line, 5-mil space, 10-inch length, as 50-Ohm striplines in FR4. Simulation set up by Mentor Graphics HyperLynx.

With the second aggressor on the other side of the victim line also switching, the noise is about twice, or 390 mV. This is 12% voltage noise from the adjacent two aggressor lines, one on either side of the victim line. Obviously, we get the same contribution of noise from each aggressor, and its noise contribution to the victim line just adds.

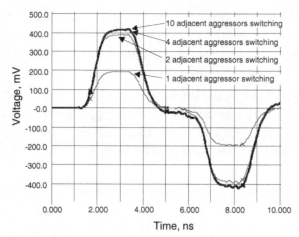

Figure 10-39 Simulated noise at the receiver of the victim line under the case of 1 aggressor, 2 aggressors, 4 aggressors, and 10 aggressors switching. Most of the noise is taken into account by including just the nearest two aggressors. Simulated with Mentor Graphics HyperLynx.

In stripline, the coupled noise drops off rapidly as the trace separation increases. We would expect the noise from the next two distant aggressor traces to be very small compared with the immediately adjacent traces. With the four nearest aggressors switching, the amount of noise at the receiver of the victim line is 410 mV, or 12.4%. Finally, the worst case is when all the aggressor lines switch. The noise is still 410 mV, indistinguishable from the noise with just the adjacent four nearest aggressors switching.

We see that the absolute worst-case noise in the bus is about 2.1 times the basic NEXT noise level. If we take into account only the adjacent traces switching, we would still be including about 95% of the total amount of noise, given this worst-case geometry. By including the switching from the two adjacent aggressor lines on both sides, we are including 100% of the coupled noise.

TIP   Using just the noise from the adjacent traces on either side of a victim line in cross-talk analysis is usually sufficient for most system-level simulations. This will account for 95% of the cross talk in a closely coupled bus.

In this example, very aggressive design rules were used. The amount of cross talk, 12%, is far above any reasonable noise budget. If, instead, the spacing equals two times the line width, the amount of noise from just one adjacent trace switching would be 1.5%. The noise from two adjacent traces switching would be 3%. When the four nearest aggressor traces switch, the coupled noise at the receiver of the quiet line would still be only 3%. It is not affected to any greater degree if all 10 adjacent traces switch. The same simulation, using this practical design rule of 5-mil-line width and 10-mil spacing, is shown in Figure 10-40.

It is only necessary to look at the adjacent trace on either side of the victim line for a worst-case analysis.

TIP   When the most aggressive design rules are used (i.e., a spacing equal to the line width), more than 95% of the noise on a victim line that is part of a bus is due to coupling from the two nearest aggressor traces, one on either side of the victim line. When conservative design rules are used and the spacing is twice the line width, virtually all of the noise from the bus comes from coupling from the most adjacent aggressor on either side of the victim line.

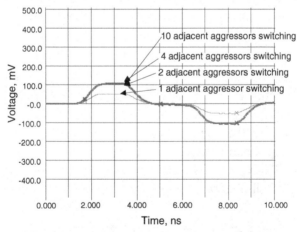

Figure 10-40   Noise on the receiver of a central trace in a bus composed of 11 50-Ohm striplines in FR4 with 5-mil-wide traces and 10-mil spacings. The number of active traces switching is changed from just 1 to 2 to 4 and then all 10. Simulated using Mentor Graphics HyperLynx.

This suggests that when establishing design rules for adequate spacings for long, parallel signal buses, the worst-case noise we might expect to find is at most 2.1 times the basic NEXT values. If the noise budget allocates 5% the voltage swing for cross talk on any victim net, the actual amount of NEXT that we should allow between adjacent nets should be 5% ÷ 2.1, which is a NEXT value of about 2%. We can use this as the design goal in establishing a spec for the closest spacing allowed in either microstrip or stripline traces.

10.15 Guard Traces

One way to decrease cross talk is by spacing the traces farther apart. Keeping the spacing twice the line width will ensure that the worst-case cross talk is less than 5%. In some cases, especially mixed-signal cases, keeping the cross talk significantly less than 5% is important. For example, a sensitive rf receiver may need as much as –100-dB isolation from any digital signals. A noise level of –100 dB is less than 0.001% of the active signal appearing on the sensitive quiet line.

It is often suggested that using guard traces will significantly decrease cross talk. This is indeed the case—but only with the correct design and configuration. A much more effective way of isolating two traces is to keep them on separate layers with different return planes. This can offer the greatest isolation of any routing alternative. However, if it is essential to route an aggressor and a victim line on the same signal layers and adjacent, then guard traces can offer an alternative way of controlling the cross talk under specific conditions.

Guard traces are separate traces that are placed between the aggressor lines and the victim lines that we want to shield. Examples of the geometry of guard traces are shown in Figure 10-41. A guard trace should be as wide as will fit between the signal lines, consistent with the fabrication design rules for spacing. Guard traces can be used in both microstrip and stripline cross sections. The value of a guard trace in microstrip is not very great, as a simple example will illustrate.

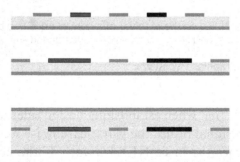

Figure 10-41   Examples of cross sections of traces that incorporate guard traces. The guard trace is hatched; all other traces are signal traces.

If we stick to design rules which specify that the smallest allowable spacing is equal to the line width, then before we can even add a guard trace, we must increase the spacing between two traces to three times the line width just to fit a trace between the aggressor and victim lines. In this geometry, we will always have less than 2% worst-case noise on the victim line. We can evaluate the added benefit of a guard trace in microstrip structures by comparing three cases:

1. Two closely spaced microstrip lines, 5-mil trace width, 5-mil spacing
2. Increasing the spacing to the minimum spacing (15 mils) that will allow a guard trace to fit
3. Increased spacing of 15 mils *and* adding the guard trace in place

These three examples are illustrated in Figure 10-42.

In addition, we can evaluate different termination strategies for the guard trace. Should it be left open, terminated at the ends, or shorted at the ends?

Figure 10-43 compares the peak noise on the receiver of the quiet line for all of these alternatives. The noise seen on the receiver is the combination of the reflections of the signal on the active line and the near- and far-end noise on the quiet line and their reflections. Combined with these first-order effects are the second-order contributions from the parasitics of the packages and resistor elements.

The peak noise with the traces close together is 130 mV. This is about 4%. Simply by pulling the traces far enough apart to fit a guard trace, the noise is reduced to 39 mV, or 1.2%. This is almost a factor-of-four reduction and is almost always a low enough cross talk, except in very rare situations. When the guard trace is inserted but left open, floating, we see the noise on the quiet line actually increases a slight amount.

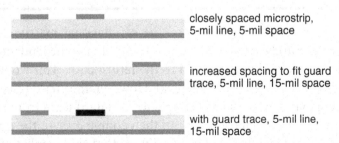

closely spaced microstrip, 5-mil line, 5-mil space

increased spacing to fit guard trace, 5-mil line, 15-mil space

with guard trace, 5-mil line, 15-mil space

Figure 10-42   Three different microstrip structures evaluated for noise at the receiver of the victim line.

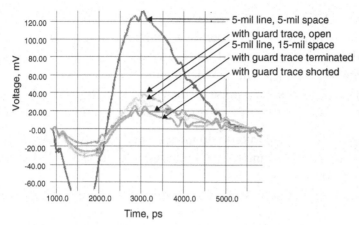

Figure 10-43   Three different geometries tested and with the guard trace left open on the ends, shorted on the ends, and terminated on the ends. Simulated with Mentor Graphics HyperLynx.

However, if the guard trace is terminated with 50-Ohm resistors on both ends, the noise is reduced to about 25 mV, or 0.75%. When the guard trace is shorted at the ends, the noise on the quiet line is reduced to 22 mV, or about 0.66%.

TIP   You will get the most benefit by increasing the spacing. The noise is reduced by a factor of four. By adding the guard trace and shorting it on both ends, the noise is reduced further by a factor of two. If left open on the ends, it will actually increase the cross talk.

TIP   The guard trace affects the electric- and magnetic-field lines between the aggressor line and the victim line, and it will always decrease both the capacitance- and inductance-matrix elements.

The reduction in magnitude of the off-diagonal matrix elements by a guard trace is purely related to the geometry. It is independent of how the guard trace is electrically connected to the return path. This would imply that a guard trace would always be beneficial. However, the guard trace will also act as another signal line. Noise will couple to the guard trace from the aggressor line. This noise on the guard trace can then couple over to the quiet line. The amount of noise generated on the guard trace that can couple to the quiet line will depend on how it is terminated.

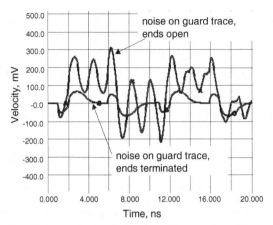

Figure 10-44   Noise at the end of a guard trace when it is open and when it is terminated. This noise will couple to the quiet line. Simulated with Mentor Graphics HyperLynx.

If the guard trace is left open, the maximum amount of noise is generated on the guard trace. If it is terminated with 50 Ohms on each end, less noise appears. Figure 10-44 shows the noise at the far end of the guard trace under the same conditions as above, with 5-mil line width, 5-mil space, and 50-Ohm microstrips with a 3.3-v, 100-MHz signal on the active line using source-series termination. It is clear that there is more noise generated on the guard trace when the guard trace is left open at the ends. This extra noise will couple more noise to the quiet line, which is why a guard trace with open ends can sometimes create more noise on the victim line than if it were not there.

The biggest benefit of a guard trace is when the guard trace is shorted on the ends. As the signal moves down the aggressor line, it will still couple noise to the guard trace. The backward-moving noise on the guard trace will hit the short at the near end and reflect with a reflection coefficient of −1. This means that much of the near-end noise on the guard trace, moving in the backward direction, will be canceled out with the coincident, negative, reflected near-end noise traveling in the forward direction on the guard trace.

TIP   Shorting the end of the guard trace will eliminate any near-end noise that would appear along the guard trace.

There will still be the buildup of the forward-moving far-end noise on the guard trace. This noise will continue to grow until it hits the shorted far end of the guard trace. At this point, it will see a reflection coefficient of –1 and reflect back. At the very far end of the guard trace, the net noise will be zero since there is, after all, a short. But the reflected far-end noise will continue to travel back to the near end of the guard trace. If all we do is short the two ends of the guard trace, the far-end noise on the guard trace will continue to reflect between the two ends, acting as a potential noise source to the victim line that is supposed to be protected. If there were no losses in the guard trace, the far-end noise would continue to rattle around the guard trace, always acting as a low-level noise source to couple back onto the victim line.

TIP   We can minimize the amount of far-end noise generated on the guard trace by adding more shorting vias, distributed down the length of the guard trace. These vias will have no impact on the noise directly coupled from the aggressor line to the victim line. They will only suppress the noise voltage generated on the guard trace.

The spacing between the shorting vias on the guard trace affects the amount of voltage noise generated on the guard trace in two ways. The far-end noise on the guard trace will only build up in the region between the vias. The closer the spacing, the lower the maximum far-end noise voltage that can build up on the guard trace. The more vias, the lower the far-end noise on the guard trace. This will mean lower voltage noise available to couple to the victim line.

Figure 10-45 shows a comparison of the coupled noise on the victim line for two coupled, 10-inch-long microstrip traces with a guard trace between them, one simulation with 1 shorting via at each end of the guard trace and a second with 11 shorting vias, spaced at 1-inch intervals along the guard trace.

The initial noise on the victim line is the same, regardless of the number of shorting vias. This noise is due to the direct coupled noise between the aggressor and victim lines. The amount of coupled noise is related to the reduction in the size of the matrix element due to the presence of the guard trace. By adding multiple vias, we limit the buildup of the noise on the guard trace, which eliminates the possibility of this additional noise coupling to the victim line.

The second role of multiple shorting vias is in generating a negative reflection of the far-end noise, which will cancel out the incident far-end noise. However, this will be canceled out only where the incident and reflected far-end noise overlaps. If the spacing between the shorting vias is shorter than the width of the far-end noise, which is the rise time, this effect will not occur.

TIP   As a rough rule of thumb, the shorting vias should be distributed along the guard trace so that there are at least three vias within the spatial extent of the signal rise time. This will guarantee overlap of far-end noise and its negative reflection, causing cancellation of noise voltage on the guard trace.

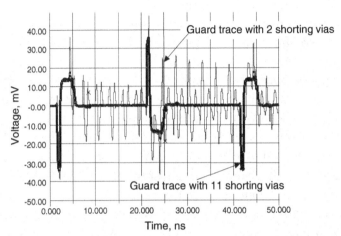

Figure 10-45 Comparing the noise on the victim line using a guard trace with only 2 shorting vias and one with 11 shorting vias. The signal is 3.3 v, with a 0.7-nsec rise time and a low clock frequency of 25 MHz, to show the effect of the noise in the guard trace with only shorting vias on its ends. The maximum noise on the victim line is 35 mV out of 3.3 v, or 1%. Simulated with Mentor Graphics HyperLynx.

If the rise time is 1 nsec, the spatial extent is 1 nsec × 6 inches/nsec = 6 inches. The spacing between shorting vias should be 6 inches/3 = 2 inches. If the rise time were 0.7 nsec, as in the previous example, the spatial extent would be 6 inches/nsec × 0.7 nsec = 4.2 inches, and the optimum spacing between shorting vias would be 4.2 inches/3 = 1.4 inches. In the previous example, vias were placed every 1 inch. Placing them any closer together would have no impact on the noise coupled to the victim line.

Of course, the shorter the signal rise time, the closer the spacing for the shorting vias required for optimum isolation. There will always be a compromise between the cost of the vias and the number that should be added. This should be an issue only when high isolation is required. In practice, there will be only a small impact on the noise in the victim line as the number of shorting vias is increased.

However, when high isolation is important, the coupled-noise to the quiet line will be much less in stripline, and stripline should always be used. In stripline structures, there will be much less far-end noise and much less need for shorting vias distributed along the length of the guard trace.

A guard trace can be very effective at limiting the isolation in a stripline structure. Figure 10-46 is the calculated near-end cross talk between an aggressor and victim, for a 50-Ohm stripline pair, with and without a guard trace between the aggressor and victim lines. In this example, the spacing between the two lines was increased and the guard trace was made as wide as would fit, consistent with the design rule of the spacing always greater than 5 mils. The guard trace, in the stripline configuration, offers a very significant amount of isolation over a configuration with no guard trace. Even isolations as large as −160 dB can be achieved using a wide guard trace in stripline. With separations of 30 mils between the active and aggressor lines, the guard trace can reduce the amount of isolation by almost three orders of magnitude.

A guard trace shields more than just the electric fields. There will also be induced currents in the guard trace from the proximity of the signal currents in the adjacent active lines. Figure 10-47 shows the current distribution in an active trace, a guard trace, and a quiet trace, with 5-mil width and spacing. Current is driven only in the active line, yet there is induced current in the guard trace, comparable to the current density in the return planes.

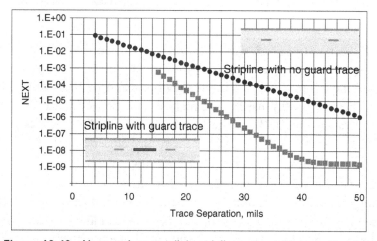

Figure 10-46 Near-end cross talk in stripline pairs, with and without a guard trace. The signal-line width is 5 mils, and each line is 50 Ohms in FR4. The guard-trace width is increased as the signal-trace separation is increased to maintain a 5-mil gap. The flattening of the near-end noise at about 10^{-9} is due to the numerical noise floor of the simulator. Simulated with Ansoft's SI2D.

This induced current will be in the opposite direction of the current in the active line. The magnetic-field lines from the induced current in the guard trace will further act to cancel the stray magnetic-field lines from the active line at the location of the quiet line. Both the electric-field shielding and magnetic-field shielding effects are fully accounted for by a 2D field solver when calculating the new capacitance- and inductance-matrix elements for a collection of conductors when one or more are used as a guard trace.

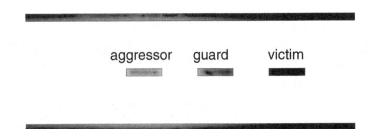

aggressor   guard   victim

Figure 10-47   Current distribution in the conductors when a 100-MHz signal is driven in the aggressor line and the guard trace is shorted to the return path. Lighter color is higher current density. Gray scale is a log scale. Simulated with Ansoft's SI2D.

10.16 Cross Talk and Dielectric Constant

Near-end noise is related to the sum of the relative capacitive and inductive coupling: $C_{12}/C_{11} + L_{12}/L_{11}$. Of course, the inductive coupling is completely unaffected by the dielectric material around the conductors.

In a 50-Ohm stripline geometry with a uniform dielectric constant everywhere, if the dielectric constant of all surrounding materials were decreased, the capacitance between the signal and return paths, C_{11}, would decrease. However, the fringe-field capacitance between the two signal lines, C_{12}, would also decrease by the same amount. There would be no change to the cross talk in a stripline.

However, if the dielectric constant were decreased everywhere, the characteristic impedance would increase from 50 Ohms. One way to get back to 50 Ohms would be to decrease the dielectric thickness. If the dielectric thickness were decreased to reach 50 Ohms, the fringe field lines would more tightly couple to the return plane, they would extend less to adjacent traces, and cross talk would decrease.

TIP   In a very subtle way, a lowered dielectric constant will decrease
cross talk—but only indirectly. The lower dielectric constant will allow a
closer spacing between the signal and return paths for the same target
impedance, which will cause the lower cross talk.

Figure 10-48 shows the characteristic impedance of a stripline trace
as the plane-to-plane separation changes, for a line width of 5 mils for two
different dielectric constants. If a design were to use a laminate with a dielectric
constant of 4.5, such as FR4, and then switch to 3.5, such as with Polyimide,
the plane-to-plane spacing to maintain 50 Ohms would have to decrease from
14 mils to 11.4 mils. If the spacing between the 5-mil traces were kept at 5 mils,
the near-end cross talk would reduce from 7.5% to 5.2%. This is a reduction in
near-end cross talk of 30%.

TIP   Using a lower-dielectric-constant material would either allow
lower cross-talk for the same routing pitch or tighter routing pitch for the
same cross-talk spec. This could result in a smaller board if it were limited
by cross-talk design rules.

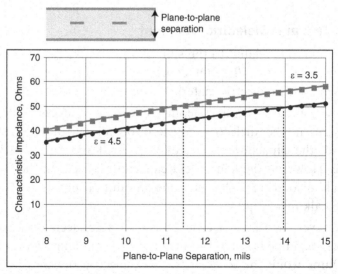

Figure 10-48 Variation in characteristic impedance for two dielectric materials in
5-mil-wide stripline traces as the total separation between the planes is changed.
A lower dielectric constant allows closer spacing between the planes for the same
impedance, which would result in lower cross talk. Simulated with Ansoft's SI2D.

10.17 Cross Talk and Timing

The time delay, TD, of a signal line depends on the length of the interconnect and the speed of the signal in the line. The speed of the signal depends on the dielectric constant of the surrounding material. In principle, cross talk from adjacent aggressor lines should not affect the time delay on a victim line. After all, how could the signal on adjacent traces affect the speed of the signal on the victim line?

> **TIP**   In stripline, cross talk from adjacent aggressor lines does not affect the time delay on a victim line. The speed of a signal on a victim line is completely independent of the signals appearing on any nearby aggressor lines, and there is no impact on timing from cross talk.

However, in microstrip traces, there is a subtle interaction between cross talk and timing. This is due to the combination of the asymmetry of the dielectric materials and the different fringe-electric fields between the signal lines, depending on the data pattern on the aggressor lines.

Consider three closely spaced, 10-inch-long microstrip signal lines, each 5 mils wide with 5-mil spacing. The outer two lines will be aggressor lines, and the center line will be a victim line. The time delay for the victim line when there is no signal on the active lines is about 1.6 nsec. This is shown in Figure 10-49.

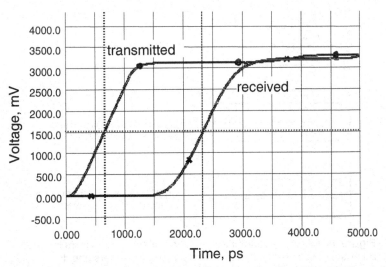

Figure 10-49 Time delay of 1.6 nsec between the signal leaving the victim driver and when it reaches the receiver on the victim line, both aggressors off. Simulated with Mentor Graphics HyperLynx.

TIP   The time delay of the signal on the victim line will depend on the voltage pattern on the aggressor lines. When the aggressors are switching opposite to the victim line, the time delay is decreased. When the aggressor lines are switching with the same signal as the victim line, the time delay of the victim line is increased.

The relationship of the voltage pattern on the aggressor lines to the time delay of the signal on the victim line is illustrated in Figure 10-50. The simulated results are shown in Figure 10-51.When the aggressor lines are off, the field lines from the signal voltage on the victim line see a combination of the bulk-material dielectric constant and the air above the line. This effective dielectric constant determines the speed of the signal.

When the aggressor lines are switching opposite from the signal on the victim line, there are large fields between the victim and aggressor traces, and many of these field lines are in air with a low dielectric constant. The effective dielectric constant the victim signal sees would have a larger fraction of air and would be reduced compared to if the aggressors were not switching. A lower effective dielectric constant would result in a faster speed and a shorter time delay for the signal on the victim line. This is illustrated in Figure 10-52.

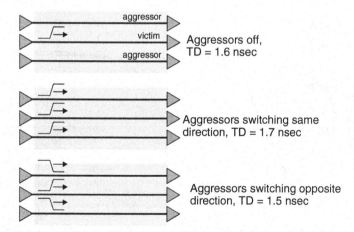

Figure 10-50   Three configurations for signals on the victim line and aggressor lines. The voltage pattern on the aggressors affects the time delay of the victim line.

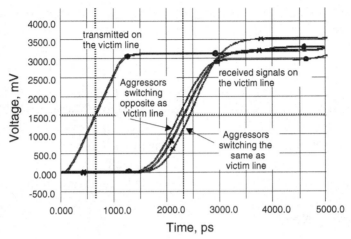

Figure 10-51 Received signal on the victim line when aggressors switch in the same direction and opposite to the signal on the victim line. Simulated with Mentor Graphics HyperLynx.

When the aggressor lines are switching in the same direction as the victim line, each trace is at the same potential, and there are few field lines in the air; most of them are in the bulk material. This means the effective dielectric constant the victim line sees will be higher, dominated by the bulk-material value. This will increase the effective dielectric constant for the victim-line signal when the aggressor lines switch in the same direction. This will decrease the signal speed and increase the time delay of the victim signal.

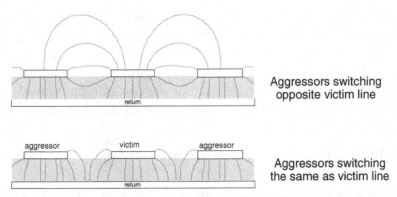

Figure 10-52 Electric-field distribution around victim and aggressor lines when they switch opposite and then when they switch in the same direction. Simulated with Mentor Graphics HyperLynx.

It is as though when the aggressors are switching opposite to the victim line, the speed of the signal is increased and the time delay is decreased. When the aggressors are switching with the same voltage pattern as the victim line, the speed of the victim signal is decreased and the time delay increased.

This impact of cross talk on time delay will happen only in tightly coupled microstrip lines where there is significant overlap of the fringe fields. If the lines are far enough apart where the cross-talk voltage is not a problem, the fringe fields will not overlap, and the time delay of the victim line will be independent of how the other traces are switching. This effect is fully taken into account when simulating with a distributed, coupled, transmission-line model.

10.18 Switching Noise

So far, we have been discussing cross talk when the return path is a wide, continuous plane, as with most transmission lines on circuit boards. In this environment, there is near- and far-end noise, and all the coupling occurs between adjacent signal lines, and virtually no coupling occurs from farther-away traces.

When the return path is not a uniform plane, the inductive coupling will increase much more than the capacitive coupling, and the noise will be dominated by the loop mutual inductance. This usually occurs in small localized regions of the interconnect, such as in packages, connectors, or local regions of the board where the return path is interrupted by gaps.

When loop mutual inductance dominates, and it occurs in a small region, we can model the coupling with a single lumped mutual inductor. The noise generated in the quiet line by the mutual inductance arises only when there is a dI/dt in the active line, which is when the edge switches. For this reason, the noise created when mutual inductance dominates is sometimes called *switching noise*, or *dI/dt noise*, or *delta I noise*. Ground bounce, as discussed previously, is a form of switching noise for the special case when the loop mutual inductance is dominated by the total inductance of the common return lead. Whenever there are shared return paths, ground bounce will occur. As pointed out, there are three ways of reducing ground bounce:

1. Increase the number of return paths so the total dI/dt in each return path is reduced.
2. Increase the width and decrease the length of the return path so it has a minimum partial self-inductance.
3. Bring each signal path in proximity to its return path to increase their partial mutual inductance with the return path.

By using a simple model, we can estimate how much loop mutual inductance between two signal/return loops is too much. As a signal passes through one pair of pins in a connector (the active path), there is a sudden change in the current in the loop right at the wavefront. This changing current causes voltage noise to be induced in an adjacent, quiet loop due to the mutual inductance between the two loops.

The voltage noise that is induced in the quiet loop is approximated by:

$$V_n = L_m \frac{dI_a}{dt} = L_m \frac{V_a}{RT \times Z_0} \qquad (10\text{-}25)$$

where:

V_n = voltage noise in the quiet loop

L_m = loop mutual inductance between the active and quiet loops

I_a = current change in the active loop

Z_0 = typical impedance the signal sees in the active and quiet loops

V_a = signal voltage in the active loop

RT = rise time of the signal (i.e., how fast the current turns on)

The only term that is really influenced by the connector or package design is the loop mutual inductance between the loops. The impedance the signal sees, typically on the order of 50 Ohms, is part of the system specification, as are the rise time and signal voltage.

The magnitude of acceptable switching noise depends on the allocated noise in the noise budget. Depending on the negotiating skills of the engineer who is responsible for selecting the connector or IC package, the switching noise would probably have to be below about 5% to 10% of the signal swing.

If the maximum acceptable switching noise is specified, this will define a maximum allowable loop mutual inductance between an active net and a quiet net. The maximum allowable loop mutual inductance is:

$$L_m = \frac{V_n}{V_a}(RT \times Z_0)$$ (10-26)

where:

L_m = loop mutual inductance between the active and quiet loops

V_n = voltage noise in the quiet loop

V_a = signal voltage in the active loop

RT = rise time of the signal, or how fast the current turns on

Z_0 = typical impedance the signal sees in the active and quiet loops

As a starting place, we will use these values:

$V_n/V_a = 5\%$

$Z_0 = 50$ Ohms

RT = 1 nsec

The maximum allowable loop mutual inductance in this example would be 2.5 nH. If the rise time were to decrease, more switching noise would be created by the dI/dt, and the loop mutual inductance would have to be decreased to keep the switching noise down.

TIP   This suggests a simple rule of thumb: To keep switching noise to an acceptable level between a pair of signal and return paths, the loop mutual inductance between them should be less than L_m < 2.5 nH × RT, with RT in nsec.

If the rise time were 0.5 nsec, the maximum allowable loop mutual inductance would be 1.2 nH. As rise times decrease, the maximum allowable mutual inductance will also decrease. This will make connector and package design more difficult. There are three primary geometrical features that decrease loop mutual inductance:

1. **Loop length**—The most important term that influences loop mutual inductance is the length of the loops. Decrease the length of the loops,

and the mutual inductance will decrease. This is why the trend in packages and connectors is to make them as short as possible, such as with chip scale packages.

2. **Spacing between the loops**—Increase the spacing, and the loop mutual inductance will decrease. There are practical limits to how far apart signal- and return-path pairs can be placed, though.

3. **Proximity of the signal to return path of each loop**—The loop mutual inductance is related to the loop self-inductance of either loop. Decreasing the loop self-inductance of either loop will decrease the loop mutual inductance between them. Bringing the signal closer to the signal path of one loop will decrease its impedance. In general, switching noise will be decreased with lower-impedance signal paths. Of course, using a value that is too low will introduce a new set of problems related to impedance discontinuities.

This analysis has assumed that the switching noise is generated between just two adjacent signal paths. If there is significant coupling between two aggressor lines and the same quiet line, keeping the same total switching noise on the victim line would require half the mutual inductance between either pair. The more aggressor lines that couple to the same victim line, the lower the allowable loop mutual inductance.

If we know how much loop mutual inductance there is between signal pairs in a package or connector, we can use the approximation in Equation 10-26 to immediately estimate their shortest usable rise time or highest usable clock frequency. For example, if the loop mutual inductance were 2.5 nH and the shortest rise time before the switching noise were more than 5%, the signal swing would be 1 nsec if there was coupling between just the two signal and return paths.

TIP This suggests a simple rule of thumb: The shortest usable rise time, in nsec, limited by switching noise is $RT > L_m/2.5$ nH. The highest usable clock frequency would be about $10 \times 1/RT = 250$ MHz/L_m. This assumes that the clock period is $10 \times RT$.

For example, if the loop mutual inductance between a pair of signal paths is 1 nH, the maximum operating clock frequency might be on the order of 250 MHz. Of course, if there were five aggressor pairs that could couple to a signal victim line, each with 1-nH loop mutual inductance, the highest operating clock frequency would be reduced to 250 MHz/5 = 50 MHz.

This is why leaded packages, with typically 1 nH of loop mutual inductance between signal paths, have maximum clock operating frequencies in the 50-MHz range.

10.19 Summary of Reducing Cross Talk

Cross talk can never be completely eliminated; it can only be reduced. The general design features that will reduce cross talk include the following:

1. Increase the pitch of the signal traces.
2. Use planes for return paths.
3. Keep coupled lengths short.
4. Route on stripline layers.
5. Decrease the characteristic impedance of signal traces.
6. Use laminates with lower dielectric constants.
7. Do not share return pins in packages and connectors.
8. When high isolation between two signal lines is important, route them on different layers with different return planes.
9. Because guard traces have little value in microstrip traces, in stripline, use guard traces with shorting vias on the ends and throughout the length.

TIP   Unfortunately, the features that reduce cross talk also invariably increase system costs. It is critically important to be able to accurately predict the expected cross talk so that the right trade-offs can be established to achieve the lowest cost at acceptable cross talk.

For cross talk in uniform transmission lines, the right tool is a 2D field solver with an integrated circuit simulator. For nonuniform transmission-line sections, the right tool is a 3D field solver, either static (if the section can be approximated by a single-lumped section) or full wave (if it is electrically long). With a predictive tool, it will be possible to evaluate how large a bang can be expected for the buck it will cost to implement the feature.

10.20 The Bottom Line

1. Cross talk is related to the capacitive and inductive coupling between two or more signal and return loops. It can often be large enough to cause problems.

2. The lowest cross talk between adjacent signal lines is when their return paths are a wide plane. In this environment, the capacitive coupling is comparable to the inductive coupling, and both terms must be taken into account.

3. Since cross talk is primarily due to the coupling from fringe fields, the most important way of decreasing cross talk is to space the signal paths farther apart.

4. The noise signature will be different on the near end and far end of a quiet line adjacent to a signal path. The near-end noise is related to the sum of the capacitively and inductively coupled currents. The far-end noise is related to the difference between the capacitively and inductively coupled currents.

5. To keep the near-end noise below 5% for the worst-case coupling in a bus, the separation should be at least two times the line width for 50-Ohm lines.

6. Near-end noise will reach a maximum value when the coupling length equals half the spatial extent of the rise time.

7. Far-end noise will scale with the ratio of the time delay of the coupling length to the rise time. For a pair of microstrip traces with a spacing equal to their line width, there will be 4% far-end noise when the time delay of the coupling length equals the rise time.

8. In a tightly coupled bus, 95% of the coupled noise can be accounted for by considering just the nearest aggressor line on either side of the victim line.

9. There is no far-end cross talk in striplines.

10. For very high isolation, signal lines should be routed on different layers with different return planes. If they have to be on the same layer and adjacent, use stripline with guard traces. Isolation greater than −160 dB is possible. In this case, watch out for ground bounce cross talk when signals change reference layers through vias.

11. Mutual inductance will dominate the coupled noise in some packages and connectors. As rise times decrease, the maximum amount of allowable mutual inductance between signal- and return-path loops must decrease. This will make it harder to design high-performance components.

End-of-Chapter Review Questions

10.1 In most typical digital systems, how much cross-talk noise on a victim line is too much?

10.2 If the signal swing is 5 V, how many mV of cross talk is acceptable?

10.3 If the signal swing is 1.2 V, how many mV of cross talk is acceptable?

10.4 What is the fundamental root cause of cross talk?

10.5 What is superposition, and how does this principle help with cross-talk analysis?

10.6 In a mixed signal system, how much isolation might be required?

10.7 What are the two root causes of cross talk?

10.8 What elements are in the equivalent circuit model for coupling?

10.9 If the rise time of a signal is 0.5 nsec, and the coupled length is 12 inches, how many sections would be needed in an n-section LC model?

10.10 What are two ways of reducing the extent of the fringe electric and magnetic fields?

10.11 Why is the total current transferred between the active line and the quiet line independent of the rise time of the signal?

10.12 What is the signature of near-end noise?

10.13 What is the signature of far-end noise in microstrip?

10.14 What is the signature of far-end noise in stripline?

10.15 What is the typical near-end cross-talk coefficient for a tightly coupled pair of striplines?

10.16 When the far end of the quiet line is open, what is the noise signature on the near end of the quiet line of two coupled microstrips?

10.17 When a positive edge is the signal on the active line, the near-end signature is a positive signal. What is the noise signature when the active signal is a negative signal? What is the noise signature for a square pulse as the active signal?

10.18 Why are the off-diagonal Maxwell capacitance matrix elements negative?

10.19 If the rise time of a signal is 0.2 nsec, what is the saturation length in an FR4 interconnect for near-end cross talk?

10.20 In a stripline, which contributes more near-end cross talk: capacitively induced current or inductively induced current?

10.21 How does the near-end noise scale with length and rise time?

10.22 How does the far-end cross talk in microstrip scale with length and rise time?

10.23 Why is there no far-end noise in stripline?

10.24 What are three design features to adjust to reduce near-end crosstalk?

10.25 What are three design features to adjust to reduce far-end cross talk?

10.26 Why does lower dielectric constant reduce cross talk?

10.27 If a guard trace were to be added between two signal lines, how far apart would the signal lines have to be moved? What would the near-end cross talk be if the traces were moved apart and no guard trace were added? What applications require a lower cross talk than this?

10.28 If really high isolation is required, what geometry offers the highest isolation with guard traces? How should the guard trace be terminated?

10.29 What is the root cause of ground bounce?

10.30 List three design features to reduce ground bounce.

CHAPTER 11

Differential Pairs and Differential Impedance

A *differential pair* is simply a pair of transmission lines with some coupling between them. The value of using a pair of transmission lines is not so much to take advantage of the special properties of a differential pair as to take advantage of the special properties of differential signaling, which uses differential pairs.

Differential signaling is the use of two output drivers to drive two independent transmission lines, one line carrying one bit and the other line carrying its complement. The signal that is measured is the difference between the two lines. This difference signal carries the information.

Differential signaling has a number of advantages over single-ended signaling, such as the following:

- The total dI/dt from the output drivers is greatly reduced over single-ended drivers, so there is less ground bounce, less rail collapse, and potentially less EMI.
- The differential amplifier at the receiver can have higher gain than a single-ended amplifier.

- The propagation of a differential signal over a tightly coupled differential pair is more robust to cross talk and discontinuities in the return path shared by the two transmission lines in the pair.
- Propagating differential signals through a connector or package will be less susceptible to ground bounce and switching noise.
- A low-cost, twisted-pair cable can be used to transmit a differential signal over long distances.

The most important downside to differential signals comes from the potential to create EMI. If differential signals are not properly balanced or filtered, and there is any common signal component present, it is possible for real-world differential signals that are driven on external twisted-pair cables to cause EMI problems.

The second downside is that transmitting a differential signal requires twice the number of signal lines as transmitting a single-ended signal. The third downside is that there are many new principles and a few key design guidelines to understand about differential pairs. Due to the non-intuitive effects in differential pairs, there are many myths in the industry that have needlessly complicated and confused their design.

Ten years ago, less than 50% of the circuit boards fabricated had controlled-impedance interconnects. Now, more than 90% of them have controlled-impedance interconnects. Today, less than 50% of the boards fabricated have differential pairs. In a few years, we may see over 90% with differential pairs.

11.1 Differential Signaling

Differential signals are widely used in the small computer system interface (SCSI) bus; in Ethernet; in USB; in many of the telecommunications optical-carrier (OC) protocols, such as OC-48, OC-192, and OC-768; and for all high-speed serial protocols. One of the popular signaling schemes is low-voltage differential signals (LVDS).

When we look at a signal voltage, it is important to keep track of where we are measuring the voltage. When a driver drives a signal on a transmission line, there is a signal voltage on the line that is measured between the signal conductor and the return path conductor. We usually refer to this as the *single-ended transmission-line signal*. When two drivers drive a differential pair, in addition to the two single-ended signals, there is also a voltage difference between the two signal lines. This is the difference voltage, or the differential signal. Figure 11-1 illustrates how these two different signals are measured in a differential pair.

In LVDS, two output pins are used to drive a single bit of information. Each signal has a voltage swing from 1.125 v to 1.375 v. Each of these signals drives a separate transmission line. The single-ended voltage between the signal and return paths of each line is shown in Figure 11-2.

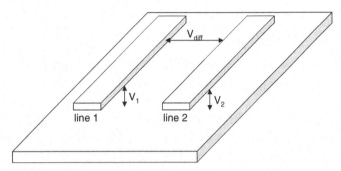

Figure 11-1   The single-ended signal is measured between the signal conductor and the return conductor. The differential signal is measured between the two signal lines in a differential pair.

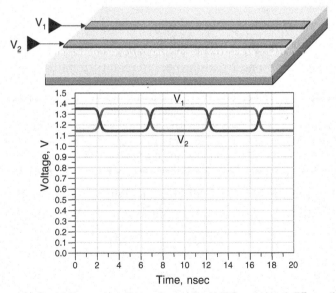

Figure 11-2   Voltage signaling scheme for LVDS, a typical differential signal.

At the receiver end, the voltage on line 1, measured between the signal line and its return path, is V_1 and the voltage on line 2 is V_2. A differential receiver measures the differential voltage between the two lines and recovers the differential signal:

$$V_{diff} = V_1 - V_2$$

(11-1)

where:

V_{diff} = differential signal
V_1 = signal on line 1 with respect to its return path
V_2 = signal on line 2 with respect to its return path

In addition to the intentional differential signal that carries information, there may be some common signal. The common signal is the average voltage that appears on both lines, defined as:

$$V_{comm} = \frac{1}{2}(V_1 + V_2)$$

(11-2)

where:

V_{comm} = common signal
V_1 = signal on line 1 with respect to its return path
V_2 = signal on line 2 with respect to its return path

TIP These definitions of differential and common signals apply universally to all signals. When any arbitrary signal is generated on a pair of transmission lines, it can always be completely and uniquely described by a combination of a common- and a differential-signal component.

Given the common and differential signals, the single-ended voltages on each line with respect to the return plane can also be recovered from:

$$V_1 = V_{comm} + \frac{1}{2}V_{diff}$$

(11-3)

$$V_2 = V_{comm} - \frac{1}{2} V_{diff} \qquad (11\text{-}4)$$

where:

V_{comm} = common signal

V_{diff} = differential signal

V_1 = signal on line 1 with respect to its return path

V_2 = signal on line 2 with respect to its return path

In LVDS, there are some differential-signal components and some common-signal components. These signal components are displayed in Figure 11-3. The differential signal swings from −0.25 v to +0.25 v. As the differential-signal propagates down the transmission line, its voltage is a 0.5-v transition. When describing the magnitude of the differential signal, we usually refer to the *peak-to-peak value.*

There is also a common signal. The average value is 1.25 v. This is more than a factor of 2 larger than the differential-signal component.

TIP   Even though an LVDS is called a differential signal, it has a very large common-signal component, which is nominally constant.

When we call an LVDS a differential signal, we are really lying. It has a differential component, but it also has a very large common component. It is not a pure differential signal. In ideal conditions, the common signal is constant. The common signal normally will not have any information content and will not affect signal integrity or system performance.

Unfortunately, as we will see, very small perturbations in the physical design of the interconnects can cause the common-signal component to change, and a changing common-signal component has the potential of causing two very important problems:

1. If the value of the common signal gets too high, it may saturate the input amplifier of the differential receiver and prevent it from accurately reading the differential signal.
2. If any of the changing common signal makes it out on a twisted-pair cable, it has the potential of causing excessive EMI.

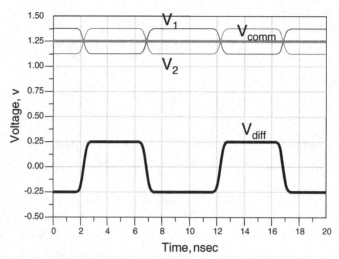

Figure 11-3 The common- and differential-signal components in LVDS. Note that there is a very large common-signal component, which, in principle, is constant.

TIP   The terms *differential* and *common* always and only refer to the properties of the signal, never to the properties of the differential pair of transmission lines themselves. Misuse of these terms is one of the major causes of confusion in the industry.

11.2 A Differential Pair

All it takes to make up a differential pair is two transmission lines. Each of the two transmission lines can be a simple, single-ended transmission line. Together, the two lines are called a *differential pair*. In principle, any two transmission lines can make up a differential pair.

Just as there are many cross sections for a single-ended transmission line, there are many cross sections for differential pair transmission lines. Figure 11-4 shows the cross sections of the most popular geometries.

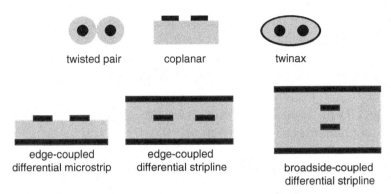

twisted pair   coplanar   twinax

edge-coupled   edge-coupled
differential microstrip   differential stripline   broadside-coupled
differential stripline

Figure 11-4   Some of the most popular cross sections for differential-pair transmission.

Although, in principle, any two transmission lines can make up a differential pair, five features will optimize their performance for transporting high-bandwidth difference signals:

1. The most important property of a differential pair is that it be of uniform cross section down its length and provide a constant impedance for the difference signal. This will ensure minimal reflections and distortions for the differential signal.
2. The second most important property is that the time delay between each line is matched so that the edge of the difference signal is sharp and well defined. Any time delay difference between the two lines, or skew between them, will also cause a distortion of the differential signal and some of the differential signal to be converted into common signal.
3. Both transmission lines should look exactly the same. The line width and dielectric spacing of the two lines should be the same. This property is called symmetry between the two lines. There should be no other asymmetries such as a test pad on one and not on the other or a neck-down on one but not the other. Any asymmetries will convert differential signals into common signals.
4. Each line in a pair should have the same length. The total length of each line should be exactly the same. This will help maintain the same delay and minimize skew.

5. There does not have to be any coupling between the two lines, but most of the noise immunity benefits of differential pairs is lost if there is no coupling. Coupling between the two signal lines will allow differential pairs to be more robust to ground bounce noise picked up from other active nets than single-ended transmission lines. The greater the coupling, the more robust the differential signals will be to discontinuities and imperfections.

Figure 11-5 shows an example of a differential pair of transmission lines with a differential signal propagating. In this example, the signal on one line is a transition from 0 v to 1 v, and the other signal is a transition from 1 v to 0 v. As the signals propagate down the transmission lines, they have the voltage distribution shown in the figure.

Of course, even though we called this a differential signal, we see right away that we are lying again. It is not a pure differential signal but contains a large common signal component of 0.5 v. However, this common signal is constant, and we will ignore it, paying attention only to the differential component.

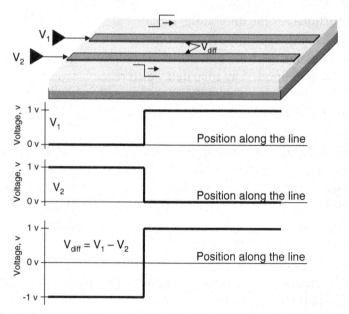

Figure 11-5   A differential pair with a signal propagating on each line and the differential signal between the two lines.

Given these voltages on each transmission line, the difference voltage can be easily calculated. By definition, it is $V_1 - V_2$. The pure differential-signal component that is propagating down the differential pair is also shown in the figure. With a 0-v to 1-v signal transition on each single-ended transmission line, the differential-signal swing is a 2-v transition propagating down the interconnect. At the same time, there is a common-signal component, $1/2 \times (V_1 + V_2) = 0.5$ v, which is constant along the line.

> **TIP** The most important electrical property of a differential pair is the impedance the differential signal sees, which we call the *differential impedance*.

11.3 Differential Impedance with No Coupling

The impedance the differential signal sees, the differential impedance, is the ratio of the voltage of the signal to the current of the signal. This definition is the basis of calculating differential impedance. What makes it subtle is identifying the voltage of the signal and the current of the signal.

The simplest case to analyze is when there is no coupling between the two lines that make up the differential pair. We will look at this case, determine the differential impedance of the pair, and then turn on coupling and look at how the coupling changes the differential impedance.

To minimize the coupling, assume that the two transmission lines are far enough apart, i.e., at least a spacing equal to twice the line width, so they do not appreciably interact with each other and so each line has a single-ended characteristic impedance, Z_0, of 50 Ohms. The current going into one signal trace and out the return is:

$$I_{one} = \frac{V_{one}}{Z_0} \qquad (11\text{-}5)$$

where:

I_{one} = current into one signal line and out its return

V_{one} = voltage between the signal line and the adjacent return path of one line

Z_0 = single-ended characteristic impedance of the signal line

For example, when a signal of 0 v to 1 v is imposed on one line and at the same time a 1-v to 0-v signal is imposed on the other line, there will also be a current loop into each line. Into the first line will be a current of $I = 1$ v/50 Ohms = 20 mA flowing from the signal conductor down to and out its adjacent return path. Into the second line will also be a current loop of 20 mA, but flowing from the return conductor up to and out its signal conductor.

The differential-signal transition propagating down the line is the differential signal between the two signal-line conductors. This differential-signal swing is twice the voltage on either line: $2 \times V_{one}$. In this case, it is a 2-volt transition that propagates down the signal-line pair. At the same time, just looking at the signal-line conductors, it looks like there is a current loop of 20 mA going into one signal conductor and coming out the other signal conductor.

By the definition of impedance, the impedance the differential signal sees is:

$$Z_{diff} = \frac{V_{diff}}{I_{one}} = \frac{2 \times V_{one}}{I_{one}} = 2 \times \frac{V_{one}}{I_{one}} = 2 \times Z_0 \qquad (11\text{-}6)$$

where:

Z_{diff} = impedance the differential signal sees, the differential impedance

V_{diff} = difference or differential-signal transition

I_{one} = current into one signal line and out its return

V_{one} = voltage between the signal line and the adjacent return path of one line

Z_0 = single-ended characteristic impedance of the signal line

The differential impedance is twice the single-ended characteristic impedance of either line. This is reasonable because the voltage across the two lines is twice the voltage across either one and its return path, but the current going into one signal and out the other is the same. If the single-ended impedance of either line is 50 Ohms, the differential impedance of the pair would be 2×50 Ohms = 100 Ohms.

If a differential signal propagates down a differential pair and reaches the end, where the receiver is, the impedance the differential signal sees at the end will be very high, and the differential signal will reflect back to the source. These multiple reflections will cause signal-quality noise problems. Figure 11-6 is an example of the simulated differential signal at the end of a differential pair. The ringing is due to the multiple bounces of the differential signal between the low impedance of the driver and the high impedance of the end of the line.

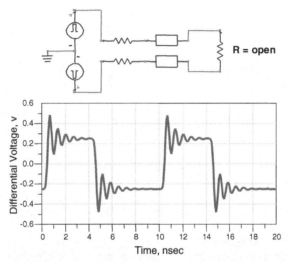

Figure 11-6 Differential circuit and the received signal at the far end of the differential pair when the interconnect is not terminated. There is no coupling in this differential pair. Simulated with Agilent ADS.

One way of managing the reflections is to add a terminating resistor across the ends of the two signal lines that matches the impedance the difference signal sees. The resistor should have a value of $R_{term} = Z_{diff} = 2 \times Z_0$. When the differential signal encounters the terminating resistor at the end of the line, it will see the same impedance as in the differential pair, and there will be no reflection. Figure 11-7 shows the simulated received differential signal with a 100-Ohm differential-terminating resistor between the two signal lines.

Differential impedance can also be thought of as the equivalent impedance of the two single-ended lines in series. This is illustrated in Figure 11-8. Looking into the front of each line, each driver sees the instantaneous impedance as the characteristic impedance of the line, Z_0. The instantaneous impedance between the two signal lines is the series combination of the impedances of each line to the return path. The equivalent impedance of the two signal lines, or the differential impedance, is the series combination:

$$Z_{diff} = Z_0 + Z_0 = 2 \times Z_0 \qquad (11\text{-}7)$$

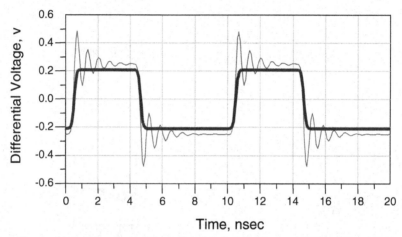

Figure 11-7   Received differential signal at the far end of the differential pair when the interconnect is terminated. There is no coupling in this differential pair. Simulated with Agilent ADS.

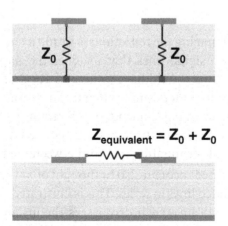

Figure 11-8   The impedance between each line and its return path in a differential pair and the equivalent impedance between the two signal lines.

where:

Z_{diff} = equivalent impedance between the signal lines, or the differential impedance

Z_0 = impedance of each line to the return path

If all we had to worry about was the differential impedance of uncoupled transmission lines, we would essentially be done. The differential impedance of a differential pair is always two times the single-ended impedance of either line, as seen by either driver. Two important factors complicate the real world. First is the impact from coupling between the two lines, and second is the role of the common signal and its generation and control.

11.4 The Impact from Coupling

If we bring two stripline traces closer and closer together, their fringe electric and magnetic fields will overlap, and the coupling between them will increase. Coupling is described by the mutual capacitance per length, C_{12}, and mutual inductance per length, L_{12}. (Unless otherwise noted, the capacitance-matrix elements always refer to the SPICE capacitance-matrix elements, not the Maxwell capacitance-matrix elements. These terms were introduced in Chapter 10, "Cross Talk in Transmission Lines.")

As the traces are brought together, both C_{11} and C_{12} will change. C_{11} will decrease as some of the fringe fields between signal line 1 and its return path are intercepted by the adjacent trace, and C_{12} will increase. However, the loaded capacitance, $C_L = C_{11} + C_{12}$, will not change very much. Figure 11-9 shows the equivalent capacitance circuit of two stripline traces and how C_{11}, C_{12}, and C_L vary for the specific case of two 50-Ohm stripline traces in FR4 with 5-mil-wide traces.

TIP   It is important to note that the coupling described by the capacitance- and inductance-matrix elements is completely independent of any applied voltages. It is purely related to the geometry and material properties of the collection of conductors.

Both L_{11} and L_{12} will change as the traces are brought together. L_{11} will decrease very slightly (less than 1% at the closest spacing) due to induced eddy currents in the adjacent trace, and L_{12} will increase. This is shown in Figure 11-10.

As the two traces are brought together, the coupling increases. However, even in the tightest spacing, the spacing equal to the line width, the maximum relative coupling, C_{12}/C_L or L_{12}/L_{11}, is less than 15%. When the spacing is more than

15 mils, or three times the line width, the relative coupling is reduced to 1%—a negligible amount. Figure 11-11 shows how this ratio, both the relative capacitive coupling and the relative inductive coupling, changes as the separation changes.

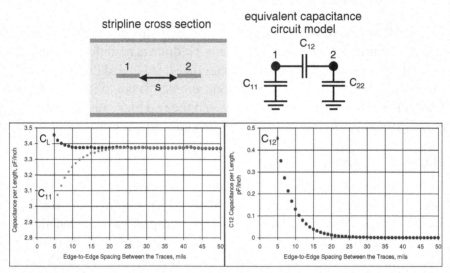

Figure 11-9 Variation in the loaded capacitance per length, C_L, and the SPICE diagonal capacitance per length, C_{11}. Also plotted is the variation in the coupling capacitance, C_{12}. Simulated with Ansoft's SI2D.

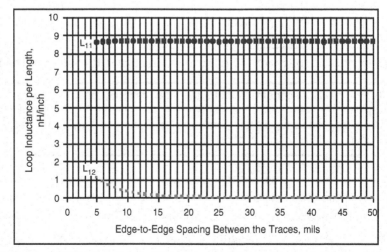

Figure 11-10 Variation in the loop self-inductance per length, L_{11}, and the loop mutual inductance per length, L_{12}. Simulated with Ansoft's SI2D.

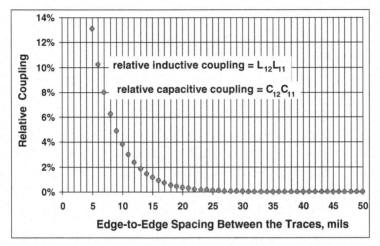

Figure 11-11 Relative mutual capacitance and mutual inductance as the spacing between two 50-Ohm, 5-mil stripline traces in FR4 changes. The relative coupling capacitance and coupling inductance are identical for a homogeneous dielectric structure such as stripline. Simulated with Ansoft's SI2D.

When the lines are far apart, the characteristic impedance of line 1 is completely independent of the other line. The characteristic impedance varies inversely with C_{11}:

$$Z_0 \sim \frac{1}{C_{11}} \tag{11-8}$$

where:

Z_0 = characteristic impedance of the line

C_{11} = capacitance between the signal trace and the return path

When the lines are brought closer together, the presence of the other line will affect the impedance of line 1. This is called the *proximity effect*. If the second line is tied to the return path—i.e., a 0-v signal is applied to line 2, and only line 1 is being driven—the impedance of line 1 will depend on its loaded capacitance in the proximity of the other line. The characteristic impedance of the driven line will be related to the capacitance per length that is being driven:

$$Z_0 \sim \frac{1}{C_{11}+C_{12}} = \frac{1}{C_L} \tag{11-9}$$

where:

Z_0 = characteristic impedance of the line

C_{11} = capacitance per length between the signal trace and the return path

C_{12} = capacitance per length between the two adjacent signal traces

C_L = loaded capacitance per length of one line

As the traces move closer together, the single-ended impedance of line 1 will decrease—but only very slightly (less than 1%). Figure 11-12 shows the change in the single-ended characteristic impedance of line 1 as the two traces are brought into proximity. The single-ended characteristic impedance is essentially unchanged if the second line is pegged low and the two traces are brought closer together.

However, suppose the second trace is also being driven, and the signal on line 2 is the opposite of the signal on line 1. As the signal on line 1 increases from 0 v to 1 v, the signal on line 2 is simultaneously dropped from 0 v to –1 v. As the driver on line 1 turns on, it will drive a current through the C_{11} capacitance due to the dV_{11}/dt between line 1 and the return path. In addition, there will be a current from line 1 to line 2, due to the changing voltage between them, dV_{12}/dt. This voltage will be twice the voltage change between line 1 and its return, $V_{12} = 2 \times V_{11}$.

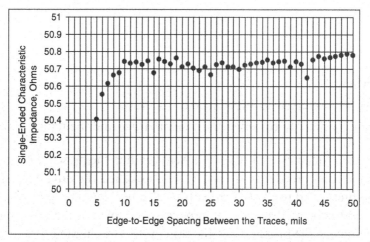

Figure 11-12 Single-ended characteristic impedance of one line when the second is shorted to the return path and the separation changes, for 5-mil-wide, 50-Ohm striplines in FR4. The small fluctuations, on the order of 0.05 Ohm, are due to the numerical noise of the simulation tool. Simulated with Ansoft's SI2D.

The current into one signal line will be related to:

$$I_{one} = v \times RT \times \left(C_{11} \frac{dV_{11}}{dt} + C_{12} \frac{dV_{12}}{dt} \right) \sim C_{11}V_{one} + 2C_{12}V_{one} = V_{one}(C_L + C_{12}) \quad (11\text{-}10)$$

where:

I_{one} = current going into one line

v = speed of the signal moving down the signal line

C_{11} = capacitance between the signal line and its return path per length

V_{11} = voltage between the signal line and the return path

C_{12} = capacitance between each signal line per length

V_{12} = voltage between each signal line

V_{one} = voltage change between the signal line and return path of one line

RT = rise time of the transition

As the two traces move together and they are both being driven by signals transitioning in opposite directions, the current from the driver into line 1 and out the return path will increase in order to drive the higher capacitance of the single-ended line.

TIP   If the current increases for the same applied voltage, the input impedance the driver sees will decrease. The characteristic impedance of the line will drop if the second line is driven with the opposite signal.

Suppose the second line is driven with exactly the same signal as the first line. The driver to line 1 will see less capacitance to drive since there will be no voltage between the two signal lines and the driver will see only the C_{11} capacitance. When the second line is driven with the same voltage as line 1, the current into signal line 1 will be:

$$I_{one} = v \times RT \times \left(C_{11} \frac{dV_{11}}{dt} \right) \sim C_{11}V_{one} = V_{one}(C_L - C_{12}) \quad (11\text{-}11)$$

where:

I_{one} = current going into one line

v = speed of the signal moving down the signal line

C_{11} = capacitance between the signal line and its return path per length

V_{11} = voltage between the signal line and the return path

C_{12} = capacitance between each signal line per length

V_{one} = voltage change between the signal line and return path of one line

RT = rise time of the transition

We see that the characteristic impedance of one line, when in the proximity of a second line, is not a unique value. It depends on how the other line is being driven. If the second line is pegged low, the impedance will be close to the uncoupled single-ended impedance. If the second line is switching opposite the first line, the impedance of the first line will be lower. If the second line is switching exactly the same as the first line, the impedance of the first line will be higher. Figure 11-13 shows how the characteristic impedance of the first line varies as the separation changes for these three cases.

This is a critically important observation. When we only dealt with single-ended signals, a transmission line had only one impedance that described it. But when it is part of a pair and coupling plays a role, it now has three different impedances that describe it. We need to identify new labels to describe which of the three different impedances we talk about when referring to "the impedance" of a transmission line when it is part of a pair. We will therefore introduce the terms *odd* and *even mode impedance* later in this chapter to provide clear and unambiguous language describing the properties of differential pairs.

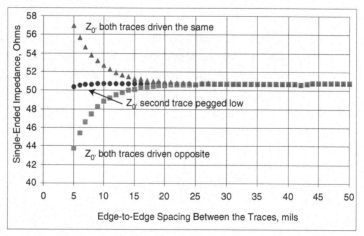

Figure 11-13   The characteristic impedance of line 1 when the second line is pegged low, switching opposite and switching the same, as the separation between the traces changes, for 5-mil-wide, 50-Ohm striplines in FR4. Simulated with Ansoft's SI2D.

TIP   When trace separations are closer than about three line widths, the presence of the adjacent trace will affect the characteristic impedance of the first trace. Its proximity as well as the manner in which it is being driven must be considered.

The differential signal drives opposite signals on each of the two lines. As we saw above, the impedance of each line, when the pair is driven by a differential signal, will be reduced due to the coupling between the two lines.

When a differential signal travels down a differential pair, the impedance the differential signal sees will be the series combination of the impedance of each line to its respective return path. The differential impedance, with coupling, will still be twice the characteristic impedance of either line. It's just that the characteristic impedance of each line decreased due to coupling.

Figure 11-14 shows the differential impedance as the spacing between the lines decreases. At the closest spacing that can be realistically manufactured (i.e., a spacing equal to the line width), the differential impedance of a pair of coupled stripline traces is reduced only about 12% from when the striplines are three line widths apart and uncoupled.

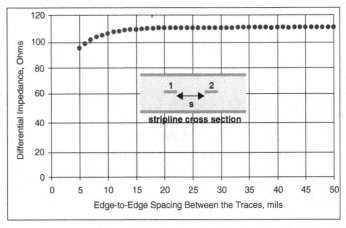

Figure 11-14   The differential impedance of a pair of 5-mil-wide, 50-Ohm stripline traces in FR4, as the spacing between them decreases. Simulated with Ansoft's SI2D.

11.5 Calculating Differential Impedance

Additional formalism must be introduced to describe the impact from coupling on differential impedance, which is an effect of 12% at most. The complexity arises when quantifying the decrease in differential impedance as the traces are brought closer together and coupling begins to play a role. There are five different approaches to this analysis:

1. Use the direct results from an approximation.
2. Use the direct results from a field solver.
3. Use an analysis based on modes.
4. Use an analysis based on the capacitance and inductance matrix.
5. Use an analysis based on the characteristic impedance matrix.

There is only one reasonably close approximation useful for calculating the differential impedance of either an edge-coupled microstrip or an edge-coupled stripline, originally offered in a National Semiconductor application note by James Mears (AN-905). These approximations are based on empirical fitting of measured data.

For edge-coupled microstrip using FR4 material, the differential impedance is approximately:

$$Z_{\text{diff}} = 2 \times Z_0 \left[1 - 0.48 \exp\left(-0.96 \frac{s}{h} \right) \right] \qquad (11\text{-}12)$$

where:

Z_{diff} = differential impedance, in Ohms

Z_0 = uncoupled single-ended characteristic impedance for the line geometry

s = edge-to-edge separation between the traces, in mils

h = dielectric thickness between the signal trace and the return plane

For an edge-coupled stripline with FR4, the differential impedance is:

$$Z_{\text{diff}} = 2 \times Z_0 \left[1 - 0.37 \exp\left(-2.9 \frac{s}{b} \right) \right] \tag{11-13}$$

where:

Z_{diff} = differential impedance, in Ohms

Z_0 = uncoupled single-ended characteristic impedance for the line geometry

s = edge-to-edge separation between the traces, in mils

b = total dielectric thickness between the planes

We can evaluate how accurate these approximations are by comparing them to the predicted differential impedance calculated with an accurate field solver. In Figure 11-15, we compare these approximations to three different cross sections for edge-coupled microstrip and edge-coupled stripline. In each case, we use the field-solver-calculated single-ended characteristic impedance in the approximation, and the approximation predicts the slight perturbation on the differential impedance from the coupling. The accuracy varies from 1% to 10%, provided that we have an accurate starting value for the characteristic impedance.

A more accurate tool that will also predict the single-ended characteristic impedance of either line is a 2D field solver. A 2D field solver requires input regarding the cross-sectional geometry and material properties. It offers output including, among other quantities, the differential impedance of the pair of lines.

The advantage of using a field solver is that some tools can be accurate to better than 1% across a wide range of geometrical conditions. They account for the first-order effects, such as line width, dielectric thickness, and spacing, as well as second-order effects, such as trace thickness, trace shape, and inhomogeneous dielectric distributions.

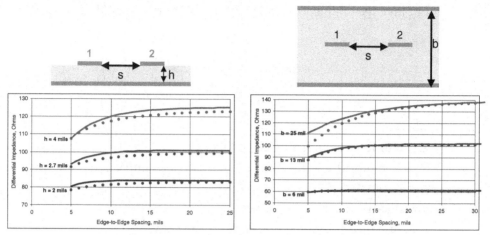

Figure 11-15 Comparing the accuracy of the differential impedance approximation with the results from a 2D field solver. In each case, the line width is 5 mils, and the material is FR4. The lines are the approximations, and the points are the simulation using Ansoft's SI2D.

T I P   When accuracy is important, such as in signing off on a drawing for the fabrication of a circuit board, the only tool that should be used is a verified 2D field solver. Approximations should never be used to sign off on a design.

However, if we are limited to using only a field solver to calculate differential impedance, we will lack the terms needed to describe the behavior of common signals, terminations, and cross talk. The following sections introduces the concepts of odd and even modes and how they relate to differential and common impedance. From this foundation, we will evaluate termination strategies for differential and common signals.

We will end up introducing the description of two coupled transmission lines in terms of the capacitance and inductance matrix and the characteristic impedance matrix to calculate the odd and even mode impedances and the differential and common impedances. This is the most fundamental description and can be generalized to n different coupled lines. Ultimately, this is what is really going on inside most field solver and simulation tools.

Even without invoking more complex descriptions at this point, we have the terminology to be able to use the results from a field solver to design for a target differential impedance and evaluate another important quality of differential pairs—the current distribution.

11.6 The Return-Current Distribution in a Differential Pair

When the spacing between two edge-coupled microstrip transmission lines is greater than three times their line width, the coupling between the signal lines will be small. In this situation, as we might expect, when we drive them with a differential signal, there is some current into each signal line and an equal amount of current in the opposite direction in the return plane. An example of the current distribution for a 100-MHz differential signal into microstrip conductors that are 1.4 mils thick, or for 1-ounce copper, is illustrated in Figure 11-16.

The direction of the current into line 1 is into the paper. The return current in the return plane below line 1 is out of the paper. Likewise, at the same time, the current into line 2 is out of the paper, and the current into its return path is into the paper. In the return plane, the return-current distributions are localized under the signal lines. When driven by a differential signal, there is virtually no overlap between the two return-current distributions in the plane.

If we pay attention only to the current in the signal traces, it looks like the same current goes into one trace and comes out the other. We might conclude that the return current of the differential signal in one trace is carried by the second trace. While it is perfectly true that the same amount of current going into one signal trace comes out of the other signal trace, it is not the complete story.

Because of the wide spacing between the differential pairs, there is no overlap of the return currents in the planes when the pair is driven with a differential signal. While it is true the net current into the plane is zero, there is still a well-defined, localized current distribution in the plane under each signal path. Anything that will distort or change this current distribution will change the differential impedance of the pair of traces.

Figure 11-16   The current distribution, at 100 MHz, in a pair of coupled 50-Ohm microstrip transmission lines, 5 mils wide and separated by 15 mils. Lighter shading means higher current density. The current density scale in the plane is 10 times more sensitive than the traces to show the current distribution more clearly. Simulated with Ansoft's SI2D.

After all, the presence of the plane defines the single-ended impedance of either trace. Increase the plane spacing, and the single-ended impedance of the trace increases, which will change the differential impedance.

Even in the extreme case for edge-coupled microstrip, bringing the signal traces as close together as is practical with a spacing equal to the line width, the degree of overlap of the currents in the return plane is very slight. The comparison of the current distributions is shown in Figure 11-17.

TIP   When the coupling between the signal line and the return plane is much stronger than the coupling between adjacent signal traces in a differential pair, there are separate and distinct return currents in the planes and very little overlap in the return-current distributions. These current distributions will strongly affect the differential impedance of the pair. Disturb the current distribution, and the differential impedance will be affected.

For any pair of single-ended transmission lines sharing a common return conductor, if the return conductor is moved far enough away, the signals' return-current distributions in the return conductor will completely overlap and cancel each other out, and the presence of the return conductor will play no role in establishing the differential impedance. In this specific situation, it is absolutely true that the return current of one line would be carried by the other line.

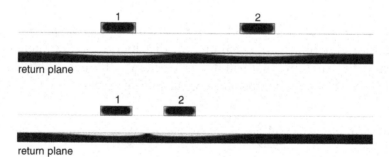

Figure 11-17   The current distribution, at 100 MHz, in a pair of coupled 50-Ohm microstrip transmission lines, 5 mils wide and separated by 5 mils compared with a separation of 15 mils. Lighter shading means higher current density. The current density scale in the planes is 10 times more sensitive than the traces to show the current distribution more clearly. Simulated with Ansoft's SI2D.

There are three such cases that should be noted:

1. Edge-coupled microstrip with return plane far away
2. Twisted-pair cable
3. Broadside-coupled stripline with return planes far away

In edge-coupled microstrip, the coupling between the two signal traces is largest when the spacing between the signal lines is as close as can be fabricated, typically equal to the line width. For the case of near-50-Ohm lines and closest spacing, as shown in Figure 11-17, there is significant return-current distribution in the return plane, and the presence of the plane affects the differential impedance. If the plane is moved farther away, the single-ended impedance of either line will increase, and the differential impedance will increase. However, as the plane is moved farther away, the return currents of the differential signal, in the plane, will increasingly overlap.

There is a point where the return currents overlap so much that there is no current in the plane, and the plane will not influence the differential impedance. This is illustrated in Figure 11-18. The single-ended impedance continues to get larger, but the differential impedance reaches a highest value of about 140 Ohms and then stops increasing. This is when the return currents completely overlap, at a trace-to-plane distance of about 15 mils.

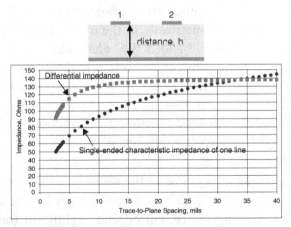

Figure 11-18   Single-ended and differential impedance of a pair of edge-coupled microstrips with 5-mil width and 5-mil spacing, as the distance to the return plane increases. Simulated with Ansoft's SI2D.

> **TIP**   As a rough rule of thumb, when the distance to the return plane is
> about equal to or larger than the total edge-to-edge span of the two signal
> conductors, the current distributions in the return path overlap, and the
> presence of the return path plays no role in the differential impedance of
> the pair. In this case, it is true that for a differential signal, the return current
> of one signal line really is carried by the other line.

After all, in this geometry of edge-coupled microstrip with the plane
far away, isn't this really like a coplanar transmission line with a single-ended
signal? The signal in each case is the voltage between the two signal lines. The
single-ended signal is the same as a differential signal, so the impedance the
single-ended signal sees will be the same impedance that a differential signal
sees. The single-ended characteristic impedance of a coplanar line with thick
dielectric on one side is the same as the differential impedance of an edge-
coupled pair with the return plane very far away.

In a shielded twisted pair, the return path of each signal line is the shield.
The spacing between the twisted wires is determined by the thickness of
the insulation around each wire. In some cables, the center-to-center pitch of
the wires is 25 mils, and the wire diameter is 16 mils, or 26-gauge wire. We
can use a 2D field solver to calculate the differential impedance as the spacing
to the shield increases.

When one wire of a twisted pair is driven single ended, with the shield as
the return conductor, the signal current flows down the wire, and the return
current flows roughly symmetrically in the outer shield. Likewise, when the
second wire is driven single ended, its return current will have about the
same current distribution, but in the opposite direction. When driven by a
differential signal and both wires are nearly at the center of the shield, their
return-current distributions flow in opposite directions in the shield and
overlap. There will be no residual-current distribution in the shield. In this
case, the shield plays no role in influencing the differential impedance of the
wires and can be eliminated.

When the shield is very close to the wires, the off-axis location of the two
wires causes their current distributions to be slightly different in the shield,
and the differential impedance will depend slightly on the location of the
shield. When the shield is far enough away for the return current to be mostly
symmetrical, the two return currents overlap, and the shield location has no
effect on the differential impedance. Figure 11-19 shows how the differential

impedance varies as the shield radius increases. When the shield radius is about equal to twice the pitch of the wires, the currents mostly overlap, and the differential impedance is independent of the shield position.

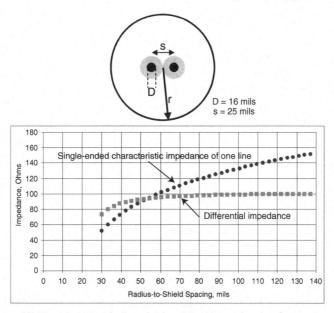

Figure 11-19 Variation in the single-ended impedance of one wire and the shield in a twisted pair and the differential impedance of the two wires as the radius of the shield expands. When the distance is more than about two times the center-to-center spacing of the twisted pair, the return currents in the shield overlap and cancel out, and the presence of the shield plays no role in the differential impedance. Simulated with Ansoft's SI2D.

An unshielded differential pair has exactly the same differential impedance as a shielded differential pair with a large radius shield. As far as the differential impedance is concerned, the shield plays no role at all. As we will see, the shield plays a very important role in providing a return path for the common current, which will reduce radiated emissions.

The same effect happens in broadside-coupled stripline. When the two reference planes are close together, there is significant, separated return current in the two planes when the transmission lines are driven by a differential signal, and the presence of the planes affects the differential impedance. As the spacing between the planes increases so the return current from each line has roughly the same current distribution in each plane, the currents cancel out in each plane, and the impact from the planes is negligible.

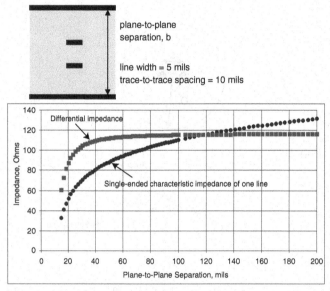

Figure 11-20   Variation in the single-ended impedance of one trace and the planes in a broadside-coupled stripline and the differential impedance of the two traces as the plane-to-plane spacing increases. Simulated with Ansoft's SI2D.

Figure 11-20 shows the differential impedance as the spacing between the planes is increased. In this example, the line width is 5 mils, and the trace-to-trace spacing is 10 mils. This is a typical stack-up for a 100-Ohm differential-impedance pair, when the plane-to-plane spacing is about 25 mils. When the distance between a trace and the nearest plane is greater than about twice the separation of the traces, or 20 mils in this case, and the plane-to-plane spacing is more than 50 mils, the differential impedance is independent of the position of the planes.

These three examples illustrate a very important principle with differential pairs. When the coupling between any one trace and the return plane is stronger compared with the coupling between the two signal lines, there is significant return current in the plane, and the presence of the plane is very important in determining the differential impedance of the pair.

When the coupling between the two traces is much stronger than the coupling between a trace and the return plane, there is a lot of overlap of the return currents, and they mostly cancel each other out in the plane. In this case, the plane plays no role and will not influence the differential signal. It can be removed without affecting the differential impedance. In this case, it is absolutely true that the second line will carry the return current of the first line.

T I P   As a rough rule of thumb, for the coupling between the two traces to be larger than the coupling between a trace and the return plane, the distance to the nearest plane must be about twice as large as the span of the signal lines.

In most board-level interconnects, the coupling between a signal trace and plane is much greater than the coupling between the two signal traces, so the return current in the plane is very important. In board-level interconnects, it generally is not true that the return current of one trace is carried in the other trace. However, if the return path is removed, as in a gap, the coupling between the traces now dominates, and in this region of discontinuity, it may be true that the return current of one line is carried by the other. In the case of a discontinuity in the return path, the change in differential impedance of the pair can be minimized by using tighter coupling between the pair in the region of the return-path discontinuity. This is discussed later in this chapter.

In connectors, the trace-to-trace coupling is usually stronger than the trace-to-return-pin coupling, so it is generally true that the return current of one pin is carried by the other. The only way to know for sure is to put in the numbers using a field-solver tool.

11.7 Odd and Even Modes

Any voltage can be applied to the front end of a differential pair, such as a microstrip differential pair. If we were to launch the voltage pattern of a 0-v to 1-v signal in line 1 and a 0-v constant signal in line 2, we would find that as we moved down the lines with the signal, the actual signal on the lines would change. There would be far-end cross talk between line 1 and line 2. Noise would be generated on line 2, and as the noise built, the signal on line 1 would decrease.

Figure 11-21 shows the evolution of the voltages on the two lines as the signals propagate. The voltage pattern launched into the differential pair changes as it propagates down the line. In general, any arbitrary voltage pattern we launch into a pair of transmission lines will change as it propagates down the line.

However, in the case of an edge-coupled microstrip differential pair, there are two special voltage patterns we can launch into the pair that will propagate down the line undistorted. The first pattern is when exactly the same signal is applied to either line; for example, the voltage transitions from 0 v to 1 v in each line.

In this case, there is no dV/dt between the signal lines, so there is no capacitively coupled current. The inductively coupled current into one line is identical to the inductively coupled current into the other line since the dI/dt is the same in each line. Whatever one line does to the other, the same action is returned. The result is that as this special voltage pattern propagates down the transmission lines, the voltage pattern on either line will remain exactly the same.

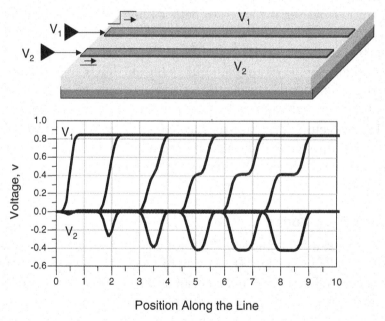

Figure 11-21   Voltage as would be measured by a scope at various positions on the two lines in an edge-coupled microstrip when one line is driven by a 0-v to 1-v transition and the other line is pegged low. Simulated with Agilent's ADS.

The second special voltage pattern that will propagate unchanged down the differential pair is when the opposite-transitioning signals are applied to each line; for example, one of the signals transitions from 0 v to 1 v and the other goes from 0 v to −1 v.

The signal in the first line will generate far-end noise in the second line that has a negative-going pulse. This will decrease the voltage on the first line as the signal propagates. However, at the same time, the negative-going signal in the second line will generate a positive far-end noise pulse in the first line. The magnitude of the positive noise generated in line 1 is exactly the same as the magnitude of the drop in the signal in line 1 from the loss due to its noise to line 2. The result is that the voltage pattern on the differential pair will propagate down the lines undistorted. Figure 11-22 shows the voltage pattern on the two signal lines as the signal propagates down the pair of lines when these two patterns are applied.

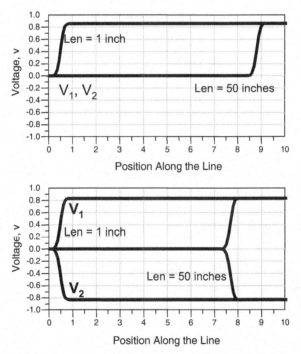

Figure 11-22 Voltage pattern on the two lines in an edge-coupled microstrip when the pair is driven in the even mode and the odd mode. The voltage pattern is the same going into the line as coming out 50 inches later. Simulated with Agilent's ADS.

These two special voltage patterns that propagate undistorted down a differential pair correspond to two special states at which the pair of lines can be excited or stimulated. We call these special states *the modes of the pair*.

When a differential pair is excited into one of these two modes, the signal will have the special property of propagating down the line undistorted. To distinguish these two states, we call the state where the same voltage drives each line *the even mode* and the state where the opposite-going voltages drive each line *the odd mode*.

TIP   Modes are special states of excitation for a pair of lines. A signal that excites one of these special states will propagate undistorted down a differential pair of transmission lines. For a two-signal-line differential pair, there are only two different special states or modes. For a three-signal conductor array of coupled lines, there are three modes. For a collection of four-signal conductors and their common return path, there are four special voltage states for which a voltage pattern will propagate undistorted.

Modes are intrinsic properties of a differential pair. Of course, any voltage pattern can be imposed on a pair of lines. When the voltage pattern imposed matches one of these special states, the voltage signal propagating down the line has special properties. The modes of a differential pair are used to define the special voltage patterns.

When the two conductors in a differential pair are geometrically symmetrical, with the same line widths and dielectric spacings, the voltage patterns that excite the even and odd modes correspond to the same voltages launched in both lines and the opposite-going voltages launched into both lines. If the conductors are not symmetric (e.g., have different line widths or dielectric thickness), the odd- and even-mode voltage patterns are not so simple. The only way to determine the specific odd- and even-mode voltage patterns is by using a 2D field solver. Figure 11-23 shows the field patterns of the odd- and even-mode states for a pair of symmetric lines.

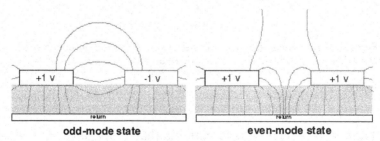

odd-mode state   **even-mode state**

Figure 11-23   Field distribution for odd- and even-mode states of a symmetric microstrip, calculated with Mentor Graphics HyperLynx.

It is important to keep separate, on one hand, the definition of modes as special, unique states into which the pair of lines can be excited based on the geometry of the pair and, on the other hand, the applied voltages that can be any arbitrary values. Any voltage pattern can be applied to a differential pair—just connect a function generator to each signal and return path.

In the special case of a symmetric, edge-coupled microstrip differential pair, the odd-mode state can be excited by the voltage pattern corresponding to a pure differential signal. Likewise, the even-mode state can be excited by applying a pure common signal.

TIP   A differential signal will drive a signal in the odd mode, and a common signal will drive a signal in the even mode for a symmetric, edge-coupled microstrip differential pair. Odd and even modes refer to special intrinsic states of the differential pair. *Differential* and *common* refer to the specific signals that are applied to the differential pair. It's important to note that 90% of the confusion about differential impedance arises from misusing these terms.

Introducing the terms *odd mode* and *even mode* allows us to label the special properties of a symmetric, differential pair. For example, as we saw above, the impedance a signal sees on one line depends on the proximity of the adjacent line and the voltage pattern on the other line. Now we have a way of labeling the different cases. We refer to the impedance of one line when the pair is driven in the odd-mode state as the *odd-mode impedance* of the line. The impedance of one line when the pair is driven in the even-mode state is the *even-mode impedance* of that line.

The odd mode is often incorrectly labeled the differential mode. If we equate the odd mode with the differential mode, then we might easily confuse the differential-mode impedance with the odd-mode impedance. If these were the same mode, why would there be any difference between the odd-mode impedance and the differential-mode impedance?

In fact, there is no such thing as the differential mode, so there is no such thing as the differential-mode impedance. Figure 11-24 emphasizes that we should remove the terms *differential mode* from our vocabulary, and then we will never be confused between odd-mode impedance and differential impedance. They are completely different quantities. There are odd-mode impedances, differential signals, and differential impedances.

TIP   The *odd-mode impedance* is the impedance of one line when the pair is driven in the odd-mode state. The *differential impedance* is the impedance the differential signal sees as it propagates down the differential pair.

Figure 11-24   There is no such thing as differential mode. Forget the words, and you will never confuse differential impedance with odd-mode impedance.

11.8 Differential Impedance and Odd-Mode Impedance

As we saw previously, the differential impedance a differential signal sees is the series combination of the impedance of each line to the return path. When there is no coupling, the differential impedance is just twice the characteristic impedance of either line. When the lines are close enough together for coupling to be important, the characteristic impedance of each line changes.

When a differential signal is applied to a differential pair, we now see that this excites the odd-mode state of the pair. By definition, the characteristic impedance of one line when the pair is driven in the odd mode is called the *odd-mode characteristic impedance*. As illustrated in Figure 11-25, the differential impedance is twice the odd-mode impedance. The differential impedance is thus:

$$Z_{diff} = 2 \times Z_{odd} \qquad (11\text{-}14)$$

where:

Z_{diff} = differential impedance

Z_{odd} = characteristic impedance of one line when the pair is driven in the odd mode

> **TIP**   The way to calculate or measure the differential impedance is by first calculating or measuring the odd-mode impedance of the line and multiplying it by 2.

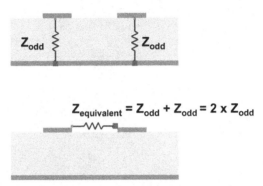

$$Z_{equivalent} = Z_{odd} + Z_{odd} = 2 \times Z_{odd}$$

Figure 11-25   The impedance between each line and its return path is the odd-mode impedance when the pair is excited with a differential signal. The differential impedance is the equivalent impedance between the two signal lines.

The odd-mode impedance is directly related to the differential impedance, but they are not the same. The differential impedance is the impedance the differential signal sees. The odd-mode impedance is the impedance of one line when the pair is driven in the odd mode.

11.9 Common Impedance and Even-Mode Impedance

Just as we can describe the impedance a differential signal sees propagating down a line, we can also describe the impedance a common signal sees propagating down a differential pair. The common signal is the average voltage between the signal lines. A pure common signal is when the differential signal is zero. This means there is no difference in voltage between the two lines, and they each have the same signal voltage.

A common signal will drive a differential pair in its even-mode state. As the common signal propagates down the line, the characteristic impedance of each line will be, by definition, the even-mode characteristic impedance. The common signal will see the two lines in parallel, as illustrated in Figure 11-26.

The impedance the common signal sees will be the parallel combination of the impedances of each line. The parallel combination of the two even-mode impedances is:

$$Z_{comm} = Z_{equiv} = \frac{Z_{even} \times Z_{even}}{Z_{even} + Z_{even}} = \frac{1}{2} Z_{even} \qquad (11\text{-}15)$$

where:

Z_{comm} = common impedance

Z_{even} = characteristic impedance of one line when the pair is driven in the even mode

Figure 11-26   The impedance between each line and its return path is the even-mode impedance when the pair is excited with a common signal. The common impedance is the equivalent impedance between the two signal lines and the return plane.

The impedance the common signal sees is, in general, a low impedance. This is because the common signal is the same voltage as applied between each signal line and the return path, but the current going into the pair of lines and coming out the return path is twice the current going into either line. If a signal sees the same voltage but twice the current, the impedance will look half as large.

TIP   For two uncoupled 50-Ohm transmission lines that make up a differential pair, the odd- and even-mode impedances will be the same—50 Ohms. The differential impedance will be 2 × 50 Ohms = 100 Ohms, while the common impedance will be 1/2 × 50 Ohms = 25 Ohms.

As we turn on coupling, the odd-mode impedance of either line will decrease, and the even-mode impedance of either line will increase. This means the differential impedance will decrease, and the common impedance will increase. The most accurate way of calculating the differential impedance or common impedance is by using a field solver to first calculate the odd-mode impedance and even-mode impedance.

Figure 11-27 shows the complete set of impedances calculated with a field solver for the case of an edge-coupled microstrip in FR4 with 5-mil-wide traces and an uncoupled single-end characteristic impedance of 50 Ohms. As the separation between the traces gets smaller, coupling increases, and the odd-mode impedance decreases, causing the differential impedance to decrease. The even-mode impedance increases, causing the common impedance to increase. As this example illustrates, even with the tightest coupling manufacturable, the differential impedance and common impedance are only slightly affected by the coupling. The tightest coupling decreases the differential impedance in microstrip or stripline by only 10%.

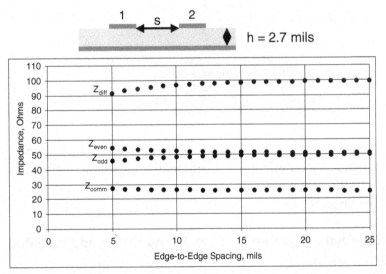

Figure 11-27 All the impedances associated with a pair of edge-coupled microstrip traces, with 5-mil-wide traces in FR4 and nominally 50 Ohms, as the separation increases. Simulated with Ansoft's SI2D.

In many microstrip traces on boards, a solder-mask layer is applied to the top surface. This will affect the single-ended impedance as well as the odd-mode impedance. Figure 11-28 shows the impedance variation for a tightly coupled microstrip differential pair as the thickness of solder mask increases. Because of the stronger field lines between the traces in the odd mode, the presence of the solder mask will affect the odd-mode impedance more than it affects the other impedances.

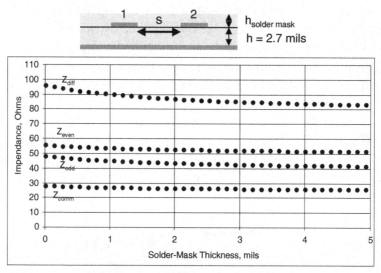

Figure 11-28   Effect on all the impedances as a solder-mask thickness applied to the top of the surface traces increases, for the case of tightest spacing of 5-mil-wide and 5-mil-spaced traces in FR4. Simulated with Ansoft's SI2D.

This is why it is so important to take into account the presence of solder mask when designing the differential impedance of a surface trace. In addition, this effect can cause the fabricated differential impedance to be off by as much as 10%.

11.10 Differential and Common Signals and Odd- and Even-Mode Voltage Components

The terms *differential* and *common* refer to the signals imposed on a line. The part of an arbitrary signal that is the differential component is the difference between the two voltages on each signal line. The part of an arbitrary signal that is the common component is the average of the two signals on the lines.

For a symmetric differential pair, a differential signal travels in the odd mode of the pair, and a common signal travels in the even mode of the pair. We can also use the terms *odd* and *even* to describe an arbitrary signal. The voltage component of an applied signal that is propagating in the even mode, V_{even}, is the common component of the signal. The component propagating in the odd mode, V_{odd}, is the differential component of the signal. These are given by:

$$V_{odd} = V_{diff} = V_1 - V_2 \qquad\qquad (11\text{-}16)$$

$$V_{even} = V_{comm} = \frac{1}{2} \times (V_1 + V_2) \qquad\qquad (11\text{-}17)$$

Likewise, any arbitrary signal propagating down a differential pair can be described as a combination of a component of the signal propagating in the even mode and a component of the signal propagating in the odd mode:

$$V_1 = V_{even} + \frac{1}{2} V_{odd} \qquad\qquad (11\text{-}18)$$

$$V_2 = V_{even} - \frac{1}{2} V_{odd} \qquad\qquad (11\text{-}19)$$

where:
V_{even} = voltage component propagating in the even mode
V_{odd} = voltage component propagating in the odd mode
V_1 = signal on line 1, with respect to the common return path
V_2 = signal on line 2, with respect to the common return path

For example, if a single-ended signal is imposed on one line of 0 v to 1 v and the other line is pegged low, at 0 v, the voltage component that propagates in the even mode will be $V_{even} = 0.5 \times (1\text{ v} + 0\text{ v}) = 0.5$ v. The voltage component that propagates in the odd mode will be $V_{odd} = 1\text{ v} - 0\text{ v} = 1$ v. Simultaneously on the differential pair, a 0.5-v signal will be propagating down the line in the even mode, seeing the even-mode characteristic impedance of each line, and a 1-v component will be propagating down the line in the odd mode, seeing the odd-mode characteristic impedance of each line. The description of this signal in terms of odd- and even-mode components is illustrated in Figure 11-29.

Any imposed voltage can be described as a combination of an even-mode voltage component and an odd-mode component. The voltage components that travel in the odd mode are completely independent of the voltage components traveling in the even mode. They propagate independently and do not interact. Each signal component will see a different impedance for each signal line to its return path, and each signal component may also travel at a different velocity.

The examples above used an edge-coupled microstrip to illustrate the different modes. When the dielectric material completely surrounding the conductors is everywhere uniform and homogeneous, there is no longer a unique voltage pattern that propagates down the differential pair in each mode. Every voltage pattern imposed on the differential pair will propagate undistorted. After all, when the dielectric is homogeneous, as in a stripline geometry, there is no far-end cross talk. Any signal launched into the front end of the pair of lines will propagate undistorted down the line. However, by convention, we still use the voltage patterns above to define the odd and even modes for any symmetrical differential pair.

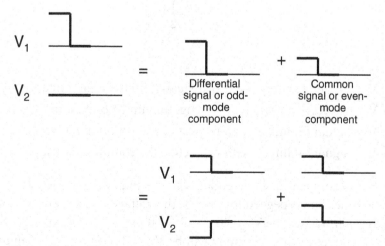

Figure 11-29 Three equivalent descriptions of the same signals on a differential pair: as the voltage on each line, as the differential and common signals, and as the components propagating in the odd mode and the even mode.

11.11 Velocity of Each Mode and Far-End Cross Talk

The description of the signal in terms of its components propagating in each of the two modes is especially important in edge-coupled microstrip because signals in each mode travel at different speeds.

The velocity of a signal propagating down a transmission line is determined by the effective dielectric constant of the material the fields see. The higher the effective dielectric constant, the slower the speed, and the longer the time delay of a signal propagating in that mode. In the case of a stripline, the dielectric material is uniform all around the conductors, and the fields always see an effective dielectric constant equal to the bulk value, independent of the voltage pattern. The odd- and even-mode velocities in a stripline are the same.

However, in a microstrip, the electric fields see a mixture of dielectric constants—part in the bulk material and part in the air. The precise pattern of the field distribution and how it overlaps the dielectric material will influence the value of the resulting effective dielectric constant and the actual speed of the signal. In the odd mode, more of the field lines are in air; in the even mode, more of the field lines are in the bulk material. For this reason, the odd-mode signals will have a slightly lower effective dielectric constant and will travel at a faster speed than do the even-mode signals.

Figure 11-30 shows the field patterns of the odd and even modes for a symmetric microstrip and stripline differential pair. In a stripline, the fields see just the bulk dielectric constant for each mode. There is no difference in speed between the modes for any homogeneous dielectric interconnect.

In an edge-coupled microstrip, a differential signal will drive the odd mode so it will travel faster than a common signal, which drives the even mode. Figure 11-31 shows the different speeds for these two signals. As the spacing between the traces increases, the degree of coupling decreases, and the field distribution between the odd mode and the even mode becomes identical. If there is no difference in the field distributions, each mode will have the same effective dielectric constant and the same speed.

In this example, at closest spacing, the odd-mode speed is 7.4 inches/ nsec, while the even-mode speed is 6.8 inches/nsec. If a signal were launched into the line with only a differential component, this differential signal would propagate down the line undistorted at a speed of 7.4 inches/nsec. If a pure common signal were launched into the line, it would propagate down the line undistorted at a speed of 6.8 inches/nsec.

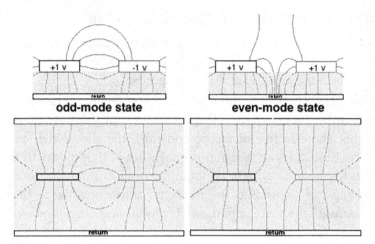

Figure 11-30   Electric-field distribution compared to the dielectric distribution for odd and even modes in microstrip and stripline. Simulated with Mentor Graphics HyperLynx.

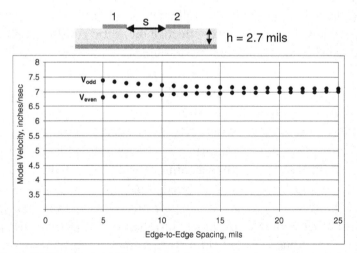

Figure 11-31   Speed of the even and odd modes for edge-coupled microstrip traces in FR4 with 5-mil-wide lines and roughly 50 Ohms.

For an interconnect that is 10 inches long, the time delay for a signal propagating in the odd mode would be TD_{odd} = 10 inches/7.4 inches/nsec = 1.35 nsec, while the time delay for an even-mode signal would be TD_{even} = 10 inches/6.8 inches/nsec = 1.47 nsec.

TIP   The difference of only 120 psec may not seem like much, but this small difference in delay between signals propagating in the even and odd modes is the effect that gives rise to far-end cross talk in single-ended, coupled, transmission lines.

If instead of driving the differential pair with a pure differential or common signal, we drive it with a signal that has both components, each component will propagate down the line independently and at different speeds. Though they start out coincident, as they move down the line, the faster component, typically the differential signal, will move ahead. The wavefronts of the two signals—the differential and common components—will separate. At any point along the differential pair, the actual voltage on each line will be the sum of the differential and common components. As the edges spread out, the voltage patterns on both lines will change.

Suppose we launch a voltage signal into a differential pair that has a 0-v to 1-v transition on one line, but we keep the voltage at 0 v on the other line. This is the same as sending a single-ended signal down line 1 and pegging line 2 low. Line 1 becomes the aggressor line, and line 2 becomes the victim line.

We can describe the voltage patterns on the two lines in terms of equal amounts of a signal propagating as a differential signal in the odd mode and as a common signal in the even mode. After all, a common signal of 0.5 v on line 1 and 0.5 v on line 2 and a differential signal of 0.5 v on line 1 and −0.5 v on line 2 will result in the same signal that is launched on the pair. This is shown in Figure 11-32.

In a homogeneous dielectric system, such as stripline, the odd- and even-mode signals will propagate at the same speed. The two modal signals will reach the far end of the line at the same time. When they are added up again, they will recompose, with no changes, the original waveform that was launched. In such an environment, there is no far-end cross talk.

In a microstrip differential pair, the differential component will travel faster than the common component. As these two independent voltage components propagate down the differential pair, their leading-edge wavefronts will separate. The leading edge of the differential component will reach the end of line 2 before the common-signal component reaches the end of line 2.

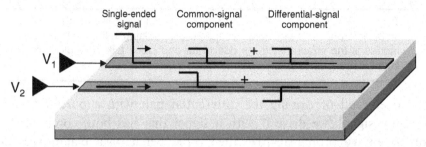

Figure 11-32 Describing a signal on an aggressor line and a quiet line as a simultaneous common- and differential-signal component on a differential pair.

The received signal at the far end of line 2 will be the recombination of the –0.5-v differential component and the delayed 0.5-v common-signal component. These combine to give a net, though transient, voltage at the far end of line 2. We call this transient voltage the *far-end noise*.

> **TIP**   Far-end noise in a pair of coupled transmission lines can be considered to be due to the capacitively coupled current minus the inductively coupled current or the sum of the shifted differential component and the common component. These two views are perfectly equivalent.

If the leading edge of the imposed differential and common signals is a linear ramp, we can estimate the expected far-end noise due to the time delay between the two components. The setup is shown in Figure 11-33, where the voltage on line 2 is a combination of a common component and a differential component. The net voltage on line 2 is just the sum of these two components. The differential-signal and common-signal magnitudes will be exactly 1/2 the voltage on line 1, $1/2 \times V_1$. The time delay between the arrival of the differential component (traveling in the odd mode) and the common component (traveling in the even mode) will be:

$$\Delta T = \frac{\text{Len}}{v_{\text{even}}} - \frac{\text{Len}}{v_{\text{odd}}} \qquad (11\text{-}20)$$

The initial part of the transient signal is the leading edge of the rise time. It reaches a peak value, the far-end voltage, related to the fraction of the rise time the time delay represents:

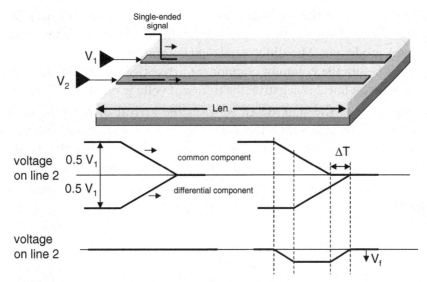

Figure 11-33   The signal on line 2 has a differential-signal component and common-signal component. The differential-signal component gets to the end of line 2 before the common component, causing a transient net signal on line 2.

$$V_f = -\frac{1}{2}V_1 \times \frac{\Delta T}{RT} = -\frac{1}{2}V_1 \frac{Len}{RT}\left(\frac{1}{v_{even}} - \frac{1}{v_{odd}}\right) = \frac{1}{2}V_1 \frac{Len}{RT}\left(\frac{1}{v_{odd}} - \frac{1}{v_{even}}\right) \quad (11\text{-}21)$$

where:

V_f = far-end voltage peak on line 2, the victim line

V_1 = voltage on line 1, the aggressor line

Len = length of the coupled region

ΔT = time delay between the arrival of the differential signal and the common signal

RT = signal rise time

v_{even} = velocity of a signal propagating in the even mode

v_{odd} = velocity of a signal propagating in the odd mode

We can interpret far-end noise as the difference in speed between the odd and even modes. If the differential pair has homogeneous dielectric and there is no difference in the speeds of the two modes, there will be no far-end noise.

If there is air above the traces and the odd mode has a lower effective dielectric constant than the even mode, the odd mode will have a higher speed than the even mode has. The differential-signal component will arrive at the end of line 2 before the common component arrives. Since the differential-signal component on line 2 is negative, the transient voltage on line 2 will be negative.

As long as the time delay between the arrival of the differential and common signals is less than the rise time, the far-end noise will increase with the coupling length. However, if the time delay is greater than the rise time, the far-end noise will saturate at the differential-signal magnitude of 0.5 V_1.

The saturation length of far-end noise is when $V_f = 0.5\ V_1$, which can be calculated from:

$$\text{Len}_{sat} = -\frac{RT}{\dfrac{1}{V_{odd}} - \dfrac{1}{V_{even}}}$$ (11-22)

where:

Len_{sat} = coupling length where the far-end noise will saturate

RT = rise time

V_{even} = velocity of a signal propagating in the even mode

V_{odd} = velocity of a signal propagating in the odd mode

For example, in the case of the most tightly coupled microstrip and a rise time of 1 nsec, the saturation length is:

$$\text{Len}_{sat} = -\frac{1\text{ns}}{\dfrac{1}{7.4\dfrac{\text{in}}{\text{ns}}} - \dfrac{1}{6.8\dfrac{\text{in}}{\text{ns}}}} = -\frac{1\text{ns}}{0.135\dfrac{\text{ns}}{\text{in}} - 0.147\dfrac{\text{ns}}{\text{in}}} = 83\text{ in}$$ (11-23)

The smaller the difference between the odd- and even-mode velocities, the greater the saturation length. Of course, long before the far-end noise saturates, the magnitude of the far-end noise can become large enough to exceed any reasonable noise margin.

11.12 Ideal Coupled Transmission-Line Model or an Ideal Differential Pair

A pair of coupled transmission lines can be viewed as either two single-ended transmission lines with some coupling that gives rise to cross talk on the two lines or as a differential pair with an odd- and even-mode characteristic impedance and an odd- and even-mode velocity. Both views are equivalent and separate.

In Chapter 10, we saw that the near-end (or backward) noise, V_b, and the far-end (or forward) noise, V_f, in a pair of coupled transmission lines was related to:

$$V_b = V_a k_b \tag{11-24}$$

$$V_f = V_a \frac{\text{Len}}{\text{RT}} k_f \tag{11-25}$$

where:
V_b = backward noise
V_f = far-end noise
V_a = voltage on the active line
k_b = backward cross-talk coefficient
k_f = forward cross-talk coefficient
Len = length of the coupled region
RT = rise time of the signal

Using the view of a differential pair, the cross-talk coefficients are:

$$k_b = \frac{1}{2} \frac{Z_{\text{even}} - Z_{\text{odd}}}{Z_{\text{even}} + Z_{\text{odd}}} \tag{11-26}$$

$$k_f = \frac{1}{2} \left(\frac{1}{v_{\text{odd}}} - \frac{1}{v_{\text{even}}} \right) \tag{11-27}$$

The near-end noise is a direct measure of the difference in the characteristic impedance of the odd and even modes. The farther apart the traces, the smaller the difference between the odd- and even-mode impedances and the less the coupling. When the traces are very far apart, there is no interaction between the traces, and the characteristic impedance of one line is independent of the signal on the other line. The odd- and even-mode impedances are the same, and the near-end cross-talk coefficient is zero.

TIP   This connection allows us to model a pair of coupled lines as a differential pair. An ideal distributed differential pair is a new ideal circuit model that can be added to our collection of ideal circuit elements. It can model the behavior of a differential pair or a pair of independent, coupled transmission lines.

Just as an ideal, single-ended transmission line is defined in terms of a characteristic impedance and time delay, an ideal differential pair is defined with just four parameters:

1. An odd-mode characteristic impedance
2. An even-mode characteristic impedance
3. An odd-mode time delay
4. An even-mode time delay

These terms fully take into account the effects of the coupling, which would give rise to near-end and far-end cross talk. This is the basis of most ideal circuit models used in circuit or behavioral simulators.

If the transmission line is a stripline, the odd- and even-mode velocities and time delays are the same, so only three parameters are needed to describe a pair of coupled stripline transmission lines.

Usually, the parameter values for these four terms describing a differential pair are obtained with a 2D field solver.

11.13 Measuring Even- and Odd-Mode Impedance

A time-domain reflectometer (TDR) is used to measure the single-ended characteristic impedance of a single-ended transmission line. The TDR will launch a voltage step into the transmission line and measure the reflected voltage. The magnitude of the reflected voltage will depend on the change in the instantaneous impedance the signal encounters in moving from the 50 Ohms of the TDR and its interconnect cables to the front of the transmission line. For a uniform line, the instantaneous impedance the signal encounters will be the characteristic impedance of the line. The reflected voltage is related to:

$$\rho = \frac{V_{\text{reflected}}}{V_{\text{incident}}} = \frac{Z_0 - 50\Omega}{Z_0 + 50\Omega}$$
(11-28)

where:

ρ = reflection coefficient

$V_{\text{reflected}}$ = reflected voltage measured by the TDR

V_{incident} = voltage launched by the TDR into the line

Z_0 = characteristic impedance of the line

50 Ohms = output impedance of the TDR and cabling system

By measuring the reflected voltage and knowing the incident voltage, we can calculate the characteristic impedance of the line:

$$Z_0 = 50\Omega \frac{1+\rho}{1-\rho}$$
(11-29)

This is how we can measure the characteristic impedance of any single-ended transmission.

In order to measure the odd-mode impedance or even-mode impedance of a line that is part of a differential pair, we must measure the characteristic impedance of one line while we drive the pair either into the odd mode or into the even mode.

To drive the pair into the odd mode, we need to apply a pure differential signal to the pair. Then the characteristic impedance we measure of either line is its odd-mode impedance. This means that if the incident signal transitions from 0 v to +200 mV between the line and its return path, into the line under test, we need to also apply another signal to the second line of 0-v to −200-mV between the second signal line and its return path. Likewise, measuring the even-mode characteristic impedance of the line requires applying a 0-v to +200-mV signal to both lines.

This requires the use of a special TDR that has two active heads. This instrument is called a *differential TDR* (*DTDR*). Figure 11-34 shows the measured TDR voltages with the two outputs connected to opens to illustrate how the two channels are driven when set up for differential-signal outputs and common-signal outputs.

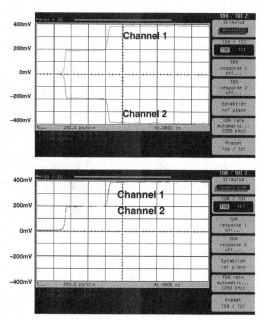

Figure 11-34   Measured voltages from the two channels of a differential TDR when driving a differential signal (top) and a common signal (bottom) into the differential pair under test. Measured with Agilent 86100 DCA with DTDR plug-in.

In a DTDR, the reflected voltages from both channels can be measured, so the odd- or even-mode impedance of both lines in the pair can be measured. Figure 11-35 shows an example of the measured odd- and even-mode characteristic impedance of one line in a pair when the lines are tightly coupled. In this example, the odd-mode impedance is measured as 39 Ohms, and the even-mode impedance of the same line is measured as 50 Ohms.

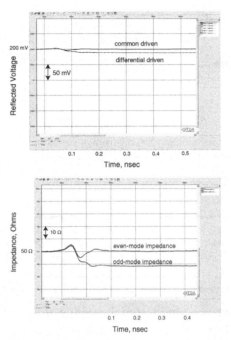

Figure 11-35   Measured voltage (top) from one channel of a differential TDR when the output is a differential signal and a common signal. These are interpreted as the impedance of the even mode (common driven) and odd mode (differential driven). The odd-mode impedance is 39 Ohms, and the even-mode characteristic impedance is 50 Ohms. Measured with an Agilent 86100 DCA with DTDR plug-in and TDA Systems IConnect software, and a GigaTest Labs Probe Station.

11.14 Terminating Differential and Common Signals

When a differential signal reaches the open end of a differential pair, it will see a high impedance and will be reflected. If the reflections at the ends of the differential pair are not managed, they may exceed the noise margin and cause excessive noise. A commonly used method of reducing the reflections is to have the differential signal see a resistive impedance at the end of the pair that matches the differential impedance.

If the differential impedance of the line is designed for 100 Ohms, for example, the resistor at the far end should be 100 Ohms, as illustrated in Figure 11-36. The resistor would be placed across the two signal lines so the differential signal would see its impedance. The differential signal would be terminated with this single resistor. But what about the common signal?

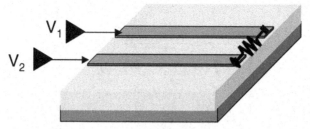

Figure 11-36 Terminating the differential signal at the far end of a differential pair with a single resistor that has a resistance equal to the differential impedance of the pair.

TIP   While it is true that the common-signal component even in LVDS levels is very high, this voltage is nominally constant, even when the drivers are switching, and may not impact the measurement of a differential signal at the receiver.

Any transient common signal moving down the differential pair will see a high impedance at the end of the pair and reflect back to the source. Even if there were a 100-Ohm resistor across the signal lines, the common signal, having the same voltage between the two signal lines, would not see it. Depending on the impedance of the driver, any common signal created will rattle back and forth and show the effect of ringing.

Is it important to terminate the common signal? If any devices in the circuit are sensitive to the common signal, then managing the signal quality of the common signal is important. If there is an asymmetry in the differential pair that converts some differential signal into common signal, if the common signal rattles around and sees the asymmetry again, some of it may convert back into differential signal and contribute to differential noise.

TIP   Terminating the common signal will not eliminate the common signal but will simply prevent the common signal from rattling around. If a common signal is contributing to EMI, it's true that terminating the common signal will help reduce EMI slightly, but it will still be important to create a design that eliminates the source of the common signal.

One way of terminating the common signal is by connecting a resistor between the end of each signal line and the return path. The parallel combination of the two resistors should equal the impedance the common signal sees. If the two lines were uncoupled and the common impedance were

25 Ohms, each resistor would be 50 Ohms, and the parallel impedance would be 25 Ohms. This is shown in Figure 11-37.

If this termination scheme is used, the common signal will be terminated, and the differential signal will also be terminated. However, if the lines were tightly coupled, and the even-mode impedance were 25 Ohms, while two 50-Ohm resistors would terminate the common impedance, the differential impedance would not be terminated.

The equivalent resistance a differential signal would see in this termination scheme is the series combination of the two resistors, $4 \times Z_{comm}$. It is only when $Z_{even} = Z_{odd}$ that this resistance is the differential impedance. As coupling increases, the common impedance will increase, and the differential impedance will decrease.

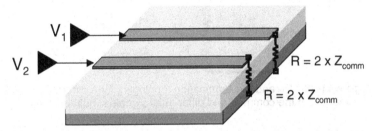

Figure 11-37 Terminating the common signal at the far end of a differential pair with two resistors, each with resistance equal to twice the common impedance of the pair.

If it is important to terminate the common signal and the differential signal simultaneously, we must use a special termination technique. This can be implemented with two different resistor topologies, each using three resistors. The "pi" and "tee" topologies are shown in Figure 11-38.

In the pi topology, the values of the resistors can be calculated based on designing the equivalent resistance the common signal sees to equal the common impedance and the equivalent resistance the differential signal sees to equal the differential impedance. The equivalent resistance the common signal sees is the parallel combination of the two R_2 resistors:

$$R_{equiv} = \frac{1}{2}R_2 = Z_{comm} = \frac{1}{2}Z_{even} \qquad (11\text{-}30)$$

where:

R_{equiv} = equivalent resistance the common signal would see

R_2 = resistance of each R_2 resistor

Z_{comm} = impedance the common signal sees on the differential pair

Z_{even} = even-mode characteristic impedance of the differential pair

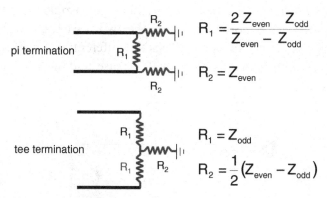

$$R_1 = \frac{2\, Z_{even}\, Z_{odd}}{Z_{even} - Z_{odd}}$$

$$R_2 = Z_{even}$$

$$R_1 = Z_{odd}$$

$$R_2 = \frac{1}{2}\left(Z_{even} - Z_{odd}\right)$$

Figure 11-38   Pi- and tee-termination topologies for a differential pair that can terminate both the common and differential signals simultaneously.

This requirement defines $R_2 = Z_{even}$.

The equivalent resistance the differential signal sees is the parallel combination of the R_1 resistor with the series combination of the two R_2 resistors:

$$R_{equiv} = \frac{R_1 \times 2R_2}{R_1 + 2R_2} = Z_{diff} = 2 \times Z_{odd} \qquad (11\text{-}31)$$

where:

R_{equiv} = equivalent resistance the differential signal would see

R_1 = resistance of each R_1 resistor

R_2 = resistance of each R_2 resistor

Z_{diff} = impedance the differential signal sees on the differential pair

Z_{odd} = odd-mode characteristic impedance of the differential pair

Since $R_2 = Z_{even}$, we can calculate the value of R_1 as:

$$R_1 = \frac{2Z_{even}Z_{odd}}{Z_{even} - Z_{odd}} \qquad (11\text{-}32)$$

When there is little coupling, and $Z_{even} \sim Z_{odd} \sim Z_0$, then $R_2 = Z_0$ and $R_1 = $ open. After all, with little coupling, this pi-termination scheme reduces to a simple far-end termination of a resistor equal to the characteristic impedance of either line. As the coupling increases, the resistor to the return path will increase to match the even-mode characteristic impedance. A high-value shunt resistor will short the two signal lines so the impedance the differential signal sees will drop as the coupling increases and will match the lower differential impedance.

For a typical tightly coupled differential pair, the odd-mode impedance might be 50 Ohms, and the even-mode impedance might be 55 Ohms. In this case, in a pi termination, the resistors would be 1 kOhm between the signal lines and 55 Ohms from each line to the return path. This combination will simultaneously terminate the 100-Ohm differential impedance and the 27.5-Ohm common impedance.

In a tee topology, the differential signal will see an equivalent resistance of just the series combination of the R_1 resistors:

$$R_{equiv} = Z_{diff} = 2R_1 = 2Z_{odd} \qquad (11\text{-}33)$$

where:

R_{equiv} = equivalent resistance the differential signal would see

R_1 = resistance of each R_1 resistor

Z_{diff} = impedance the differential signal sees on the differential pair

Z_{odd} = odd-mode characteristic impedance of the differential pair

This defines $R_1 = Z_{odd}$.

The equivalent resistance the common signal sees is the series combination of the two R_1 resistors that are in parallel with the R_2 resistor:

$$R_{equiv} = Z_{comm} = \frac{1}{2}R_1 + R_2 = \frac{1}{2}Z_{even} \qquad (11\text{-}34)$$

The value of R_2 can be extracted as:

$$R_2 = \frac{1}{2}(Z_{even} - Z_{odd}) \tag{11-35}$$

In a tee termination, when there is little coupling, so $Z_{even} \sim Z_{odd} \sim Z_0$, the termination is simply the two R_1 resistors in series between the signal lines, each equal to the odd-mode characteristic impedance. In addition, there is a center tap connection to the return path that is a short. The tee termination reduces to the pi termination in the case of no coupling. As coupling increases, the differential impedance decreases, and the R_1 values decrease to match. The common impedance increases, and the R_2 term increases to compensate.

If the odd-mode impedance were 50 Ohms and the even-mode impedance 55 Ohms, the tee termination would use two series resistors between the signal lines of 50 Ohms each, and it would use a 2.5-Ohm resistor between the center of the resistors and the return path.

The most important consideration when implementing a pi- or tee-termination strategy is the potential DC load on the drivers. In both cases, there is a resistance between either signal line to the return path on the order of the even-mode impedance. The lower the even-mode impedance, the more current flow from the driver. Typical differential drivers may not be able to handle low-DC resistance to the low-voltage side, so it may not be practical to terminate the common signal. Rather, care should be used to minimize the creation of common signals right from the beginning and simply to terminate the differential signals.

An alternative configuration to terminate both the differential and common signals is to use a tee topology with a DC-blocking capacitor. This circuit topology is shown in Figure 11-39. In this topology, the resistor values are the same as with a tee topology, and the value of the capacitor is chosen so that the time constant the common signal will see is long compared to the lowest-frequency component in the signal. This will ensure a low impedance for the capacitor compared with the impedance for the resistors, at the lowest-frequency component of the signal. As a first-order estimate, the capacitor value is initially chosen from:

$$RC = 100 \times RT \tag{11-36}$$

$$C = \frac{100 \times RT}{Z_{comm}} \qquad (11\text{-}37)$$

where:

R = equivalent resistance the common signal sees

C = capacitance of the blocking capacitor

RT = rise time of the signal

Z_{comm} = impedance the common signal sees on the differential pair

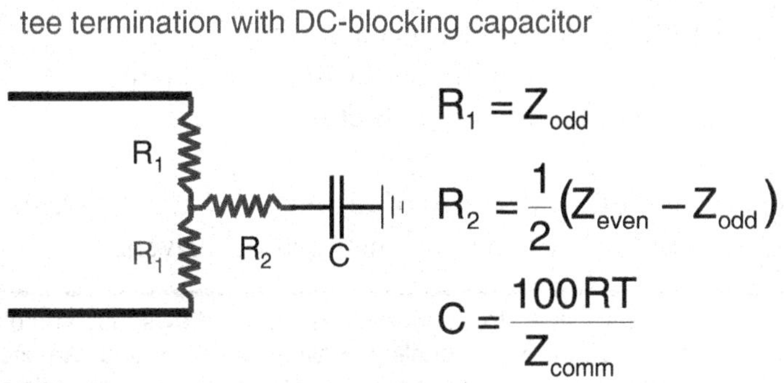

tee termination with DC-blocking capacitor

$$R_1 = Z_{odd}$$

$$R_2 = \frac{1}{2}(Z_{even} - Z_{odd})$$

$$C = \frac{100\,RT}{Z_{comm}}$$

Figure 11-39   Tee termination with DC-blocking capacitor to terminate common signals but minimize the DC-current drain.

For example, if the common impedance is about 25 Ohms, and the rise time is 0.1 nsec, the blocking capacitor should be about 10 nsec/25 Ohms = 0.4 nF. Of course, whenever an RC termination is used, a simulation should be performed to verify the optimum value of the capacitor.

Another alternative termination scheme, which works well for on-die termination, is to just terminate each signal line with a Vtt termination to a separate Vtt voltage supply. For 50-Ohm lines, this means connecting a 50-Ohm resistor to the Vtt supply. This effectively terminates each line as a single-ended transmission line.

When there is no coupling between the two signal lines, this terminates both the differential signal and the common signal, with half the power consumption of a termination to ground. As coupling increases and the differential impedance is kept to 100 Ohms, the differential signal will be terminated, and the common signal will be mostly terminated. It is not a perfect termination but can be a match to 90%.

The advantages of this approach are that it gives good differential signal termination, adequate common signal termination, and good power consumption, and it can be impedance on-die.

11.15 Conversion of Differential to Common Signals

The information content in differential signaling is carried by the differential signal. First, it is important to maintain the signal quality of the differential signal. This is done by using the following guidelines:

- Use controlled differential impedance lines.
- Minimize any discontinuities in the differential pair.
- Terminate the differential signals at the far end.

There is an additional source of distortion of the differential signal, due to asymmetries in the lines and skew between the drivers.

> **TIP** Distortion due to asymmetries and skews is completely independent of the degree of coupling between the lines and can occur in a differential pair with no coupling or the tightest coupling, both microstrip and stripline.

A skew between the transitioning of the two differential drivers will distort the differential signal. Figure 11-40 shows the edge of the differential signal as the skew between the drivers is increased from 20% of the rise time to two times the rise time. Most high-speed serial links specify that a line-to-line skew in a channel should be less than 20% the unit interval (UI). A line-to-line skew will directly affect the signal edges and reduce the horizontal opening of an eye.

For example, in a 5-Gbps signal, the UI is 200 psec, and 20% of the UI is 40 psec, the maximum acceptable line-to-line skew. With a typical speed of a signal as 6 inches/nsec, the length difference between the two lines that make up the differential pair could be 0.04 nsec × 6 inches/nsec = 240 mils.

The line-to-line skew becomes noticeable when the skew is longer than the signal rise time. The rising and falling edges of the differential signal will be distorted.

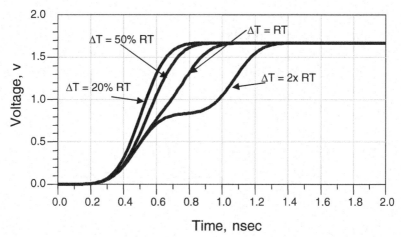

Figure 11-40 Received differential signal with driver skew varying from 20% of the rise time to 200% of the rise time. Simulated with Agilent's ADS.

To keep the maximum line-to-line skew below 20%, the unit interval requires:

$$\Delta L = 0.2 \times UI \times v = 0.2 \times \frac{v}{BR} \qquad (11\text{-}38)$$

where:

ΔL = maximum length skew between paths to keep skew less than 20% of the unit interval

UI = unit interval

BR = bit rate

v = speed of the differential signal

If the speed of a signal is roughly 6 inches/nsec and the UI is 1 nsec, then the maximum length mismatch allowed in a differential pair would have to be less than 0.2×6 inches/nsec $\times 1$ nsec ~ 1.2 inch. This is relatively easy to accomplish.

If the UI were 100 psec, the lengths would have to be matched to better than 120 mils. As the skew budget allocated to line-length differences decreases, it becomes more and more important to match the line lengths.

Other sources of asymmetry will potentially distort the differential signal. In general, if something affects one line and not the other, the differential signal will be distorted. For example, if one of the traces sees a test pad, so there is a capacitive load and the other trace does not, the differential signal will be distorted. Figure 11-41 shows the simulated differential signal at the end of a differential pair with a capacitive load on one line.

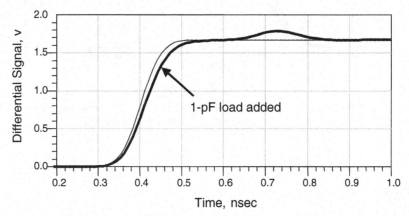

Figure 11-41   Received differential signal with and without a 1-pF capacitive load on one line of a differential pair. The rise time is 100 psec. Simulated with Agilent's ADS.

There is another impact from skew and distortions. Any asymmetry can convert some of the differential signal into common signal. In general, if the drivers and receivers are not sensitive to common signals, the amount of common signal created may not be important. After all, differential receivers typically have a good common-mode rejection ratio (CMRR). However, if the common signal were to get out of the box on twisted pair, for example, it would dramatically contribute to EMI. It is critical to minimize any common signal that may have an opportunity, either intentionally or unintentionally, to escape outside the product enclosure through an aperture or on any cables.

TIP   Any asymmetry will convert differential signal into common signal. This includes cross talk, driver skew, length skew, and asymmetrical loading. An important motivation to keep skews to a minimum is to minimize the conversion of differential signal into common signal.

A small skew that may not affect the differential-signal quality at all can have considerable impact on the common signal. Figure 11-42 shows the voltages on the signal lines when there is a skew of just 20% of the rise time and the received differential and common signal when only the differential signal is terminated. The differential-signal component is terminated by the resistor at the far end, but the common-signal component sees an open at the far end and reflects back to the low impedance of the source. This creates the ringing.

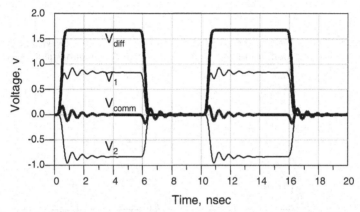

Figure 11-42 Signals at the far end of a differential pair with a differential terminating resistor. The driver skew is only 20% of the rise time of the signal. Note the good quality of the differential signal, even with the voltage on each signal line distorted. Simulated with Agilent's ADS.

Even if both the common and differential signals are terminated, there will still be common signal generated by any asymmetry. Figure 11-43 shows the voltages at the far end with a tee-termination network, where both differential and common signals are simultaneously terminated. The ringing is gone, but there is still a common signal, which could contribute to EMI if it were to get out of the box.

As the skew between the two signal lines increases, the magnitude of the common signal will increase, even with the common signal terminated. Figure 11-44 shows the common signal for skews of 20%, 50%, 100%, and 200% of the rise time.

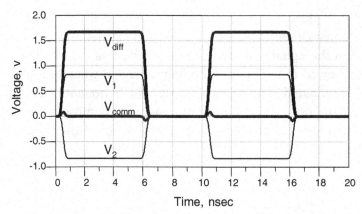

Figure 11-43 Received signals at the far end with a 20% skew and both differential- and common-signal termination. Simulated with Agilent's ADS.

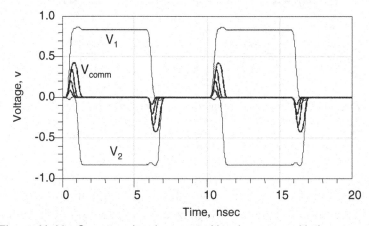

Figure 11-44 Common signal generated by skew even with the common signal terminated. Skew of 20%, 50%, 100%, and 200% of the rise time. Simulated with Agilent's ADS.

TIP   A very small driver skew can create significant common signal. This is why it is important to try to minimize all asymmetries.

To minimize the creation of common signals, care should be used to try to keep the paths as symmetrical as possible. Imperfections should be modeled to anticipate the amount of common signal created. This will allow some prediction of the severity of potential EMI problems.

11.16 EMI and Common Signals

If an unshielded twisted-pair cable, such as category-5 cable, is connected to a differential pair on a circuit board, both a differential signal and a common signal may be transmitted onto the cable. The differential signal is the intended signal that carries information content. A twisted-pair cable is a very poor radiator of electromagnetic energy from the differential signals. However, any common current on the cable will radiate and contribute to EMI.

If there is common current on the twisted pair, where is the return of the common signal? On the circuit board where the differential signal is created, the common signal will return on the return plane of the board. When the signal transitions from the circuit board to the twisted pair, there is no direct connection to the return plane of the circuit board.

This poses no problem for the differential signal. It is possible to design the differential impedance of the circuit-board interconnects to match the differential impedance of the twisted pair. The differential signal may not see any impedance discontinuity in transitioning from the board to the twisted pair.

> **TIP**   In an unshielded twisted pair, the return path for the common signal is literally the ground or floor, or any other conductor that is nearby. The coupling to the nearest conducting surface is usually less than the coupling between adjacent signal paths, so the common impedance is usually high—on the order of a few hundred Ohms.

There may be a large impedance mismatch between what the common signal sees in the board and in the twisted pair. The connection between the two common return paths is through whatever path the current will find. At high frequency, the return path is typically dominated by stray capacitances between the circuit-board ground to the box chassis to the floor. A capacitance of only 1 pF will have an impedance at 1 GHz of about 160 Ohms. It is even lower at higher frequency, and this is for just 1 pF of stray capacitance.

As the common signal propagates down the twisted pair, the return current is continuously coupled to the nearest conductor by the stray capacitance between the twisted pair and the conductor. This is illustrated in Figure 11-45.

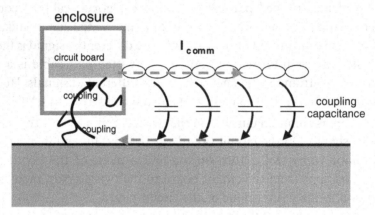

Figure 11-45   Schematic of the common current path when a twisted-pair cable is connected to a circuit board. The coupling paths are typically capacitive for high-frequency common-signal components.

The amount of common current on the twisted pair will depend on the common-signal voltage launched into the cable and on the impedance the common signal sees in the cable:

$$I_{comm} = \frac{V_{comm}}{Z_{comm}}$$

(11-39)

This common current will radiate. The amount of common current that will fail a certification test depends on the specific test spec, length of cable, and frequency. As a rough rule of thumb, it takes only 3 microAmps of common current, at 100 MHz, in a 1-meter-long external cable to fail an FCC Class B certification test. This is a tiny amount of common current.

If the radiated field strength exceeds the allowed limits of the EMI certification regulation, the product will not pass certification, and this may delay the ship date of the product. In most countries, it is illegal to sell a product for revenue unless it has passed the local certification regulations. In the United States, the Federal Communications Commission (FCC) specifies the acceptable radiated field levels in two categories of products. Class A products are those that will be used in an industrial or manufacturing environment. Class B products are those that will be used in a home or office environment. Class B requirements limit acceptable radiated emissions to lower levels than Class A requirements.

Freq (MHz)	μ V/m
30–88	100
88–216	150
216–960	200
> 960	500

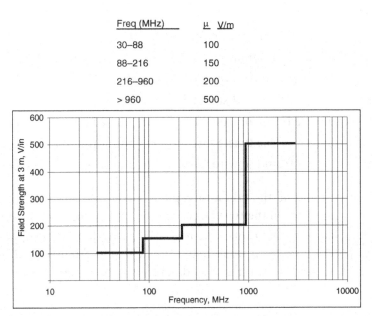

Figure 11-46 FCC Class B certification limits for the maximum allowable far-field electric-field strengths at a distance of 3 m.

Electric-field strength is measured in units of volts/m at a specified frequency. For Class B certification, the maximum radiated field strengths are measured 3 m from the product. The maximum electric-field strengths vary within different frequency ranges. They are listed and plotted in Figure 11-46. As a reference point, the maximum acceptable field strength at 3 m for 100 MHz is 150 microV/m.

We can estimate the radiated electric-field strength from a common signal in a twisted pair by approximating the twisted pair as an electric monopole antenna. The far field exists at a distance about one-sixth of the wavelength of the radiation. The FCC test condition of 3 m is in the far field for frequencies larger than about 16 MHz. The far-field electric-field strength from an electric monopole antenna is:

$$E = 4\pi 10^{-7} \times f \times I_{comm} \times \frac{Len}{R}$$
(11-40)

where:

I_{comm} = common current on the twisted pair, in Amps

V_{comm} = common signal that is launched into the twisted pair

Z_{comm} = impedance the common signal sees

E = field strength at a distance R from the wire, in volts/m

f = sine-wave frequency of the common current component, in Hz

Len = length of the twisted wire that is radiating, in m

R = distance from the wire to the point where the field is measured, in m

For example, if the common signal is 100 mV and the impedance the common signal sees in the twisted pair is 200 Ohms, the common current may be as large as 0.1 v/200 Ohms = 0.5 mA. If the length of the twisted pair is 1 m and the distance at which we measure the field strength is 3 m, corresponding to the FCC Class B test, the radiated field strength at 100 MHz is:

$$E = 4\pi 10^{-7} \times 10^{8} \times 5 \times 10^{-4} \times \frac{1}{3} = 20000 \frac{microV}{m}$$
(11-41)

The radiated field strength from a common current that might easily be generated in an unshielded twisted pair is more than a factor of 100 larger than the FCC Class B certification limit.

TIP   Even small amounts of common currents getting onto an unshielded twisted-pair cable can cause a product to fail an EMI certification test. Common currents larger than about 3 microAmps will fail FCC Class B Certification.

There are generally three techniques to reduce the radiated emissions from common currents in twisted-pair cables:

1. Minimize the conversion of differential signal into common signal by minimizing any asymmetries in the differential pairs and skews in the drivers. This minimizes the problem at the source.
2. Use a shielded twisted pair so the return path of the common current in the twisted pair is carried in the shield. Using a shielded cable may even increase the amount of common current since the common impedance may be lowered with the return path brought closer to the signal path. If the shield is connected to the chassis, the return current of the common signal may have an opportunity to flow on the inside of the shield. The common signal would then be flowing in a coax geometry—going out on the cable on the twisted-pair center and back on the inside of the shield. In this geometry, there would be no external magnetic or electric field, and the common current would not radiate. This requires using a low-inductance connection between the shield and the chassis so that the common-return current can maintain its coaxial distribution.
3. Increase the impedance of the common-current path by adding common chokes. Common signal chokes are of two forms. Virtually all cables used by peripherals have cylinders of ferrite material on the outside of the cable. The placement of a ferrite around a cable is shown in Figure 11-47. The high permeability of the ferrite will increase the inductance and the impedance of any net currents going through the ferrite. In addition, many RJ45 connectors used in Ethernet links have common-mode ferrite chokes built in.

If a differential signal flows through the ferrite, there will be no net magnetic field through the ferrite. The currents in each line of the differential pair are equal and in opposite directions, so their external magnetic and electric fields mostly cancel each other. Only common-signal currents going through the ferrite, with return currents on the outside, would have magnetic-field line loops that would go through the ferrite material and see a higher impedance. The ferrite on the outside of the cable increases the impedance the common signal sees, reducing the common current, which decreases the radiated emissions. This type of ferrite choke can be used on the outside of any cable—twisted pair or shielded.

The second type of common-signal choke is primarily useful for twisted-pair cable and is often built into the connector. Connectors with a built-in choke are sometimes called "magnetics." The goal is to dramatically increase the impedance a common signal would see but not affect the impedance a differential signal would see. To radically increase the impedance of the common signal, a twisted-pair wire is wound into a coil, sometimes with a ferrite core.

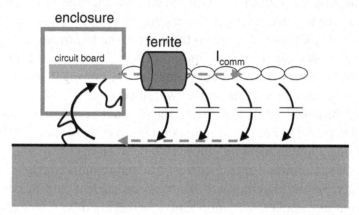

Figure 11-47   Placement of ferrite around the differential currents but enclosing only the common signal path, not the common return path.

Any common current flowing down the twisted pair will see a much higher inductance from the coiling and the high-permeability ferrite core. However, the differential signal on the twisted pair will have very few magnetic-field lines outside the twisted pair, and the differential current will not see the coil or the ferrite. The differential signal will be nearly unaffected by the coiling.

A twisted-pair coil can be built into a connector. Many RJ-45 connectors, used with Ethernet cables, for example, have built-in common-signal chokes.

TIP   The higher impedance seen by the common signal can reduce the common current that would get onto the twisted-pair cable by more than 99%, or –40 dB. They are essential components to minimize radiated emissions from any common currents created by asymmetries in the circuit board.

11.17 Cross Talk in Differential Pairs

If we bring a single-ended transmission line close to a differential pair, some of the signal voltage will couple from the active, single-ended line to both lines of the differential pair. This is illustrated in Figure 11-48. The noise that couples to each line of the differential pair will be the same polarity, just of different magnitudes.

The closer line of the pair will end up with more noise than the farther line. The more tightly coupled the differential pair, the more equal the noise generated on the two lines, and the less the differential noise.

TIP   In general, the more identically we can arrange the noise that is picked up on each line of the pair, the less the differential noise. This generally means a configuration of tightly coupled lines, with a large space between the pair and the aggressor line.

Figure 11-49 shows the differential noise at the receiver generated on the differential pair constructed in stripline, with the far end differentially terminated and the near end having the low impedance of a typical driver.

In this configuration, the aggressor line is spaced a distance equal to the line width away from the nearest victim trace. Two different coupling levels are evaluated. Tight coupling has a spacing between lines in the differential pair equal to the line width, while weak coupling has spacing twice the line width. In the pair, the line that is farther away from the aggressor will have slightly less noise coupled to it than the line close to the aggressor line. The differential noise is the difference between the noise levels on the lines. In this example, there is about 1.3% differential noise on the weakly coupled victim differential pair, while there is about half this value on the tightly coupled victim differential pair. Tighter coupling can decrease the differential noise by about 50%.

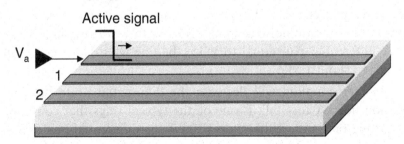

Figure 11-48   Cross talk from a single-ended signal to a differential pair.

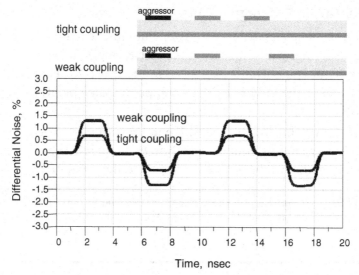

Figure 11-49 Differential noise in a differential pair from an adjacent single-ended aggressor line.

Though the differential noise from the active line to the differential line may be only 1.3% in the worst case, this may sometimes cause problems. If the aggressor line is a 3.3-v signal level, the differential noise on the differential pair may be as large as 40 mV. If there is another active line on the other side of the victim lines, switching in the opposite direction of the first aggressor line, the differential noise from the two active lines will add to the victim line. This can contribute as much as 80 mV of noise, which may approach the allocated noise budget for some low-voltage differential signals.

The common noise on the victim differential pair is the average voltage on each line. Figure 11-50 shows the common noise for the two levels of coupling. The common noise will not be strongly affected as the coupling between the differential pair changes. With tight coupling (i.e., a spacing equal to the line width), the common noise is about 2.1%. With weak coupling (i.e., a spacing in the differential pair equal to twice the line width), the common noise is reduced to about 1.5%.

Tighter coupling will decrease the differential noise but increase the common noise. Cross talk is one of the typical ways that common currents are created on a differential pair. Even if the differential pair is designed for perfect symmetry, cross talk can cause the creation of common voltages on a differential pair. This is why it is always important to provide common-signal chokes for any externally connected twisted-pair cables.

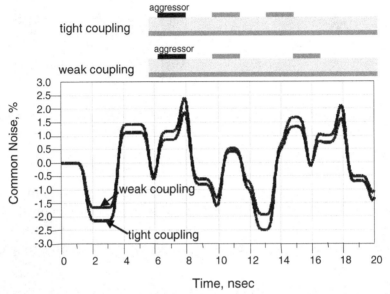

Figure 11-50 Common noise in a differential pair from an adjacent single-ended aggressor line. The ringing and distortion are due to the common signal not being terminated.

T I P   This suggests the general rule that to minimize the differential noise on a differential pair from a single-ended aggressor line, the coupling in the differential pair should be as tight as possible. Of course, the spacing between the aggressor and the victim pair should also be as large as possible to minimize the coupled noise. Tight coupling will not eliminate cross talk but will just reduce it in some cases.

The differential noise coupling between two differential pairs is slightly less than the differential noise coupling from a single-ended line. Figure 11-51 shows the differential and common noise between two differential pairs. In this case, the edge-to-edge spacing between the two pairs is equal to the line width. The difference between tight and weak coupling is not very large. As a rough estimate, the differential noise can be less than 1%, and the common noise can be less than 2% of the differential signal on the aggressor pair.

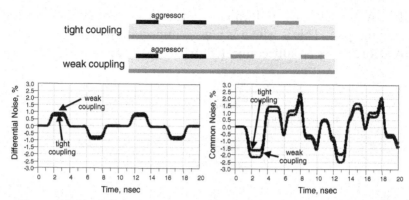

Figure 11-51   Differential (left) and common (right) noise on a differential pair when the aggressor line is also a differential pair for the case of tight and weak coupling within the victim pair.

Generally, the coupling in a differential pair, especially in stripline, has only a small effect on the differential cross talk picked up on the pair from another trace. It is only when there is not adjacent return plane, such as in a connector or leaded package, that tight coupling will have a strong impact on reducing cross talk.

11.18 Crossing a Gap in the Return Path

Gaps in the return path are often used to isolate one region of a board from another. They also occur when a power plane is used as the reference layer and when a split power layer is used. Sometimes, a gap in the return path occurs unintentionally, as when the clearance holes in a return plane are over-etched and overlap. In this case, any signal lines that pass through the via field will see a gap in the return path.

A single-ended signal will see a potentially disastrous discontinuity if the gap it encounters is wide. This will look like a large inductive discontinuity. Figure 11-52 shows the simulated reflected and transmitted single-ended signal passing over a 1-inch gap in the return path of an otherwise uniform 50-Ohm line with terminations at each end. The 100-psec initial rise time is dramatically increased due to the series inductive discontinuity.

While it may be possible in some situations to use a low-inductance capacitor to span the gap and provide a lower-impedance path for the return, it is difficult to get good high-frequency performance.

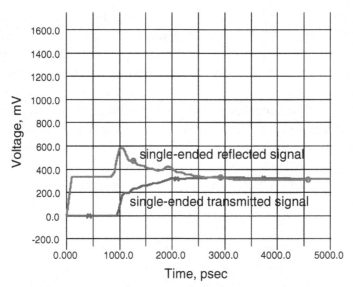

Figure 11-52   Reflected and transmitted 100-psec signal through a 50-Ohm transmission line with a 1-inch-wide gap in the return path in the middle. Simulated with Mentor Graphics HyperLynx.

An alternative approach to transport a signal across a gap in the return path with acceptable performance is using a differential pair. Figure 11-53 illustrates a typical case where the signal starts in one region of a board as an edge-coupled microstrip differential pair, in a second region where the return plane is far away, and in a third region where the interconnect is again an edge-coupled microstrip.

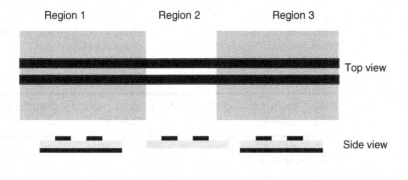

Figure 11-53   Region 2 of a circuit board is where there is a gap in the return path. This can be modeled as a region where the return path is very far away.

In regions 1 and 3, the differential pair has a differential impedance of about 90 Ohms. In the middle region, the return path is removed. With a 5-mil line and space and a 2.7-mil dielectric beneath the traces, but no other conductor, the differential impedance is about 160 Ohms. If the return plane is at least as far away as the span of the signal conductors, the differential impedance will be independent of the position of the return plane, and it is as though it is not there. This trick allows us to build a circuit model for this discontinuity and to explore the impact of the discontinuity on the differential-signal quality. Figure 11-54 shows the simulated reflected and transmitted differential signal passing through the same gap as before. The rise time is preserved in the transmitted differential signal.

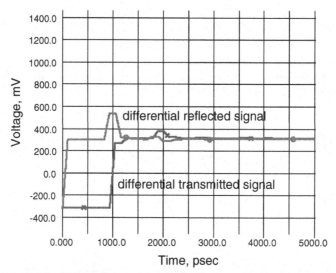

Figure 11-54   Differential signal with a 100-psec rise time transmitted and reflected across a 1-inch-wide gap in the return plane. The differential signal is minimally distorted across the gap. Simulated with Mentor Graphics HyperLynx.

TIP   Using a tightly coupled differential pair is one way of transmitting high-bandwidth signals in regions with poorly defined return planes.

While the 160-Ohm differential impedance in the region of the gap is higher than the 90-Ohm differential impedance of the rest of the interconnect, it is a uniform transmission line. This will cause a degradation of signal quality.

More importantly, single-ended signals, in passing over the gap in the return path, will create ground bounce as their return currents will see the common inductance of the gap. However, differential signals will have their return currents overlap in the common inductance and mostly cancel. With virtually no net return current crossing the gap, the ground bounce from differential signals will be dramatically less than for single-ended signals.

11.19 To Tightly Couple or Not to Tightly Couple

A differential pair can transport a differential signal just as well if the coupling between the pairs is tight or loose, provided that the differential impedance is the same. Of course, in fabricating for a target differential impedance, the cross section and stack-up must take into account the coupling. With an accurate 2D field-solver tool, the stack-up for any target differential impedance and any level of coupling is equally easy to design.

This means that from a differential signal's perspective, as long as the instantaneous differential impedance is constant, it doesn't matter what the coupling is. The two lines that make up the pair can come close together and then move far apart, as long as the line width is adjusted to always maintain 100 Ohm differential impedance.

The most important advantage of using tightly coupled differential pairs is the higher interconnect density with tight coupling. This means either fewer layers or potentially a smaller board, both of which could contribute to lower cost. If there are no other important considerations, tightly coupled differential pairs should always be the first choice because they can result in the lowest-cost board.

In some cases, the differential cross talk to a tightly coupled pair might be a little less than if loosely coupled. However, this is not universally true so should be considered on a case-by-case basis.

The other advantage of tightly coupled differential pairs is that the signal is robust to imperfections in the return path. As a general rule, whenever the return planes are disrupted, such as in twisted-pair cables, ribbon cables, connectors, and some IC packages, tight coupling should always be used. For these reasons, tight coupling should always be the first choice. However, tight coupling may not be the best choice in every design.

In comparing the line width of a loosely coupled and tightly coupled 100-Ohm differential pair, the line width has to decrease about 30% to maintain the 100 Ohms. This means a loosely coupled pair will have about 30% lower series-resistive loss than a similar-impedance tightly coupled pair. When loss is important, this can be an important reason to use loosely coupled pairs. Generally, above 10 Gbps, when loss is the most important performance term, loosely coupled differential pairs will enable the widest lines and the lowest loss and should be the first choice.

The most significant advantage of using loosely coupled lines is the opportunity to use a wider line width. If we start with a differential pair that is loosely coupled and bring the traces closer together, keeping everything else the same, the differential impedance will decrease. To achieve the target impedance, we have to make the lines narrower. This means higher resistance lines.

TIP   When cost is the driving force, tightly coupled differential pairs should be used. When loss is important, loosely coupled differential pairs should be used.

11.20 Calculating Odd and Even Modes from Capacitance- and Inductance-Matrix Elements

The first-order model of a single-ended transmission line is an n-section lumped-circuit model. This is described in terms of a capacitance per length and a loop inductance per length. The characteristic impedance and time delay of a single-ended transmission line are given by:

$$Z_0 = \sqrt{\frac{L_L}{C_L}} \qquad (11\text{-}42)$$

$$TD = \sqrt{L_L C_L} \qquad (11\text{-}43)$$

where:

Z_0 = single-ended characteristic impedance of the line

L_L = loop inductance per length of the line

C_L = capacitance per length of the line

TD = time delay of the line

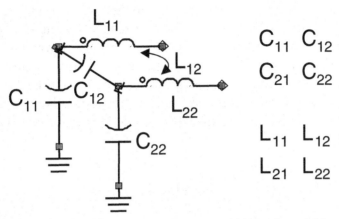

Figure 11-55 One section of an n-section lumped-circuit model for a pair of coupled transmission lines, with the SPICE capacitance matrix and a loop-inductance matrix that defines all the model elements.

We can extend this model to include the coupling between two lines. Figure 11-55 shows the equivalent circuit model of a section of the two coupled transmission lines. The capacitor elements are defined in terms of the SPICE capacitor matrix, and the inductance elements are defined in terms of the loop-inductance matrix. Of course, the values of the C and L matrix elements are obtained directly from the stack-up geometry of the differential pair by using a 2D field solver tool. We can use this model to determine the even- and odd-mode characteristic impedance values based on the values of the matrix elements.

When the pair is driven in the odd-mode state, the impedance of one line is the odd-mode characteristic impedance. The equivalent capacitance of one line is:

$$C_{odd} = C_{11} + 2C_{12} = C_{load} + C_{12} \tag{11-44}$$

where:

C_{odd} = capacitance per length from the signal to the return path of one line when the pair is driven in the odd mode

C_{11} = diagonal element of the SPICE capacitance matrix

C_{12} = off-diagonal element of the SPICE capacitance matrix

C_{load} = loaded capacitance of the signal line = $C_{11} + C_{12}$

In the odd mode, current is going into the signal line of trace 1 and back out its return path. At the same time, current is coming out of line 2 and into its return path. In this configuration, around transmission line 1 will be mutual magnetic-field lines from the signal in transmission line 2; these mutual magnetic-field lines are going opposite to the self-field lines of line 1. The equivalent loop inductance of line 1 will be reduced due to the current in line 2. The equivalent loop inductance of line 1 when the pair is driven in the odd mode is:

$$L_{odd} = L_{11} - L_{12} \tag{11-45}$$

where:

L_{odd} = loop inductance per length from the signal to the return path of one line when the pair is driven in the odd mode

L_{11} = diagonal element of the loop-inductance matrix

L_{12} = off-diagonal element of the loop-inductance matrix

Looking into line 1 of the pair, we see a higher capacitance and lower loop inductance as the coupling between the lines increases. From these two terms, the odd-mode characteristic impedance and time delay can be calculated as:

$$Z_{odd} = \sqrt{\frac{L_{odd}}{C_{odd}}} = \sqrt{\frac{L_{11} - L_{12}}{C_{load} + C_{12}}} \tag{11-46}$$

$$TD_{odd} = \sqrt{L_{odd}C_{odd}} = \sqrt{(L_{11} - L_{12})(C_{load} + C_{12})} \tag{11-47}$$

When the pair is driven in the even mode, the capacitance between the signal and the return is reduced due to the shielding from the adjacent trace driven at the same voltage as trace 1. The equivalent capacitance per length is:

$$C_{even} = C_{11} = C_{load} - C_{12} \tag{11-48}$$

where:

C_{even} = capacitance per length from the signal to the return path of one line when the pair is driven in the even mode

C_{11} = diagonal element of the SPICE capacitance matrix

C_{12} = off-diagonal element of the SPICE capacitance matrix

C_{load} = loaded capacitance of the signal line = $C_{11} + C_{12}$

When driven in the even mode, current goes into signal line 1 and back out its return path. At the same time, current also goes into signal line 2 and back out its return path. The magnetic mutual-field lines from line 2 are in the same direction as the self-field lines in line 1. The equivalent loop inductance of line 1 when driven in the even mode is:

$$L_{even} = L_{11} + L_{12} \qquad (11\text{-}49)$$

where:

L_{even} = loop inductance per length from the signal to the return path of one line when the pair is driven in the even mode

L_{11} = diagonal element of the loop-inductance matrix

L_{12} = off-diagonal element of the loop-inductance matrix

From the capacitance per length and loop inductance per length looking into the front of line 1 when the pair is driven in the even mode, we can calculate the even-mode characteristic impedance and time delay:

$$Z_{even} = \sqrt{\frac{L_{even}}{C_{even}}} = \sqrt{\frac{L_{11} + L_{12}}{C_{load} - C_{12}}} \qquad (11\text{-}50)$$

$$TD_{even} = \sqrt{L_{even}C_{even}} = \sqrt{(L_{11} + L_{12})(C_{load} - C_{12})} \qquad (11\text{-}51)$$

These relationships based on the capacitance- and inductance-matrix elements are used by all field solvers to calculate the odd- and even-mode characteristic impedances and the time delays for any transmission, with any degree of coupling and any stack-up configuration. In this sense, the capacitance- and inductance-matrix elements completely define the electrical properties of a pair of coupled transmission lines. This is a more fundamental description that shows the underlying effect of how the off-diagonal terms, C_{12} and L_{12}, that describe coupling affect the characteristic impedance and time delay for each mode. As coupling increases, the off-diagonal terms increase, the odd-mode impedance decreases, and the even-mode impedance increases.

11.21 The Characteristic Impedance Matrix

There is an alternative description of two or more coupled transmission lines that uses an impedance matrix. This is just as fundamental as the use of the capacitance and inductance matrices; it's just different. In this description, an impedance matrix is defined that relates the current and voltage into either line of a differential pair. Though the following analysis illustrates the impedance matrix for two coupled lines, it can be generalized to n coupled lines. It applies no matter what the stack-up geometry, symmetry, or materials distribution. These features will affect the parameter values but not the general approach.

Two transmission lines are shown in Figure 11-56. Any arbitrary signal into either line can be described by the voltage between each line and the current into each signal line and out its return path. If there were no coupling between the lines, the voltage on each line would be independent of the other line. In this case, the voltages on each line would be given by:

$$V_1 = Z_1 I_1 \qquad\qquad (11\text{-}52)$$

$$V_2 = Z_2 I_2 \qquad\qquad (11\text{-}53)$$

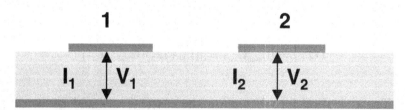

Figure 11-56   The voltage on each line and the current into each signal line and back out the return are labeled as shown.

However, if there is some coupling, this cross talk will cause the voltage on one line to be influenced by the current on the other line. We can describe this coupling with an impedance matrix. Each element of the matrix defines how the current into one line and out its return path influences the voltage in the other line. With the impedance matrix, the voltages on line 1 and 2 become:

$$V_1 = Z_{11}I_1 + Z_{12}I_2 \qquad (11\text{-}54)$$

$$V_2 = Z_{21}I_1 + Z_{22}I_2 \qquad (11\text{-}55)$$

The diagonal elements of the impedance matrix are the impedances of each line when there is no current into the other line. Of course, when the coupling is turned off, the diagonal elements reduce to what we have called the characteristic impedance of the line.

The off-diagonal elements of the impedance matrix describe the amount of coupling but in a non-intuitive way. These are matrix elements. They are not the actual impedance between line 1 and line 2. Rather, they are the amount of voltage generated on line 1 per Amp of current in line 2. In this sense, they are really *mutual* impedances:

$$Z_{12} = \frac{V_1}{I_2} \qquad (11\text{-}56)$$

$$Z_{21} = \frac{V_2}{I_1} \qquad (11\text{-}57)$$

When there is little coupling, no voltage is created on one line for any current in the other line, and the off-diagonal element between the lines is nearly zero. The smaller the off-diagonal term compared to the diagonal term, the smaller the coupling.

From this description, we can identify the odd- and even-mode impedances in terms of the characteristic impedance matrix. When a pure differential signal is applied to the two lines, the odd mode is driven. The odd mode is defined as when the current into one line is equal and opposite to the current into the other line, or $I_1 = -I_2$. Based on this definition, the voltages are:

$$V_1 = Z_{11}I_1 - Z_{12}I_1 = I_1(Z_{11} - Z_{12}) \tag{11-58}$$

From this, we can calculate the odd-mode impedance of line 1:

$$Z_{odd1} = \frac{V_1}{I_1} = Z_{11} - Z_{12} \tag{11-59}$$

Likewise, the even mode is when the currents into each line are the same, $I_1 = I_2$. The voltage on line 1 in the even mode is:

$$V_1 = Z_{11}I_1 + Z_{12}I_1 = I_1(Z_{11} + Z_{12}) \tag{11-60}$$

The even-mode impedance of line 1 is:

$$Z_{even1} = \frac{V_1}{I_1} = Z_{11} + Z_{12} \tag{11-61}$$

In the same way, the odd- and even-mode impedance of the other line can be found. In this definition, the odd-mode impedance of one line is the difference between the diagonal and off-diagonal impedance-matrix elements. The larger the coupling, the larger the off-diagonal element and the smaller the differential impedance. The even-mode impedance of one line is the sum of the diagonal and off-diagonal impedance-matrix elements. As coupling increases, the off-diagonal element increases, and the even-mode impedance gets larger.

Most 2D field solvers report the odd- and even-mode impedance, the capacitance and inductance matrices, and the impedance matrix.

From a signal's perspective, the only things that are important are the differential and common impedances, which can be described in three equivalent ways by:

1. Odd- and even-mode impedances
2. Capacitance- and inductance-matrix elements
3. Impedance matrix

These are all separate and independent ways of describing the electrical environment seen by the common and differential signals.

11.22 The Bottom Line

1. A differential pair is any two transmission lines.
2. Differential signaling has many signal-integrity advantages over single-ended signals, such as contributing to less rail collapse, less EMI, better noise immunity, and less sensitivity to attenuation.
3. Any signal on a differential pair can be described by a differential-signal component and a common-signal component. Each component will see a different impedance as it propagates down the pair.
4. The differential impedance is the impedance the differential signal sees.
5. A mode is a special state in which a differential pair operates. A voltage pattern that excites a mode will propagate down the line undistorted.
6. A differential pair can be fully and completely described by an odd-mode impedance, an even-mode impedance, and a time delay for the odd mode and for the even mode.
7. The odd-mode impedance is the impedance of one line when the pair is driven in the odd mode.
8. Forget the words *differential mode*. There is only *odd mode*, differential signal, and differential impedance.
9. Coupling between the lines in a pair will decrease the differential impedance.
10. The only reliable way of calculating the differential or common impedance is with an accurate 2D field solver.

11. Tighter coupling will decrease the differential cross talk picked up on a differential pair and will minimize the discontinuity the differential signal sees when crossing a gap in the return plane.

12. One of the most frequent sources of EMI is common signals getting onto an external twisted-pair cable. The way to reduce this is to minimize any asymmetries between the two lines in a differential pair and add a common-signal choke to the external cable.

13. All the fundamental information about the behavior of a differential pair is contained in the differential and common impedance. These can be more fundamentally described in terms of the even and odd modes, in terms of the capacitance and inductance matrices, or in terms of the characteristic impedance matrix.

End-of-Chapter Review Questions

11.1 What is a differential pair?

11.2 What is differential signaling?

11.3 What are three advantages of differential signaling over single-ended signaling?

11.4 Why does differential signaling reduce ground bounce?

11.5 List two misconceptions you had about differential pairs before learning about differential pairs.

11.6 What is the voltage swing in LVDSsignals? What is the differential signal?

11.7 Two single-ended signals are launched into a differential pair: 0 V to 1 V and 1 V to 0 V. Is this a differential signal? What is the differential signal component, and what is the common signal component?

11.8 List four key features of a robust differential pair.

11.9 What does differential impedance refer to? How is this different from the characteristic impedance of each line?

11.10 Suppose the single-ended characteristic impedance of two uncoupled transmission lines is 45 Ohms. What would a differential signal see when propagating on this pair? What common impedance would a signal see on this pair?

11.11 Is it possible to have a differential pair with the two signal lines traveling on opposites ends of a circuit board? What are three possible disadvantages of this?

11.12 As the two lines that make up a differential pair are brought closer together, the differential impedance decreases. Why?

11.13 As the two lines that make up a differential pair are brought closer together, the common impedance increases. Why?

11.14 What is the most effective way of calculating the differential impedance of a differential pair? Justify your answer.

11.15 Why is it a bad idea to use the terms differential mode impedance and common mode impedance?

11.16 What is odd-mode impedance, and how is it different from differential impedance?

11.17 What is even-mode impedance, and how is it different from common impedance?

11.18 If two lines in a differential pair are uncoupled and their single-ended impedance is 50 Ohms, what are their differential and common impedances? What happens to each impedance as the lines are brought closer together?

11.19 If two lines in a differential pair are uncoupled and their single-ended impedance is 37.5 Ohms, what are their odd mode and even mode impedances?

11.20 If the differential impedance in a tightly coupled differential pair is 100 Ohms, is the common impedance higher, lower, or the same as 25 Ohms?

11.21 In a tightly coupled differential microstrip pair, where is most of the return current?

11.22 List three different interconnect structures in which the return current of one line of a differential pair is carried by the other line.

11.23 In stripline, how does the speed of the differential signal compare to the speed of the common signal as the two lines are brought closer together? How is this different in stripline?

11.24 Why is it difficult to engineer a perfectly symmetrical differential pair in broadside-coupled stripline?

11.25 What happens to the odd mode loop inductance per length as the two lines in a microstrip differential pair are brought closer together? What about the loop inductance of the even mode? How is this different in microstrip?

11.26 If you only used a 100-Ohm differential resistor at the end of a 100-Ohm differential pair, what would the common signal do at the end of the line?

11.27 Why is on-die Vtt termination of each line separately an effective termination strategy for a differential pair? If each resistor is 50 Ohms and the differential impedance is 100 Ohms, what might be the worst-case reflection coefficient of the common signal if the two lines are tightly coupled?

CHAPTER 12

S-Parameters for Signal-Integrity Applications

A new revolution hit the signal-integrity field when signal bandwidths exceeded the 1 GHz mark. The rf world of radar and communications has been dealing with these frequencies and higher for more than 50 years. It is no wonder that many of the engineers entering the GHz regime of signal integrity came from the rf or microwave world and brought with them many of the analysis techniques common to this world. One of these techniques is the use of S-parameters.

TIP   Though S-parameters are a technique with origins in the frequency domain, some of the principles and formalism can also be applied to the time domain. They have become the new universal standard to describe the behavior of any interconnect.

12.1 S-Parameters, the New Universal Metric

In the world of signal integrity, S-parameters have been relabeled a *behavioral model* as they can be used as a way of describing the general behavior of any linear, passive interconnect, which includes all interconnects with the exception of some ferrites.

In general, a signal is incident to the interconnect as the stimulus and the behavior of the interconnect generates a response signal. Buried in the stimulus–response of the waveforms is the behavioral model of the interconnect.

The electrical behavior of every sort of interconnect, such as that shown in Figure 12-1, can be described by S-parameters. These interconnects include:

- Resistors
- Capacitors
- Circuit board traces
- Circuit board planes
- Backplanes
- Connectors
- Packages
- Sockets
- Cables

That's why S-parameters have become and will continue to be such a powerful formalism and in such wide use.

Figure 12-1   The formalism of S-parameters applies to all passive, linear interconnects such as the ones shown here.

12.2 What Are S-Parameters?

Fundamentally, a behavioral model describes how an interconnect interacts with a precision incident waveform. When describing in the frequency domain, the precision waveform is, of course, a sine wave. However, when describing the behavior in the time domain, the precision waveform can be a step edge or even an impulse waveform. As long as the waveform is well characterized, it can be used to create a behavioral model of the interconnect or the device under test (DUT).

In the frequency domain, where sine waves interact with the DUT, the behavioral model is described by the S-parameters. In the time domain, we use the labeling scheme of S-parameters but interpret the results differently.

TIP   Fundamentally, the S-parameters describe how precision waveforms, like sine waves, scatter from the ends of the interconnect. The term *S-parameters* is short for *scattering parameters*.

When a waveform is incident on an interconnect, it can scatter back from the interconnect, or it can scatter into another connection of the interconnect. This is illustrated in Figure 12-2.

For historical reasons, we use the term *scatter* when referring to how the waveform interacts. An incident signal can "scatter" off the front of the DUT, back into the source, or it can "scatter" into another connection. The "S" in S-parameters stands for *scattering*.

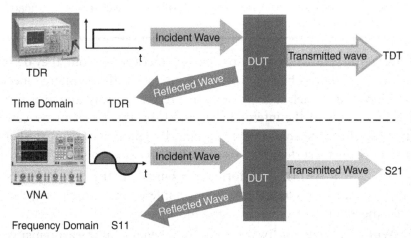

Figure 12-2   S-parameters are a formalism to describe how precision waveforms scatter from an interconnect or a device under test (DUT).

We also call the wave that scatters back to the source the *reflected wave* and the wave that scatters through the device the *transmitted wave*. When the scattered waveforms are measured in the time domain, the incident waveform is typically a step edge, and we refer to the reflected wave as the time-domain reflection (TDR) response. The instrument used to measure the TDR response is called a *time-domain reflectometer (TDR)*. The transmitted wave is the *time-domain transmitted (TDT)* wave.

In the frequency domain, the instrument used to measure the reflected and transmitted response of the sine waves is a *vector-network analyzer (VNA)*. *Vector* refers to the fact that both the magnitude and phase of the sine wave are being measured. On the complex plane, each measurement of magnitude and phase is a vector. A *scalar-network analyzer* measures just the amplitude of the sine wave, not its phase.

The frequency-domain reflected and transmitted terms are referred to as specific S-parameters, such as S11 and S21, or the return and insertion loss.

This formalism of describing the way precision waveforms interact with the interconnect can be applied to measurements or the output from simulations, as illustrated in Figure 12-3. All electromagnetic simulations, whether conducted in the time domain or the frequency domain, also use the S-parameter formalism.

> **TIP**   S-parameters, a formalism developed in the rf world of narrow-band carrier waves, has become the de facto standard format to describe the wide-bandwidth, high-frequency behavior of interconnects in signal-integrity applications.

Regardless of where the S-parameter values come from, they describe the way electrical signals behave when they interact with the interconnect. From this behavioral model, it is possible to predict the way any arbitrary signal might interact with the interconnect, and from this behavior we can predict output waveforms, such as an eye diagram. This process of using the behavioral model to predict a system response is called *emulation* or *simulation*.

There is a wealth of information buried in the S-parameters, which describe some of the characteristics of an interconnect, such as its impedance profile, the amount of cross talk, and the attenuation of a differential signal.

With the right software tool, the behavioral measurements of an interconnect can be used to fit a circuit topology–based model of an interconnect, such as a connector, a via, or an entire backplane. With an accurate circuit model that matches physical features with performance, it is possible to "hack" into the model and identify what physical features contribute to the limitations of the interconnect and suggest how to improve the design. This process is popularly called *hacking interconnects*.

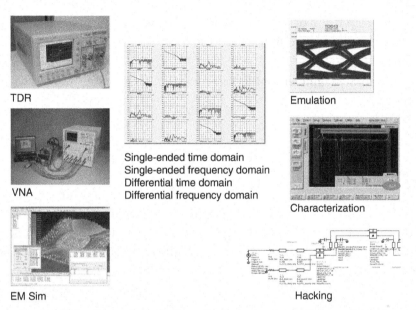

TDR

Emulation

Single-ended time domain
Single-ended frequency domain
Differential time domain
VNA   Differential frequency domain

Characterization

EM Sim   Hacking

Figure 12-3 Regardless of where the behavioral model comes from, it can be used to emulate system performance, characterize the interconnect, or hack into the performance limitations of the interconnect.

12.3 Basic S-Parameter Formalism

S-parameters describe the way an interconnect affects an incident signal. We use the term *port* to describe an end where signals enter or exit a device under test (DUT). Ports are connections for both the signal and return paths of the DUT. The easiest way of thinking about a port is as a small coaxial connection to the DUT.

Unless otherwise stated, the impedance the signal sees inside the connection leading up to the DUT is 50 Ohms. In principle, the port impedance can be made any value.

> **TIP**   S-parameters are confusing enough without also arbitrarily changing the port impedances. Unless there is a compelling reason to change them, the port impedances should be kept at 50 Ohms.

Each S-parameter is the ratio of a sine wave scattered from the DUT at a specific port to the sine wave incident to the DUT at a specific port.

For all linear, passive elements, the frequency of the scattered wave will be exactly the same as the incident wave. The only two qualities of the sine wave that can change are the amplitude and phase of the scattered wave.

To keep track of which port the sine wave enters and exits, we label the ports with consecutive index numbers and use these index numbers in each S-parameter.

Each S-parameter is the ratio of the output sine wave to the input sine wave:

$$S = \frac{\text{output sine wave}}{\text{input sine wave}} \tag{12-1}$$

The ratio of two sine waves is really two numbers. It is a magnitude that is the ratio of the amplitudes of the output to the input sine waves, and it is the phase difference between the output and the input sine waves. The magnitude of an S-parameter is the ratio of the amplitudes:

$$\text{mag(S)} = \frac{\text{amplitude(output sine wave)}}{\text{amplitude(input sine wave)}} \tag{12-2}$$

While the magnitude of each S-parameter is just a number from 0 to 1, it is often described in dB. As discussed in Chapter 11, the dB value *always* refers to the ratio of two powers. Since an S-parameter is the ratio of two voltage amplitudes, the dB value must relate to the ratio of the powers within the voltage amplitudes. This is why when translating between the dB value and the magnitude value, a factor of 20 is used:

$$S_{dB} = 20 \times \log\left(S_{mag}\right) \tag{12-3}$$

where:

S_{dB} = value of the magnitude, in dB

S_{mag} = value of the magnitude, as a number

The phase of the S-parameter is the phase difference between the output wave minus the input wave:

$$\text{phase(S)} = \text{phase(output sine wave)} - \text{phase(input sine wave)} \qquad (12\text{-}4)$$

As we will see, the order of the waveforms in the definition of the phase of the S-parameter will be important when determining the phase of the reflected or transmitted S-parameter and will contribute to a negative advancing phase.

TIP   While the port assignment of a DUT can be arbitrary, and there is no industry-standardized convention, there should be. If we are using the indices to label each S-parameter, then changing the port assignment index labels will change the meaning of specific S-parameters.

For multiple coupled transmission lines, such as a collection of differential channels, there is one port assignment scheme that is convenient and scalable and should be adopted as the industry standard. It is illustrated in Figure 12-4.

The ports are assigned as an odd port on the left end of a transmission line and the next higher number on the other end of the line. This way, as the number of coupled lines in the array increases, additional index numbers can be added by following this rule. This is a very convenient port assignment approach. S-parameters are confusing enough without creating more confusion with mix-ups in the port assignments.

TIP   You should always try to adopt a port assignment for multiple transmission lines that has port 1 connected to port 2 and port 3 adjacent to port 1, feeding into port 4. This is scalable to *n* additional transmission lines.

In order to distinguish to which combination of ports each S-parameter refers, two indices are used. The first index is the output port, and the second index is the input port.

For example, the S-parameter for the sine wave going into port 1 and coming out port 2 would be S21. This is exactly the opposite of what you would expect. It would be logical to use the first index as the going-in port and the second number as the coming-out port. However, part of the mathematical formalism for S-parameters requires this reverse-from-the-logical convention. This is related to the matrix math that is the real power behind S-parameters. By using the reverse notation, the S-parameter matrix can transform an array of stimulus-voltage vectors, represented by the vector a_j, into an array of response-voltage vectors, b_k:

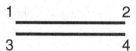

Figure 12-4 Recommended port assignment labeling scheme for multiple transmission-line interconnects. The pattern continues as additional lines are added.

$$b_k = S_{kj} \times a_j \qquad (12\text{-}5)$$

Using this formalism, the transformation of a sine wave into and out of each port can be defined as a different S-parameter, with a different pair of index values. The definition of each S-parameter element is:

$$S_{kj} = \frac{\text{sine wave out port k}}{\text{sine wave into port j}} \qquad (12\text{-}6)$$

This basic definition applies no matter what the internal structure of the DUT may be, as illustrated in Figure 12-5. A signal going into port 1 and coming out port 1 would be labeled S11. A signal going into port 1 and coming out port 2 would be labeled S21. Likewise, a signal going into port 2 and coming out port 1 would be labeled S12.

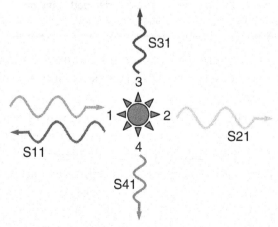

Figure 12-5 Index labeling definition for each S-parameter.

12.4 S-Parameter Matrix Elements

With only one port on a DUT, there is only one S-parameter, which would be identified as S11. This one element may have many data points associated with it for many different frequency values. At any single frequency, S11 will be complex, so it will really be two numbers. It could be described by a magnitude and phase or by a real component and an imaginary component. This single S-parameter value, at one frequency, could be plotted on either a polar plot or a Cartesian plot.

In addition, S11 may have different values at different frequencies. To describe the frequency behavior of S11, the magnitude and phase could be plotted at each frequency. An example of the measured S11 from a short-length transmission line, open at the far end, is shown in Figure 12-6.

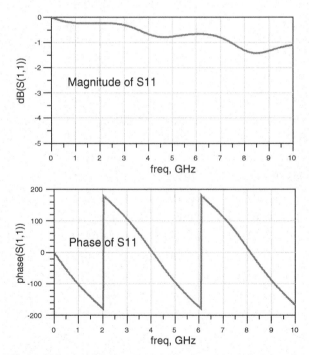

Figure 12-6 Measured S11 as a magnitude (top) and phase (bottom) for a transmission line, open at the far end. Connection was made to both the signal and return path at port 1. Measured with an Agilent N5230 VNA and displayed with Agilent's ADS.

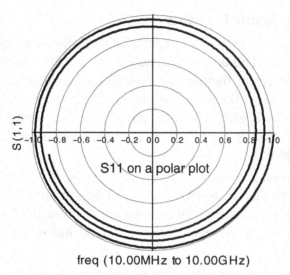

freq (10.00MHz to 10.00GHz)

Figure 12-7   The same measured S11 data as in Figure 12-6 but replotted in a polar plot. The radius position is the magnitude, and the angular position is the phase of S11.

Alternatively, the S11 value at each frequency can be plotted in a polar plot. The radial position of each point is the magnitude of the S-parameter, and the angle from the real axis is the phase of S11. The positive direction of the angle is always in the counterclockwise (CCW) direction. An example of the same measured S11 but plotted in a polar plot is shown in Figure 12-7.

The information is exactly the same in each case but just looks different, depending on how it is displayed. When displayed in a polar plot, it is difficult to determine the frequency values of each point unless a marker is used.

A two-port device would have four possible S-parameters. From port 1, there could be a signal coming out at port 1 and coming out at port 2. The same could happen with a signal going into port 2. The S-parameters associated with the two-port device can be grouped into a simple matrix:

$$\begin{matrix} S11 & S12 \\ S21 & S22 \end{matrix}$$

(12-7)

In general, if the interconnect is not physically symmetrical when looking in from each end, S11 will not equal S22. However, for all linear, passive devices,

S21 = S12—always. There are only three unique terms in the four-element
S-parameter matrix. An example of the measured two-port S-parameters of a
stripline transmission line is shown in Figure 12-8.

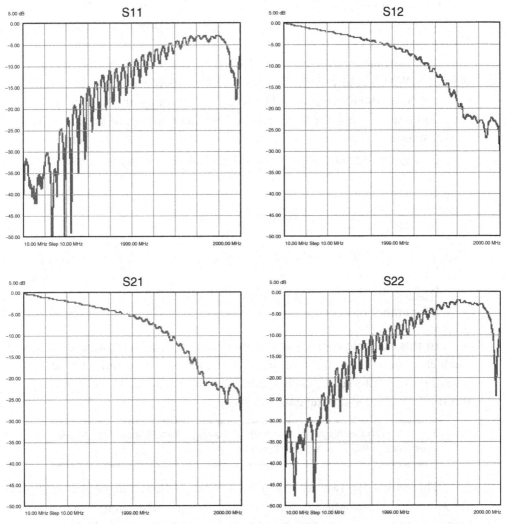

Figure 12-8 Measured two-port S-parameters of a simple 5-inch-long nearly
50-Ohm stripline transmission line, from 10 MHz to 20 GHz. Shown here are only
the magnitude values. There are also phase values for each matrix element and
at each frequency point, not shown. Measured with an Agilent N5230 VNA and
displayed with Agilent's PLTS.

This example illustrates part of the power behind the S-parameter formalism. So simple a term as the S-parameter matrix represents a tremendous amount of data. Each of the three unique elements of the 2×2 matrix has a magnitude and phase at each frequency value of the measurement, from 10 MHz to 20 GHz, at 10-MHz intervals. This is a total of $2000 \times 2 \times 3 = 12,000$ specific, unique data points, all neatly and conveniently catalog by the S-parameter matrix elements.

This formalism can be extended to include an unlimited number of elements. With 12 different ports, for example, there would be $12 \times 12 = 144$ different S-parameter elements. However, not all of them are unique. For an arbitrary interconnect, the diagonal elements are unique, and the lower half of the off-diagonal elements is unique. This is a total of 78 unique terms.

In general, the number of unique S-parameter elements is given by:

$$N_{unique} = \frac{n(n+1)}{2} \qquad (12\text{-}8)$$

where:

N_{unique} = number of unique S-parameter elements
n = number of ports

In this example of 12 ports, there are $(12 \times 13)/2 = 78$ unique elements. Each element has two different sets of data: a magnitude and a phase. This is a total of 156 different plots. If there are 1000 frequency values per plot, this is a total of 156,000 unique data points.

TIP   With this huge amount of data, having a simple formalism to just keep track of the data is critically important.

There are really two aspects of S-parameters. First and most important is the analytical information contained in the S-parameter matrix elements. But there is also information that can be read from the patterns the S-parameter values make when plotted either in polar or Cartesian coordinates. A skilled eye can pick out important properties of an interconnect just from the pattern of the plots.

These S-parameter matrix elements and the data each contains really specify the precise behavior of the interconnect. Everything you ever wanted to know about the behavior of an interconnect is contained in its S-parameter matrix elements. Each S-parameter matrix element tells a different story about the 12-port device. The analytical information these elements contain can be immediately accessed with a variety of simulation tools.

12.5 Introducing the Return and Insertion Loss

In a two-port device, there are three unique S-parameters: S11, S22, and S21. Each of these matrix elements contains complex numbers that may vary with frequency.

The S11 term is also called the *reflection coefficient*. The S21 term is also called the *transmission coefficient*.

For historical reasons, the absolute value of the magnitude of the S11, in dB, is called the *return loss*, and the absolute value of the magnitude of S21, in dB, is called the *insertion loss*. For example, if S11 = –40 dB, the return loss is 40 dB. If S21 = –15 dB, the insertion loss is 15 dB.

The historical origin of the terms *return* and *insertion loss* are based on how these terms were measured before the days of modern VNAs. A fixture would be prepared that could be separated to insert the device under test (DUT). When closed, and with nothing between the two ports, the received signal at port 2 would be measured. The fixture would be separated and the DUT inserted. The loss in what was measured for the through reference, when the DUT was insert, was called the *insertion loss*—the loss in the transmitted signal when the DUT was inserted.

A large value for insertion loss means much less signal gets to port 2, and the interconnect is not very transparent. Since insertion loss is considered a "loss," a larger value means there is more loss when the DUT is inserted, and less gets through.

Return loss is measured by first opening the fixture so that port 1 sees an open. The signal reflected from the open is the reference. After the DUT is inserted in the fixture, connecting between port 1 and port 2, the returned signal is measured. The loss in the returned signal, compared with the open, is the return loss.

The better matched the DUT and fixture, the less reflects and the more loss in the reflected signal compared with the reference open. A large return loss means a good match compared to an open. A small return loss means a lot of signal is reflecting, and it is looking like a really big impedance mismatch from the 50-Ohm port, so that it looks more like an open or a short.

Even though these terms were introduced long before VNAs, they are still used when referring to S11 and S21, though they carry a lot of confusion. Many of us in the industry have adapted the bad habit of using *return loss* as a substitute for S11 and the *insertion loss* as a substitute for S21. This is not technically correct, even though it's incredibly convenient.

When insertion loss *increases*, for example, and there is less received signal, the transmission coefficient *decreases*, and S21 *decreases*. The S21 value becomes a larger, more negative dB value. When we say the insertion loss is *increasing*, do we mean there is a *decreasing* transmission coefficient or an *increasing* transmission coefficient? Likewise, for return loss. When return loss is large, this means there is very little signal reflected, and the DUT is a good match to the fixture. If return loss *increases*, does this mean the reflection coefficient *increases* or the reflection coefficient *decreases*?

Some in the industry advocate changing the definition of *return* and *insertion loss* and have them mean precisely the magnitude of S11 and S21, in dB. This would remove the ambiguity. In this context, a *larger* return loss means a *larger* reflection coefficient and a *larger* S11 and more signal reflects. A *smaller* insertion loss means a *smaller* transmission coefficient and a *smaller* S21 and fewer signal transmits. While technical, if not historically, correct, this usage dramatically reduces the confusion associated with S-parameters. Given how many confusing aspects there are, this is not a bad idea.

> **TIP**   Throughout this book, we adopt the growing consensus definition that the insertion loss is the same as S21 and the return loss is the same as S11. Although doing so removes one source of confusion, it introduces another.

> **TIP**   It is always unambiguous to refer to S11 as the reflection coefficient and S21 as the transmission coefficient. A more transparent interconnect will have a smaller reflection coefficient and a larger transmission coefficient. This is exactly the opposite directions from the historically correct return and insertion loss terms.

When the interconnect is symmetrical from one end to the other, the return losses, S11 and S22, are equal. In an asymmetric two-port interconnect, S11 and S22 will be different.

In general, calculating by hand the return and insertion loss from an interconnect line is complicated. It depends on the impedance profile and time delays of each transmission-line segment that makes up the interconnect and the frequency of the sine waves.

In the frequency domain, any response to the sine waves is at steady state. The sine wave is on for a long time, and the total reflected or transmitted response is observed. This is an important distinction between the frequency domain and the time domain.

In the time domain, the instantaneous voltage reflected or transmitted by the transmission line is observed. The reflected response, the TDR response, can map out the spatial impedance profile of the interconnect by looking at when a reflection occurred and where the exciting edge must have been when the reflection happened. Of course, after the first reflection, an accurate interpretation of the impedance profile can be obtained only by post processing the reflected signal, but a good first-order estimate can always be read directly from the TDR response.

In the frequency domain, the spatial information is intermixed throughout the frequency-domain data and not directly displayed. Consider an interconnect with many impedance discontinuities down its length, as shown in Figure 12-9.

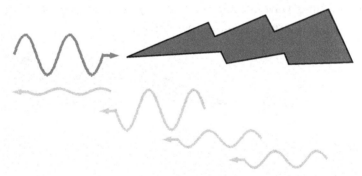

Figure 12-9 Each arbitrary impedance discontinuity in the irregularly shaped interconnect causes a reflected sine wave back to the incident port, each with a different magnitude and phase.

As the incident sine wave encounters each discontinuity, there will be a reflection, and some of the sine wave will head back to the port. Some of the reflected waves will bounce multiple times between the discontinuities until they either are absorbed or eventually make it out to one of the ports where they are recorded.

What is seen as the reflected signal from port 1, for example, is the combination of all the reflected sine waves from all the possible discontinuities.

For an interconnect 1 meter long, the typical time for a reflection down and back is about 6 nsec. In the 1 msec, a typical time for a VNA to perform one frequency measurement, a sine wave could complete more than 100,000 bounces, far more than would ever occur.

> **TIP**   For any single frequency, the reflected signal or transmitted signal is a steady-state value. It represents all the possible combinations of reflections at all the different impedance interfaces. This is very different from the behavior in the time domain.

If the frequency incident to the port is fixed during the 1-msec measurement or simulation time, the frequency of every wave reflecting back from every discontinuity will be exactly the same frequency. However, the amplitude and phase of each resulting wave exiting the interconnect at each port will be different.

Coming out of each port will be a large number of sine waves, all with the same frequency but with an arbitrary combination of amplitudes and phases, as illustrated in Figure 12-10. Surprisingly, when we add together an arbitrary number of sine waves, each with the same frequency but with arbitrary amplitude and phase, we get another sine wave.

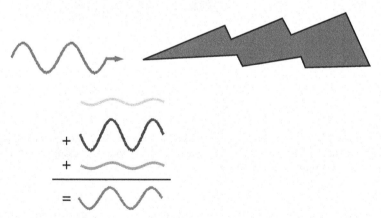

Figure 12-10   The sum of a large number of sine waves, all with the same frequency but different amplitudes and phase, is another sine wave.

When a sine wave is incident into one of the ports, the resulting signal that comes back out one of the ports is also a sine wave of exactly the same frequency but with a different amplitude and phase. This is what is captured in the S-parameters. Unfortunately, other than in a few special cases, the behavior of the S-parameters is a complicated function of the impedance profile and sine wave frequency. It is not possible to take the measured magnitude and phase of any S-parameter and back out each individual sine wave element that created it.

In general, other than in a few simple cases, it is not possible to calculate the S-parameters of an interconnect by hand with pencil and paper. A simulator must be used. Though there are a number of sophisticated, commercially available simulators that can simulate the S-parameters of arbitrary structures, any SPICE simulator can calculate the return and insertion loss of any arbitrary structure with a simple circuit, as illustrated in Figure 12-11. An example of an interconnect with a few discontinuities is also shown.

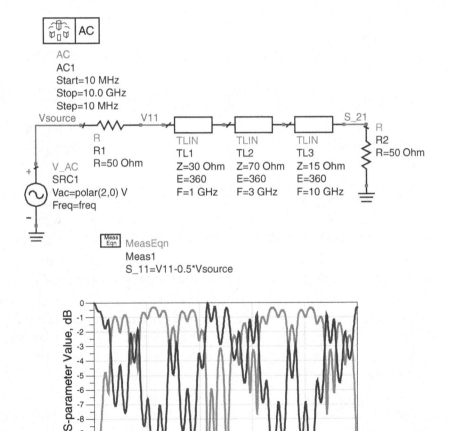

Figure 12-11   SPICE circuit to calculate the S11 and S21 of any interconnect circuit between the ports. Circuit set up in Agilent's ADS version of SPICE.

While any arbitrary interconnect can be simulated, there are a few important patterns in the S-parameters that are indicative of specific features in the interconnect. Learning to recognize some of these patterns will enable the trained observer to immediately translate S-parameters into useful information about the interconnect.

One important pattern to recognize is that of a *transparent interconnect*. The pattern of the return and insertion loss can immediately indicate the "quality" of the interconnect as "good" or "bad." In this context, we assume that "good" means the interconnect is transparent to signals, and "bad" means it is not transparent to signals.

12.6 A Transparent Interconnect

There are three important features of a transparent interconnect:

1. The instantaneous impedance down the length matches the impedance of the environment in which it is embedded.
2. The losses through the interconnect are low, and most of the signal is transmitted.
3. There is negligible coupling to adjacent traces.

These three features are clearly displayed at a glance in the reflected and transmitted signals, which correspond to the S11 and S21 terms, when the ports are attached to opposite ends of the interconnect, labeled port 1 and port 2.

Examples of the port configuration and measured return and insertion loss for a nearly transparent interconnect are shown in Figure 12-12.

When the impedance throughout the interconnect closely matches the port impedances, very little incident signal reflects, and the reflected term, S11, is small. When displayed in dB, a smaller return loss is a larger, more negative dB value. The 50-Ohm impedance of port 2 effectively terminates the interconnect.

Of course, in practice, it is almost impossible to achieve a perfect match to 50 Ohms over a large bandwidth. Typically, the measured reflection coefficient of an interconnect gets larger at higher frequency, as shown in Figure 12-12 as a smaller negative dB value at higher frequency.

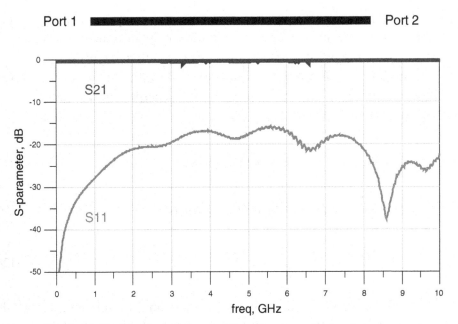

Figure 12-12 Measured return and insertion loss of a nearly transparent interconnect. Measured with an Agilent N5230 VNA and displayed with Agilent's ADS.

TIP When displayed as a dB value, a transparent interconnect will have a large negative dB reflection coefficient and a small S11. Using the modern term, we would say the return loss is small.

The worse the interconnect and the larger the impedance mismatch to the port impedances, the closer the return loss will be to 0 dB, corresponding to 100% reflection.

The insertion loss is a measure of the signal that transmits through the device and out port 2. The larger the impedance mismatch, the less transmitted signal. However, when there is close to a good match, the insertion loss is very nearly 0 dB and is insensitive to impedance variations.

There is a specific connection between the reflection coefficient and the transmission coefficient. Always keep in mind that the S-parameters are ratios of voltages. There is no law of conservation of voltage, but there is a law of conservation of energy.

If the interconnect is low loss, and there is no coupling to adjacent traces, and there is no radiated emissions, then the energy into the interconnect must be the sum of the reflected energy and the transmitted energy.

The energy in a sine wave is proportional to the square of its amplitude. This condition of energy in being equal to the energy reflected plus the energy transmitted is described by:

$$1 = S11^2 + S21^2 \tag{12-9}$$

Given the return loss, the insertion loss is:

$$S21 = \sqrt{1 - S11^2} \tag{12-10}$$

For example, if the impedance somewhere on the interconnect is 60 Ohms in an otherwise 50-Ohm environment, the worst-case return loss would be:

$$S11 = \frac{(60-50)}{(60+50)} = \frac{10}{110} = 0.091 = -21\,\text{dB} \tag{12-11}$$

The impact on insertion loss is:

$$S21 = \sqrt{1 - 0.091^2} = \sqrt{0.992} = 0.996 = -0.04\,\text{dB} \tag{12-12}$$

Even though the return loss is as high as −20 dB, the impact on insertion loss is very small and indistinguishable from 0 dB. Figure 12-13 illustrates this for a wide range of return loss values.

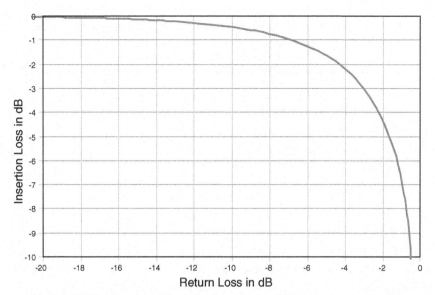

Figure 12-13 Impact on the insertion loss for different return losses. The return loss must be as high as −10 dB for there to be even a 0.5-dB impact on the insertion loss if there are no losses in the interconnect.

T I P   Only for a return loss higher than −10 dB will there be a noticeable impact on the insertion loss.

12.7 Changing the Port Impedance

The industry-standard port impedance is 50 Ohms. However, in principle, it can be made any value. As the port impedance changes, the behavior of the displayed return and insertion loss changes. To first order, moving the port impedance farther from the interconnect's characteristic impedance will increase the return loss. Other than this simple pattern, the specific values of the return loss and insertion loss are complicated functions of the port impedance.

Given the S-parameters at one port impedance, the S-parameters with any other port impedance can be calculated using matrix math. They can also be calculated directly in a SPICE circuit simulation.

TIP   To describe the S-parameters of an interconnect, it is not necessary
to match the port impedance to the impedance of the device. Unless there
is a compelling reason otherwise, 50 Ohms should always be used.

Regardless of the port impedance, an analytical analysis of each
S-parameter element can be equally well performed. The only practical reason
to switch the port impedance from 50 Ohms is to be able to qualitatively
evaluate, from the front screen, the quality of the device in a non-50-Ohm
environment.

If the device, like a connector or cable, is designed for a non-50-Ohm
application, like the 75-Ohm environment of cable TV applications, its return
loss will look "bad" when displayed with 50-Ohm port impedances.

The reflections from the impedance mismatches at the ends will cause
ripples in the return loss. The excessive return loss will show ripples in the
insertion loss. To the trained eye, the behavior will look complicated and will be
difficult to interpret, other than that the impedance is way off from 50 Ohms.

But if the application environment were 75 Ohms, the interconnect might
be acceptable. By changing the port impedance to 75 Ohms, the application
impedance, the behavior of the device in this application environment can be
visually evaluated right from the front screen.

An example of the return loss and insertion loss of a nominally 75-Ohm
transmission line with 50-Ohm connectors attached is shown in Figure 12-14
for the case of 50-Ohm and 75-Ohm port impedances.

In a 75-Ohm environment, the 75-Ohm cable looks nearly transparent at
low frequency, below about 1 GHz. Above 1 GHz, the connectors dominate
the behavior, and the interconnect is not very transparent, no matter what the
port impedance. The behavioral model contained in the measured S-parameter
response is exactly the same, regardless of the port impedances. It is merely
redisplayed for different port impedances.

S-parameters are confusing enough without also leaving the port
impedance ambiguous. Of course, when the S-parameters are stored in the
industry-standard touchstone file format, the port impedances of each port
to which the data is referenced are specifically called out at the top of the file.
From one touchstone file, the S-parameters for any port impedance can easily
be calculated.

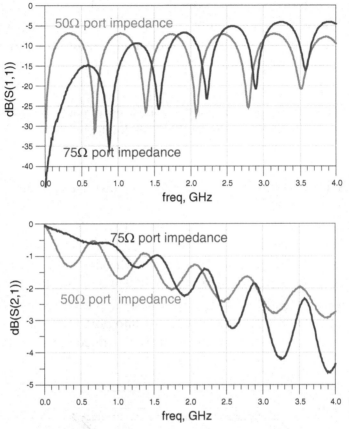

Figure 12-14 Measured return and insertion loss for a 75-Ohm transmission line with 50-Ohm connectors using port impedances of 50 Ohms and 75 Ohms. Measured with an Agilent N5230 VNA and displayed with Agilent's ADS.

12.8 The Phase of S21 for a Uniform 50-Ohm Transmission Line

The simplest interconnect to evaluate is when the impedance of the line is 50 Ohms, matched to the impedance of the ports. In this case, there are no reflections, and the magnitude of S11 is 0. In dB, this is a large, negative dB, usually limited by the noise floor of the instrument or simulator, on the order of −100 dB.

All of the sine wave would be transmitted so the magnitude of S21 would be 1, which is 0 dB at each frequency. The phase of S21 would vary depending on the time delay of the transmission line and the frequency. The behavior of the phase is one of the most subtle aspects of S-parameters.

The definition of the S-parameters is that each matrix element is the ratio of the sine wave that comes out of a port to the sine wave that goes into a port.

For S21, the ratio of the sine wave coming out of port 2 to the sine wave going into port 1, we get two terms:

$$\text{mag(S21)} = \frac{\text{sine wave amplitude out of port 2}}{\text{sine wave amplitude intoport 1}} \qquad (12\text{-}13)$$

$$\text{phase(S21)} = \text{phase(sine wave out of port 2)} - \text{phase(sine wave into port 1)} \quad (12\text{-}14)$$

When we send a sine wave into port 1, it doesn't come out of port 2 until a time delay, TD, later. If the phase is 0 degrees when the sine wave went into port 1, it will also have a phase of 0 degrees when it comes out of port 2. It has merely been transmitted from one end of the line to the other.

However, when we compare the phase of the sine wave coming out of port 2 to the phase of the sine wave going into port 1, it is the phases that are present at the same instant of time. This is illustrated in Figure 12-15.

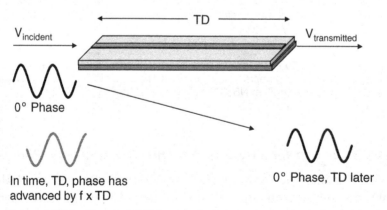

Figure 12-15 Phase of S21 is negative due to the incident phase advancing while the sine wave is transmitting through the transmission line.

When we see the 0-degree phase sine wave coming out of port 2 and then immediately look at the phase of the wave going into port 1, we are looking not at the 0-degree phase when the sine wave entered the transmission line TD nsec ago but at the current phase of the sine wave entering port 1 now.

During the time the 0-degree wavefront has been propagating through the transmission line, the phase of the sine wave entering port 1 has advanced. The phase of the sine wave entering port 1 is now $f \times TD$.

When we calculate the phase of S21 as the difference between the phase coming out of port 2 minus the phase going into port 1, the phase coming out of port 2 may be 0 degrees, but the phase going into port 1 *now* has advanced to $f \times TD$. This means the phase of S21 is:

$$phase(S21) = 0° - f \times TD \qquad (12\text{-}15)$$

where:

f = frequency of the sine wave going into port 1

TD = time delay of the transmission line

TIP   The phase of S21 will start out negative and will become increasingly negative with frequency. This is the most bizarre and confusing aspect of S-parameters: The phase of S21 through a transmission line is increasingly negative.

We see that this behavior is based on two features. First, the definition of the phase of S21 is that it is the phase of the sine wave coming out of port 2 minus the phase of the sine wave going into port 1. Second, it is the difference in phase between the two waves at the same instant in time. The phase of S21 will always be negative and will become increasingly negative as frequency increases. An example of the measured phase of S21 for a stripline transmission line is shown in Figure 12-16.

At low frequency, the phase of S21 starts out very nearly 0 degrees. After all, if the TD is small compared to a cycle, then the phase does not advance very far during the transit time of the wave through the interconnect. As frequency increases, the number of cycles the sine wave advances during the transit time increases. Since the phase of S21 is the negative of the incident phase, as the incident phase advances, the phase of S21 gets more and more negative.

Figure 12-16   Measured phase of S21 for a uniform, 50-Ohm stripline, about 5 inches long, measured from 10 MHz to 5 GHz. Measured with an Agilent N5230 VNA and displayed with Agilent's PLTS.

We usually count phase from −180 degrees to +180 degrees. When the phase advances to −180, it is reset to +180 and continues to count down. This gives rise to the typical sawtooth pattern for the phase of S21.

12.9 The Magnitude of S21 for a Uniform Transmission Line

The magnitude of the insertion loss is a measure of all the processes that prevent the energy being transmitted through the interconnect. The total energy flowing into an interconnect must equal the total energy coming out. There are five ways energy can come out:

1. Radiated emissions
2. Losses in the interconnect turning into heat
3. Energy coupled into adjacent traces, whether measured or not
4. Energy reflected back to the source
5. Energy transmitted into port 2 and measured as part of S21

In most applications, the impact on S21 from radiated losses is negligible. Though radiated emissions are an important factor in causing FCC certification test failures, the amount of energy typically radiated is so small a fraction of the signal as to be difficult to detect in S21.

The losses in the interconnect that turn into heat are accounted for by the conductor loss and dielectric loss. As shown in Chapter 11, both of these effects increase monotonically with frequency. When all other mechanisms that affect S21 are eliminated, when measured in dB, the insertion loss is a direct measure of the attenuation in the line and will increase as a more negative dB value with increasing frequency. Figure 12-17 shows the measured insertion loss of a stripline transmission line about 5 inches long.

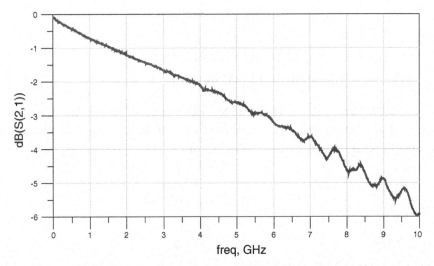

Figure 12-17 Measured insertion loss of 5-inch-long stripline. Measured with an Agilent N5230 VNA and displayed with Agilent ADS.

There is often some ambiguity around the sign of the attenuation. Usually, when describing the attenuation, the sign chosen is positive because a larger attenuation should be a larger number. In this formalism, attenuation is not the same as the transmission coefficient. There is a sign difference between them. If the only energy loss mechanism is from attenuation, the historical term *insertion loss* would be the same as *attenuation*.

The transmission coefficient is the negative of the attenuation. When attenuation is described by the insertion loss and measured in dB, it is critically important to be absolutely consistent and always make the insertion loss the correct—negative—sign.

TIP   Insertion loss is a direct measure of the attenuation from both the conductor and dielectric losses. Given the conductor and dielectric properties of the transmission line, as described in Chapter 9, the insertion loss can be easily calculated.

The insertion loss due to just the losses is:

$$S21 = -(A_{diel} + A_{cond}) \qquad (12\text{-}16)$$

where:

S21 = insertion loss, in dB

A_{diel} = attenuation from the dielectric loss, in dB

A_{cond} = attenuation from the conductor loss, in dB

When there are no impedance discontinuities in the transmission lines, and conductor loss is small compared to the dielectric loss, the insertion loss is a direct measure of the dissipation factor:

$$S21 = -2.3 \times f \times Df \times \sqrt{Df} \times Len \qquad (12\text{-}17)$$

where:

S21 = insertion loss, in dB

f = frequency, in GHz

Df = dissipation factor

Dk = dielectric constant

Len = interconnect length, in inches

A measurement of the insertion loss, when scaled appropriately, is a direct measure of the dissipation factor of the laminate material:

$$Df = \frac{-S21 \,[\text{in dB}]}{2.3 \times \sqrt{Dk} \times Len \times f} \qquad (12\text{-}18)$$

where:

S21 = insertion loss, in dB

f = frequency, in GHz

Df = dissipation factor

Dk = dielectric constant

Len = interconnect length, in inches

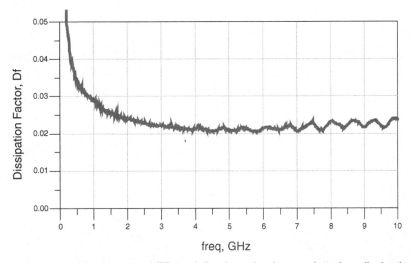

Figure 12-18 Measured FR4 stripline insertion loss replotted as dissipation factor, showing the value of about 0.022. The higher values at low frequency are due to conductor losses not taken into account. Measured with an Agilent N5230 VNA and displayed with Agilent ADS.

When the dielectric constant is close to 4, which is typical for most FR4 type materials, the insertion loss is approximated as $-5 \times$ Df dB/inch/GHz. Figure 12-18 shows the same measured insertion loss for the stripline case above but scaled by 5 and by the length and frequency to plot the dissipation factor directly.

At low frequency, conductor loss can be a significant contributor to loss and is not taken into account in this simple analysis of insertion loss. When the approximation above is used and all the attenuation is assumed to be dielectric loss, the extracted dissipation may be artificially higher, compensating for the conductor loss. At higher frequency, the impedance discontinuities, typically from connectors or vias, cause ripples in the insertion loss.

While the slope of the insertion loss per length is a rough indication of the dissipation factor of an interconnect, it is only a rough indication. In practice, the discontinuities from impedance discontinuities of both the impedance of the line and from the connectors, as well as from contributions to the attenuation from the conductor losses will complicate the interpretation of the insertion loss. In general, for accurate measurements of the dissipation factor, these features must be taken into account.

However, the behavior of the insertion loss is a good indicator of dissipation factor of the materials. Figure 12-19 shows an example of the measured insertion loss for two nearly 50-Ohm transmission lines, each about 36 inches long. One is made from a Teflon substrate, while the other is made from FR5, a very lossy dielectric. The much larger slope of the FR5 substrate is an indication of this much higher dissipation factor.

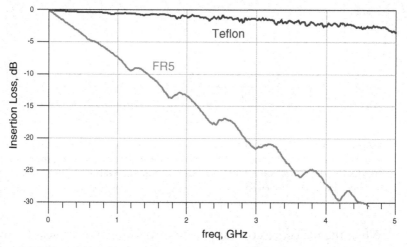

Figure 12-19 Measured FR5 and Teflon transmission lines, showing the impact on the insertion losses from the dissipation factors of the laminates. Each line is a 50-Ohm microstrip and 36 inches long. Measured with an Agilent N5230 VNA and displayed with Agilent ADS.

When plotted as a function of frequency, the insertion loss shows a monotonic drop in magnitude. At the same time, the phase is advancing at a constant rate in the negative direction. When S21 is plotted in a polar plot, the behavior is a spiral, as shown in Figure 12-20.

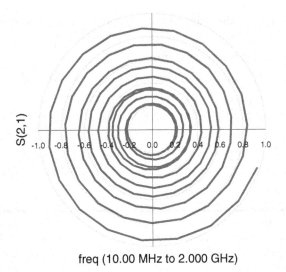

freq (10.00 MHz to 2.000 GHz)

Figure 12-20   Measured insertion loss of the 36-inch-long FR5 micro-strip from Figure 12-19 but plotted in a polar plot. The frequency range is 10 MHz to 2 GHz, measured every 10 MHz. Measured with an Agilent N5230 VNA and displayed with Agilent ADS.

The lowest frequency has the largest magnitude of S21 and a phase close to zero. As frequency increases, S21 rotates around in the clockwise direction, with smaller and smaller amplitude. It spirals into the center.

The spiral pattern of the insertion loss of a uniform transmission line, or the return loss when looking at the one-port measurement of a transmission line open at the end, is a pattern indicative of losses in the line that increase with frequency, which is typical of conductor and dielectric losses.

12.10 Coupling to Other Transmission Lines

Even if the cross-talk noise into adjacent lines is not measured, it will still happen, and its impact will be the magnitude reduction of S11 and S21. The simplest case is a uniform microstrip line with an adjacent microstrip closely coupled to it.

The insertion loss of an isolated microstrip will show the steady drop-off due to the dielectric and conductor loss. As an adjacent microstrip is brought closer, some of the signal in the driven microstrip will couple into the adjacent one, giving rise to near- and far-end cross talk. In microstrip, the far-end noise can be much higher than the near-end noise.

If the ends of the adjacent microstrip are also terminated at 50 Ohms, then any cross talk will effectively be absorbed by the terminations and not reflect in the victim microstrip line.

As frequency increases, the far-end cross talk will increase, and more signal will be coupled out of the driven line, reducing S21. In addition to the drop in S21 from attenuation, there will also be a drop in S21 from cross talk. As the two traces are brought closer together and coupling increases, more energy flows from the active line to the quiet line, and less signal makes it out as S21. Figure 12-21 shows that S21 will decrease with frequency and with coupling as the spacing between the lines decreases.

TIP   Just looking at the S21 response, it is difficult to separate out how much of S21 is due to attenuation and how much is due to coupling to other interconnects, unless the coupled signal into the other interconnects is also measured.

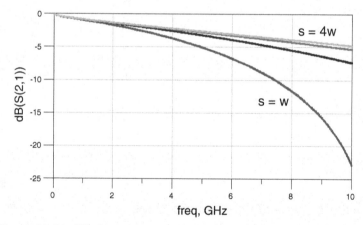

Figure 12-21   The insertion loss of one line in a microstrip pair. As the coupling increases, energy is coupled out of the line, and S21 decreases more than just due to attenuation. Simulated with Agilent's ADS.

Using the standard port-labeling convention, two adjacent microstrip lines can be described by four ports. The insertion loss of one line is S21, while the insertion loss of the other line is S43. The near-end noise is described by S31 and the far-end noise by S41. As the coupling increases and S21 drops, we can see the corresponding increase in the far-end noise, S41, as shown in Figure 12-22.

In this example, the reduction in magnitude of S21 from coupling increased with frequency but was relatively slowly varying. When the coupling is to an interconnect that is not terminated but is floating, the Q of the isolated floating line can be very high. If it were excited, the noise injected would rattle around between the open ends, attenuating slightly with each bounce. It could rattle around for as many as 100 bounces before eventually dying out from its own losses.

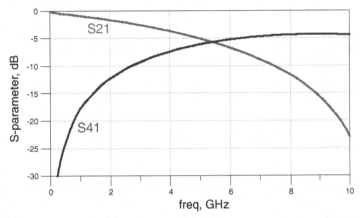

Figure 12-22 Insertion loss, S21, of one line in a microstrip pair and the far-end cross talk, S41, showing the insertion loss decreasing as the far-end noise increases. Simulated with Agilent ADS.

The impact from coupling to high-Q resonators is very narrow-band absorption in S21 or S11. Figure 12-23 shows the S21 of the same microstrip as above, but now the adjacent victim line is floating, open at each end. It has very narrow frequency resonances, which suck out energy in the active line in a very narrow frequency range at the line resonance frequency.

Coupling to high-Q resonators is also apparent when measuring the return loss in one-port configurations. In principle, when one port is connected to a transmission line with its other end open, S11 should have a magnitude of 1, or 0 dB. In practice, as we have seen, all interconnects have some loss, so S11 is always smaller than 0 dB, and it continues to get smaller as frequency increases.

If there is coupling between the transmission line being measured and adjacent transmission lines that are unterminated, the lines will act as high-Q resonators and show very narrow absorption lines.

TIP   Narrow, sharp dips in the return or insertion loss are almost always indications of coupling to high-Q resonant structures. When the resonating structure has a complex geometry, the multiple resonant frequency modes may be difficult to calculate without a full-wave field solver.

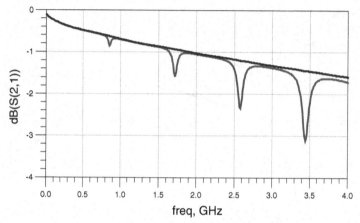

Figure 12-23   The insertion loss of a microstrip line when a floating adjacent microstrip line is very far away and when it is closely spaced. When close, the high-Q resonance of the floating line couples energy in narrow frequency bands. Simulated with Agilent ADS.

In the case of a single adjacent transmission line, the frequency of the dip is the resonant frequency of the quiet line. The width of the resonance is related to the Q of the resonator. By definition, the Q is given by:

$$Q = \frac{f_{res}}{FWHM} \tag{12-19}$$

where:

Q = value of the quality of the resonance

f_{res} = resonant frequency

FWHM = full width, half minimum (the frequency width across the dip at half the minimum value)

The higher the Q, the narrower the frequency dip. The depth of the dip is related to how tight the coupling is. With a larger coupling, the dip will be deeper. Of course, as the dip gets deeper, the coupling to the floating line increases, and its damping generally increases, causing the Q to decrease.

The resonator to which a signal line couples does not have to be another uniform transmission line but can be the cavity made up of two or more adjacent planes. For example, when a signal transitions from one layer to another and its return plane also changes, the return current, transitioning between the return planes, can couple into the cavity, creating high-Q resonant couplings. An example of this layer transition in a four-layer board is shown in Figure 12-24.

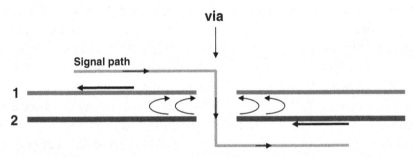

Figure 12-24 A signal transitions two reference planes, layer 1 and layer 2. The return current, flowing between the two planes, couples into the plane cavity resonance.

The resonant frequency of a cavity composed of two planes is the frequency where an integral number of half-wavelengths can fit between the two open ends of the cavity. It is given by:

$$f_{res} = n\frac{11.8}{\sqrt{Dk}}\frac{1}{2 \times Len} = n\frac{2.95\,\text{GHz}}{Len} \qquad (12\text{-}20)$$

where:

f_{res} = resonant frequency, in GHz

Dk = dielectric constant of the laminate inside the cavity = 4

Len = length of a side of the cavity, in inches

n = index number of the mode

11.8 = speed of light in vacuum in inches/nsec

For example, when the length of a side is about 1 inch, the resonant frequency with an FR4 laminate begins at about 3 GHz and increases from there for higher modes. In real cavities, the resonant frequency spectrum can be more complex due to cutouts in the planes or rectangular shapes.

In typical board-level applications, the length of a side can easily be 10 inches, with a resonant frequency starting at about 300 MHz. In this range, the decoupling capacitors between power and ground planes can sometimes suppress resonances or shift them to higher frequency, and the resonances may not be clearly visible.

However, in multilayer packages, typically on the order of 1 inch on a side, the first resonances are in the GHz range, and the decoupling capacitors are not effective at suppressing these modes since their impedance at 1 GHz is high compared to the plane's impedance. When a signal line transitions from the top layer to a bottom layer, going through the plane cavity, the return current will excite resonances in the GHz range.

This can easily be measured by using a one-port network analyzer. Figure 12-25 shows the measured return loss of six different leads in a four-layer BGA. In each case, the signal line is measured at the ball end, and the cavity end—where the die would be—is left open. Normally, the return loss should be 0 dB, but at the resonant frequencies of the cavity, a considerable amount of energy is absorbed.

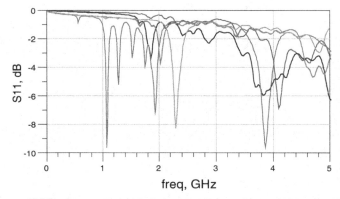

Figure 12-25 The measured return loss of six different leads in a BGA package, each open at the far end. The dips starting at 1 GHz are resonances in the cavity formed by the power and ground planes, as well as coupling into adjacent signal lines. Measured with an Agilent N5230 VNA and displayed with Agilent ADS.

In this example, the return loss drops off very slowly from 0 dB as frequency increases due to the dielectric loss in the laminate. Above about 1 GHz, there are very large, narrow bandwidth dips. These are the absorptions from coupling into the power and ground cavity of the package.

If the return loss from a reflection at the open source is −10 dB, which is from a round-trip path, the drop in the one-way insertion loss might be on the order of −5 dB. When the magnitude is −5 dB, this means that only 50% of the signal amplitude would get through. This is a huge reduction in signal strength and would result in distortions of the signal as well as excessive cross talk between channels.

The return loss is really a measure of the absorption spectrum of the package, similar to an infrared absorption spectrum of an organic molecule. Infrared spectroscopy identifies the distinctive resonant modes of specific atomic bonds. In the same way, these high-Q package resonances identify specific resonant modes of the package or component.

Not only can the plane cavities absorb energy, but other adjacent traces can act as resonant absorbers. The resonant absorptions of a package will often set the limit to its highest usable frequency. One of the goals in package design is to push these resonances to higher frequency or reduce the coupling of critical signal lines to the resonant modes. This can be accomplished in a number of ways:

- Don't transition the signal between different return planes.
- Use return vias adjacent to each signal via to suppress the resonance.
- Use low-inductance decoupling capacitors to suppress the resonances.
- Keep the body size of the package very small.

12.11 Insertion Loss for Non-50-Ohm Transmission Lines

When the losses are small and coupling out of the line to adjacent lines is small, the dominant mechanism affecting the insertion loss is reflections from impedance discontinuities. The most common source of discontinuity is when the transmission line is different from the 50 Ohms of the ports. The impedance mismatch at the front of the line and the end of the line will give rise to a type of resonance, causing a distinctive pattern to the return and insertion loss.

Consider the case, as illustrated in Figure 12-26, of a short length of lossless transmission line. It has a characteristic impedance Z0, different from

50 Ohms, and a time delay, TD. The port impedances are 50 Ohms. As the sine wave hits the transmission line, there will be a reflection at the front interface, sending some of the incident sine wave back into port 1, contributing to the return loss. However, most of the incident sine wave will continue through the transmission line to port 2, where it will reflect from that interface.

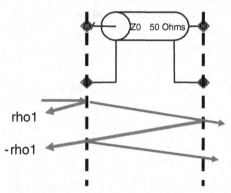

Figure 12-26 Multiple reflections from the interfaces through a non-50-Ohm, uniform transmission line.

The reflection coefficient for a signal traveling from port 1 into the transmission line, rho_1, is:

$$rho_1 = \frac{(Z_0 - 50\,\Omega)}{(Z_0 + 50\,\Omega)} \qquad (12\text{-}21)$$

Once the signal is in the transmission line, when it hits port 2 or port 1, the reflection coefficient back into the line, rho, will be:

$$rho = \frac{(50\,\Omega - Z_0)}{(Z_0 + 50\,\Omega)} = -rho_1 \qquad (12\text{-}22)$$

where:

rho_1 = reflection coefficient from port 1 to the transmission line

rho = reflection coefficient from the line into port 1 or port 2

Z_0 = characteristic impedance of the transmission line

Whatever the phase shift off the reflection from the first interface, the phase shift from the reflection off the second interface will be the opposite. The phase shift propagating a round-trip distance down the interconnect and back is:

$$\text{Phase} = 2 \times \text{TD} \times f \times 360 \text{ degrees} \qquad (12\text{-}23)$$

where:

Phase = phase shift of the reflected wave, in degrees

TD = time delay for one pass through the transmission line

f = sine wave frequency

If the round-trip phase shift is very small, the reflected wave from the end of the line when it enters port 1 will have nearly equal magnitude but opposite phase from the reflected signal off port 2. As the wave heads back into port 1, the signals will cancel, and the net reflected signal into port 1 will be 0.

TIP   At low frequency, the return loss of all lossless interconnects will always start at very little reflecting, or a very large negative dB value.

If nothing reflects back into port 1, then everything must be transmitted into port 2. and the insertion loss of all lossless interconnects will start at 0 dB. This is due to the second reflected wave back toward port 2 having the same phase as the first wave into port 2, and they both add.

TIP   At low frequency, the insertion loss of all lossless interconnects will always start at 0 dB.

As the frequency increases, the round-trip phase shift of the transmitted signal will increase until it is exactly half a cycle. At this point, the reflected signal from the front interface heading into port 1 will be exactly in phase with the reflected signal from the back interface, heading into port 1. They will add in phase, and the return loss will be a maximum. The return loss will be:

$$S11 \sim 2 \times \text{rho}_1 \qquad (12\text{-}24)$$

If S11 is a maximum, the insertion loss will be a minimum. When the round-trip phase shift is 180 degrees, the second reflected wave heading back to port 2 will be 180 degrees out of phase with the first wave heading into port 2, and they will partially cancel out.

As frequency increases, the round-trip phase will cycle between 0 and 180, causing the return and insertion losses to cycle between minimum and maximum values.

Dips in return loss and peaks in insertion loss will occur when the round-trip phase is a multiple of 360 degrees, or Phase = n × 360. This occurs when:

$$n \times 360 = 2 \times TD \times f \times 360 \text{ degrees} \qquad (12\text{-}25)$$

or:

$$f = \frac{n}{2} \times \frac{1}{TD} \qquad (12\text{-}26)$$

The longer the time delay, TD, the shorter the spacing between frequency intervals for a 180-degree phase shift. An example of the return and insertion loss for two 30-Ohm transmission lines, with a TD of 0.5 and 0.1 nsec, is shown in Figure 12-27. The frequency intervals between dips should be 1 GHz and 5 GHz, respectively.

A glance at the frequency interval between the ripples in the return or insertion loss can give a good indication of the physical length between the discontinuities of a transmission-line interconnect. The TD of the interconnect is roughly:

$$TD = \frac{1}{2 \times \Delta f} \qquad (12\text{-}27)$$

where:

TD = interconnect time delay

Δf = frequency interval between dips in the return loss or peaks in insertion loss

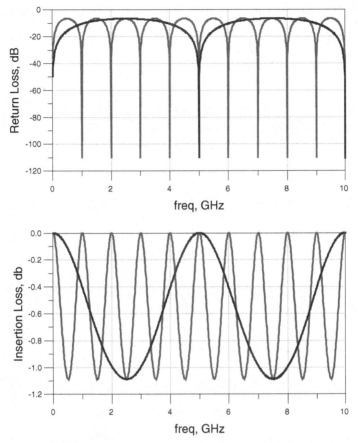

Figure 12-27 The return and insertion loss for two 30-Ohm, lossless, uniform transmission lines with TD of 0.5 nsec and 0.1 nsec. The longer the time delay, the shorter the interval between dips. Simulated with Agilent's ADS.

For example, in the return-loss plot at the top of Figure 12-27, the frequency spacing between dips in one transmission line is 1 GHz. This corresponds to a time delay of the interconnect of about $1/(2 \times 1) = 0.5$ nsec.

TIP The best way to make a transparent interconnect is first to match the interconnect impedance to 50 Ohms. If you can't make the impedance 50 Ohms, then the next most important design guide is to keep it short.

Interposers between a semiconductor package and a circuit board are designed to be transparent. When their impedance is far off from 50 Ohms, the design guideline to keep them transparent is to keep them short. This condition is that:

$$2 \times TD \times f \times 360 \text{ degrees} \ll 360 \text{ degrees} \qquad (12\text{-}28)$$

If the wiring delay of an interconnect is roughly 170 psec/inch, then the maximum length of a transparent interposer is given by:

$$2 \times Len \times 170 \text{ psec/inch} \times f_{max} \ll 1 \qquad (12\text{-}29)$$

or:

$$Len \ll \frac{3}{f_{max}} \quad \text{and} \quad f_{max} \ll \frac{3}{Len} \qquad (12\text{-}30)$$

If we translate the $\ll$ condition to be 10x, then this rough rule of thumb becomes:

$$Len \ll \frac{0.3}{f_{max}} \quad \text{and} \quad f_{max} \ll \frac{0.3}{Len} \qquad (12\text{-}31)$$

where:

Len = interposer length in inches

f_{max} = maximum usable frequency where the interconnect is still transparent, in GHz

For example, if the operating frequency is 1 GHz, an interposer or a connector should be shorter than about 0.3 inches to still be transparent. Likewise, if an interposer is 10 mils long, it could have a usable bandwidth of 30 GHz without any other special design conditions.

An example of a compliant interposer from Paricon, about 10 mils thick and with a 30-GHz bandwidth, is shown in Figure 12-28.

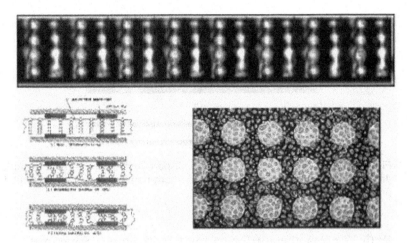

Figure 12-28   Cross section of the Pariposer interposer from Paricon, which is about 10 mils thick and has a bandwidth in excess of 30 GHz.

Of course, if the interconnect is designed as a controlled-impedance path with a characteristic impedance near 50 Ohms, it can have a much higher bandwidth and can be much longer.

12.12 Data-Mining S-Parameters

The interpretation of the S-parameter elements depends on the port assignments. For example, a semiconductor package could have 12 different traces, one end of each trace connected to a port, and every trace open at its far end. Or there could be one net on a board with a fanout of 11 and a port at each end of the single net. Or the port assignment could be as shown in Figure 12-29, with six different straight-through interconnects in some proximity. These six different transmission lines could be grouped in pairs to create three different differential channels.

The precise internal connections to the device will influence how to interpret each S-parameter. The most common case is with the six different through connections with port assignments as shown. Always remember that if the port assignments change, the interpretation of each specific S-parameter will change as well.

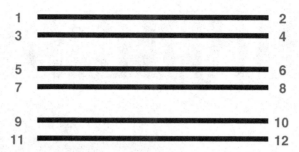

Figure 12-29 The index labeling for a 12-port device, which could be three differential channels. Not shown are the return paths, but they are assumed to be present and connected to the ports.

With 12 ports, there will be a total of 78 unique S-parameter elements, each having a magnitude and phase and each varying with frequency.

The diagonal elements are the return losses of each transmission line and have information about impedance changes in the interconnect. If the lines are all similar and symmetrical, all of the diagonal elements could be the same.

The six unique, direct through signals, the insertion losses S21, S43, S65, S87, S10,9, and S12,11 have information about the losses, the impedance discontinuities, and even resonances from stubs. All the other S-parameter elements represent coupling terms, containing information about cross talk. For example, S51, the ratio of a sine wave coming out of port 5 to the sine wave going into port 1, is related to the near-end cross talk between lines two away. The S61 term has information about the far-end cross talk between lines far away.

Of course, the near-end noise between the first line and each of the adjacent signal lines will drop off with spacing. Figure 12-30 is an example of the simulated near-end noise from one line to five other adjacent lines, each 5 mils in width and spaced 7 mils to the adjacent trace, each being 10 inches long.

Though the near-end noise between adjacent lines, S31, may be –25 dB, the near-end noise to the line five lines away, S11,1, is less than –55 dB.

TIP   Everything you ever wanted to know about the electrical behavior of interconnects is contained in their S-parameters.

So far, we have considered the signals into the ports as single-ended sine wave signals. Two other types of signals provide an alternative description of the stimulus-response behavior, which can often provide valuable insight into the behavior of the interconnects. Depending on the specific question being asked, these other forms may offer a faster route to the correct answer.

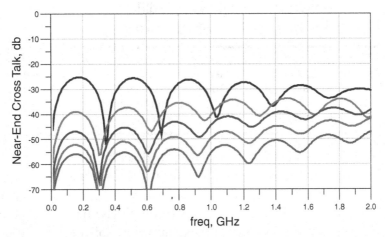

Figure 12-30 Near-end cross talk from one line to five adjacent microstrip lines, each 10 inches long, with spacing between them of 7 mils. The farther away the line, the lower the near-end cross talk. Simulated with Agilent's ADS.

The two other forms of the S-parameters are differential and time domain.

12.13 Single-Ended and Differential S-Parameters

Two independent transmission lines in proximity and with coupling can be described in two equivalent ways. On the one hand, they are two independent transmission lines, each with independent properties. For example, if we use the labeling scheme suggested in the preceding section, where port 1 goes through to port 2 and port 3 goes through to port 4, each transmission line would have reflected elements, S11 and S33, and each would have a transmitted element, S21 and S43.

In addition, there would be cross talk between the two lines, the unique near end would be S31 and S42, and the unique far-end terms would be S41 and S32. The near- and far-end noise signatures would vary depending on the spacing, coupling lengths, and whether the topology were stripline or microstrip. All the electrical properties of these two transmission lines are completely described by these 10 unique S-parameter elements across the frequency range.

These same two lines can also be described as one differential pair. No assumptions have to be made about the lines; this description as a differential pair is a complete description. But, the words we use and the sorts of behavior we describe are very different for a single differential pair description than for two independent single-ended transmission lines with coupling.

> **TIP** When the S-parameters are describing the differential properties of the interconnects, we refer to the S-parameters as the *differential S-parameters* or the *mixed-mode S-parameters* or the *balanced S-parameters*. These terms are used interchangeably in the industry. The preferred term is mixed-mode S-parameters.

Using mixed-mode S-parameters, we describe the four-port interconnect in terms of being a *differential pair*, in which case we refer to the ports as *differential ports*. With one differential pair and a differential port on either end, the only types of signals that exist are differential signals and common signals. Any arbitrary waveform going into either of the differential ports can be described by a combination of differential and common signals.

These signals are often called the *differential mode* and *common mode* signals. Using this terminology is a bad habit. There is no need to invoke the word *mode*. The signals entering the interconnect as stimuli or leaving the interconnect as responses are either differential or common signals. If you call them differential *mode* signals, it is too easy to confuse the signals with the even and odd propagation modes, which refer to the state of the interconnect. This is discussed in great detail in Chapter 11.

> **TIP** To avoid confusion, it is strongly recommended that you never use the terms *differential mode signals* and *common mode signals*. There is no need. Always refer to signals as *differential signals* or *common signals* to minimize confusion about mixed-mode S-parameters.

The differential S-parameters describe how these differential and common signals interact with the interconnect. With just two differential ports, one on either end of the differential pair, as shown in Figure 12-31, there are only three unique S-parameters with different index numbers, S11, S22, and S21, which describe how signals enter and come out of the differential pair. These are defined in the conventional way: Each element describes the ratio of the sine wave coming out of one port to the sine wave going into another port. Each differential S-parameter has a magnitude and a phase.

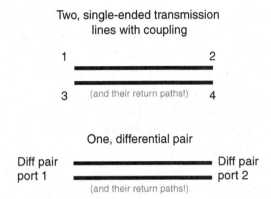

Figure 12-31   The same two transmission lines can be equivalently described as two single-ended lines with coupling or one differential pair. As one differential pair, there is one differential port on each end.

However, we must keep track of what type of signals enter and come out of the differential pair. In a differential pair, there are only differential and common signals. These interact with the differential pair in four possible ways:

1. A differential signal can enter a port and come out as a differential signal.
2. A common signal can enter a port and come out as a common signal.
3. A differential signal can enter a port and come out as a common signal.
4. A common signal can enter a port and come out as a differential signal.

With single-ended S-parameters, each S-parameter describes the ratio of single-ended sine waves coming out of a port, compared with the single-ended amplitude and phase going into a port. With differential S-parameters, we have to include a labeling system to describe not only what port the sine waves enter and come out but also what type of signal it is.

We use the letter D to refer to a differential signal and C to refer to a common signal. To designate a differential signal going in and a differential signal coming out, we use SDD. For a common signal in and common signal out, we use SCC. We also use the backward convention for the order of the type of signal, with the coming-out port signal first, and the going-in port signal second. So, for a differential signal in and a common signal out, we use SCD. And for a common signal in and a differential signal out, we use SDC.

Each differential S-parameter must contain information about the in and out ports and the in and out signals. By convention, we use the signal letters first and then the port indices. For example, SCD21 would describe the ratio of the differential signal going in at port 1 to the common signal coming out at port 2.

The port impedances of the single-ended S-parameters are all 50 Ohms. When two ports drive a differential signal, the outputs are in series, and the differential port impedance for the differential signal is 100 Ohms.

When two ports drive a common signal, the single-ended ports are in parallel, and the port impedance for the common signal is the parallel combination, or 25 Ohms. This means that a differential signal looking into one of the differential ports will always see a 100-Ohm termination impedance, while a common signal will always see the 25-Ohm common impedance port termination.

When displaying the mixed-mode S-parameters as a matrix, again by convention, we usually place the pure differential behavior in the upper-left quadrant and the pure common signal in the lower-right quadrant. The lower left is the conversion of differential into common signal, and the upper right is the conversion of common signal into differential signal. The mixed-mode matrix using this notation is shown in Figure 12-32.

			Stimulus			
			Differential Signal		Common Signal	
			port 1	port 2	port 1	port 2
Response	Differential Signal	port 1	SDD11	SDD12	SDC11	SDC12
		port 2	SDD21	SDD22	SDC21	SDC22
	Common Signal	port 1	SCD11	SCD12	SCC11	SCC12
		port 2	SCD21	SCD22	SCC21	SCC22

Figure 12-32 Mixed-mode S-parameter labeling scheme for each matrix element. For mixed-mode S-parameters, the ports are always differential ports.

This same matrix orientation can be used to display the 16 measured mixed-mode S-parameters. Using this formalism makes it much easier and less prone to introducing errors. An example of the measured mixed-mode or differential S-parameters for a differential channel in a backplane is shown in Figure 12-33.

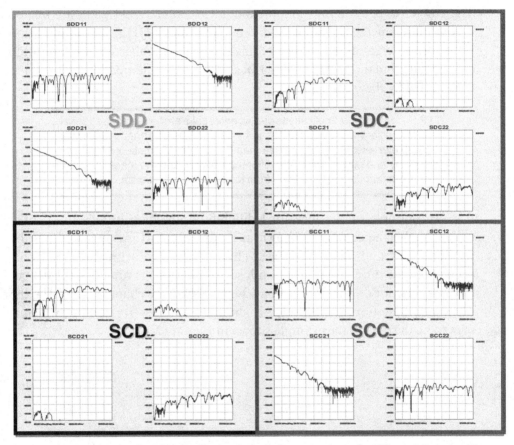

Figure 12-33 Measured mixed-mode S-parameters of a differential channel in a backplane, arranged in the same order as the matrix elements. Only the magnitude of each element is displayed; there is also a corresponding phase plot for each element, not shown. Measured with an Agilent N5230 VNA and displayed with Agilent's PLTS.

The CC terms contain information about how common signals are treated by the interconnect. The reflected common signal, SCC11, has information about the common impedance profile of the interconnect. The transmitted common signal, SCC21, describes how a common signal is transmitted through the interconnect. While the CC terms are important in the overall complete characterization of the differential pair, in most applications, the common signal properties of the interconnect are not very important.

12.14 Differential Insertion Loss

The DD terms contain information about how differential signals are treated by the interconnect. The reflected differential signal, SDD11, has information about the differential impedance profile of the interconnect. The transmitted differential signal, SDD21, describes how well differential signals are transmitted by the interconnect.

> **TIP**   Since most applications of differential pairs are for high-speed serial links, the differential insertion loss is by far the most important differential S-parameter element. The phase has important information about the time delay and dispersion of the differential signal, and the magnitude has information about the attenuation from losses and other factors.

An example of the measured differential insertion loss, SDD21, of a typical backplane channel is shown in Figure 12-34. It is dominated first by the conductor and dielectric losses. This gives rise to a general monotonic decrease in the differential insertion loss. In FR4, this is roughly −0.1 dB per inch per GHz for dielectric loss. In a 40-inch backplane channel, at 5 GHz, the differential insertion loss expected is about $-0.1 \times 40 \times 5 = -20$ dB, which is seen to be the case in the example shown in Figure 12-34.

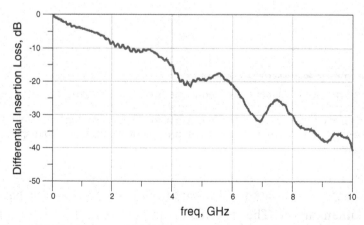

Figure 12-34 Measured SDD21 of a backplane trace. The general drop is due to losses. The ripples are due to impedance discontinuities from connectors and vias.

The second factor affecting SDD21 is impedance mismatches from connectors, layer transitions, and vias. These give rise to some ripples in the SDD21 plot. Two other factors can sometimes dominate SDD21: resonances from via stubs and mode conversion.

Figure 12-35   Measured differential return loss of two backplane channels, one with a 0.25-inch-long via stub and one with the via stub backdrilled. Measured with an Agilent N5230 VNA and displayed with Agilent's PLTS. Data courtesy of Molex Corporation.

Figure 12-35 shows an example of the measured differential insertion loss, SDD21, for two different differential channels in a backplane. In one channel there is a long via stub, approximately 0.25 inches long. In the other channel this via stub has been removed by backdrilling.

The sharp resonant dip at about 6 GHz is due to the *quarter wave stub resonance* and is a prime factor limiting the usable bandwidth of high-speed serial links.

The origin of the resonance is very easy to understand by referring to the illustration in Figure 12-36. In this illustration, the return path is not included but is present as planes on adjacent layers.

When the signal flows down the via to the connected signal layer, it splits. Part of the signal continues on the signal layer, and another part of the signal heads down to the end of the via stub, which is open.

When the signal hits the end of the via, it reflects and again splits at the layer transition. Some of the signal goes back to the source, while the rest of it continues in the same direction as the initial signal but with a phase shift.

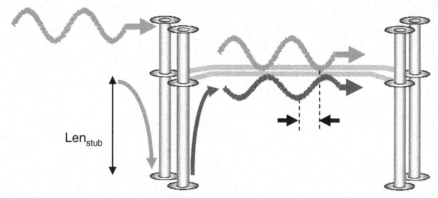

Figure 12-36   The insertion loss is due to the combination of the incident wave and reflected wave from the bottom of the via stub. When the stub length is one-quarter a wavelength, the two waves arriving at the receiver are 180 degrees out of phase and cancel out.

The phase shift of the reflected wave is the round-trip path length going down to the stub and back up to the signal layer. When this round-trip path length is half a wave, the two components—the initial signal and the reflected signal—heading to port 2 will be 180 degrees out of phase and offer the most cancellation. The condition for maximum cancellation, the dip in the differential insertion loss, is that the round-trip time delay for the stub is half a cycle, or:

$$2\frac{\text{Len}_{stub}}{v} = \frac{1}{2}\frac{1}{f_{res}}$$
(12-32)

Or, for the case of Dk = 4:

$$f_{res} = \frac{1}{4}\frac{12 \text{ inches/nsec}}{\sqrt{Dk} \times \text{Len}_{stub}} = \frac{1.5}{\text{Len}_{stub}} \text{ GHz}$$
(12-33)

where:

Len_{stub} = length of the stub, in inches

v = speed of the signal, in inches/nsec

f_{res} = resonant frequency for the stub, in GHz

Dk = dielectric constant of the laminate around the stub = 4

For example, if the stub is 0.25 inches long, the resonant frequency is 1.5 GHz/0.25 inches = 6 GHz, which is very close to what is observed.

The condition for maximum cancellation and the minimum differential insertion loss is that the round-trip length of a via stub be one-half a wavelength or that the one-way length be one-quarter of a wavelength. This is why this resonance is often called a *quarter-wave stub resonance*.

As a rough guideline, for the large resonant absorption in the stub to not affect the high-speed signal, the resonant frequency should be engineered to be a frequency that is at least twice the bandwidth of the signal. In the worst case, when the channel is low loss and short, the bandwidth of the signal would be roughly the fifth harmonic of the Nyquist. The Nyquist frequency is one-half the bit rate, so the signal bandwidth is:

$$BW = 5 \times 0.5 \times BR = 2.5 \times BR \qquad (12\text{-}34)$$

where:

BW = bandwidth of the signal, in GHz

BR = bit rate, in Gbps

The condition for an acceptable via stub length is:

$$f_{res} > 2 \times BW = 5 \times BR \qquad (12\text{-}35)$$

or:

$$\frac{1.5}{Len_{stub}} > 5 \times BR \qquad (12\text{-}36)$$

or:

$$\text{Len}_{\text{stub}} < \frac{300}{\text{BR}} \qquad (12\text{-}37)$$

where:

Len_{stub} = maximum acceptable stub length, in mils

BR = bit rate, in Gbps

TIP   For example, to engineer the stub resonant frequency to be far above the Nyquist frequency of a 5-Gbps signal so that no signal frequency components would see the resonant dip, the maximum length of any via stub in the channel should be shorter than 300/5 = 60 mils. This defines the design space for acceptable via stubs for 5-Gbps signals.

The via stub length can be engineered by restricting layer transitions between the top few layers and the bottom few layers, using thinner boards, backdrilling the long stubs, or using alternative via technologies such as blind and buried vias or microvias.

12.15 The Mode Conversion Terms

The off-diagonal quadrants of the differential S-parameter matrix are the most confusing, as they have information about how the differential pair interconnect converts differential signals into common signals and vice versa. They are also very important when it comes to hacking into the interconnect to determine possible root causes of behavior.

Even though it is incorrect to call the signal entering a port the differential mode signal, the industry has adopted a convention that is very confusing. These two off-diagonal quadrants are called the *mode conversion quadrants*, as they describe how one signal mode is converted into another signal mode.

The SCD quadrant describes how a differential signal enters the differential pair but comes out as a common signal. It can enter at one port and can come out that port or can be transmitted to the other port.

TIP   Only asymmetry between one line and the other line in a differential pair will convert some differential signal into common signal or vice versa. In a perfectly symmetrical differential pair, there is never any mode conversion. As long as what is done to one line is done to the other line, no matter how large the discontinuity, there will be no mode conversion, and the SCD terms will be zero.

Any asymmetry between the two lines that make up the differential pair will cause mode conversion. This can be a length difference, a local variation in the dielectric constant between the two lines, a test pad on one line but not the other, a difference in the rise time of the drivers, a skew between the drivers of the two channels, a variation in the clearance holes in a via field, or a line width difference. How much signal appears in the SCD11 term compared to the SCD21 term depends on how much common impedance discontinuity the common signal encounters to reflect some of the common signal back to the originating port.

The same asymmetry that converts a differential signal into a common signal will convert a common signal into a differential signal. Mode conversion creates three possible problems for differential pairs used in high-speed serial link applications.

The first problem is that if there is significant mode conversion of the differential signal, the differential signal amplitude can drop. This drop might increase the bit error rate. An example of the impact on the differential signal from a small length difference between the two lines in a differential pair is shown in Figure 12-37.

The second problem with mode conversion is that the common signal created might reflect from the unterminated ends of the differential pair, and each time it passes through the asymmetry, some of it might convert back into a differential signal, but asynchronous with the data stream, and distort the original differential signal. This would also increase the bit error rate by collapsing the eye.

The third problem with the common signal generated by mode conversion arises if the common signal gets out of the box and on an unshielded twisted-pair cable. A pure differential signal on a twisted-pair cable does not radiate EMI and has no problem passing an FCC or similar EMC certification test. However, if more than 3 microAmps of common signal at 100 MHz or higher frequency is created and gets out on 1 meter of twisted-pair cable, it will radiate and cause a failure in an FCC test.

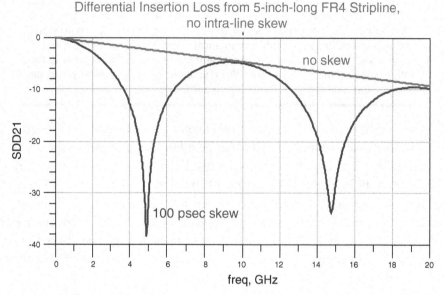

Differential insertion loss with 0.6-inch-length skew (100 psec)

Figure 12-37 Differential insertion loss through a uniform differential pair, with no intra-line skew and with 100 psec of skew between the two lines. The skew causes a dip in the differential signal when the skew is one-half of a cycle. Simulated with Agilent's ADS.

If the common impedance is 300 Ohms, it takes only about 1 mV of common signal generated from mode conversion driving an unshielded, external twisted-pair cable to cause an FCC failure. Depending on the differential signal swing, this might be on the order of 1% mode conversion.

Figure 12-38 shows an example of the measured mode conversion in three different, differential channels routed adjacent to each other in a backplane. In each case, the converted common signal would exceed 1 mV. If any of this escaped onto external, unshielded twisted-pair cable, it could cause an FCC failure. However, if this common signal stays inside the box, it will not contribute to EMI. It may still cause a problem in distorting the differential signal, though.

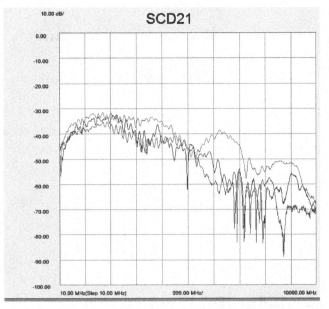

Figure 12-38 Measured-mode conversion in the transmitted signal, SCD21, for three different channels in a backplane. The peak mode conversion is about −35 dB, which is slightly above 1%. If the differential signal were 100 mV, the common signal would be above 1 mV, and if this appeared on unshielded twisted pair, it would probably fail an FCC test. Measured with an Agilent N5230 VNA and displayed with Agilent's PLTS.

12.16 Converting to Mixed-Mode S-Parameters

Virtually all instruments measure only the single-ended S-parameters. To display the mixed-mode S-parameters, the single-ended S-parameters are mathematically transformed into the mixed-mode S-parameters. For all linear, passive interconnects, the mixed-mode S-parameters are just a linear combination of the single-ended S-parameters.

Exactly how the 10 unique, single-ended S-parameters are transformed into the 10 unique differential S-parameters requires some complicated matrix math but is very straightforward to implement. The matrix math is summarized in a simple matrix equation:

$$S_d = M^{-1} S_s M \qquad (12\text{-}38)$$

where:

S_d = mixed-mode or differential S-parameter matrix

M = transform matrix

S_s = single-ended S-parameter matrix that is typically measured

The transform matrix relates how single-ended matrix elements combine to create the mixed-mode matrix. When the port assignments are as described in this chapter, $1 \rightarrow 2$ and $3 \rightarrow 4$, the transform matrix is:

$$M = \frac{1}{\sqrt{2}} \begin{bmatrix} 1 & 0 & -1 & 0 \\ 0 & 1 & 0 & -1 \\ 1 & 0 & 1 & 0 \\ 0 & 1 & 0 & 1 \end{bmatrix} \qquad (12\text{-}39)$$

For example, the SDD11 element is given by:

$$SDD11 = 0.5 \times (S11 + S33 - 2 \times S31) \qquad (12\text{-}40)$$

and the SDD21 element is given by:

$$SDD21 = 0.5 \times (S21 + S43 - S41 - S23) \qquad (12\text{-}41)$$

This matrix manipulation allows us to take any single-ended S-parameter matrix of measured or simulated S-parameters and routinely transform it into the differential or mixed-mode elements. This is another reason it is so important to use the correct port assignment because otherwise the transform matrix will change.

12.17 Time and Frequency Domains

For linear, passive interconnects, the same information contained in an individual S-parameter element can be displayed either in the frequency domain or in the time domain. After all, the frequency-domain S-parameters describe how any sine wave voltage waveform is treated by the interconnect. We can synthesize any time-domain waveform by combinations of different frequency components.

A linear system will have no interaction between the frequency components. If we know how each frequency component interacts with the interconnect, we can literally add up the right combinations of frequency components and evaluate any time-domain waveform. This is essentially the inverse Fourier transform process.

There are two commonly used synthesized time-domain waveforms that provide immediately useful information about an interconnect: a *step edge* and an *impulse response*. The reflected and transmitted behavior is interpreted slightly differently.

A step response is the same waveform used in a TDR and is often referred to as the *TDR response*. S11, when viewed in the time domain as a step response, is identical to a TDR response and has information about the single-ended impedance profile of an interconnect. The S21 term, when viewed in the time domain as a step edge, has information about how the leading or falling edge of a signal is distorted by the interconnect. This is usually dominated by losses in the interconnect.

When there are multiple responses to evaluate in the time domain, we usually use the S-parameter notation for each matrix element to signify which ports are involved, but instead of using the letter S, we use the letter T. For example, T11 is the TDR response, and T21 is the TDT response.

> **TIP**   Each domain has the same information but displayed in a different format. They obviously look different visually, and each term has a different sensitivity to interconnect properties. Depending on the question being asked, one or the other format might be the right one to use to get to the answer faster.

For example, if the question is what is the characteristic impedance of the line, the T11 element would get us the answer the fastest. If the question is what are the losses in the line, the S21 element would get us the answer the fastest.

This is illustrated in Figure 12-39, where the same measured insertion and return loss of a line in a microstrip pair is displayed in the frequency and time domains. Included are examples of two microstrip lines: one in a tightly coupled pair and one in a weakly coupled pair.

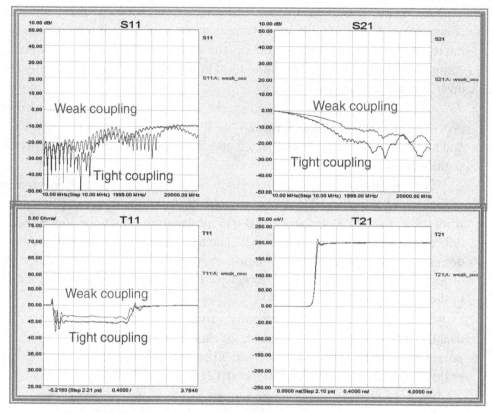

Figure 12-39 Measured single-ended return and insertion loss for a single line in a microstrip pair. One pair is tightly coupled, and the other pair is loosely coupled. Exactly the same measured response is shown in the frequency domain and the time domain. Measured with an Agilent N5230 VNA and displayed with Agilent PLTS.

TIP   Even though the information content in the time domain and the frequency domain is exactly the same, the information that is immediately available from the front screen is different. To get to the answer the fastest, it is important to have flexibility in converting the S-parameters between the time and frequency domains and between single-ended and differential domains.

It is much easier to interpret the impedance profile of an interconnect from the step edge time-domain form of S11 than from the frequency-domain form. Likewise, the frequency-domain form of S21 is a more sensitive measure of the interconnect losses than the step time-domain response. Depending on the question being asked, one element in a particular domain might get you to the answer more quickly than another. This is why having flexibility in data mining all the S-parameters is valuable.

The impulse time-domain response is particularly useful when using the S-parameters as a behavioral model in circuit simulators or to emulate the interconnect's response to any arbitrary waveform.

The impulse response is also sometimes called *Green's function of the interconnect*. It describes how any tiny incident voltage is treated by the interconnect. If we know how the interconnect treats an impulse response, we can take any waveform in the time domain and describe it as a combination of successive impulse responses. Using a convolution integral, we can combine how each of the successive, scaled impulse responses is treated by the interconnect to get the output waveform.

This technique allows us to synthesize any arbitrary time-domain waveform, convolve it with the impulse response, and simulate the output waveform. One application is to emulate eye diagrams of interconnects, as diagrammed in Figure 12-40.

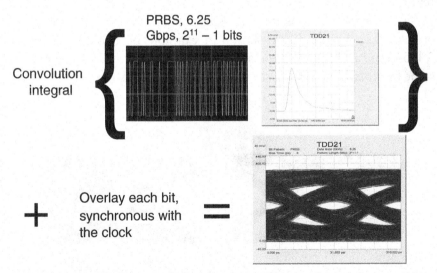

Figure 12-40   Process of simulating an eye diagram for an interconnect by convolving the impulse response with a synthesized pseudorandom bit sequence and overlaying each bit synchronously with the clock. Measured with an Agilent N5230 VNA and simulated and displayed with Agilent's PLTS.

The SDD21 frequency-domain waveform is converted into the time-domain impulse-response waveform. A pseudorandom bit sequence (PRBS) is synthesized using an ideal square wave with some rise time. This time-domain waveform is convolved with the impulse response of the interconnect. Essentially, the impulse response says how each voltage point will be treated by the interconnect. The convolution integration takes each voltage point, multiplies it by the interconnect's response, and adds up all the time-domain responses, walking along the time-domain waveform.

The result is how the interconnect would treat the synthesized waveform. It is the simulation of the output real-time waveform. This is an example of using the S-parameters as a behavioral model of the interconnect. Its measured performance in the frequency domain is used to simulate the expected performance in the time domain without knowing anything about the internal workings of the interconnect. Many commercially available tools do this simulation automatically.

The simulated real-time, differential responses from the single-ended measurements can be used to generate eye diagrams. The clock waveform is used to slice out one, two, or three consecutive bits, and all the bits in the real-time response are superimposed with infinite persistence. The resulting superposition of all possible bit signals looks like an eye and is called an *eye diagram*. Figure 12-41 shows an example of two eye diagrams for a 3.125-Gbps and a 6.25-Gbps serial signal as they would appear through the same backplane channel based on S-parameter measurements of the channel.

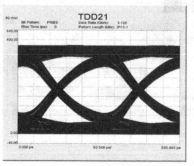 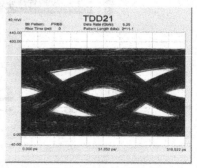

3.125 Gbps   **6.25 Gbps**

Figure 12-41   Emulated eye diagrams for a backplane channel at two different bit rates, based on the measured S-parameters. Measured with an Agilent N5230 VNA and simulated and displayed with Agilent's PLTS.

Two important features of an eye diagram can often predict bit error rates: the vertical opening and the horizontal opening. The vertical opening is often called the *collapse* of the eye, and the horizontal opening is related to the bit period minus the deterministic jitter.

The center-to-center spacing of the crossovers is always the unit interval bit time. The width of the crossovers is a measure of the jitter. In the measured SDD21 response of an interconnect, the random jitter is related to the quality of the measurement instrument and is almost always negligible. All the jitter in a simulated eye diagram is deterministic in the sense that it is *nonrandom* and predictable from just the interconnect behavior. It arises from losses, impedance discontinuities, and other factors.

By synthesizing PRBS waveforms with different bit rates, a family of eye diagrams can be simulated to indicate the performance limits of an interconnect.

12.18 The Bottom Line

1. As a universal description of any interconnect, even though they are new to the signal-integrity field, S-parameters are quickly becoming an industry standard.
2. Each S-parameter is the ratio of an output sine wave to an input sine wave. It represents the behavior of the interconnect to sine waves across a defined frequency range.
3. The reflected S-parameters, S11 and S22, have information about impedance discontinuities of the interconnect.
4. The transmitted S-parameters, S21 and S43, have information about losses, discontinuities, and couplings to other lines.
5. Other terms may describe the cross talk between interconnects.
6. While the port impedance is generally 50 Ohms, the same S-parameter response can be transformed to any port impedance. If there is no compelling reason to use something else, 50 Ohms should always be used.
7. The single-ended S-parameters obtained in the frequency domain can be transformed into differential S-parameters or into the time domain.

8. S-parameters in the frequency domain are a measure of the overall integrated, steady-state response of the interconnect; while transformed in the time domain, S-parameters can provide spatial information about the properties of an interconnect.

9. Narrow dips in the return or insertion loss are indictors of coupling to high-Q resonating structures.

10. The differential insertion loss describes the most important property of a differential channel.

11. Mode conversion is described by two of the differential S-parameter matrix elements, SCD11 and SCD21. They are useful in debugging the root cause of drops in the insertion loss.

12. Everything you ever wanted to know about the behavior of an interconnect is contained in the S-parameters, obtained either through measurement or by simulation.

End-of-Chapter Review Questions

12.1 Fundamentally, what do S-parameters really measure?

12.2 What is another name for an S-parameter model of an interconnect structure?

12.3 In the frequency domain, what do S-parameters describe about the properties of an interconnect?

12.4 What is different about the S-parameter model of an interconnect when displayed in the time domain?

12.5 List four properties of an interconnect that could be data mined from their S-parameter model.

12.6 What is the significance of using a 50-Ohm port impedance?

12.7 Why is a there a factor of 20 instead of 10 when converting an S-parameter, as a fraction, into a dB value?

12.8 Which S-parameter term is the impedance of an interconnect most sensitive to?

12.9 Which S-parameter term is the attenuation in an interconnect most sensitive to?

12.10 What are the S-parameters of an ideal transparent interconnect?

12.11 What is the recommended port labeling scheme for a pair of transmission lines?

12.12 Which S-parameter term contains information about near-end cross talk between two transmission lines?

12.13 Which S-parameter term contains information about far-end cross talk between two transmission lines?

12.14 How many individual numbers are contained in the S-parameter model of a 4-port interconnect, measured from 10 MHz to 40 GHz at 10-MHz steps?

12.15 Using the historically correct definition of return loss, if the return loss increases, what happens to the reflection coefficient? Is this in the direction of a more transparent or less transparent interconnect?

12.16 Using the modern, conventional definition of insertion loss, if the insertion loss increases, what does the transmission coefficient do? Is this in the direction of a more or less transparent interconnect?

12.17 Suppose the reflection coefficient of a cable with 50-Ohm port impedances has a peak value of −35 dB. What would you estimate its characteristic impedance to be? Suppose the port impedance were changed to 75 Ohms. Would the peak reflection coefficient increase or decrease?

12.18 As two of the most important consistency tests for the S-parameter of a through interconnect, what should the return and insertion loss be at low frequency? How will this change if the impedance of the interconnect is increased?

12.19 What causes ripples in the reflection coefficient?

12.20 Above what magnitude of reflection coefficient will the return loss ripples begin to appear in the transmission coefficient?

12.21 Why does the phase of S21 increase in the negative direction?

12.22 Why does the polar plot of the S21 spiral inward with increasing frequency?

12.23 As a consistency test, what insertion loss would you expect for an FR4 interconnect at 10 GHz that is 20 inches long?

12.24 Why does S21 of one line in a microstrip show a deep dip at some frequency? What would be an important consistency test to test your explanation?

12.25 What is the most common source of sharp, narrow dips in the insertion loss of through connections?

12.26 What are the two types of signal that can appear on a differential pair?

12.27 What is the only feature that causes mode conversion in a differential pair? Which two S-parameter terms are strong indicators of mode conversion?

12.28 In an uncoupled differential pair, how would the differential insertion loss compare with the single-ended insertion loss of either line?

12.29 When displaying the TDR response of an interconnect, which S-parameter is displayed, and how do you interpret the display?

12.30 What information will be in the TDT response of a through interconnect?

CHAPTER 13

The Power Distribution Network (PDN)

The power distribution network (PDN; sometimes also called the power delivery network) consists of all the interconnects from the voltage-regulator module (VRM) to the pads on the chip and the metallization on the die that locally distributes power and return current. This includes the VRM itself, the bulk decoupling capacitors, the vias, the traces, the planes on the circuit board, the additional capacitors added to the board, the solder balls or leads of the packages, the interconnects in the packages mounted to the board, the wire bonds or C4 solder balls, and the interconnects on the chips themselves.

The primary difference between the PDN and signal paths is that there is just one net for each voltage rail in the PDN. It can be a very large net that can physically span the entire board and can have many components attached.

TIP   As we will see, the PDN is like an ecosystem. If one small part of the PDN changes, the performance of the entire system can be affected. This makes generalizations very difficult.

13.1 The Problem

Figure 13-1 shows an example of a motherboard with all the interconnects mentioned in the preceding section.

The first and primary role of the PDN is to keep a constant supply voltage on the pads of the chips and to keep it within a narrow tolerance band, typically on the order of 5%. This voltage has to be stable, within the voltage limits, from DC up to the bandwidth of the switching current, typically above 1 GHz.

Figure 13-1   Example of typical motherboard, showing all the interconnects of the PDN.

TIP   The purpose of the PDN is threefold: Keep the voltage across the chip pads constant, minimize ground bounce, and minimize EMI problems.

In most designs, the same PDN interconnects that are used to transport the power supply are also used to carry the return currents for signal lines. The second role of the PDN interconnects is to provide a low-impedance return path for the signals.

The easiest way of doing this is by making the interconnects wide, so the return currents can spread out as much as they want and by keeping the signal traces physically separated so that the return currents do not overlap. If these conditions are not met, the return current is constricted, and the return currents from different signals overlap. The result is ground bounce, also called *simultaneous switching noise* (*SSN*) or just *switching noise*.

Finally, because the PDN interconnects are usually the largest conducting structures in a board, carry the highest currents, and sometimes carry high-frequency noise, they have the potential of creating the most radiated emissions and causing failure of an EMC certification test. When done correctly, the PDN interconnects can mitigate many potential EMI problems and help prevent EMC certification test failures.

The consequence of not designing the PDN correctly is that there will be excessive noise on the voltage rails of the chips. This can cause a bit failure directly, or it can mean the clock frequency of the chip can't be met, and timing errors result.

Figure 13-2 shows an example of the voltage noise on the pads of a processor chip. In this example, the nominally constant 2.5-v rail to the core of the chip, referred to as Vdd, shows voltage noise of as much as 125 mV on some pads. As the Vdd supply drops, the propagation delay of the core gates will increase, and timing problems can cause bit failures.

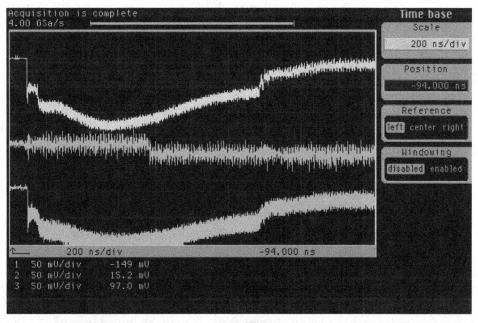

Figure 13-2 Example of the measured voltage between three different pairs of Vdd and Vss pads of a processor chip, showing as much as a 125-mV drop. The initial step down is when the processor comes out of an idle state. The three traces are three different locations on the die. The precise shape is related to the microcode running on the processor.

13.2 The Root Cause

If the problem is a voltage drop or a droop on the power supply rails on the pads of the chip, why not just use a "heftier" regulator—one that can supply a more rock stable voltage? Why not pay extra for a regulator with a 1% regulation or even 0.1% regulation? This way, the voltage from the regulator will be absolutely stable, no matter what, right?

What the chip cares about is the voltage on its pads. If there were no current flow in the PDN interconnects from the regulator pads to the chip pads, there would be no voltage drop in this path, and the constant regulator voltage would appear as a constant rail voltage on the chip pads.

If there were a constant DC current draw by the chip, this DC current would cause a voltage drop in the PDN interconnects due to the series resistance of the interconnects. This is commonly referred to as the *IR drop*. As the current from the chip fluctuates, the voltage drop in the PDN would fluctuate, and the voltage on the chip pads would fluctuate.

Now add not just the resistive impedance of the PDN but also the reactive components, including the inductive and capacitive qualities of the PDN interconnects. The impedance of the PDN, as seen by the pads on the chip, in general, is some complex impedance verses frequency, $Z(f)$. This is diagrammed in Figure 13-3.

As fluctuating currents with some spectrum, $I(f)$, pass through the complex impedance of the PDN, there will be a voltage drop in the PDN:

$$V(f) = I(f) \times Z(f) \qquad (13\text{-}1)$$

where:

$V(f)$ = voltage amplitude as a function of frequency

$I(f)$ = current spectrum drawn by the chip

$Z(f)$ = impedance profile of the PDN, as seen by the chip pads

This voltage drop in the PDN means that the constant voltage of the regulator is not seen by the chip but is changed. In order to keep the voltage drop on the chip pads less than the voltage noise tolerance, usually referred to as the ripple, given the chip current fluctuations, the impedance of the PDN needs to be below some maximum allowable value. This is referred to as the *target impedance*:

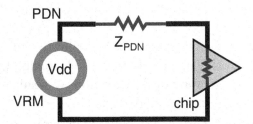

Figure 13-3 Diagram of the connections from the VRM through the PDN to the chip pads and the voltage drop across the PDN interconnects due to the impedance of the PDN.

$$V_{ripple} > V_{PDN} = I(f) \times Z_{PDN}(f) \tag{13-2}$$

$$Z_{target}(f) = Z_{PDN}(f) < \frac{V_{ripple}}{I(f)} \tag{13-3}$$

where:

V_{ripple} = voltage noise tolerance for the chip, in volts

V_{PDN} = voltage noise drop across the PDN interconnects, in volts

$I(f)$ = current spectrum drawn by the chip, in Amps

$Z_{PDN}(f)$ = impedance profile of the PDN as seen by the chip pads, in Ohms

Z_{target} = maximum allowable impedance of the PDN, in Ohms

As mentioned many times so far in this book, the most important step in solving a signal integrity problem is to identify the root cause of the problem. The root cause of rail collapse or voltage noise on the PDN conductors is that a voltage drop in the PDN interconnects is created by the chip's transient current flowing through the complex impedance of the PDN.

TIP   If we want to keep the voltage stable across the pads of the chip, given the chip's current fluctuations, we need to keep the impedance of the PDN below a target value. This is the fundamental guiding principle in the design of the PDN.

13.3 The Most Important Design Guidelines for the PDN

The goal of the PDN is to deliver clean, stable, low-noise voltage to the pads of all the active devices that require power. We often translate this performance metric into designing the PDN interconnects to bring their impedance, as viewed from the chips' pads, below a target value from DC to high frequency. In general, this will be accomplished by following three important design principles. Though it may not always be possible to push these to the limit, it is always important to be aware of the directions to head, even if the real-world constraints keep you from the ultimate path.

The three most important guidelines in designing the PDN are:

1. Use power and ground planes on adjacent layers, with as thin a dielectric as possible, and bring them as close to the surface of the board stack-up as practical.
2. Use a surface trace that is as short and wide as possible between the decoupling capacitor pads and the vias to the buried power and ground plane cavity and place the capacitors where they will have the lowest loop inductance.
3. Use SPICE to help select the optimum number of capacitors and their values to bring the impedance profile below the target impedance.

Unfortunately, in the real world of practical product design, you may not always have the luxury of power and ground planes on adjacent layers or placed near the top of the board stack-up. There may be multiple voltage rails, and they may have odd and irregular shapes, with many antipad clearance holes.

You may not be able to use as many capacitors as you think you need, nor place them in proximity to the devices they are decoupling. Even if you do the best you can, it will still be important to know if it is "good enough" before you build the product. The last place you want to find a problem is when you are making 100,000 units and finding that 1% of them are failing due to excessive noise in the PDN. The time to find this out is as close to the beginning of the design process as possible, and the only way to determine this is by using analysis tools that allow you to explore design space.

TIP   It is essential to try to follow the three important design guidelines above and, at the same time, to use combinations of rules of thumb, approximations, and numerical simulation tools to predict the impedance profiles and voltage noise under typical and worst-case conditions.

The most important principle to follow for cost-effective design is to add appropriate analysis as early in the design cycle as possible. This will reduce the surprises as the design progresses and result in a product with acceptable performance at the lowest cost and that works the first time.

13.4 Establishing the Target Impedance Is Hard

The first step in designing the PDN is to establish the target impedance. This must be done separately and independently for each voltage rail to all the chips on the board. Some designs may use as many as 10 different voltages. In each one, the target impedance may vary with frequency due to the specific current spectrum of the chip.

Suppose that the current from the chip on one rail is a sine wave, with a peak-to-peak value of 1 A. The amplitude of the sine wave of current will be 0.5 A. This current from the chip is shown in Figure 13-4 in both the time domain and the frequency domain.

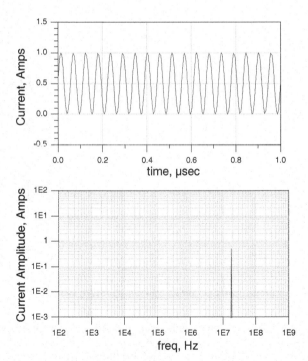

Figure 13-4 Example of the current waveform in the time domain (top) and the frequency domain (bottom) for a sine-wave current draw.

When this frequency component of the current flows through the specific impedance profile of the PDN, a voltage noise will be generated in the PDN. An example of an impedance profile with the frequency component of the current, and the resulting time-domain voltage noise across the chip pads, is shown in Figure 13-5.

When the sine-wave current passes through an impedance that is too large, the voltage generated is above the ripple spec, which is typically ±5%, shown as the reference lines.

There is the potential for the current draw through a chip to be at almost any frequency from DC to above the clock frequency. This means that unless the precise current spectrum from the chip is well known, for all the possible microcode that could be running through it, we have to assume that the peak current could be anywhere from DC to the bandwidth of the signals.

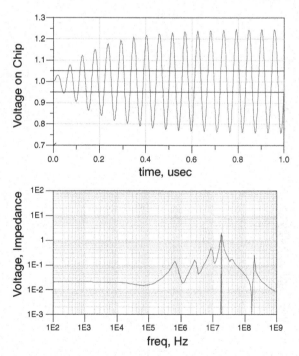

Figure 13-5 Voltage noise on the chip pads (top) as the sine-wave current flows through the PDN impedance profile of simulated PDN (bottom). The spike in the PDN profile at about 20 MHz shows where the sine-wave-frequency component of the current is with respect to the PDN impedance peaks.

In a few rare cases, if it is known that the chip processing will have a high-frequency fall off the current draw above some frequency, it may be possible to put some constraints on the current spectrum. This should always be done whenever possible.

While the current draw from a chip is rarely a pure sine wave, there are always sine-wave-frequency components in the current. The precise spectrum of the current amplitudes will interact with the impedance profile of the PDN completely independently of each other, but the resulting voltage waves will add together. Sometimes they can add and still meet the ripple spec, while at other times, they can add and exceed the ripple spec, depending on the precise overlap of current peaks and impedance peaks.

Figure 13-6 shows an example of a 1-A peak-to-peak square wave current draw for two slightly different modulation frequencies. The square wave current will have sine-wave-frequency components at odd multiples of the square wave frequency. Above roughly the fifth harmonic, depending on the rise time, the amplitude of the sine-wave harmonics will drop off much faster than 1/f. As the modulation frequency changes, the frequency distribution of the harmonics shifts and interacts differently with the PDN impedance profile.

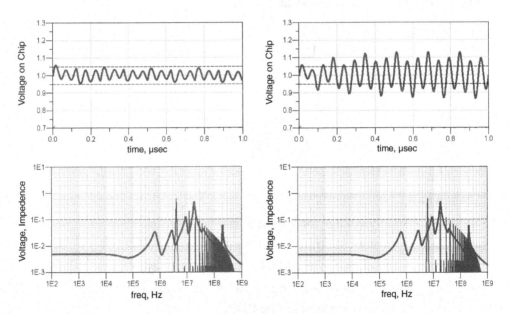

Figure 13-6 The resulting ripple noise when the same 1 A current draw has a slightly different modulation frequency, where one of the harmonic components overlaps an impedance peak of the PDN impedance profile.

A very slight frequency shift in the modulation of the current can mean the difference between acceptable performance and failure. Unfortunately, the engineer designing the PDN has very little control of the current draw spectrum of the chip. It is whatever the chip is going to do, depending on its operation.

This means that unless there is precise information about the specific, worst-case spectrum of the current draw of the chip, a conservative design must assume a worst-case current that could occur at any frequency from DC to the bandwidth of the clock, which is a few times the clock frequency.

In practice, it is not the peak current but the maximum transient current that interacts with the higher frequencies of the PDN. If there is a steady-state DC current draw from the chip, the sense lines of the VRM can usually compensate to keep the rail voltage close to the specified voltage value. It's when the current changes from the DC value, either increasing or decreasing, at frequencies above the response frequency of the VRM, that the current will interact with the PDN impedance.

The maximum impedance for the PDN, the target impedance, is established based on the highest impedance that will create a voltage drop still below the acceptable ripple spec. This is given by:

$$Z_{PDN} \times I_{transient} = V_{noise} < V_{dd} \times ripple\% \qquad (13\text{-}4)$$

or:

$$Z_{target} < \frac{V_{dd} \times ripple\%}{I_{transient}} \qquad (13\text{-}5)$$

where:

V_{dd} = supply voltage for a specific rail

$I_{transient}$ = worst-case transient current

Z_{PDN} = impedance of the PDN at some frequency

Z_{target} = target impedance, the maximum allowable impedance of the PDN

V_{noise} = worst-case noise on the PDN

ripple% = ripple allowed, assumed to be ±5% in this example

TIP   The optimum PDN impedance should be below the target impedance but not too far below the target impedance.

If the PDN impedance is kept below the target impedance at every frequency, the worst-case voltage noise generated across it as the worst case, maximum transient current flows through it will be less than the ripple spec. If the PDN impedance is much below the target impedance, it means the PDN was overdesigned and costs more than it needs to.

TIP   Whenever possible, the peak transient current should be used in estimating the target impedance. When the peak transient current is not available, it can be roughly estimated from the maximum current draw or from the power consumption of the chip.

While the worst-case transient current is what is important, rarely is this provided in a spec sheet. Rather, it is the worst-case peak current per rail that is on spec sheets. After all, this is important to estimate how large a voltage regulator module is needed. It must be capable of supplying the maximum current draw at the rated voltage.

The peak current could be mostly DC with a 10% transient current, or it could be a very low quiescent current with most of the peak current being transient that would last only for a few microseconds. Without special knowledge of the behavior of the chips in each application, the conservative design has to plan for the worst case.

What fraction of the maximum current is transient? Obviously, it depends on the function of the chip. It could vary from 1% to 99%, depending on the application. As a rough, general rule of thumb, without any special knowledge, the transient current can be estimated as being half of the maximum current:

$$I_{transient} \sim \frac{1}{2} \times I_{max} \qquad (13\text{-}6)$$

where:

$I_{transient}$ = worst-case transient current from the chip
I_{max} = maximum total current from the chip

Alternatively, the worst-case power dissipation of the chip will always be provided in a chip's specs since this is critical information when designing the thermal management approach for the package. It is not usually separated by voltage rail, so some assumptions would have to be made on the power consumption of each rail.

However, given the power consumption per voltage rail, the peak current draw of a chip can be estimated from:

$$I_{peak} = \frac{P_{max}}{V_{dd}} \tag{13-7}$$

And from this, the target impedance can be evaluated as:

$$Z_{target}(f) < \frac{V_{dd} \times ripple\%}{I_{transient}} = 2\frac{V_{dd} \times ripple\% \times V_{dd}}{P_{max}} \tag{13-8}$$

where:

I_{peak} = worst-case peak current, in Amps
P_{max} = worst-case power dissipation, in watts
V_{dd} = voltage rail, in volts
ripple% = ripple spec, in %
2 = comes from the transient current being ½ the peak current

For example, if the ripple spec is 5%, the target impedance is:

$$Z_{target}(f) = 0.1 \times \frac{V_{dd}^2}{P_{max}} \tag{13-9}$$

In the case of a 1-volt rail and 1-watt power dissipation device, the target impedance would be about 0.1 Ohm:

$$Z_{target}(f) < 0.1 \times \frac{1^2}{1} = 0.1\Omega \tag{13-10}$$

Some chip vendors, especially FPGA vendors, also provide calculation tools that allow simple estimates of the current draw of specific voltage rails, depending on the gate utilization. These can be used to estimate the target impedance specs of the rails. An example of the results of using one such analysis for the Altera Stratix II GX FPGA is shown in Figure 13-7.

Power rail	Voltage (v)	ripple%	Max current (A)	Transient current amplitude (A)	Z_{target} (Ohms)
VCCT/R	1.2	2.5%	1.2	0.6	0.05
VCCH	1.5	2%	0.17	0.085	0.35
3.3 v Analog	3.3	3%	0.274	0.137	0.72
VCCP	1.2	2%	1.03	0.51	0.047

Figure 13-7 Example of a calculation of the target impedance of different voltage rails based on gate utilization of an Altera FPGA.

Finally, the current requirements of the I/O voltage rails, typically referred to as either the Vcc or Vddq rails, can be estimated based on the number of gates that are switching.

If each output gate drives a transmission line with some characteristic impedance, then the load it sees, if only for a round-trip time, is the same as the characteristic impedance of the line it drives.

If n gates could switch simultaneously, the transient current draw could be:

$$I_{transient} = n \frac{V_{cc}}{Z_0} \qquad (13\text{-}11)$$

And the target impedance of the VCC rail would be:

$$Z_{target}(f) < 0.05 \times \frac{1}{n} Z_0 \qquad (13\text{-}12)$$

where:

$I_{transient}$ = worst-case transient current

n = number of I/Os that could switch simultaneously

V_{cc} = voltage rail

Z_0 = characteristic impedance of the transmission lines

For example, if the lines are all 50 Ohms, and there are 32 bits switching simultaneously, the target impedance for the Vcc rail would be:

$$Z_{target}(f) < 0.05 \times \frac{1}{32} 50 = 0.08\Omega \qquad (13\text{-}13)$$

Even with the peak current and the target impedance established, the current could be fluctuating at almost any frequency due to the specific microcode or application running. This means that unless there is information to the contrary, it must be assumed this is the target impedance, flat from DC to very high frequency.

TIP   The goal in designing the PDN is to keep the impedance of the PDN interconnects below this target value over a very wide bandwidth. A PDN above the target impedance may result in excessive ripple. A PDN impedance much below the target impedance may be overdesigned and more expensive than it needs to be.

When the impedance profile is kept below the target value, the worst-case voltage rail noise will be less than the ripple spec. An example of a successful impedance profile is shown in Figure 13-8.

However, if there is a peak in the impedance profile that exceeds the target spec, and if the worst-case current happens to fall on top of this peak impedance, there is a chance that the ripple spec may be exceeded. This is shown in Figure 13-9. In this example, the step current change excites the peak impedance, and we see the characteristic ringing at the peak frequency.

Peak impedances in the PDN impedance profile are an important design feature to watch out for. Many aspects of PDN impedance design, especially the selection of capacitor values, are driven by the desire to reduce the peak impedances in the PDN.

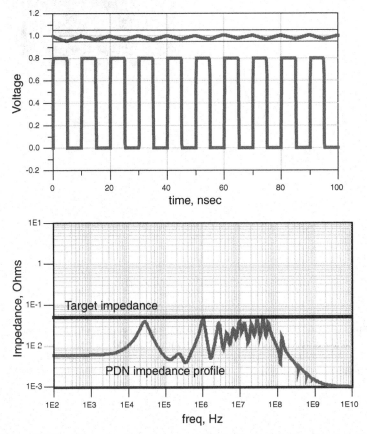

Figure 13-8   When the impedance profile (bottom) is below the target impedance, the worst-case voltage noise (top) is below the ripple spec. The square wave is the current draw by the chip, while the flat curve is the voltage on the supply rail.

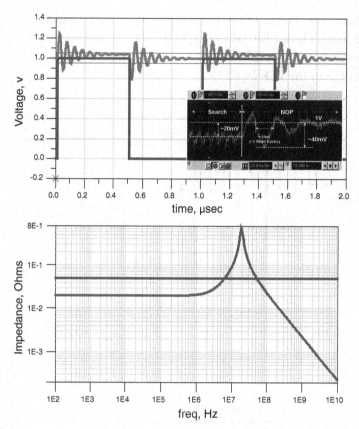

Figure 13-9 When the impedance exceeds the target spec (bottom) and the current's peak frequency hits this impedance peak, excess ripple can result (top). The square wave is the current draw through the chip. The ringing wave is the voltage on the supply rail. Inset is an example of the measured voltage noise on a PDN, showing the typical ringing response of a peak in the impedance profile.

13.5 Every Product Has a Unique PDN Requirement

One of the greatest sources of confusion in PDN design is created by taking the PDN design features of one product and blindly applying them to another product.

TIP   Unlike with the design of signal paths, where the design rules in one product can often be applied to other products of similar bandwidth, the behavior of the PDN depends on the interactions of all of its parts, and the goals and constraints vary widely from product to product.

The PDN is one giant net rather than a large number of individual nets, with only a small amount of local coupling between them. In this respect, the PDN net is like an ecosystem of interconnects. While it may be possible to suggest an optimized design based on optimizing individual elements, the most cost-effective designs are based on optimizing the entire ecology of all the elements, across the entire frequency range.

The voltage level of the rails can vary from 5 v to less than 1 v, depending on the chip type and technology node. The ripple spec may be as large as 10% in some devices or as low as 0.5% in others, such as phase locked loop (PLL) supplies or the analog-to-digital converter (ADC) reference voltage rails.

The current draw from chips can vary from more than 200 A in high-end graphics chips and processors to as low as 1 mA for some low-power microcontrollers. This means that the target impedance values can vary from below 1 mOhm for high-end chips to more than 100 Ohms. This is five orders of magnitude in impedance.

There could be as many as 10 different voltage rails in some designs, many sharing the same layer, while other designs may have just one voltage and ground. Some of the planes may be solid; others may be irregularly shaped and full of clearance holes.

TIP   This wide variety of applications and board constraints means no one solution is going to fit all. Instead, each design must be treated as a custom design.

It is dangerous to blindly apply the specific features of one design to another design. However, a general strategy can be followed to arrive at an acceptable impedance profile.

13.6 Engineering the PDN

It is remarkable that so complex a structure as the PDN interconnects can be partitioned in the frequency domain into just five simple regions. Figure 13-10 diagrams these five regions, based on the frequency ranges they can influence.

At the lowest frequency, the VRM dominates the impedance the chip sees when looking into the PDN. Of course, the series resistance of the interconnects can also set a limit on the lowest impedance of the PDN if it is larger than the VRM impedance. The VRM performance dominates from DC to about 10 kHz.

The next higher frequency regime, roughly in the 10-kHz to 100-kHz range, is dominated by the bulk decoupling capacitors. These are typically electrolytic and tantalum capacitors that provide a low impedance beyond where the VRM can go.

The highest-frequency impedance is set by the on-die capacitance. This is the only feature in the PDN that the chip sees in the GHz regime. It generally has the lowest loop inductance associated with it and offers the lowest impedance at the highest frequency of any element in the PDN.

TIP   Every chip interfaces to the board it is mounted to through some mounting inductance. Usually, this is dominated by the package, the board vias, and the spreading inductance of the via contacts into the power, and ground planes of the board.

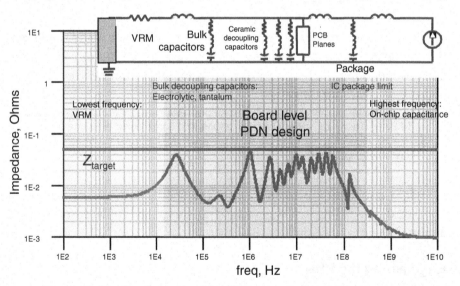

Figure 13-10   The five parts of the PDN separated by the frequency range they influence.

The PDN interconnects in the package are generally inductive. This means that at high frequency, they will act as a high-impedance path. Even if the board on which the package is mounted were designed with an impedance of a dead short, the chip would be looking at this short through the chip attach and package attach inductance, and it would see an impedance limited by these inductances.

The series inductance of the package's PDN will always limit the highest frequency at which the chip will see the board-level PDN. This acts as a high-frequency limit to the board-level PDN design, and above this package-limited frequency, the impedance the chip sees will be determined by the on-die capacitance and any capacitance in the package. This limit is generally in the 10- to 100-MHz range. Above this frequency, the impedance the chip sees is all about the package and the chip.

TIP   The frequency region that board-level PDN design can influence is roughly from the 100-kHz range up to about 100-MHz. This is where the planes of the board and the multilayer ceramic chip capacitors (MLCC) can play a role.

These capacitors—typically in sizes of 60 mils × 30 mils and referred to as 0603 or 40 mils × 20 mils and referred to as 0402—are called *chip capacitors* because they look like small "chips" of something on a board. Figure 13-11 is a close-up of some typical multilayer ceramic chip (MLCC) capacitors on a small memory board.

Figure 13-11   Close-up of typical 0402 MLCC capacitors mounted to a small memory board.

13.7 The VRM

The low-frequency impedance is set by the VRM. All VRMs, regardless of regulator type, have an output impedance profile. This can easily be measured using a two-port impedance analyzer.

An example of the measured impedance profile of a typical VRM is shown in Figure 13-12. In this example, the impedance looking into the output leads of the VRM was measured when the regulator was turned off and when it was turned on and providing regulation. In addition, the impedance of a simple two-capacitor model is superimposed.

This illustrates that when the regulator is off, the behavior seen at the output leads is almost exactly as predicted by a two-capacitor model, each capacitor being modeled as an RLC circuit.

This behavior corresponds to the two bulk decoupling capacitors associated with the VRM. The 910 µF capacitor is an electrolytic capacitor, while the 34 µF capacitor is a tantalum capacitor. This impedance profile is for the passive network of the leads and the two capacitors.

When the regulator is turned on, its output impedance drops by orders of magnitude at low frequency. This is exactly what is expected from a regulator. The output voltage is kept constant, independently of the current load. A large change in current produces a small change in voltage, the behavior of a low impedance. However, we see that in the actual behavior of the VRM, this low impedance is maintained from DC only up to the kHz range.

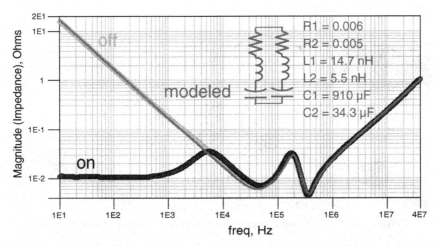

Figure 13-12 Measured impedance profile, from 10 Hz to 40 MHz, using an Ultimetrix Impedance Analyzer for a typical VRM, showing the impedance when on and when turned off and showing the modeled impedance based on a two-capacitor model.

Above about 1 kHz, the impedance is seen to increase, until it matches the impedance of the bulk capacitor at about 4 kHz, at which frequency the impedance is brought down by the passive capacitor network on the regulator.

Above about 1 kHz, the output impedance of the VRM is completely due to the passive capacitors, and the active regulation plays no role at all. Whether the regulator is on or off, the impedance is the same.

This is a slight exaggeration because the regulator fights with the capacitance of the passive network, and when the regulator is turned on, its impedance is actually higher than if it were literally turned off.

TIP The output impedance of most VRMs is low up to the kHz regime. Beyond this, what brings the impedance down is the bulk capacitors associated with the regulator.

The total amount of capacitance needed on a board in the form of electrolytic or tantalum capacitors can be estimated based on achieving the target impedance at the frequency where the VRM is no longer able to maintain the low impedance.

The capacitance is chosen so that its impedance at 1 kHz is less than the target impedance. The minimum capacitance needed is given by:

$$C_{bulk} > \frac{1}{Z_{target} \times 2\pi \times 1\,kHz} = \frac{160\,\mu F}{Z_{target}} \tag{13-14}$$

where:

C_{bulk} = minimum bulk capacitance needed, in μF

Z_{target} = target PDN impedance, in Ohms

1 kHz = frequency at which the VRM is no longer able to provide low impedance

For example, if the target impedance is 0.1 Ohm, the minimum bulk capacitance needed is about 1600 μF. Of course, this is only a rough estimate, but it is a good starting place. When it comes to establishing the actual target values of the capacitance, the interactions of the VRM effective inductance and the capacitor's capacitance must be taken into account with a SPICE simulation.

The low-frequency model of a VRM can be easily approximated by a simple RL model with a voltage source. The equivalent circuit model of the

VRM and bulk decoupling capacitor is shown in Figure 13-13. This circuit can be used to optimize the capacitor value to keep the impedance below the target value at low frequency.

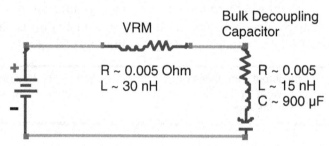

Figure 13-13 Typical equivalent circuit model of a VRM and bulk decoupling capacitor with typical values for each element.

13.8 Simulating Impedance with SPICE

Simulating the impedance profiles of different circuit models is essential in PDN design. Luckily, most of the simple circuits that need to be analyzed can be simulated with free versions of SPICE, such as QUCS, that can be downloaded from the Internet.

TIP   The secret to using SPICE to perform impedance simulations is to build an impedance analyzer as a SPICE circuit. This is done with a single element in SPICE: a constant-current AC current source.

This element is defined as a constant-current sine-wave source, outputting a sine wave of current with a constant amplitude. The output voltage of this element will be whatever it needs to be to always output a constant-amplitude sine wave of current. The frequency of the current is set by the frequency of the frequency-domain simulation. An example of a SPICE impedance analyzer is shown in Figure 13-14.

The amplitude of the constant current source is set to 1 A with a phase of 0. The voltage across the constant current source will depend on the impedance of whatever is connected across the leads. This voltage generated will be given by:

$$V = I(f) \times Z(f) = 1 \times Z(f) = Z(f) \tag{13-15}$$

where:

V = voltage generated across the current source, in volts

$I(f)$ = current from the source, a constant 1-A amplitude sine wave

Z(f) = impedance of the device connected across the current source, in Ohms

We set the current amplitude to be exactly 1 A. This means the voltage generated across the current source is numerically the impedance in Ohms. The impedance of the circuit connected may vary with frequency. As the 1-A constant amplitude sine wave flows through it, a voltage will be generated that is numerically equal to the impedance. The phase of the voltage will even track the phase of the impedance.

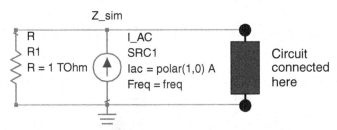

Figure 13-14   A SPICE impedance analyzer consisting of a constant-current AC sine-wave source.

A large shunt resistor, in this case 1 TOhm, is connected across the current source. This is to keep SPICE from halting due to an error. SPICE wants to see a DC path to ground for all nodes. Without the resistor, an open across the constant-current source could result in an infinite voltage, causing an error.

With this circuit, the impedance of any circuit model can be simulated. It's actually the voltage across the current source that is simulated, but this is equal to the impedance of the circuit. The impedance of the two-capacitor model in the VRM was simulated using this SPICE impedance analyzer.

13.9 On-Die Capacitance

The impedance at the highest frequency is established by the on-die decoupling capacitance. This arises from three general sources: the capacitance between the power and ground rail metallization, the gate capacitance from all the p and n junctions, and any added capacitance.

The largest component is from the gate capacitance distributed over the die. Figure 13-15 shows a typical CMOS circuit, found by the millions on most chips, and by the billions for some chips. At any one time, one of the gates is on, and the other is off.

This means that the gate capacitance of one of the gates, either the p channel or the n channel, is connected between the power and ground rails on the die. The capacitance per area associated with the gate is simply approximated by:

$$\frac{C}{A} = \frac{8.85 \times 10^{-12} \frac{F}{m} \times Dk}{h} \tag{13-16}$$

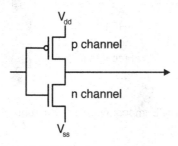

Figure 13-15   Typical CMOS circuit model for the transistors on a chip.

where:
C/A = capacitance per area, in F/m²
Dk = dielectric constant of the oxide ~3.9 for SiO_2
h = dielectric thickness, in meters

In general, the shorter the channel length, the thinner the gate oxide. As a rough rule of thumb, the gate oxide thickness is about 2 nm per 100 nm of channel length. However, below about 100-nm channel length, the scaling of h flattens out due to higher leakage currents, but then the dielectric constant is increased with the use of "high-Dk" gate insulator materials. This keeps the rule of thumb a good approximation even below 100 nm channel lengths.

For the 130-nm channel length node, the capacitance per area is about:

$$\frac{C}{A} = \frac{8.85 \times 10^{-12} \frac{F}{m} \times 3.9}{0.02 \times 130 \times 10^{-9}} = \frac{8.85 \times 10^{-12} \frac{F}{m} \times 3.9}{2.6 \times 10^{-9}} = 1.3 \frac{\mu F}{cm^2} \tag{13-17}$$

Of course, not all the die is gate area. If we assume that 10% of the surface of the die is gate capacitance, then we see that as a rough rule of thumb, the on-die decoupling capacitance on a 130-nm technology chip due to its p and n junctions is about:

$$\frac{C}{A} = 130 \frac{nF}{cm^2} \tag{13-18}$$

As the technology node advances and the channel length decreases, the gate capacitance per area will increase, but the total gate area on a die will stay about the same. This means the capacitance per unit of die surface area will increase inversely with the technology node.

The capacitance of 65-nm chips is about 260 nF/cm². This estimate suggests that for a die that could be 2 cm × 2 cm, close to the largest mask size in volume production, at 65-nm channel length, the on-die decoupling capacitance could easily be in excess of 1000 nF.

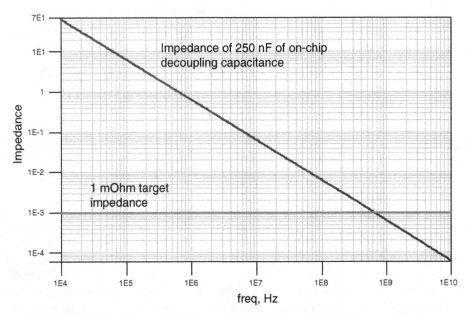

Figure 13-16 Impedance, in Ohms, provided by 250 nF of on-chip decoupling capacitance, typical of a die that is 65 nm and 1 cm on a side.

A typical chip in many embedded processors, only 1 cm × 1 cm, would have as much as 260 nF of capacitance. If the gate utilization on the die were larger, the on-die capacitance could be higher as well.

TIP   At high frequency, it is the on-die capacitance that provides the low impedance.

The impedance profile of a capacitance that is 250 nF is shown in Figure 13-16. In this example, the on-die capacitance provides an impedance below 1 mOhm at frequencies above 800 MHz. All high-frequency decoupling is provided by this mechanism.

If the target impedance were 10 mOhms, the on-die capacitance would provide significant decoupling for frequencies above about 100 MHz.

13.10 The Package Barrier

Between the pads on the chip and the pads on the circuit board is typically the IC package. Styles range from lead frame-based packages to miniature circuit board–based packages to minimalist or chip-scale packages.

The loop inductance of the package leads in the power/ground distribution path is in series with the pads of the chip to the pads on the circuit board. This series inductance creates an impedance barrier. The impedance of an inductance is given by:

$$Z = 2\pi fL \qquad\qquad (13\text{-}19)$$

where:

Z = impedance, in Ohms

f = frequency, in Hz

L = inductance, in H

For example, at 100 MHz, the impedance of a 0.1-nH inductor is about 0.06 Ohm. Even if the impedance of the PDN on the board were implemented as a dead short, the chip, looking through the package, would see a PDN impedance of 0.06 Ohm at 100 MHz. Of course, this is why on-die and on-package capacitance is so important.

Low-cost packages are often leaded, either as a stamped lead frame or as a two-layer printed circuit board. The loop inductance of adjacent leads is roughly about 20 nH/inch of length. For package leads 0.25 inches long, the loop inductance of a single power and ground lead pair can be as much as 5 nH. In the case of chip-scale packages, the loop inductance of a pair of leads may be on the order of 2 nH.

In multilayer BGA packages with at least four layers, a dedicated power and ground plane is often used. The loop inductance can be reduced to less than 1 nH per power and ground pair, limited by the roughly 50 mil total path length solder ball and its associated package via.

In small packages, there may be only a few power and ground pairs. In large BGA packages, there can be hundreds of pairs. This means the effective package lead inductance can vary from 1 nH to as low as 1 pH.

In addition to the package leads, there is also the loop inductance of the vias into the circuit board and the spreading inductance launching current into the power and ground planes of the board. When the package lead inductance is small, the board via and spreading inductance can limit the loop inductance as seen by the chip.

When the interactions of the on-die capacitance are added to the package inductance, the behavior is even more complicated. Figure 13-17 shows the impedance profile the chip sees looking into a board that has a short for the PDN. The impedance profile is limited by the package inductance.

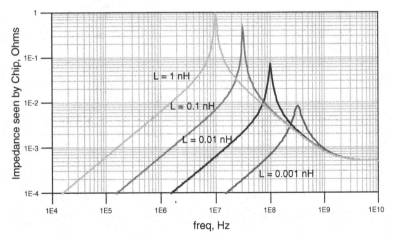

Figure 13-17   Impedance seen by the chip when the board is a dead short for different package lead inductances.

This suggests that no matter what the board-level PDN does, it can never reduce the impedance the chip sees below the package lead impedance. When the package equivalent lead inductance is 0.1 nH, the board cannot influence the impedance the chip sees to below 10 mOhms at frequencies above 10 MHz.

Of course, in this example, there is a large parallel resonance impedance peak due to the interactions of the package inductance and on-die capacitance. Many times, these can be suppressed by using on-package decoupling capacitors.

For example, Figure 13-18 shows the reduction in peak impedance for the case of 0.1 nH of lead inductance with the addition of 10 different 700-nF capacitors, each with 50 pH of ESL mounted to the package.

TIP   When establishing the design goals of the board-level PDN, the high-frequency limit to where the board-level impedance can be effective to the chip is determined by the frequency at which the impedance from the combination of the package leads, board vias, and spreading inductance exceeds the target impedance.

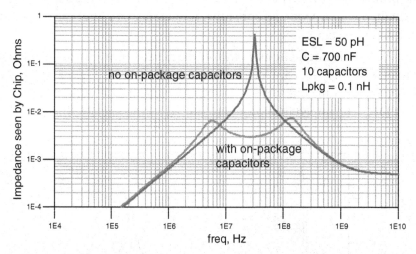

Figure 13-18 Suppression of package and on-die capacitance parallel resonances with on-package decoupling capacitors, as seen by the chip, if the board-level impedance were a dead short.

The relationship between the package lead inductance, maximum effective frequency and target impedance corresponds to:

$$Z_{target} < 2\pi \, L_{pkg} \, f_{max} \qquad (13\text{-}20)$$

where:

Z_{target} = target impedance, in Ohms

L_{pkg} = equivalent lead inductance of all the PDN paths in the package

f_{max} = highest useful frequency for the board-level PDN

As a starting place, Figure 13-19 shows the map of the target impedance and package inductance for a specific maximum frequency of 100 MHz. If a product design falls below the line—for example, if the target impedance is very low and the package lead inductance is very high—the maximum frequency for the board to be effective is below 100 MHz. In this case, the package severely limits the PDN's performance.

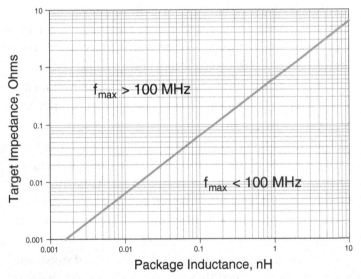

Figure 13-19   Map of the combination of target impedance and package lead inductance that has a maximum board-level frequency limit of 100 MHz. If a design is above the line, the board-level impedance will play a role at frequencies higher than 100 MHz; if the combination falls below the line, the board-level impedance will play a role at less than 100 MHz.

If a design falls above the line—for example, if the lead inductance is very low and the target impedance is high—the maximum frequency range for the board to still be effective is above 100 MHz.

As a rough rule of thumb, with about 20 nH/inch of loop inductance in a package lead and 0.05 inch of package lead length in a CSP package, the loop inductance per power and ground lead pair is about 1 nH. With 10 power and ground pin pairs in parallel, this is about 0.1 nH of equivalent lead inductance in a typical package that might have 10 pairs of PDN leads. If the target impedance were below about 0.6 Ohm, the board would not be effective much above 100 MHz.

Though it is difficult to generalize, as we've said throughout this book, sometimes an okay answer *now* is better than a good answer late. In general, the combination of packages and target impedance limits the board-level impedance to be effective under about 100 MHz. This is why the board-level PDN design goal is typically set to no higher than 100 MHz unless there is other information to the contrary.

While it is possible to set the high-frequency limit higher, achieving higher-frequency design limits is often much more expensive and should be done only when it is known to be important.

When on-package decoupling capacitors are provided, the maximum frequency at which the board-level impedance can be effective is often lower than 100 MHz.

The lead inductance also acts as a filter to keep high-frequency noise from the chip's PDN off the board. When the core gates switch, the PDN rail voltage is kept low by the on-die capacitance. After all, if there is excessive noise on the chip's PDN pads, this will cause its own problem. Any voltage noise on the chip rails will be further filtered by the lead inductance before it gets to the board.

Figure 13-20 shows an example of the simulated noise rejection from the chip pads to the board for different package lead inductances and a board-level impedance of 10 mOhms.

When the target impedance is 10 mOhms, and the package lead inductance is 0.1 nH, the noise rejection is about 0.1 or −20 dB at 100 MHz. Less than 10% of the on-chip noise is coupled into the board. The higher the package lead inductance, the less on-chip noise gets on the board. This is why very little noise above about 100 MHz gets onto the board-level PDN from the chip.

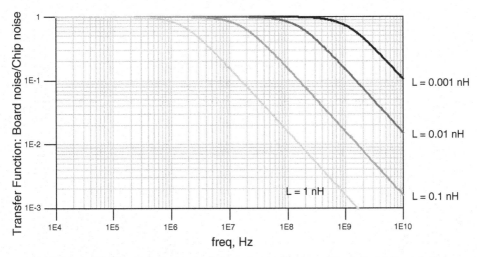

Figure 13-20 Relative noise injected onto the board from the chip pads for different package lead inductances. This is for the special case of the board-level impedance at or below 10 mOhms.

TIP In the absence of a complete package model including the PDN paths, it is difficult to do much more than roughly estimate the impact of the package on the PDN path.

13.11 The PDN with No Decoupling Capacitors

At low frequency, the VRM and the bulk decoupling capacitors provide the low impedance in the PDN. At high frequency, the on-die capacitance and on-package capacitance provide the low impedance to the PDN. We can see what the complete impedance profile might look like for this simple case using typical model parameter values.

Figure 13-21 is the simulated impedance profile for the case of power and ground planes in the board, with no added decoupling capacitors. It includes a simple VRM with bulk decoupling capacitor and 50 nF of on-die capacitance.

If the target impedance were 1 Ohm, this board would work just fine, with no added decoupling capacitors. It would not matter how many or what value capacitors were added to the board; the PDN would still have acceptable noise. Even if the target value were as low as 0.2 Ohm, as long as the current spectrum did not have any worst-case amplitude spikes in the 5 MHz to 20 MHz range, this board might work just fine.

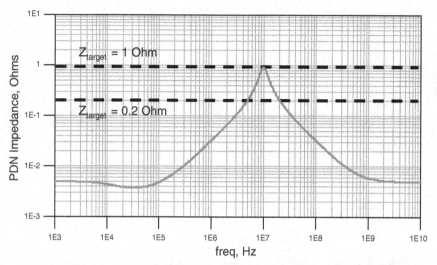

Figure 13-21   A typical impedance profile with just on-die capacitance and the VRM included.

This is why many boards work no matter what is done at the board level: because of the on-die capacitance and large bulk decoupling capacitors that are part of the VRM. This is also why it is sometimes reported that decoupling capacitors have been removed from the board, and it works just fine, thereby starting the myth that decoupling capacitors aren't very important. However, there is no guarantee that this condition will apply to your specific product application. Different chips with different current requirements and different on-die capacitance with different packages and the same board can have very different performance.

> **TIP**   In order to have confidence in a PDN design, the board-level designer must have information about the package model and the on-die capacitance, as well as the current spectrum of the chip.

While the package model and the on-die capacitance, as well as the current spectrum of the chip, are important, it is also difficult to get this information from most semiconductor suppliers. We still have to design the board-level decoupling in the absence of all the important information. In such cases, it's important to make some reasonable assumptions to base the board-level design around.

TIP   The two most common board-level design assumptions are that the package lead inductance will limit the frequency where the board-level impedance is important to below 100 MHz and that the current draw and target impedance can be estimated based on the worst-case power dissipation of the chips.

When the target impedance is 1 Ohm or above, the board design and decoupling capacitors may not play a very important role. However, achieving target impedances below 1 Ohm requires careful selection of capacitors and their integration into boards to optimize their performance.

With the correct number, value, and implementation of decoupling capacitors and power and ground planes to connect them to the VRM and the package leads, we can engineer the PDN impedance down below the mOhm range.

TIP   Knowing the behavior of individual capacitors, combinations of capacitors, and how capacitors interact with planes will lay the foundation for the most cost-effective PDN designs.

13.12 The MLCC Capacitor

An ideal capacitor has an impedance that drops off inversely with increasing frequency, given by:

$$Z = \frac{1}{2\pi fC}$$
(13-21)

where:

Z = impedance, in Ohms

f = frequency, in Hz

C = capacitance, in F

For example, the impedance profile of four ideal capacitors is shown in Figure 13-22. It is easy to believe that if this is the behavior of a capacitor, then why can't we just add a single, large capacitor to a board and use it to provide low impedance at ever higher frequencies?

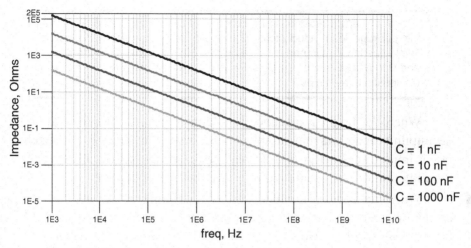

Figure 13-22   Impedance profile of ideal capacitors.

The problem with this approach is that the behavior of a real capacitor is not quite the same as that of an ideal capacitor. An example of the measured impedance of a real 0603 capacitor is shown in Figure 13-23. While the impedance starts out like that of an ideal capacitor, unlike an ideal capacitor, a real capacitor reaches a lowest impedance and then begins to increase in value.

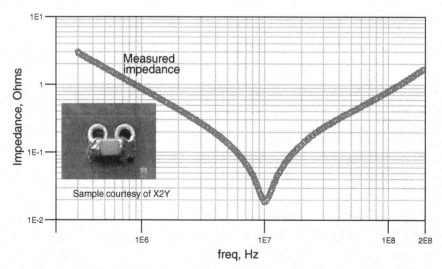

Figure 13-23   Measured impedance profile of an 0603 capacitor mounted to a test board.

A real capacitor can be approximated by a simple RLC circuit model to very high frequency. The simulated impedance of an ideal RLC circuit is an excellent match to this measured performance. Figure 13-24 shows the comparison of the measured and simulated impedance for the specific values:

R = 0.017 Ohm

C = 180 nF

L = 1.3 nH

In this model, the R, L, and C parameter values are absolutely constant with frequency. They are each ideal elements. However, when connected together in series, the resulting impedance profile of the combination of ideal elements is remarkably close to the actual measured impedance of the capacitor.

TIP The fact that an ideal RLC circuit matches the behavior of a real capacitor makes this RLC circuit model incredibility useful for modeling real capacitors, even up to very high bandwidth, above 1 GHz.

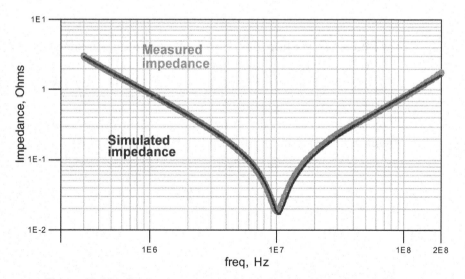

Figure 13-24 Comparing measured and simulated impedance of the 0603 MLCC capacitor.

The composite behavior of an RLC model is different than the behavior of any single element. These are compared in Figure 13-25.

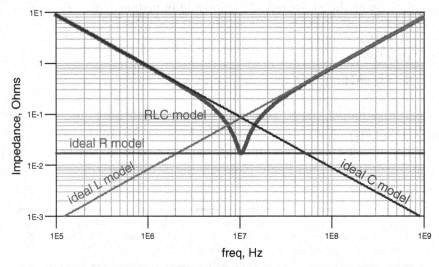

Figure 13-25 Impedance profile of the individual RLC elements that make up the RLC model.

At low frequency, the impedance of the RLC circuit is related to the ideal capacitance. At high frequency, the impedance of the RLC circuit is related to the ideal inductance. The lowest impedance of the RLC circuit is limited by the ideal resistance.

The frequency at which the impedance is the lowest is called the *self-resonant frequency* (*SRF*) and is given by:

$$f_{SFR} = \frac{1}{2\pi} \frac{1}{\sqrt{L \times C}} = \frac{159 \, \text{MHz}}{\sqrt{L \times C}} \tag{13-22}$$

where:

f_{SRF} = self-resonant frequency, in MHz
L = equivalent series inductance, in nH
C = capacitance, in nF

For example, for the real capacitor shown earlier, the self-resonance frequency is estimated to be:

$$f_{SFR} = \frac{159\,\text{MHz}}{\sqrt{1.3\,\text{nH} \times 180\,\text{nF}}} = 10.4\,\text{MHz} \qquad (13\text{-}23)$$

As can be seen in the earlier example, this is very close to the measured SRF of this capacitor.

Near the SRF, the impedance profile of the RLC circuit is not the same as the ideal L or C. It differs in a complicated way that also depends on the R value. This makes it difficult to perform simple analytical estimates but can be easily simulated with any free version of SPICE (see www.beTheSignal.com).

> **TIP**   Above the SRF, the impedance is dominated by the inductance. Reducing the high-frequency impedance is about reducing the inductance. This is the most important engineering term to adjust in the selection of capacitors and their integration into the board.

> **TIP**   Change the way you think of a capacitor. An MLCC capacitor is not a capacitor; it is an inductor with a DC block. Everything about implementing a capacitor is about the mounting inductance design, not about its capacitance.

The R is related to the series resistance of the metallization in the planes that make up the capacitors. The C is about the number of layers in the capacitor, the area of the internal planes, their separation, and dielectric constant.

13.13 The Equivalent Series Inductance

The L, often referred to as the *equivalent series inductance (ESL)*, is more about how the capacitor is mounted to the board or test fixture than the capacitor itself.

Even though many capacitor vendors offer an "intrinsic" inductance for their capacitor components, the inductance they provide is absolutely worthless and has no value in determining the performance of real capacitors. Instead, we will see how the ESL is affected by the mounting geometry of the capacitor.

Some capacitors are capable of achieving lower ESL for the same mounting features by nature of their design. It is not that they have lower intrinsic ESL but that they enable lower mounted inductance because of design features. For example, an X2Y capacitor, a type of interdigitated capacitor, will have a lower ESL under typical mounting conditions than an 0603. Figure 13-26 compares the measured impedance profile of an 0603 capacitor and an X2Y capacitor on the same board.

The impedance at low frequency for these two different capacitors is nearly the same, but their high-frequency impedance is very different. This is primarily due to the fact that the single X2Y capacitor with four terminals is really like four separate capacitors in parallel. The parallel combination of their loop inductances reduces the equivalent loop inductance of the whole capacitor. This can be a significant advantage in some designs.

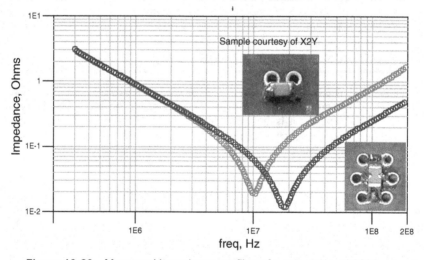

Figure 13-26   Measured impedance profiles of a conventional 0603 capacitor and an X2Y interdigitated capacitor on the same board. They have exactly the same value of capacitance, seen at low frequency, but very different ESL.

The complete path of the power and return currents from the pads of the BGA package to the capacitor is shown in Figure 13-27. The ESL of the capacitor is, to first order, related to design features in this path.

The ESL associated with the capacitor and its path to the package can be divided into four regions:

1. The loop inductance of the surface traces and top of the plane's cavity
2. The loop inductance of the vias from the capacitor pads to the top of the plane cavity
3. The spreading inductance from the capacitor vias to the vias of the BGA
4. The loop inductance from the cavity under the package to the leads or solder balls of the package

---

T I P Different design techniques should be applied to each region in order to engineer the lowest ESL possible.

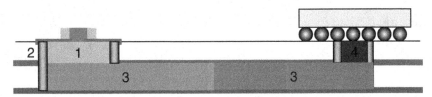

Figure 13-27 The ESL associated with a capacitor can be separated into four distinct regions.

When only a few capacitors are used on a board, and the current distributions in the planes from the capacitors to the pins of the package do not substantially overlap, the ESL of each capacitor is the loop inductance of the entire path. In this case, each capacitor behaves independently, and it is possible to accurately simulate the impedance profile of the parallel combinations of the capacitors on the board by using a simple SPICE model and simulation. The capacitors are independent.

However, when the current distributions overlap, such as when the capacitors are clustered in one region of the board or when many capacitors surround a package, the spreading inductance in the cavity of the power and ground cavity will be a complicated function of the location of the capacitors, their values, and the location of the package pins.

This is why it is useful to separate the ESL of a capacitor into the mounting inductance and the spreading inductance in the cavity. When the capacitors do not interact with each other, the cavity spreading inductance can be combined with the mounting inductance into the ESL. When the capacitors' spreading inductances interact, the only accurate way of estimating the impedance profile seen by the package is with a 3D simulator, which takes into account the current distribution of each capacitor. In this case, the location of the capacitors and the location of the power and ground pins in the package will be important.

TIP   It's always a good practice to separate the mounting inductance and the cavity spreading inductance. They can be combined when needed into one number to estimate the ESL.

13.14 Approximating Loop Inductance

There are only a few geometries for which there are simple approximations for loop inductance:

- Any uniform transmission line
- The special case of two round rods
- A pair of long, wide conductors with a thin dielectric between them
- The special case of edge-to-edge connections to planes
- Spreading inductance from a via to a distant ring
- Spreading inductance between two via contacts in a plane

The loop inductance of any uniform transmission line, assuming that the signal and return paths are shorted at the far end, is given by:

$$L_{loop} = Z_0 \times TD = \frac{Z_0 \times Len}{v}$$

(13-24)

where:

L_{loop} = loop inductance, in nH

Z_0 = characteristic impedance, in Ohms

TD = time delay of the transmission line, in nsec

Len = length of the transmission line, in inches

v = speed of light in the material, in inches/nsec

When the impedance of the line is 50 Ohms, such as a surface microstrip trace that is 10 mils wide and dielectric spacing in FR4 to the return path of 5 mils, the loop inductance is roughly:

$$L_{loop} = Z_0 \times TD = \frac{Z_0 \times Len}{v} = \frac{50 \times Len}{6} = 8.3 \frac{nH}{in} \times Len \qquad (13\text{-}25)$$

For a surface trace that is 0.2 inches long, the loop inductance of the surface trace can be as large as 1.7 nH.

This simple relationship suggests the two important design guidelines for engineering the lowest loop inductance possible for any structure that sort of looks like a uniform transmission line:

• Design the lowest a characteristic impedance possible.
• Keep the lengths as short as possible.

A special structure for which there is an analytical relationship between the geometry and loop inductance is two round rods, as illustrated in Figure 13-28.

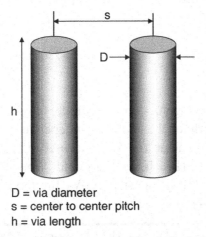

D = via diameter
s = center to center pitch
h = via length

Figure 13-28 Geometry for two round rods, similar to two vias.

The loop inductance from the end of one rod, down the rod, shorting across the end of the other rod and back again to the front is related to only the three geometry terms in Figure 13-28. If the length is increased, the loop inductance will increase. If the rods are brought closer together, their partial mutual inductance will help to cancel some of the total field lines, and the loop inductance will be reduced. If the diameter of the rods is increased, the loop inductance will be decreased.

There are a number of analytical approximations for the loop inductance of these two rods. The simplest approximation is:

$$L_{loop} = 10 \times h \times \ln\left(\frac{2s}{D}\right) pH \qquad (13\text{-}26)$$

where:

L_{loop} = loop inductance, in pH

h = length of the rods, in mils

s = center-to-center pitch of the rods, in mils

D = diameter of each rod, in mils

For example, for 2 vias, 10 mils in diameter, on 50-mil centers and 100 mils long going through an entire board, the loop inductance is roughly:

$$L_{loop} = 10 \times 100 \times \ln\left(\frac{2 \times 50}{10D}\right) pH = 2300 \ pH = 2.3 \ nH \qquad (13\text{-}27)$$

The uniform transmission-line model gives the same constant loop inductance per length for the two rods, independent of the rod length. For the case of 10-mil via diameter and 50-mil centers, the loop inductance per pair-length is roughly 23 nH/inch, or 23 pH/mil. When the center-to-center pitch is 40 mils, typical of high-density BGAs, the loop inductance per length is 21 nH/inch, or 21 pH/mil.

TIP   As a rough rule of thumb, if you want to carry around one value for the loop inductance of a pair of vias, a rough estimate is about 21 pH/mil. This is a reasonable estimate for the loop inductance contribution from vias.

When the two conductors that make up the loop are wide and closely spaced, such as with two plane segments shown in Figure 13-29, the loop inductance is approximated by:

$$L_{loop} = \left(32\frac{pH}{mil} \times h\right) \times \frac{Len}{w} pH \qquad (13\text{-}28)$$

where:

L_{loop} = loop inductance between the planes, in pH
Len = length of the planes, in inches
w = width of the planes, in inches
h = thickness between the two planes in mils

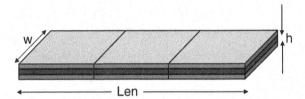

Figure 13-29   Geometry for the loop inductance of two plane segments.

For example, if the planes are 2 inches long and 0.5 inches wide, with 4 mils between them, the loop inductance would be:

$$L_{loop} = \left(32\frac{pH}{mil} \times 4\right) \times \frac{2}{0.5} = 512\ pH = 0.5\ nH \qquad (13\text{-}29)$$

When the length of the trace is equal to the width, the structure looks like a square, and the ratio of Len/w is always 1. The loop inductance of this square section of plane is the first part of the equation and is called the *loop inductance* per square, or the *sheet inductance*:

$$L_{square} = \left(32\frac{pH}{mil} \times h\right) \qquad (13\text{-}30)$$

Any square piece of a pair of planes has the same loop inductance. The thinner the dielectric between them, the lower the sheet loop inductance.

This approximation assumes that the currents flow in a uniform sheet down the top trace and back to the bottom, uniformly distributed along both sheets. When the contacts are spread along the edge of the strip, this is a good approximation. However, in via contacts to planes, the current does not flow uniformly. Instead, it spreads out from sources and constricts into sinks. An example of the current flow map in a plane between two via contacts is shown Figure 13-30.

The spreading inductance in planes is the most important property of planes and is discussed in detail in Chapter 6. It contributes to the additional loop inductance between point contacts in planes over their sheet inductance when contacts are at vias rather than at an edge of the plane.

The narrow contact regions of vias increase the current density and increase the local loop inductance. In general, spreading inductance is complicated to calculate and usually requires a 3D field solver, as the current flow is difficult to calculate by approximation.

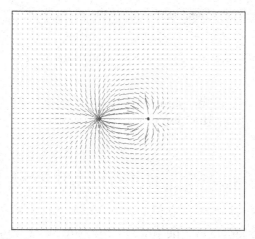

Figure 13-30   Current flow pattern in the top plane from a via source point to a via sink point into the bottom plane. Simulated with HyperLynx.

There is one special case for which there is an accurate approximation for spreading inductance. This is the case of current flowing from a central-ring contact to an outer, symmetrical-ring contact, where it flows into the bottom plane and then reverses back, constricting to an inner ring contact on the bottom plane. This is diagrammed in Figure 13-31.

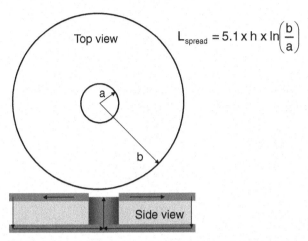

Figure 13-31 Inner and outer contact regions on the top plane, with similar regions on the bottom plane. Spreading inductance calculation is the loop inductance from the top contact point, radially outward to the edge, down the edge, and back in to the center contact.

In this geometry, the loop spreading inductance is:

$$L_{spread} = 5.1 \times h \times \ln\left(\frac{b}{a}\right) pH \qquad (13\text{-}31)$$

where:

L_{spread} = loop spreading inductance between the planes, in pH
a = radius of the inner contact region, in inches
b = radius of the outer contact region, in inches
h = thickness between the two planes, in mils

This assumes that the current is flowing from the center via contact to the bottom plane and returns to the inside edge of the clearance hole, and the clearance hole is just slightly larger than the via contact diameter in the bottom plane. For example, if the inner radius is 5 mils, corresponding to a 10-mil-diameter via, and the outer radius is 1 inch, corresponding to the perimeter of a package, and the dielectric thickness between the planes is 10 mils, the loop spreading inductance is:

$$L_{spread} = 5.1 \times 10 \times \ln\left(\frac{1}{0.005}\right) = 270 \text{ pH} \qquad (13\text{-}32)$$

This relationship of spreading inductance has the same form as the sheet loop inductance of a path, if we use as the number of squares:

$$n = \frac{1}{2\pi} \ln\left(\frac{b}{a}\right) \qquad (13\text{-}33)$$

then:

$$L_{spread} = \left(32\frac{\text{pH}}{\text{mil}} \times h\right) \times n \text{ pH} \qquad (13\text{-}34)$$

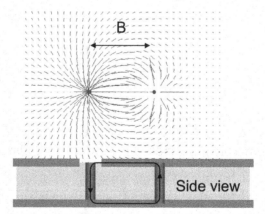

Figure 13-32   Spreading current from one via contact to another via contact in a pair of planes. There is spreading loop inductance between the two locations.

For typical cases, b/a could be on the order of 100, and the number of squares is of order 1.

In the special case of the current flow between the via contacts to a buried plane pair from a capacitor and BGA pins on the surface of a board, the loop inductance is more complicated to calculate. There are no exact analytical equations that describe this loop spreading inductance. However, by making a few assumptions, a simple approximation can be developed for the loop inductance in a pair of planes with round contact points.

Figure 13-32 illustrates the example of two via contacts positioned a distance B apart in a pair of planes and the current flow between them, spreading out and constricting in the planes.

The spreading loop inductance between these two via contacts is approximated by:

$$L_{via-via} = 21 \times h \times \ln\left(\frac{B}{D}\right) pH \tag{13-35}$$

where:

$L_{via-via}$ = loop spreading inductance in the planes between the two via contacts, in pH

h = dielectric thickness between the vias, in mils

B = distance between the via centers, in mils

D = diameter of the vias, in mils

For example, if the via diameters are 10 mils and they are spaced 1 inch apart, in a pair of planes with h = 10 mils, the spreading inductance in the planes between the contacts is about:

$$L_{via-via} = 21 \times 10 \times \ln\left(\frac{1000}{10}\right) = 967 \, pH \sim 1 \, nH \tag{13-36}$$

The contribution of the spreading inductance in the planes between the vias can be as much as 1 nH. The thinner the dielectric, the lower the spreading inductance.

> **TIP**   The lower spreading inductance in ultra-thin laminates is the real
> reason they offer a performance advantage over conventional FR4 for the
> dielectric between power and ground planes. The higher capacitance plays
> little role as there is far more capacitance in the on-die capacitance than in
> the power and ground planes.

If the connections between the capacitors and the pads of the package can be routed in planes with a cavity thickness that is not 4 mils but 1 mil or 0.5 mil, the spreading inductance in this path can be reduced from 0.4 nH with 4 mils down to 0.05 nH with a 0.5-mil-thick dielectric. An example of the cross section of a board with a 0.5-mil-thick dielectric in the power and ground planes is shown in Figure 13-33.

The predicted values of this approximation can be compared to the results predicted by a 3D field solver. Figure 13-34 shows the estimates of these approximations to the simulated via to via spreading inductance using the HyperLynx PI tool for two planes separated by 3 mils.

These various approximations can be used to roughly estimate the impact of physical design features and the resulting ESL of a capacitor mounted to a board. Using these approximations, we can explore design space to determine what general design guidelines to follow.

> **TIP**   Since each design is custom, care must be taken when applying
> an observation for one case and blindly applying it to another without
> putting in the numbers.

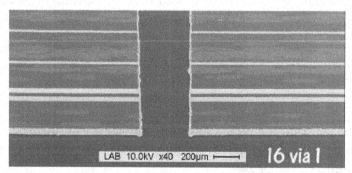

Figure 13-33   Cross section of a board with a 0.5-mil-thick layer of DuPont Interra HK04 laminate between the power and ground planes, close to the bottom surface of the board.

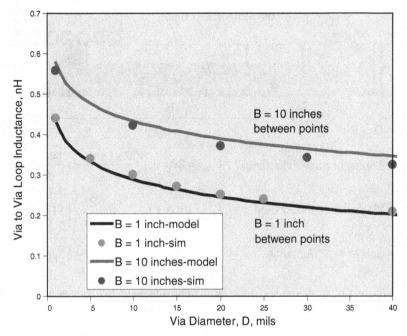

Figure 13-34 The comparison of the approximation (solid lines) and the simulated loop inductance (single dots), using HyperLynx for the case of a pair of planes separated by 3 mils.

13.15 Optimizing the Mounting of Capacitors

The three most useful approximations for loop inductance are summarized in one place in Figure 13-35.

These approximations describe the important design trade-offs. If you want to reduce the loop inductance associated with the traces from the pads of the capacitor to the vias, there are three important design knobs to adjust:

1. Keep the depth to the top of the power/ground cavity thin.
2. Use wide surface traces.
3. Keep the length of the surface traces short.

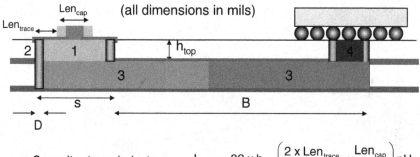

- Capacitor trace inductance $\quad L_{trace} = 32 \times h_{top} \left(\dfrac{2 \times Len_{trace}}{W_{trace}} + \dfrac{Len_{cap}}{W_{cap}} \right) pH$

- Via pair loop inductance $\quad L_{vias} = 10 \times h_{top} \times \ln\left(\dfrac{2s}{D} \right) pH$

- Spreading inductance $\quad L_{spread} = 21 \times h_{planes} \times \ln\left(\dfrac{B}{D} \right) pH$

Figure 13-35   Summary of the three approximations to estimate the ESL of a capacitor.

To reduce the inductance of the vias, there are three design knobs to adjust:

1. Keep the depth to the top of the power/ground cavity short.
2. Use large-diameter vias.
3. Keep the via pitch as close as possible.

To reduce the spreading-loop inductance in the planes, there are three knobs to adjust:

1. Keep the dielectric thickness of the power/ground cavity thin.
2. Use large-diameter vias or multiple vias in contact to the cavity.
3. Place the capacitor close to the package it is decoupling. (This is only a weak dependence.)

While these are important design guidelines to be aware of, some are more important than others.

TIP   The terms that affect the total loop inductance the most should always be optimized first.

These are the terms that affect the total loop inductance the most:

1. Keep the depth to the top of the power/ground cavity short.
2. Keep the dielectric thickness of the power/ground cavity thin.
3. Use wide surface traces.
4. Keep the length of the surface traces short.

The other design features are of second- and third-order importance and can sometimes be a distraction from the first-order concerns. In general, the only way to know what is important is to put in the numbers for specific cases. Integrating these approximations in a spreadsheet allows us to easily explore design space and identify what is really important and what is not.

In the example shown in Figure 13-36, three cases are explored. In each case, an 0603 capacitor is supplying current to one power and ground pin pair in a BGA package, located some distance away. The vias are 13 mil in diameter. This estimate is for the ESL of the capacitor as though it were not interacting with other capacitors. Case 1 is the starting place, with long and narrow surface traces. The total ESL is found to be about 6.1 nH.

In case 2, the surface traces are shortened and widened. The resulting ESL is 3.7 nH. Finally, in case 3, the capacitor is moved closer to the package, and the cavity thickness is decreased. The resulting loop inductance is reduced to 1.8 nH.

TIP   This example clearly shows that in typical cases, the loop inductance of the vias is negligible. In most typical cases, especially with thick spacing between the planes, the spreading inductance can be as significant as the surface trace inductance. By careful design of the stack-up, it is possible to routinely achieve less than 2-nH loop inductance.

Surprisingly, the board stack-up plays a significant role in the ESL of the capacitor—in two respects. By moving the top of the cavity closer to the capacitor, the loop inductance of the capacitor and the surface traces is reduced. By making the dielectric thickness of the cavity between the power and ground planes thinner, the spreading inductance is reduced. Adjusting these two design features can bring the ESL from 6 nH to 1 nH in some cases.

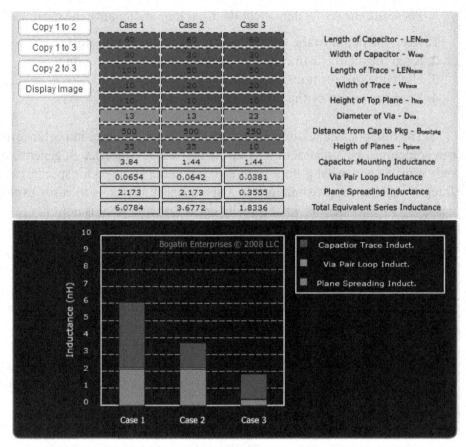

	Case 1	Case 2	Case 3	
Copy 1 to 2	60	60	60	Length of Capacitor - LEN_cap
Copy 1 to 3	30	30	30	Width of Capacitor - W_cap
Copy 2 to 3	100	50	50	Length of Trace - LEN_trace
Display Image	10	20	20	Width of Trace - W_trace
	10	10	10	Height of Top Plane - h_top
	13	13	23	Diameter of Via - D_via
	500	500	250	Distance from Cap to Pkg - B_cap2pkg
	35	35	10	Heigth of Planes - h_plane
	3.84	1.44	1.44	Capacitor Mounting Inductance
	0.0654	0.0642	0.0381	Via Pair Loop Inductance
	2.173	2.173	0.3555	Plane Spreading Inductance
	6.0784	3.6772	1.8336	Total Equivalent Series Inductance

Figure 13-36   Analysis of three typical mounting geometries for an 0603 capacitor, analyzed with an online tool at www.beTheSignal.com.

Both of these design features are first order and linear in the thickness. Changing via the diameter and moving the capacitor closer to the BGA are log-dependent factors and are of only slight (second- or third-order) importance.

If the surface trace length is also reduced and widened, ESL values as low as 0.5 nH can be achieved. An example of three similar cases is shown in Figure 13-37.

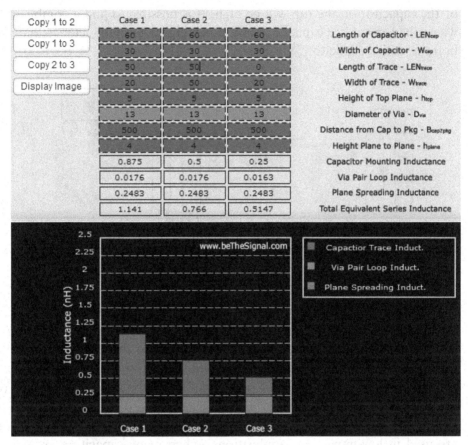

	Case 1	Case 2	Case 3	
Copy 1 to 2	60	60	60	Length of Capacitor - LEN$_{cap}$
Copy 1 to 3	30	30	30	Width of Capacitor - W$_{cap}$
Copy 2 to 3	50	50	0	Length of Trace - LEN$_{trace}$
Display Image	20	50	20	Width of Trace - W$_{trace}$
	5	5	5	Height of Top Plane - h$_{top}$
	13	13	13	Diameter of Via - D$_{via}$
	500	500	500	Distance from Cap to Pkg - B$_{cap2pkg}$
	4	4	4	Height Plane to Plane - h$_{plane}$
	0.875	0.5	0.25	Capacitor Mounting Inductance
	0.0176	0.0176	0.0163	Via Pair Loop Inductance
	0.2483	0.2483	0.2483	Plane Spreading Inductance
	1.141	0.766	0.5147	Total Equivalent Series Inductance

Figure 13-37 Three examples of thin cavity, close to the surface with three different surface traces, resulting in an ESL as low as 0.5 nH, analyzed with an online tool at www.beTheSignal.com.

This model can also be used to assess important design questions such as Is it better to add capacitors under the BGA or on the same surface as the BGA? Figure 13-38 illustrates the two options.

Of course, the most common answer to all signal integrity questions is "it depends," and the only way to get a firm answer is by putting in the numbers.

The right place to put the capacitor is where it will have the lowest loop inductance. Clearly, if the total board thickness is thin, the via loop inductance will be low. If the cavity is far from the surface and thick, the loop inductance of the capacitor on the top will be high. It is possible to find a combination where the top surface capacitor has a much higher loop inductance than the bottom surface capacitor.

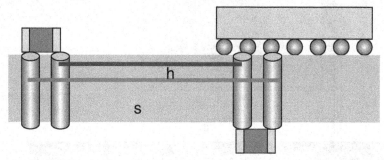

Figure 13-38   Where should the capacitor go: on the same surface as the BGA or directly under the BGA?

However, if the board is thick and the cavity is near the top surface and the cavity is thin, the capacitor on the bottom will have the higher loop inductance. Figure 13-39 summarizes three cases. It shows that placing the capacitor on the bottom can have a loop inductance on the order of 2 nH.

If it is possible to achieve lower loop inductance by placing capacitors on the top surface, this is preferred, but as a general rule, if there is the option of doing both, both locations should be used, especially when many capacitors are used in low-impedance applications. When many capacitors are placed around the periphery of the package, their currents can overlap, and the cavity spreading inductance can increase. Placing some of the capacitors under the BGA minimizes the increase in cavity spreading inductance.

TIP   The combination of short, wide surface traces—or via-in-pad technologies—and thin dielectric between the power and ground planes close to the surface can result in typical ESL values from 0.5 to 2 nH. By going to extreme efforts and utilizing interdigitated capacitors, it is possible to achieve loop inductances below 0.5 nH.

If the capacitor mounting inductance is known, based on the design constraints, it will be possible to predict the impedance profile of a collection of capacitors using a 3D field solver. If the mounting inductance changes, as from a stack-up change or a surface-mounting design change, the loop inductance will change, and the impedance profile of the collection of capacitors will change. This is why every PDN design is custom.

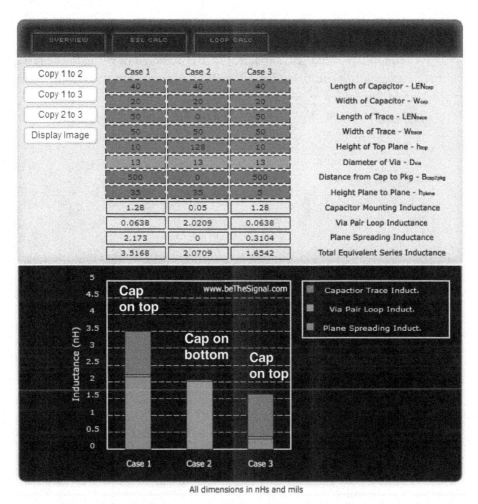

	Case 1	Case 2	Case 3	
	40	40	40	Length of Capacitor - LEN_{cap}
	20	20	20	Width of Capacitor - W_{cap}
	50	0	50	Length of Trace - LEN_{trace}
	50	50	50	Width of Trace - W_{trace}
	10	128	10	Height of Top Plane - h_{top}
	13	13	13	Diameter of Via - D_{via}
	500	0	500	Distance from Cap to Pkg - $B_{cap2pkg}$
	35	35	5	Height Plane to Plane - h_{plane}
	1.28	0.05	1.28	Capacitor Mounting Inductance
	0.0638	2.0209	0.0638	Via Pair Loop Inductance
	2.173	0	0.3104	Plane Spreading Inductance
	3.5168	2.0709	1.6542	Total Equivalent Series Inductance

All dimensions in nHs and mils

Figure 13-39 Analysis of capacitors on the top and on the bottom of the board. Analyzed with an online tool at www.beTheSignal.com.

TIP The PDN impedance profile of the combination of capacitors depends very much on the details of the board stack-up, capacitor mounting geometry, and location on the board.

13.16 Combining Capacitors in Parallel

The strategy in engineering the PDN impedance profile is to select the right number and value of capacitors to keep the peak impedance below the target value from where the VRM and bulk capacitors no longer provide low impedance, up to about 100 MHz.

When multiple identical capacitors are connected in parallel, the resulting impedance matches the behavior of an RLC circuit, but the circuit elements values are different.

The equivalent R, L, and C of n capacitors in parallel are:

$$C_n = nC \tag{13-37}$$

$$ESR_n = \frac{1}{n}ESR \tag{13-38}$$

$$ESL_n = \frac{1}{n}ESL \tag{13-39}$$

where:

C_n = equivalent capacitance of n identical real capacitors in parallel

C = capacitance of each individual capacitor

n = number of identical capacitors in parallel

ESR_n = equivalent series resistance of n identical real capacitors in parallel

ESR = equivalent series resistance of each individual capacitor

ESL_n = equivalent series inductance of n identical real capacitors in parallel

ESL = equivalent series inductance of each individual capacitor

Figure 13-40 shows an example of the impedance profile of multiple identical capacitors in parallel, showing the same general RLC profile but with lower impedance at all frequencies. We are approximating the problem by assuming that the capacitors are independent and their currents do not overlap.

The SRF stays the same; it's the entire impedance profile that scales lower. This is one way of decreasing the impedance profile of a capacitor: Add more of them in parallel.

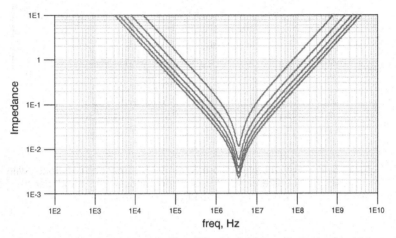

Figure 13-40   Impedance profile of five identical capacitors added in parallel. With each additional capacitor, the impedance decreases at all frequencies.

However, if the two capacitors have a different value of capacitance or ESL, when they are added in parallel, the behavior is not so simple. Figure 13-41 shows the impedance profiles of two different capacitors with the same ESL and the same ESR. The behavior of the two capacitors in parallel has the same low-impedance dips at the self-resonant frequencies of the individual RLC models. The larger capacitor has the lower SRF. The smaller capacitor has the higher SRF. They each occur when the impedance of the ideal capacitor matches the impedance of the ideal inductance associated with each capacitor. The SRF seen in the parallel combination of capacitors is the same as each individual capacitor's.

In addition, there is a new feature between the self-resonant frequencies: a peak in the impedance, called the *parallel resonant peak*, that occurs at the *parallel resonant frequency (PRF)*.

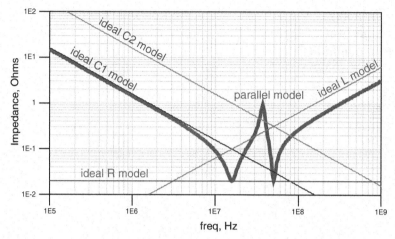

Figure 13-41   The impedance profile of two RLC circuits in parallel, with the same R and L values but different C values. Superimposed is the impedance of the two ideal capacitors and the ideal L and ideal R of both capacitors.

The value of the PRF is difficult to calculate accurately, as it depends on the ESL of the larger capacitor, the C of the smaller capacitor, and the ESR of both of them. If the SRF values are far apart, the PRF is roughly related to:

$$PRF \approx \frac{1}{2\pi} \frac{1}{\sqrt{C_2 \times ESL_1}} = \frac{160 \text{ MHz}}{\sqrt{C_2 \times ESL_1}} \qquad (13\text{-}40)$$

where:

PRF = parallel resonant frequency, in MHz

C_2 = capacitance of the smaller capacitor, in nF

ESL_1 = equivalent series inductance of the first capacitor, in nH

For example, if ESL_1 = 2 nH and C_2 = 10 nF, then the PRF is:

$$PRF \approx \frac{160 \text{ MHz}}{\sqrt{10 \times 2}} = 36 \text{ MHz} \qquad (13\text{-}41)$$

However, when the SRF values are within a factor of 10 of each other, the impedance profile of the parallel combination is distorted from the impedance of the ideal L. The PRF is a more complicated function of the elements, and can more easily be calculated using a SPICE simulation.

TIP   The PRF is one of the most important features of parallel combinations of capacitors because it denotes where there are peaks in the impedance. When few capacitors are used, it's the parallel resonant impedance that always sets the limit to the PDN performance and must be engineered to lower values.

The peak impedance at the PRF is roughly related to:

$$Z_{peak} \sim \frac{L_1}{C_2}\left(\frac{1}{R_1+R_2}\right) \tag{13-42}$$

where:

Z_{peak} = peak impedance at the PRF, in Ohms

L_1 = equivalent series inductance of the larger capacitor

C_2 = capacitance of the smaller capacitor

R_1 = equivalent series resistance of the larger capacitor

R_2 = equivalent series resistance of the smaller capacitor

This is only approximate and is less accurate as the SRFs of the capacitors are brought closer together. However, it points out the important ways of engineering a reduction in the peak impedance:

- Reduce the ESL of the larger capacitor.
- Increase the capacitance of the smaller capacitor.
- Increase the ESR of both capacitors.

TIP   Where there is the option to use higher-ESR capacitors—referred to as *controlled resistance* capacitors—they should be considered. A low enough ESR should be selected so that the equivalent ESR of all the capacitors in parallel is just below the target impedance.

The ESR of a capacitor is related to the structure of the parallel plates that make it up and the metallization between the layers. In general, the higher the capacitance, the more plates in parallel and the lower the ESR.

Looking at the specifications of a variety of 0402 capacitors can provide a simple generalization for the series resistance of capacitors by capacitor value. Figure 13-42 shows the plotted ESR for various capacitor values, taken off the AVX data sheets.

From the specified ESR, it is possible to derive a simple empirical relationship between the ESR and the capacitance of a capacitor. One empirical approximation is given by:

$$ESR \approx \frac{180 \text{ m}\Omega}{2.5^{\log(C)}} \tag{13-43}$$

where:

ESR = equivalent series resistance of the capacitor, in mOhms

C = capacitance of the capacitor, in nF

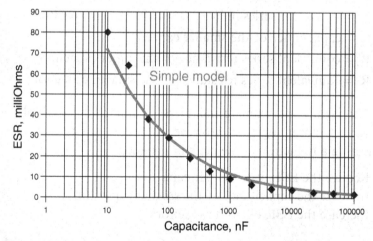

Figure 13-42   ESR and capacitance for 0402 capacitors, taken from AVX data sheets of capacitors.

This simple model is compared with the specified values of ESR in Figure 13-42, and the agreement is seen to be very good.

This suggests that it may be possible to select for higher ESR and lower parallel resonant peak heights if smaller-value capacitors are used. This is especially true when one of the capacitors is the power and ground cavity's capacitance.

Another important design feature to engineer to decrease the peak impedance value is decreasing the ESL of the larger capacitor or increasing the capacitance of the smaller capacitor. Figure 13-43 shows the impact on the peak impedance as the ESL of the larger capacitor is changed from 10 nH down to 0.1 nH.

In this example, the larger capacitor is 100 nF, and the smaller one is 10 nF, with an ESL of 3 nH. As the ESL of the larger capacitor is reduced from 10 nH, the peak impedance at the PRF decreases until the SRF of the larger capacitor matches the SRF of the smaller capacitor, in which case there is no peak.

TIP   Reducing the ESL is a significant method of reducing peak impedances.

Unfortunately, due to the complex interactions of the circuit elements, it is not possible to do a simple and accurate analytical analysis of the features of the impedance profile of multiple capacitors. This is especially true as more capacitors are added. Instead, it is critical to use SPICE for such analysis. Luckily, there are many free versions of SPICE readily available on the Internet that can routinely perform this sort of analysis. For examples of these tools, visit beTheSignal.com.

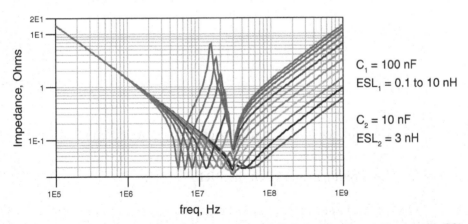

Figure 13-43   Impedance profile of a 100-nF and a 10-nF capacitance with ESL of 3 nH in parallel while changing the ESL of the larger capacitor from 10 nH down to 0.1 nH. As the ESL is reduced, the peak impedance drops.

In addition to reducing the ESL of the capacitors to bring the peak impedance values down, there is another way of reducing the peak impedances: Add more capacitors. These can be identical capacitors, or they can be different value capacitors. Both approaches can work.

13.17 Engineering a Reduced Parallel Resonant Peak by Adding More Capacitors

When two capacitors with different SRFs are added in parallel, they create a parallel resonant peak impedance between their self-resonant dips. The peak impedance can be reduced by adding a third capacitor with an SRF between them. What is the optimum value of the SRF of the third capacitor?

The optimum SRF of the third capacitor to give the lowest peak impedance depends on the capacitance values, the ESL values, and the ESR values of all three capacitors. It is difficult to evaluate without a SPICE simulation. There are two obvious algorithms to choose from: Select the third capacitor so that its SRF matches the PRF, or select its SRF so that it is midway between the two other capacitors' SRFs. The choice depends on the ESL values of the capacitors, their ESR values, and how far apart are the capacitances. Consider the simple case of two capacitors, a 10-nF capacitor and a 100-nF capacitor, both with the same ESL of 3 nH. When combined in parallel, they have a PRF at 21 MHz.

Option 1 is to add a capacitor with its SRF at 21 MHz. The capacitance value is given by:

$$C_3 = \left(\frac{160}{f_{peak}}\right)^2 \frac{1}{ESL} = \left(\frac{160}{21}\right)^2 \frac{1}{3} = 19.3 \, nF \qquad (13\text{-}44)$$

where:

C_3 = capacitance of the third capacitor to be added, in nF

ESL = 3 nH, assumed to be the same for each of the three capacitors

21 = SRF required to match the PRF, in MHz

Option 2 is to select the value of the third capacitor so its SRF is midway (on a log scale) between the SRFs of the other two capacitors. Since their ESL values are the same, this translates to a capacitance of the third capacitor that is the geometric mean of the other two:

$$C_3 = \sqrt{C_1 C_2} = \sqrt{100 \times 10} = 33 \, nF \qquad (13\text{-}45)$$

Figure 13-44 shows the resulting simulation of the original combination of two capacitors and the impedance profile of the three capacitors, with the value of the third capacitor chosen based on the two options.

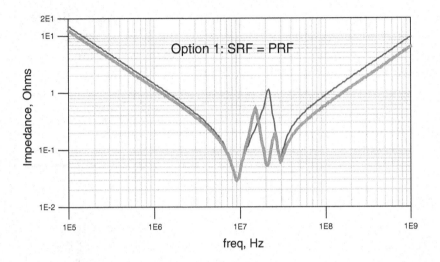

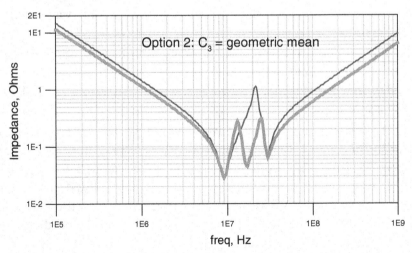

Figure 13-44   Comparing the two options for selecting the third capacitor value. The approach that gives the lowest peak impedance is Option 2: Choose the geometric mean.

This example illustrates that when the ESL of the capacitors is the same, the lowest peak impedance can be engineered by choosing a third capacitor that is the geometric mean of the other two capacitors. This is why it is often recommended to select capacitor values spread over a decade scale. If the values are distributed uniformly in a log scale, they will provide the lowest peak impedance.

The only way to know the optimum capacitor values that result in the lowest peak impedances, given the ESL and ESR values, is with a SPICE simulation. When the SRF values for the two capacitors are far apart, using a third capacitor with an SRF near the PRF may be a better choice.

TIP   Whenever different-value capacitors are brought together in parallel, there will always be parallel resonant peaks that must be managed. This will occur at low frequency with the bulk capacitors and at high frequency with the capacitance of the planes and the on-die capacitance.

13.18 Selecting Capacitor Values

Many application notes recommend that all you need to do is add three capacitors per power and ground pin in the package. Half of them recommend using all three capacitors the same value, and the other half of them recommend using different capacitor values. Which is right? The only way to tell is to put in the numbers.

Figure 13-45 compares the impedance profiles for three capacitors all with the same value of 1 μF and three capacitors with values of 1 μF, 0.1 μF, and 0.01 μF. In each case, the ESL is the same 3 nH, and the ESR is chosen based on the values specified by the manufacturer.

At first glance, the conclusion would be that the three capacitors all with the same value give the lowest impedance, but, at 100 MHz, the impedance limit they both set is the same, at about 0.6 Ohm. However, this analysis neglects two important effects: at the low-frequency end the interaction with the VRM and bulk capacitor and at the high-frequency end the interactions with the planes of the board, or the on-die capacitance and package lead inductance.

The argument for using three different capacitor values is often that they produce very low impedance at specific frequency regions. While this is true, it's totally irrelevant. What's important in the PDN is not how low the impedance profile goes but how high it goes. Peaks in the impedance profile cause failures, and the PDN should be designed to handle this.

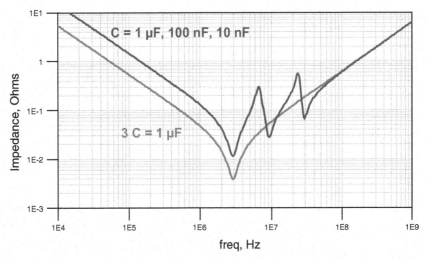

Figure 13-45 Impedance profiles for the case of three capacitors, each with a different value, and the case of three capacitors all having the same value.

TIP Dips in the impedance profile from self-resonant frequencies of capacitors are irrelevant. Peaks in the impedance profile cause failures and are therefore important. The PDN should therefore be engineered to control peaks.

The resulting impedance profiles for these two sets of capacitors, a VRM with a bulk decoupling capacitor and the planes for a board 5 inches on a side and 4 mils thick, is shown in Figure 13-46.

TIP At the low frequency, the interactions of the bulk capacitor and the small ceramic capacitors cause the impedance peak at about 1 MHz. The peak impedance is primarily due to the inductance of the bulk capacitor and the capacitance of the MLCC capacitors.

An important way to drop this peak down is to reduce the inductance of the bulk capacitors. In this example, it was assumed to be 15 nH, typical of an electrolytic capacitor. If this can't be reduced by design, one way it can be dropped is by adding more capacitors in parallel. As long as their SRF is lower than the peak impedance at roughly 1 MHz, their ESL in parallel with the electrolytic capacitor will reduce the peak impedance.

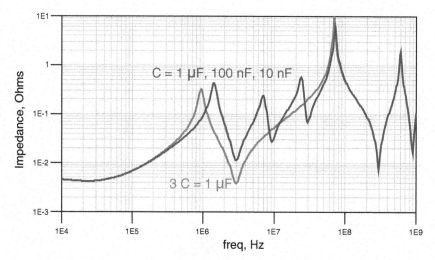

Figure 13-46 The parallel resonances at the boundaries cause peak impedances in both combinations of capacitors.

The minimum capacitance needed can be found from a simple estimate. If we assume that we use a tantalum capacitor, with an ESL on the order of 5 nH, in order to have SRF < 1 MHz, the condition is:

$$C_3 = \left(\frac{160}{f_{PRF}}\right)^2 \frac{1}{ESL} = \left(\frac{160}{1}\right)^2 \frac{1}{5} \sim 5\,\mu F \tag{13-46}$$

By adding an additional capacitor of more than 5 μF and less than 5 nH ESL, the peak impedance at low frequency can be reduced. The precise value of the capacitance is not important; the ESL value is important.

The new impedance profile with a 10 μF tantalum capacitor added is shown in Figure 13-47. The high-frequency impedance peak is created by the interactions of the capacitance of the planes and the ceramic capacitors. The capacitance in a pair of planes is:

$$C_{planes} = 0.225 \times Dk \frac{A}{h} \tag{13-47}$$

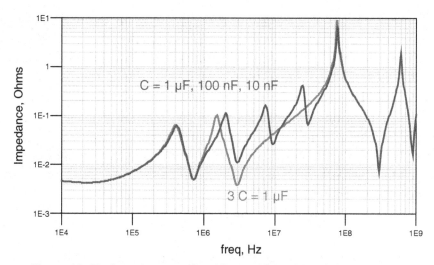

Figure 13-47 Impedance profile with an additional 10-μF bulk decoupling capacitor added with an ESL of 5 nH, reducing the peak impedance at low frequency to under 0.1 Ohm.

where:

C_{planes} = capacitance in the planes, in nF

Dk = dielectric constant of the laminate materials, typically 4 for FR4

A = area of the planes, in square inches

h = dielectric thickness, in mils

For example, in this case with A = 5 inches × 5 inches = 25 square inches and h = 4 mil and Dk = 4, the capacitance of the planes is:

$$C_{planes} = 0.225 \times 4 \frac{25}{4} = 5.6\,nF \tag{13-48}$$

The parallel resonant frequency is expected at roughly:

$$f_{PRF} = \frac{160\,MHz}{\sqrt{\frac{1}{n}ESL \times C_{planes}}} = \frac{160\,MHz}{\sqrt{\frac{1}{3}3 \times 5.6}} = 67\,MHz \tag{13-49}$$

The simulated PRF is 70 MHz.

The peak impedance at the PRF is related to the inductance of the capacitors and the capacitance of the planes. In this case, the inductance of the three capacitors is identical, independent of the value of their capacitance. This is why the peak impedance is exactly the same whether we use three capacitors of the same value or three capacitors with different values.

This peak impedance limits the PDN impedance at the board level to about 10 Ohms. If no worst-case current amplitudes are near 70 MHz, this impedance peak may not be a problem. But if you are designing the PDN assuming that you need a target impedance below 10 Ohms, this peak impedance needs to be brought down.

There are six ways of reducing the peak impedance for frequencies below 100 MHz:

1. Increase the capacitance in the planes a lot to push its SRF to very low frequency with a lower peak impedance.
2. Decrease the capacitance of the planes so the PRF is well above 100 MHz.
3. Reduce the inductance of the decoupling capacitors.
4. Increase the ESR of the capacitors.
5. Adjust a capacitor value so its SRF is closer to the PRF.
6. Add an additional capacitor with an SRF near the PRF.

We rarely can adjust the capacitance of the planes. It is what it happens to be. One of the reasons all PDN designs are custom is that the plane capacitance will vary depending on the board size and stack-up. This will shift the PRF over a wide range of frequencies.

TIP   Reducing the ESL of the bulk capacitors should always be at the top of the list of actions. Everything should always be done to reduce the ESL. Selecting lower-value-capacitance capacitors may provide higher ESR and more damping.

If we are limited to using just three capacitors, it may be possible to find a value of one of the capacitors so that its SRF is closer to the PRF of the planes. This would reduce the peak impedance by a closer SRF and by a higher ESR, increasing damping.

We would want to adjust the third capacitor, C3, so that its SRF is close to the PRF. The condition is:

$$SRF = PRF = \frac{160 \text{ MHZ}}{\sqrt{ESL \times C_3}} = \frac{160 \text{ MHz}}{\sqrt{\frac{1}{3}ESL \times C_{planes}}}$$

(13-50)

This reduces to:

$$C_3 = \frac{1}{3}C_{planes} = \frac{1}{3}5.6 \text{ nf} = 1.9 \text{ nf}$$

(13-51)

Figure 13-48 shows the impedance profile of the capacitors, VRM, and planes with the third capacitor changed from 10 nF to 2 nF.

TIP   It is counterintuitive that by decreasing the capacitance of one of the capacitors, the impedance profile actually improves. By optimizing the capacitor value, we reduce the peak impedance from 10 Ohms down to 2.5 Ohms. This reduces the PDN noise by four times in the 50-MHz to 100-MHz range.

Which is better: three capacitors all with the same value or three capacitors with different values? If you randomly select the three capacitor values or blindly use values of 1 µF, 0.1 µF, and 0.01 µF, then it probably doesn't matter which approach you choose. They each have the same chance of success or failure. However, if you can optimize the capacitor value to minimize the peak impedance at the PRF with the plane's capacitance, using different-value capacitors results in a lower peak impedance profile.

In this example, the lowest peak impedance that could be obtained below 100 MHz using just three capacitors, even if their values are optimized, is still limited to about 2 Ohms. This can be dramatically improved with more capacitors.

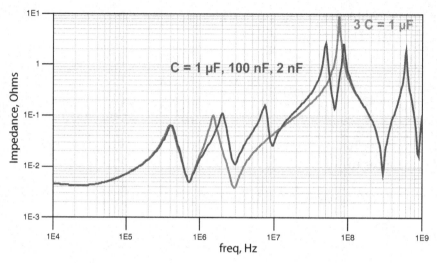

Figure 13-48   Impedance profile with the 10-nF capacitor changed to 2 nF with a lower peak impedance.

13.19 Estimating the Number of Capacitors Needed

In the absence of more detailed information, the goal of the board-level PDN design is to engineer the impedance peaks below the target value up to about 100 MHz, or roughly where the package limits the impedance that the chip will see, which could be at lower frequency.

At the low-frequency end, the number and values of the bulk capacitors can be adjusted to keep the peak impedances below the target value.

At the high-frequency end, the absolute lowest maximum impedance a collection of capacitors can theoretically have is set by their parallel combination of equivalent series inductance. The best case is if there is no parallel resonance with the plane's capacitance, and all the inductances are in parallel. The design condition is that:

$$Z_{capacitors} < Z_{target} \text{ at } F_{max} \tag{13-52}$$

where:

$Z_{capacitors}$ = impedance of the capacitors in parallel, in Ohms

Z_{target} = target in impedance, in Ohms

F_{max} = highest frequency where the board-level impedance can play a role

If the impedance of the capacitors at the high-frequency end is all due to the parallel combination of inductances and they all have the same value of ESL, this condition translates to:

$$2\pi F_{max}\left(\frac{ESL}{n}\right) < Z_{target} \tag{13-53}$$

where:

Z_{target} = target in impedance, in Ohms

F_{max} = highest frequency where the board-level impedance can play a role, in GHz

ESL = equivalent series inductance of each capacitor, in nH

n = number of capacitors needed in parallel to meet the target impedance

This establishes the theoretical minimum number of capacitors needed in parallel to meet this impedance target as:

$$n > 2\pi F_{max}\left(\frac{ESL}{Z_{target}}\right) \tag{13-54}$$

For example, if the target impedance is 0.1 Ohm and F_{max} is 100 MHz and each capacitor has 2 nH of ESL, then the theoretical minimum number of capacitors needed is:

$$n > 2\pi \times 0.1\left(\frac{2}{0.1}\right) = 13 \tag{13-55}$$

TIP   In order to reduce the number of capacitors needed, regardless of their value, the ESL must be reduced. This is why ESL is such an important number.

Figure 13-49 shows how the theoretical minimum number of capacitors varies based on the ESL and the target impedance.

The minimum number of capacitors possible to achieve a target impedance is a good figure of merit to evaluate how well optimized a design might be. This would be the case when the PDN is not delivering power to a Vdd core but providing a low impedance for return currents, for example. More capacitors in parallel would act as shorting inductors to reduce the impedance of the power and ground planes with a DC block.

In the example in the preceding section, we achieved a target impedance of 2 Ohms with three capacitors, each with an ESL of 3 nH. The theoretical minimum number that could be used, as shown in the chart in Figure 13-7, is one. Using three to get there is not very efficient due to the complication of the PRF of the planes.

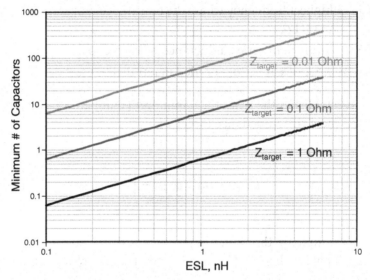

Figure 13-49   Minimum number of capacitors required to achieve a target impedance at 100 MHz based on an ESL value.

13.20 How Much Does a nH Cost?

The cost of a small ceramic decoupling capacitor is almost negligible. Its largest direct costs are in the assembly operation and in the indirect costs of more vias to drill, the surface real estate taken up, the potential of blocked routing channels, and the impact on the board layer count.

Using a rough estimate of $0.01 for the total direct material cost and assembly cost per capacitor, we can estimate how much an nH is worth. For every fraction of a nH reduction in the ESL, fewer capacitors need to be used, and there is a direct cost savings.

The cost per nH is derived from the expression above for the theoretical minimum number of capacitors needed:

$$TotalCost = \$0.01 \times n = \$0.01 \times 2\pi F_{max}\left(\frac{ESL}{Z_{target}}\right) \tag{13-56}$$

This is displayed in Figure 13-50 for different ESL and target impedance values for a maximum frequency where the board plays a role of 100 MHz.

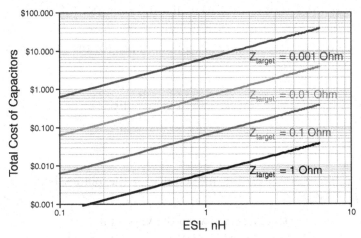

Figure 13-50 Total cost of all the capacitors, assuming $0.01 each, as the ESL is reduced.

The cost per nH can be estimated based on:

$$\frac{TotalCost}{ESL} = \frac{\$0.01 \times 2\pi F_{max}}{Z_{target}} = \frac{0.006}{Z_{target}} \frac{\$}{nH} \tag{13-57}$$

where:

TotalCost/ESL = cost per nH, in $/nH

F_{max} = highest frequency at which the board-level impedance can play a role, in GHz, assuming ~ 0.1 GHz

Z_{target} = target impedance, in Ohms

TIP   This is a simple result. It suggests that the lower the target impedance, the more expensive every nH becomes and the more valuable a reduction in the ESL becomes.

For example, with a target impedance of 0.01 Ohm, the cost per nH is $0.6/nH. If the mounted inductance is 2 nH, the total cost of the capacitors is about $1.20 for the board. If the ESL can be reduced from 2 nH to 1 nH just by changing the surface traces or bringing the power and ground plane cavity closer to the board surface, the cost savings from the reduction in the number of capacitors used would be $0.60, with no sacrifice in performance. If this were a high-volume board, with 1 million units per month, the cost savings would be $600k per month, or $7.2 million per year.

The contribution of spreading inductance for the planes is roughly:

$$L_{via-via} = 21 \times h \times \ln\left(\frac{B}{D}\right) pH \tag{13-58}$$

where:

$L_{via-via}$ = loop-spreading inductance in the planes between the two via contacts, in pH

h = dielectric thickness between the vias, in mils

B = distance between the via centers, in mils

D = diameter of the vias, in mils

For a typical case of B = 1 inch and D = 10 mils, the spreading inductance in the planes is roughly about:

$$L_{via-via} = 21 \times h \times \ln\left(\frac{1}{0.01}\right) pH \sim 0.1 \times h \; nH \tag{13-59}$$

When a conventional thickness of 4 mils is used, the spreading inductance contribution is on the order of 0.4 nH per capacitor. If an ultra-thin laminate, such as the DuPont Interra HK04 material, with a thickness of 0.5 mils, were used, the spreading inductance would be on the order of 0.05 nH. This is a reduction of about 0.35 nH and translates into a potential cost reduction of:

$$CostReduction = \frac{0.006}{Z_{target}} \frac{\$}{nH} \times 0.35 \, nH = \frac{\$0.002}{Z_{target}} \tag{13-60}$$

When the cost reduction is greater than the cost premium for the thinner dielectric, using the more expensive ultra-thin laminate becomes a reduction in the total cost of ownership. The premium is rated as an extra price per square foot of board area, or:

$$CostReduction > Premium \times area \tag{13-61}$$

$$\frac{\$0.002}{Z_{target}} > Premium \times area \tag{13-62}$$

$$\frac{\$0.002}{Premium} > Z_{target} \times area \tag{13-63}$$

If the premium were about $3/square foot extra cost, then the condition for total cost reduction with a thin laminate would be:

$$\frac{\$0.002}{\$3} = \$0.0007 > Z_{target} \times area \tag{13-64}$$

where:

CostReduction = cost reduction in the number of capacitors not needed, in $

Z_{target} = target impedance, in Ohms

Premium = added cost per square foot of thin laminate over conventional laminate, in $/square foot

area = area of the board surface in the specific application, in square feet

If the area is described in square inches, this relationship becomes:

$$\$0.1 > Z_{target} \times area \qquad (13\text{-}65)$$

This suggests that if the board area is 10 square inches, a thinner laminate is a cost reduction if the target impedance is lower than 0.01 Ohm.

13.21 Quantity or Specific Values?

To first order, the impedance at high frequency of a collection of capacitors is related to the parallel combination of their inductance. However, if a parallel resonance with the capacitance of the board planes exists near the maximum frequency where the board impedance plays a role, it will artificially increase the impedance profile of the capacitors. In this regime, the impedance of the collection of capacitors can be brought down by carefully selecting the values of capacitor to "sculpt" the impedance profile and compensate for the parallel resonance.

The impact from the parallel resonance on the impedance profile for a collection of capacitors is illustrated in Figure 13-51 for the specific case of:

$Z_{target} = 0.1$ Ohm
$F_{max} = 0.1$ GHz
$ESL = 2$ nH
$n = 13$ capacitors
$A = 65$ square inches and 6.5 square inches

In this example, the theoretical minimum number of capacitors needed to achieve the 0.1 Ohm at 0.1 GHz is:

$$n > 2\pi F_{max}\left(\frac{ESL}{Z_{target}}\right) = 2\pi 0.1\left(\frac{2}{0.1}\right) = 13 \qquad (13\text{-}66)$$

When the PRF is close to the F_{max}, it artificially increases the impedance profile of the capacitors and the planes' capacitance. This increase can be more than a factor of two or three.

However, if the PRF can be engineered to be a higher frequency—by decreasing the area of the planes, for example—the parallel resonance does not interact with the impedance of the capacitors near the F_{max}, and the impedance can be close to the theoretical impedance of the n inductors in parallel. In this case, achieving the target impedance at the F_{max} with the minimum number of capacitors does not depend on the specific value of capacitors; it just depends on their number and their ESL.

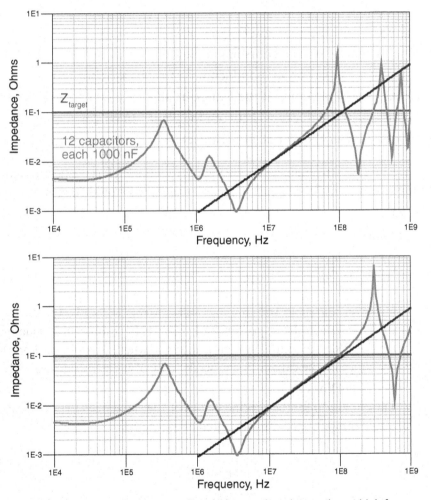

Figure 13-51   Impedance profile of 13 capacitors interacting at high frequency with the 65 and 6.5 square inches of board capacitance, compared to the impedance of the ideal inductance of 13 capacitors. Top: PRF = 100 MHz; bottom: PRF = 3 × 100 MHz.

The PRF of the capacitors' inductance interacting with the planes' capacitance is given by:

$$PRF = \frac{160 \text{ MHz}}{\sqrt{\frac{1}{n}(ESL\ C_{planes})}} = \sqrt{\frac{h}{\frac{1}{n}ESL \times A}}\,160 \text{ MHz} \qquad (13\text{-}67)$$

where:

PRF = parallel resonant frequency, in MHz

n = number of capacitors in parallel

ESL = equivalent series inductance of each capacitor, in nH

C_{planes} = capacitance of the planes, in nF

h = dielectric thickness between the planes, in mils, assuming Dk = 4

A = area of the planes, in square inches

The goal is to engineer conditions so that the PRF is pushed to frequencies above the maximum frequency. To first order, this would suggest:

• Large n
• Thicker h
• Small ESL
• Small A

However, the dielectric thickness also affects the ESL. Increasing h will increase ESL. Given the importance of lower ESL, thinner h is usually better. To push the PRF to a high enough frequency where it is not interacting with the impedance of the capacitors, it needs to be at least three times higher than the max frequency.

The condition for the specific values of the capacitors to not be significant is roughly approximated, as illustrated above, to be:

$$PRF > 3 \times F_{max}$$

$$\sqrt{\frac{h}{\frac{1}{n}ESL \times A}}\,160 \text{ MHz} > 3 \times F_{max} \qquad (13\text{-}68)$$

In addition, if the number of capacitors is adjusted to meet the target impedance at the maximum frequency, a further condition is:

$$Z_{target} = \frac{1}{n} ESL \times 2\pi F_{max} \qquad (13\text{-}69)$$

These two relationships can be combined to result in the condition where the impedance at F_{max} is independent of the specific values of the capacitors selected as:

$$\frac{h}{Z_{target} A} > 56 \times F_{max} \qquad (13\text{-}70)$$

where:

n = number of capacitors in parallel

Z_{target} = target impedance, in Ohms

A = area of the planes, in square inches

F_{max} = highest frequency where the board-level impedance is important, in GHz

ESL = equivalent series inductance of each capacitor, in nH

h = dielectric thickness between the planes, in mils, assuming Dk = 4

For the best conventional case of h = 4 mils and the typical F_{max} of 0.1 GHz, this condition reduces to:

$$Z_{target} A < 0.7 \qquad (13\text{-}71)$$

where:

Z_{target} = target impedance, in Ohms

A = area of the planes, in square inches

TIP   This suggests that to engineer a condition where the values of the capacitors are not important and where it is still possible to use the theoretical minimum number of capacitors, the area of the power planes should be kept to a minimum and the target impedance low.

In general, the area of the planes is always less than the area of the board. When split planes are used, the actual board area can be more than a factor of three larger than the power planes. If well engineered, the area of the power plane can be kept to a minimum to support all the capacitors and connections to the VRM. When power planes are mixed on signal layers as copper fill areas, the small power planes are sometimes referred to as a *copper puddle*.

TIP   Whenever split planes are used, it is always important to keep signal layers from crossing underneath them. Signal layers crossing under split planes has the potential of generating noise in the power plane and excessive coupling between adjacent signal lines.

In many applications with dedicated power planes, the plane area and board area may be close. This design space is mapped in Figure 13-52.

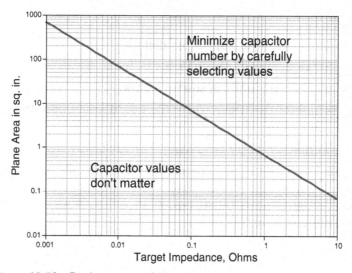

Figure 13-52   Design space of when the capacitor values selected are important and when they are not important for the special case of dielectric thickness of 4 mils.

For this specific case of a dielectric thickness of 4 mils, if the target impedance is 0.1 Ohm, then as long as the area of the planes is less than roughly 7 square inches, the selection of capacitor values is not important. They can all be the same value.

If, however, the plane area is larger than 7 square inches, the capacitance of the board will push the PRF close to 0.1 GHz and increase the impedance of the combination of capacitors. In this case, to use the minimum number of capacitors, their value is important and should be selected to "sculpt" the impedance profile.

This design space of target impedance higher than 0.1 Ohm and plane area larger than 7 square inches encompasses many boards. In this regime, to use the smallest number of capacitors and achieve the lowest cost, the precise values of capacitors should be carefully selected. Their values and number are chosen to bring the peak impedances below the target value up to the maximum frequency.

This is why for many common board applications, using a distribution of capacitor values will enable the lowest impedance with the fewest capacitors rather than all the same value of capacitance. Of course, the minimum number of capacitors is based on using the right distribution of values.

If the plane area can be kept to less than 2 square inches, then all designs with a target impedance less than 0.3 Ohm can use the same value capacitors and the minimum number of capacitors. This is a rather small plane area and not very common. However, the package is about this size.

TIP   The package can act as a small board. If enough capacitors are added to the package, the impedance may be reduced to the level where the impedance of the board is not important.

While there is rarely enough on-package decoupling capacitance to supplement all the board-level needs, it is possible to add all the necessary capacitors using a small size interposer rather than in the package. Such an alternative approach is available from Teraspeed Consulting Group.

An example of a small island of low-impedance power and ground provided in the PowerPoser is shown in Figure 13-53. This small board fits underneath the package and has multiple layers of thin laminate with all the very low inductance decoupling capacitors to bring the impedance below the target value from low frequency to very high frequency.

By using an interposer, very thin dielectric layers near the surface can be used without paying a large-area price penalty, and low inductance capacitors can be placed in proximity to the package. The area of the planes can be kept small so the parallel resonant frequency is well above the package limit, and all capacitor values can be the same.

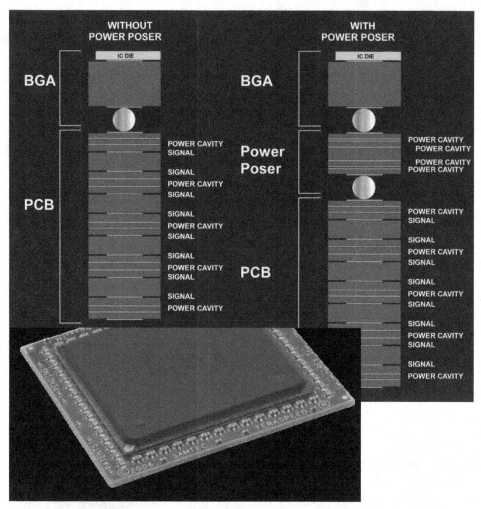

Figure 13-53 The role of the PowerPoser in a board stack-up and a close-up view of an FPGA chip mounted on a PowerPoser.

13.22 Sculpting the Impedance Profiles: The Frequency-Domain Target Impedance Method (FDTIM)

Parallel resonances in the PDN cause impedance peaks, which ultimately are the source of failures in the PDN. These parallel resonances are caused by a parallel combination of a capacitor and an inductor—somewhere.

In the case of the Vdd supply, as viewed by the pads of the die, the capacitor is the on-die capacitance, and the inductor is the package lead inductance. It's this parallel resonance that can cause excess Vdd noise. At the board level, the best we can achieve is a flat impedance response, which will damp this parallel resonance.

The impedance of the power and ground plane cavity, as seen on the board, is important for signal-return-path noise and noise coupling from the planes to other components. This noise is also related to peaks in the cavity PDN. This is primarily from the parallel resonance between the cavity capacitance and the equivalent-series inductance of all the MLCC capacitors in parallel.

The solution for this PDN is to reduce the total parallel inductance and reduce any peaks. Both of these problems can be minimized by engineering a PDN profile that is flat, without significant peaks.

By optimizing the values of the capacitors, a flat impedance profile can be engineered with the smallest number of capacitors. In this design regime, it is possible to minimize the number of capacitors used to achieve the target impedance by careful selection of their values. This process is called *sculpting* the impedance profile.

TIP   No matter what, it is always important to use the lowest possible ESL for all decoupling capacitors. This will always result in the smallest number of capacitors and the lowest-cost system.

The precise number and optimized values of capacitors needed will depend on:

- The bulk decoupling capacitors associated with the VRM
- The capacitance in the board
- The target impedance
- The maximum frequency
- The ESL of each capacitor

The combination of these terms varies dramatically from product to product, so it is not possible to give one capacitor distribution that will always work. However, the methodology can be applied to many designs.

This methodology was pioneered by Larry Smith while he was at Sun Microsystems and has been termed the *Frequency-Domain Target Impedance Method (FDTIM)*. The process leverages the simulated impedance profile of a collection of capacitors, including their ESL and ESR, with the capacitance in the planes at high frequency and the bulk capacitors associated with the VRM at the low-frequency end.

Capacitor values are selected from the values available from vendors. Not all values are available; rather, the common values are in decade steps of multiples of 1.0, 1.5, 2.2, 3.3, 4.7, and 6.8. When the ESL of each capacitor in a collection is the same, the minimum parallel resonant impedance peak is obtained when each capacitor value is the geometric mean of the capacitor on either side of it.

The optimum distribution would be using values of decade multiples of 1, 2.2, and 4.7, for example. The largest capacitance value of an 0402 capacitor easily available is about 1 μF. When higher values are needed, a 1206 capacitor can provide as much as 100 μF.

The lowest-value capacitor needed is about ⅓ × the capacitance in the planes. The board capacitance is on the order of 10 nF for a board with 40 square inches, for example. The smallest-value capacitor needed would be about 2.2 nF.

The selection of possible values might range from 1 μF to 2.2 nF and may include nine different values from which to select:

    1000 nF
    470 nF
    220 nF
    100 nF
    47 nF
    22 nF
    10 nF
    4.7 nF
    2.2 nF

A typical set of parameters might be:

• 50 square inches of board area with ~20 nF of capacitance
• Target impedance of 0.1 Ohm
• ESL of 2 nH for each capacitor
• Maximum frequency of 0.1 GHz

In this example, the board area of 50 square inches and target impedance of 0.1 Ohm puts the design in the upper part of the design space, where the value of the capacitors matters. Of course, if the target impedance were low enough or the board capacitance were small enough, it would not matter what value capacitors were used; they could all be 1 μF. In that case, parallel resonances would not play a role.

The theoretical minimum number of capacitors required to meet the target value for the above condition is:

$$n > 2\pi F_{max}\left(\frac{ESL}{Z_{target}}\right) = 2\pi 0.1\left(\frac{2}{0.1}\right) = 13 \tag{13-72}$$

where:

n = minimum number of capacitors needed

F_{max} = highest frequency for board-level impedance, in GHz

ESL = equivalent series inductance of the capacitor, including the mounting inductance and some of the cavity spreading inductance, in nH

Z_{target} = target impedance, in Ohms

Starting at the low-frequency end, the largest capacitor value is selected and simulated. Enough capacitors of each value are added to bring the peak impedance below the target value. Capacitors are added, along with enough quantity of this value, until the target impedance is reached. Specific capacitor values are skipped, especially at the low-frequency end, where the low target impedance can be reached without them. The ESR for each capacitor value is used in the simulation.

Figure 13-54 shows an impedance profile based on the following capacitor selection with a total of 14 capacitors:

C	n
470	1
100	1
47	1
22	1
10	3
4.7	3
2.2	4
Total	14

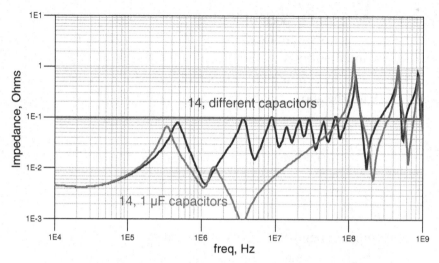

Figure 13-54 Impedance profiles for two distributions of 14 capacitors. One distribution uses all 1-µF capacitors. The other distribution was chosen to sculpt the profile. At 100 MHz, the sculpted profile meets the target impedance, but the other distribution does not.

In Figure 13-54, the impedance profile meets the target impedance up to 100 MHz, using 14 capacitors. This is close to the theoretical minimum of 13. However, using 14 capacitors all with the same value of capacitance is not able to achieve the same low impedance. If all the same value capacitors were used, more than 14 would be required. It would not be as low a cost as using the FDTIM.

TIP   Of course, if any of the initial conditions of this specific problem were changed—for example, if the ESL were not 2 nH but really 3 nH—this combination would no longer work.

Figure 13-55 shows the impedance profile for these capacitors with an ESL of 3 nH, exceeding the target impedance at a number of frequencies. This is another example of the importance of reducing the ESL for capacitors and how custom the selection of capacitors becomes when so many system parameters affect the impedance profile.

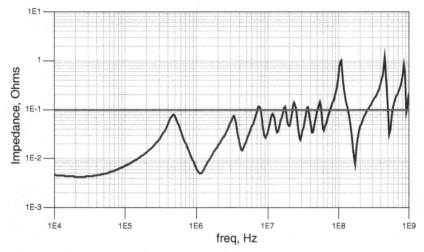

Figure 13-55 Impedance profile for the same distribution of capacitor values as in Figure 13-54 but each with an ESL of 3 nH rather than 2 nH.

TIP   Of course, there are many right distributions. The most cost-effective solutions use a total number of capacitors that is close to the theoretical minimum.

Another example of a sculpted-impedance profile for a target impedance of 0.05 Ohm is shown in Figure 13-56. In this case, the theoretical minimum with an ESL of 2 nH is 26. This distribution used 33, slightly above the theoretical minimum. The capacitor values used were:

C	n
1000 nF	1
470 nF	1
220 nF	1
100 nF	1
47 nF	2
22 nF	3
10 nF	5
4.7 nF	6
2.2 nF	13
Total	33

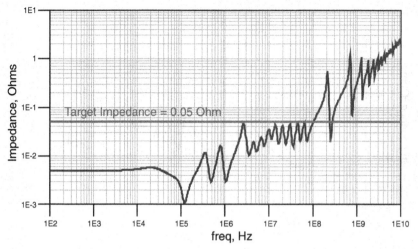

Figure 13-56   Impedance profile for a target impedance of 0.05 Ohm using 33 capacitors.

13.23 When Every pH Counts

The four most important design "habits" to follow are:

1. Use the shortest and widest possible surface trace that is consistent with the assembly design rules. In other words, use as few squares of surface interconnect between the capacitors and the vias as possible.
2. Place the capacitors in proximity to the package—some below it on the bottom side of the board and some on the same layer—to avoid saturating the spreading inductance with all peripheral capacitors.
3. When power and ground planes are on adjacent layers, use the thinnest possible dielectric that does not cost extra. This is usually 2.7 to 4 mils, depending on the vendor.
4. When possible, place the power and ground cavity as close to the surface of the board as possible.

There is rarely a reason not to do these habits, and they will always result in lower ESL.

The cost impact for an ESL reduction is:

$$\frac{\text{TotalCost}}{\text{ESL}} = \frac{0.006}{Z_{\text{target}}} \frac{\$}{\text{nH}} \qquad (13\text{-}73)$$

When the target impedance is 0.1 Ohm or higher, this is less than $0.06 per nH per voltage rail on a board. There usually isn't enough cost savings potential to justify paying extra for low-inductance features. However, when the target impedance is 0.001 Ohm, the cost savings is $6/nH for each voltage rail on the board. Every pH reduction is a half-cent cost reduction.

> **TIP**   The lower the mounting loop inductance of each capacitor, the fewer capacitors are required to achieve the low target impedance at the high frequency. Every free option should be used to reduce the ESL of all decoupling capacitors.

As pointed out earlier in this chapter, sometimes paying extra for thinner dielectric between the power and ground planes is worth the extra cost of the lower spreading inductance.

In addition, alternative capacitor technologies can offer a lower mounted ESL than conventional capacitors. Most capacitors are designed with their terminals along their long axis. The capacitor body is a minimum of two squares. Even with via-in-pad, there are still two squares of surface trace.

An alternative design uses *reverse aspect ratio* capacitors, with terminal pads along the long side of the capacitor. Mounted onto a board, these capacitors could be implemented with as low as 0.5 square. An example of these capacitors is shown in Figure 13-57.

Figure 13-57   Capacitor technologies. Left: Conventional-aspect-ratio capacitors with minimum number of surface trace squares of n = 2. Right: Reverse-aspect-ratio capacitors available from AVX, with a minimum n = 0.5.

If the depth in the board stack-up to the power and ground cavity is 5 mils below the top surface, the sheet inductance of surface traces would be about 32 pH/mil × 5 mil = 160 pH/sq. The trace loop inductance of a via-in-pad, best-case standard capacitor is about 320 pH, while for a reverse-aspect-ratio capacitor, the best case could be as low as 160 × 0.5 = 80 pH. This is a reduction of 240 pH.

Interdigitated capacitor technologies can offer even lower capacitance. These are constructed as multilayer ceramic capacitors with multiple interleaved terminals on each end. Examples are illustrated in Figure 13-58.

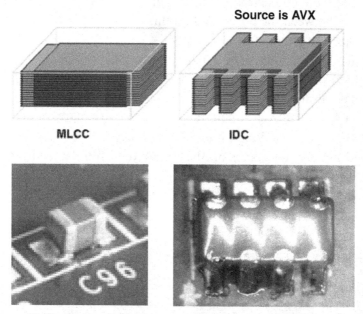

Figure 13-58 Comparison of conventional MLCC capacitors and interdigitated capacitors (IDC).

An IDC is effectively multiple capacitors in parallel, with the ESL of each current path in parallel. The equivalent ESL of the four capacitors in one IDC would be one-quarter the ESL of any one of them. In addition, since the current flow is in opposite directions in adjacent capacitor segments and they are in close proximity, the effective ESL of each one is further reduced. An IDC can have less than 20% the ESL of a conventional capacitor.

Another type of IDC is provided by X2Y Attenuators. These are multilayer ceramic capacitors with alternating plates coming out to each of four different electrode terminals. An example of these four-terminal capacitors and their internal structure is shown in Figure 13-59.

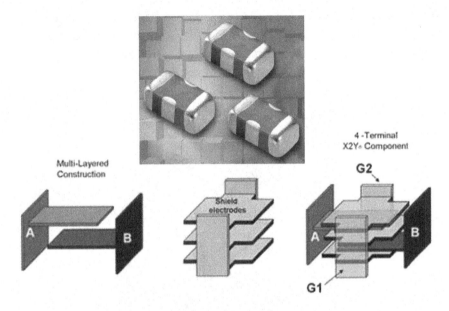

Figure 13-59 Examples of X2Y capacitors and their internal structure, as described by X2Y Attenuators.

The A and B plates are both connected in parallel to the power plane, while the central two G1 and G2 plates are connected in parallel to ground. In this configuration, the capacitor behaves like four capacitors in parallel, with current flows illustrated in Figure 13-60.

While there are similar performance advantages using either IDC technology, an advantage of the X2Y capacitors is the ease of integration with conventional through-hole circuit board technology.

An 0805 IDC with four terminals on a side will have a pad footprint with 20-mil centers between pads. This is difficult to connect to circuit boards using conventional through-hole technology and requires via in pad. An 0805 X2Y capacitor can use via holes on 40-mil centers, as shown earlier in this chapter, and can use conventional through-hole technology, leaving a routing channel through the holes available for multiple 5-mil-wide tracks.

TIP   The combination of IDC, minimum surface trace size, and cavity close to the surface can enable a total ESL from the capacitor to the package pin of less than 250 pH.

However, if the surface traces are long and the cavity is not near the surface, the same X2Y capacitor can show an ESL that is more than 1 nH—three times larger than it needs to be—just from very slight design variations. Figure 13-61 shows an example of the measured impedance profile of two X2Y capacitors with the comparison of an RLC model.

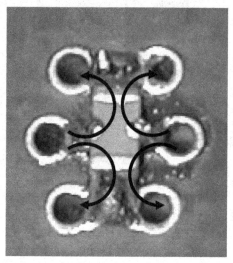

Figure 13-60   X2Y capacitor mounted to a circuit board with the top and bottom terminals tied to power and the two central terminals tied to ground. It behaves like four capacitors in parallel.

For each capacitor RLC model, the same value of C = 180 nF was used in the model, the same value of R = 0.013 Ohm was used, but two different inductance values were used. For the best case, L = 260 pH gave an excellent fit, while for the other case, L = 900 pH was the best fit. The peaks near 200 and 300 MHz are the parallel resonances with the circuit board to which the capacitors were mounted.

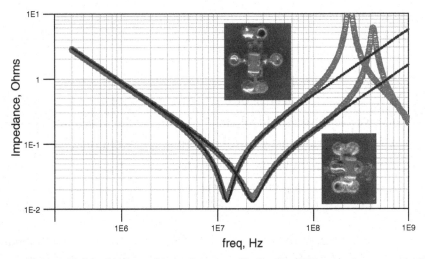

Figure 13-61 Measured impedance profile for two X2Y capacitors mounted to a test board with slightly different conditions. The measured data is compared to the simulated impedance of an ideal RLC circuit. The insets show the capacitor for each measurement.

TIP When every pH counts, the right capacitors and the optimum mounting inductance can make all the difference.

13.24 Location, Location, Location

At low frequency, below the parallel resonant frequency of the capacitors' ESL and the planes' capacitance, the planes will interact with the capacitors as a lumped element. However, when the length of an edge of the planes is comparable to a fraction of the wavelength, the resonant behavior of the board will show up in the impedance profile.

When the probe point is located at the edge of the board, the first resonant frequency will be when:

$$\text{Len} = \frac{1}{2}\lambda = \frac{1}{2\sqrt{Dk}}\frac{c}{f_{res}} = \frac{1}{2}\frac{v}{f_{res}} = \frac{3}{f_{res}} \qquad (13\text{-}74)$$

$$f_{res} = \frac{3}{Len} \qquad\qquad (13\text{-}75)$$

where:

Len = length of an edge of the board, in inches

λ = wavelength of light where the first resonance shows up, in inches

Dk = dielectric constant of the laminate between the planes

c = speed of light in air, 12 inches/nsec

f_{res} = resonant frequency, in GHz

v = speed of light in the material, assumed v = 6 inches/nsec, for FR4

For example, if the board is 10 inches on a side, the first resonance will be about 300 MHz. If the probe point is in the middle of the board, the first resonance will be at twice this frequency, or 600 MHz.

Figure 13-62 is an example of the simulated impedance profile of a 10 inch × 10 inch bare board, probed in the center, showing the capacitive behavior at low frequency, its self-resonant frequency, and the onset of board resonances at 600 MHz.

While the board will appear as a lumped capacitor below the resonant frequency, and a simple SPICE simulation will accurately reflect the impedance profile of the capacitor on the board, the spreading inductance the capacitor sees between its location on the board and the device it is decoupling will depend on location.

The farther away the capacitor is from the package it is decoupling, the higher its total ESL due to the spreading inductance. When the spreading inductance is small compared to the mounted inductance of the capacitor, the location is not important. Changing the capacitor position will change the spreading inductance, but this will have minimal impact on the total ESL of the capacitor.

TIP However, when the spreading inductance is a significant fraction of the total capacitor's ESL, the location will have a significant impact on the ESL of the capacitor, and moving the capacitors closer to the device is important.

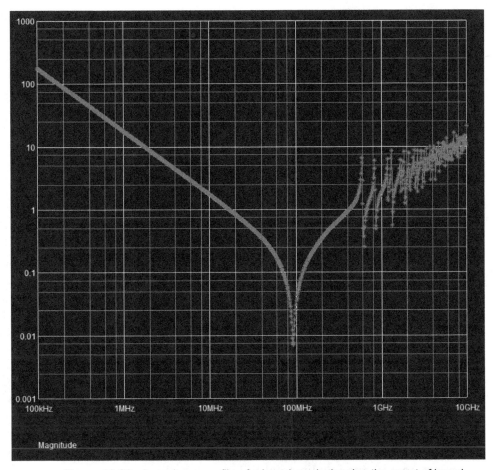

Figure 13-62 Impedance profile of a bare board, showing the onset of board resonances at 600 MHz. Simulated with HyperLynx 8.0.

The condition for when location is important is based on the amount of spreading inductance compared with the mounted inductance of the capacitor. The condition is:

When spreading inductance ~ mounting inductance, location matters.
When spreading inductance << mounting inductance, location does not matter.

For a via about 10 mil in diameter and a capacitor 1 inch away, the spreading inductance is roughly:

$$L_{via-via} = 21 \times h \times \ln\left(\frac{B}{D}\right) pH = 21 \times h \times \ln\left(\frac{1}{0.01}\right) pH = 100 \times h \ pH \qquad (13\text{-}76)$$

When h is thin, spreading inductance is small, and location is important only for very low values of mounting-inductance capacitors.

When h is thick, spreading inductance is larger. When the mounting inductance of capacitors is low, location can be important. Due to the very rough approximate nature of the spreading-inductance estimate, when the spreading inductance is on the same order as the mounting inductance, it is appropriate to use a 3D field solver tool to estimate the impact on the impedance profile of the mounted capacitors from location.

Figure 13-63 shows an example of the impedance profile of two board configurations. In each case, the same four capacitors are mounted to the board, close together. The mounting inductance of each is 5 nH, so the four of them have an equivalent inductance of 1.25 nH.

In the first example, the cavity thickness is 30 mils. The spreading inductance is on the order of 3 nH, large compared to the mounting inductance of the four capacitors. The impedance profile of these four capacitors is simulated when they are located far from the package pin and when they are close. The large difference in impedance in these two positions shows the impact from location on the total ESL.

In the second example, the cavity thickness is 4 mils, and the spreading inductance is on the order of 0.4 nH. This is small compared to the 1.25-nH inductance of the capacitors. When the position of the capacitors is changed from near to far, there is little change in the simulated impedance. The spreading inductance is not an important contributor to the capacitor's inductance, and location is not important.

In low-impedance designs, where the ESL has been optimized to less than 0.25 nH for each capacitor, the spreading inductance can be a significant contributor and should be taken into account with a 3D field solver. The spreading inductance will increase the contribution to the inductances of the individual capacitors in a complicated way that depends on the position of the capacitors and the location of the package pins. It can only be analyzed with a 3D field solver.

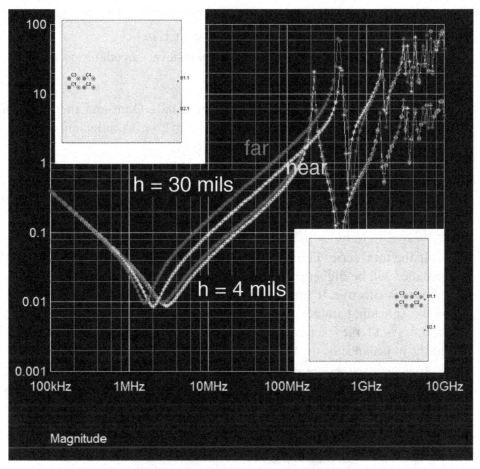

Figure 13-63 Simulated impedance of four capacitors mounted to a 30-mil-thick cavity and a 4-mil-thick cavity located close to the package pin and far from the package pin. Simulated with HyperLynx 8.0.

13.25 When Spreading Inductance Is the Limitation

For a given target impedance and maximum frequency, the total, maximum allowable series inductance that could be in the path, including the capacitors and plane spreading inductance, must be:

$$L_{max} < \frac{Z_{target}}{2\pi F_{max}} \qquad (13\text{-}77)$$

where:

L_{max} = maximum allowable series inductance, in nH

F_{max} = maximum frequency where board-level impedance is important, in GHz

For example, if the target impedance is 0.01 Ohm and the maximum frequency is 100 MHz, the maximum allowable series inductance before it dominates the impedance of all the capacitors is:

$$L_{max} < \frac{Z_{target}}{2\pi F_{max}} = \frac{0.01}{2\pi \times 0.1} = 0.016\,\text{nH} = 16\,\text{pH} \qquad (13\text{-}78)$$

If the total series inductance to the capacitors exceeds 16 pH, the PDN impedance will be higher than the target impedance at the highest frequency the board is effective. If the vias in the board from the package to the planes and the spreading inductance in the planes from the package pins to the capacitors is a large fraction of this inductance, the spreading inductance will limit the impedance of the board.

If the capacitors are uniformly distributed around the package with the power and ground pins also distributed around the perimeter of the package, as shown in Figure 13-64, the spreading inductance in the planes can be estimated.

The spreading inductance in the planes is approximately:

$$L_{spread} = 5.1 \times h \times \ln\left(\frac{b}{a}\right)\text{pH} \qquad (13\text{-}79)$$

where:

L_{spread} = spreading inductance in the planes, in pH

h = dielectric thickness between the planes, in mils

b = distance to the capacitors, in inches

a = radius of the power/ground pins in the package, in inches

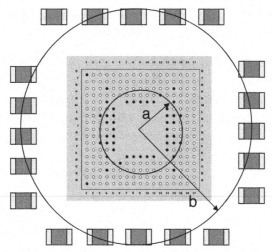

Figure 13-64 Estimating the best-case spreading inductance in the planes as current flows from the distribution of capacitors to the pins in the package.

For example, if h = 4 mils, b = 1 inch, a = 0.25 inches, then the spreading inductance is:

$$L_{spread} = 5.1 \times h \times \ln\left(\frac{b}{a}\right) = 5.1 \times h \times \ln\left(\frac{1}{0.25}\right) = 28 \text{ pH} \qquad (13\text{-}80)$$

It is possible that the equivalent inductance of the vias in the board from the package pins may also add to this limiting inductance. If the power and ground cavity is 10 mils below the top surface, there will be about 210 pH per power and ground via. For 10 power and ground pin pairs, this is an equivalent via inductance of 21 pH, comparable to the spreading inductance. The combination may almost double the limiting inductance when the package pins look into the board to the capacitors.

TIP The vias in the board to the cavity and the spreading inductance in the cavity to the capacitors can easily dominate the impedance of the collection of capacitors when the target impedance is below 0.05 Ohm.

Based on the root cause of the spreading inductance, there are only a few board-level design features that can be engineered to reduce it:

- Use thinner dielectric layers between the planes.
- Use multiple planes in parallel.
- Move the power and ground cavity closer to the top of the board.
- Spread the capacitors farther out around the perimeter of the package. This has a limit once the capacitors are uniformly distributed around the perimeter.
- Mount some of the capacitors under the package on the bottom of the board.

Some design features can be implemented in the package design if planned ahead of time including the following:

- Mount some of the capacitors within the ring of the power and ground pins of the package.
- Spread the power and ground pins of the package to the outer perimeter of the package.
- Add decoupling capacitors on the package.
- Use more power and ground pin pairs in parallel.

Figure 13-65 shows an example of adjusting the chip attach footprint of the package to allow decoupling capacitors inside the power/ground ring of pads.

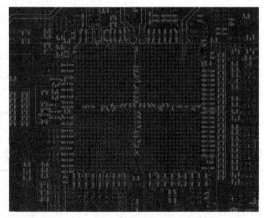

Figure 13-65   Package attach footprint of an Altera Stratix II GX FPGA, showing the capacitor attach pads around the periphery and inside the BGA footprint.

When spreading inductance limits the series inductance, the worst thing to do would be to cluster the capacitors or cluster the power and ground pins into the core region of the package.

Because of the 3D nature of the current flows in the planes, the only way to accurately estimate the series inductance contribution from the planes, given the capacitor locations and power pin locations, is with a 3D field solver.

13.26 The Chip View

Through most of this chapter, we have considered the impedance profile seen by the board. Once the board-level impedance is designed, what impact does this have on the chip-package combination?

In the following example, the conditions are:

- Target impedance of 0.01 Ohm
- F_{max} of 100 MHz
- Board area of 25 square inches
- h of 4 mils
- ESL per capacitor of 1 nH
- On-die capacitance of 250 nF

This combination puts us in the regime where the precise capacitor values don't matter, and they can all be the same value. However, where possible, the lowest value of capacitors should be used so that the capacitors have the highest possible ESR and contribute to damping of parallel resonances. The theoretical minimum number of capacitors needed is:

$$n > 2\pi F_{max}\left(\frac{ESL}{Z_{target}}\right) = 2\pi 0.1\left(\frac{1}{0.01}\right) = 63 \tag{13-81}$$

The maximum allowable spreading inductance to the package would be:

$$L_{max} < \frac{Z_{target}}{2\pi F_{max}} = \frac{0.01}{2\pi \times 0.1} = 0.016\,nH = 16\,pH \tag{13-82}$$

As we saw above, it would be impossible to achieve this with a conventional BGA package and 4-mil-thick layers with the capacitors on the top surface. The capacitors would have to be attached under the BGA, and the laminate thickness would have to be reduced or multiple planes would need to be added to the board. Only a 3D field solver would be able to confirm that these modifications provided low enough spreading inductance.

Given these modifications, the board-level decoupling can be achieved with 63 0402 1 µF capacitors, each with an ESL below 1 nH.

What will the chip see? Figure 13-66 shows the impedance profile from the board level and the chip level for this condition.

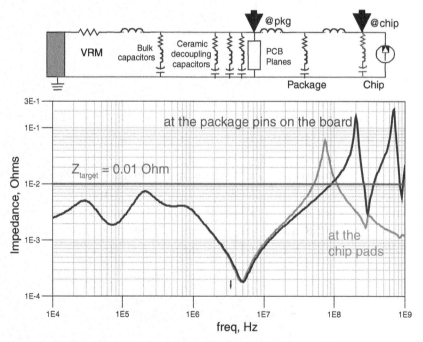

Figure 13-66   Impedance profile, as seen at the package pins on the board and the chip pads.

The low-frequency impedance is optimized by designing the VRM and bulk decoupling capacitors. The board impedance is set by adding 63 MLCC capacitors to achieve the 0.01-Ohm impedance up to 100 MHz. This meets the condition for an optimized board-level PDN design.

However, as viewed by the chip pads, looking through the package into the board, the on-die capacitance and the series package pin inductance and capacitor equivalent inductance create a parallel resonance that results in a large impedance peak near 80 MHz in this case. As with any other parallel peak impedance, it is composed from a capacitance sloping down and an inductance sloping up. An important term that affects the peak impedance is the equivalent series resistance of the parallel resonant circuit.

One important driving force for using a flat impedance profile on the board to which the package and on-die capacitor are attached is that it acts as a damping resistance to reduce the peak height. Using the FDTIM, the capacitor values can be selected to engineer a flat-impedance profile. If this flat region extends under the chip-package parallel resonance, it will damp the peak as seen by the Vdd pads on die. This is not the only way to provide damping resistance, but it can be an effective method.

If the FDTIM is not used, but all the capacitor values are the same value, there will still be some damping series resistance from the parallel combination of the ESR of the capacitors and the series resistance in the package leads and on the die itself.

When the equivalent series resistance is low, the peak impedance can be high. For capacitors near 1 μF, their ESR is on the order of 10 mOhms. With 63 in parallel, the equivalent resistance is reduced to about 0.0002 Ohm, well below the target impedance level. If 10 nF capacitors were used, their ESR would be about 70 mOhms each, with a total series resistance about 0.001 Ohm. This is better than using all higher values of capacitors, but it is still not enough to damp out the parallel resonances. The series resistance of the package traces will be below 0.001 Ohm. This results in a series resistance that might be dominated by the on-die metallization.

TIP   The resistance from on-die interconnects is an important term that can be used to engineer a lower peak impedance.

Figure 13-67 shows the impact that different on-die resistance can have. The resistance is changed from 1 milliOhm, to 3 milliOhms, and 10 milliOhms. It has to be lower than 10 mOhms to meet the target impedance spec, but some on-die resistance is a good thing.

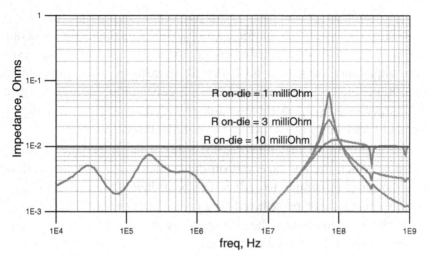

Figure 13-67   The impact on the impedance profile as seen by the chip, from different on-die series resistance in the PDN.

Even if the on-die resistance is optimized, the peak impedance at the parallel resonance will still be above the target. This is where on-package decoupling capacitors can play a role in suppressing the peak impedance.

Figure 13-68 shows the impact on the impedance profile the chip sees looking into the PDN as the number of on-package decoupling capacitors is increased from none to 10. Each has an ESL of 0.1 nH, which is typically implemented using IDC.

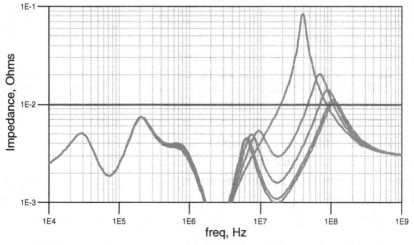

Figure 13-68   The impedance profile seen by the chip as 3, 6, 8, and 10 on-package decoupling capacitors are added.

In this example, when 10 IDC, each with an ESL of 0.1 nH, are added to the package, the impedance profile seen by the chip is as shown in Figure 13-69, meeting the 0.01-Ohm target impedance from DC to very high bandwidth.

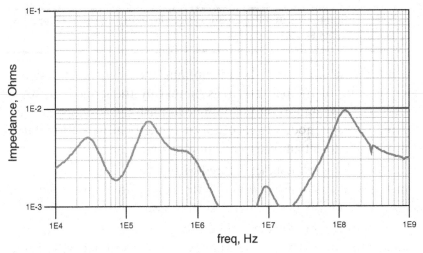

Figure 13-69 Impedance profile as seen by the chip pads, meeting the 0.01-Ohm target impedance.

13.27 Bringing It All Together

The most important feature of the PDN is how low it keeps the voltage noise on the pads of the chip. This is fundamentally related to the current draw of the chip and the impedance profile from the chips' pads to the VRM. Unfortunately, this depends very strongly on factors that are often beyond the control of the board-level designer, such as:

• Current spectrum drawn by the chip operations
• On-die capacitance
• On-die resistance
• Chip attach inductance
• Package attach inductance
• On-package decoupling capacitors

In the highest-performance systems, where the target impedance is in the milliOhm range—such as for high-end processors, servers, and graphics chips—system co-design, simultaneously optimizing the chip features, package features, and board features, is critically important for a robust, cost-effective design.

TIP   Companies that implement system-level PDN design will end up being the most successful.

But for most designs, there is nothing the board-level designer can do about the chip or the package. It is limited by the semiconductor provider. While it is always important to continue to push the envelope and ask for the chip and package models to incorporate in the board-level PDN design, it is rarely possible to get all the important information required.

The board-level designer is faced with having to follow the guideline that "sometimes an okay answer *now* is better than a good answer later" and base the board-level PDN design on reasonable assumptions.

If enough on-package decoupling capacitance is provided, the requirements for on-board decoupling can be dramatically reduced. For example, in the earlier example, the 10 on-package decoupling capacitors provide an equivalent inductance of $1/10 \times 100$ pH $= 10$ pH. This interacts with the on-die capacitance to keep the peak impedance below the target impedance in the 100-MHz region and above.

This inductance is on the chip side of the interconnects, between the package and the board. The equivalent package-attach inductance to the board could be large and may not affect the high-frequency impedance the chip sees. In fact, it could be as large as 0.1 nH. This is illustrated in Figure 13-70.

If the package inductance is 0.1 nH, the highest frequency at which the board can affect the impedance the chip sees is:

$$F_{max} = \frac{Z_{target}}{2\pi L_{pkg}} = \frac{0.01}{2\pi \times 0.1} = 0.016 \, \text{GHz} = 16 \, \text{MHz} \tag{13-83}$$

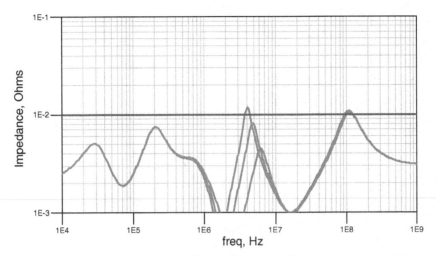

Figure 13-70 Impedance profile the chip sees with on-package capacitors providing the low impedance at high frequency and changing the package-attached inductance from 0.05 nH, to 0.1 nH, and 0.15 nH.

where:

Z_{target} = target impedance, in Ohms

L_{pkg} = inductance of the package attach to the circuit board, in nH

F_{max} = highest frequency at which the board-level impedance can affect the chip-pad impedance, in GHz

The use of package-level decoupling capacitors and higher package attach inductance can mean that the requirements for the board-level decoupling are reduced. If the maximum target impedance is no longer 100 MHz but 16 MHz, fewer capacitors are required to meet the same impedance level. Of course, the same considerations about parallel resonance peak impedances at the interfaces must be taken into account. Figure 13-71 shows that using the lower-ESL capacitors on package can reduce the number of board-level capacitors required from 63 to 15 and still meet the target impedance.

TIP When there are on-package capacitors, the board-level requirements can be reduced. Even then, selecting the capacitors is not about the capacitor value but about the ESR and ESL they provide.

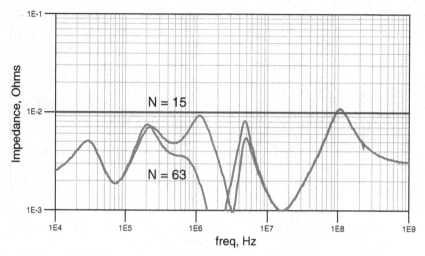

Figure 13-71 Impedance that the profile chip sees with on-package capacitors and reducing the board-level capacitors from 63 to 15 while still maintaining the impedance target.

When semiconductor vendors incorporate on-package decoupling capacitors, they often specify the recommended board-level decoupling requirements in terms of how many capacitors and what value should be added to the board. This is rarely useful information, as the number of capacitors to add to the board is not about their capacitance but about their ESL.

13.28 The Bottom Line

1. PDN design is confusing, contradictory, complicated, and complex. This is fundamentally due to the complex interactions of many features that are beyond the control of the board designer and that are often poorly documented.

2. The goal of the PDN is to provide a low impedance from the chip pads to the VRM.

3. The VRM and bulk-decoupling capacitors provide low impedance at low frequency.

4. Do everything possible to integrate the bulk capacitors with low loop inductance.

5. The chip and package design have the most influence on the impedance in the 100 MHz and above region. The most important design

guidelines are to add more on-die decoupling capacitance, to keep the chip-attach inductance low, and to add low-inductance decoupling capacitors to the package.

6. The most important starting place in PDN design is establishing a target impedance. This can be approximated based on the worst-case power consumption of the device.

7. The package lead inductance and circuit board via to the power/ground cavity will fundamentally limit the high-frequency design for the board-level impedance.

8. At the board level, everything possible should be done to reduce the loop inductance from the capacitors to the packages.

9. The most important design guidelines are to use power and ground planes on adjacent layers, with thin dielectric, placed close to the surface of the board; to use short and wide surface traces between the capacitors and their vias to the cavity; and to place the capacitors in proximity to the package on the top surface and, when spreading inductance is saturated, to place some directly beneath the packages on the bottom of the board.

10. The most important terms that influence the number of capacitors to use is their ESL and the maximum frequency for the board-level decoupling.

11. When the board planes' capacitance creates a parallel resonance with the decoupling capacitors at a low frequency, you can use the smallest number of capacitors by carefully selecting their values using SPICE to simulate the impedance profile.

12. For very low-impedance designs, the values of the capacitors are not as important as their ESL and the number used. All the same value capacitors can work well.

13. When parallel resonant peaks are involved, the lowest-capacitance capacitors will have the highest ESR and provide some damping to reduce the peak impedance.

14. For very low impedance, on-package capacitors are essential. Their use may decrease the on-board capacitor requirements.

15. For the very lowest-impedance PDN designs, co-design of the chip, package, and board-level PDN will provide the best cost–performance solutions.

End-of-Chapter Review Questions

13.1 What are five elements that are part of the PDN?

13.2 What are two examples of interconnect structures that are not part of the PDN?

13.3 What are three potential problems that could arise with a poorly designed PDN?

13.4 What is the most important design principle for the PDN?

13.5 What performance factors influence the target impedance selection?

13.6 What are the three most important design guidelines for the PDN?

13.7 A 2-V power rail has a ripple spec of 5% and draws a maximum transient current of 10 A. What is the estimated target impedance?

13.8 What is one downside to implementing a PDN impedance that is well below the target impedance?

13.9 Why is the PDN easier to engineer by looking in the frequency domain?

13.10 What are the five frequency regions of the impedance of the PDN in the frequency domain, and what are the physical features that affect it?

13.11 What feature in the PDN provides the lowest impedance at the highest frequency?

13.12 What is the only design feature you need to adjust to reduce the impedance at the highest frequency, and why is this difficult to achieve?

13.13 What are the three most important metrics that describe a VRM?

13.14 What is the key SPICE element that enables simulation of an impedance analyzer?

13.15 Why is reducing the package lead inductance so important?

13.16 A 2-layer BGA has leads that run from one side to the other side. If adjacent power and ground leads are 0.5 inches long, and there are 20 pairs, what is the equivalent package lead inductance of the PDN?

13.17 What package style would have the lowest loop inductance? What three package features would you design to reduce the package lead inductance?

13.18 Why is it sometimes noticed that if you take off all the decoupling capacitors on a board, the product will sometimes still work? Why is this not a good test?

13.19 What are the three most important features of an MLCC capacitor, and what physical features affect each term?

13.20 Why is reducing inductance in the PDN elements on a board so important?

13.21 What are the three most important design guidelines for reducing the mounting inductance of an MLCC capacitor?

13.22 What three design features will reduce the spreading inductance in a power and ground cavity, and which is the most important?

13.23 In what combination of design features will the position of a decoupling capacitor not be important? In what case is location important?

13.24 A capacitor really does not behave like an ideal capacitor on a board at the higher frequencies. What is a more effective way of thinking of a capacitor?

13.25 A surface trace on a board from a capacitor to the vias to the cavity is 10 mils wide and 30 mils long. The top of the cavity is 10 mils below the surface. What is the mounting inductance of the 0603 capacitor? What would be the loop inductance of the vias if they were 10 mils in diameter?

13.26 What happens to the self-resonant frequency when 10 identical capacitors are added to a board? What are the features that change?

13.27 What new impedance feature is created when two different-value capacitors are added in parallel? Why is this a very important feature for PDN design?

13.28 What are three design guidelines for reducing the parallel peak impedance between two capacitors?

13.29 What happens to the ESR of a capacitor as the value of the capacitance increases?

13.30 To reduce the number of capacitors needed to achieve a target impedance, what is the most important design feature to adjust?

13.31 What is one advantage of engineering a flat PDN impedance profile?

13.32 What is the FDTIM process and why is it a powerful technique?

100+ General Design Guidelines to Minimize Signal-Integrity Problems

Never follow a rule blindly. Always understand its purpose and put in the numbers to evaluate the costs and benefits for your specific design.

0. Always use the longest rise time you can.

A.1 Minimize Signal-Quality Problems on One Net

Strategy: Keep the instantaneous impedance the signal sees constant throughout its entire path.

Tactics:

1. Use controlled-impedance traces.
2. Ideally, all signals should use the low-voltage planes as their return planes.
3. If different voltage planes are used as signal returns, there should be tight coupling between them. Use the thinnest dielectric you can afford and use multiple low-inductance decoupling capacitors between the different voltage planes.

4. Use a 2D field solver to calculate the stack-up design rules for the target characteristic impedance. Include the effects of solder-mask and trace thickness.

5. Use series termination for point-to-point topologies, whether single-directional or bidirectional.

6. Terminate both ends of the bus in a multidrop bus.

7. Keep the time delay of stubs less than 20% of the rise time of the fastest signals.

8. Place the terminating resistors as close to the package pads as possible.

9. Don't worry about corners unless 10 fF of capacitance is important.

10. Follow the return path of each signal and keep the width of the return path under each signal path at least as wide—and preferably at least three times as wide—as the signal trace.

11. Route signal traces around rather than across return-path discontinuities.

12. Avoid using engineering change wires in any signal path.

13. Keep all nonuniform regions as short as possible.

14. Do not use axial-lead terminating resistors for system rise times less than 1 nsec. Use SMT resistors and mount them for minimum loop inductance.

15. When rise times are less than 150 psec, do everything possible to minimize the loop inductance of the terminating SMT resistors or consider using on-die termination (ODT) or integrated or embedded resistors.

16. Vias generally look capacitive. Minimizing the capture pads and increasing the antipad clearance diameter will help make a via look transparent.

17. Consider adding a little capacitance to the pads of a low-cost connector to compensate for its typically higher inductance.

18. Route all differential pairs with a constant differential impedance, independent of their coupling.

19. Avoid all asymmetries in a differential pair. Whatever you do to one trace, do the same to the other.

20. If the spacing between the traces in a differential pair has to change, adjust the line width to keep a constant differential impedance.

21. If a delay line is to be added to one leg of a differential pair, add it near the beginning of the trace and keep the traces uncoupled in that region.

22. It is okay to change the coupling in a differential pair as long as the differential impedance is maintained.

23. In general, route differential-pair traces with the tightest coupling practical to achieve the highest interconnect density and lowest cost.

24. Decide on edge- versus broadside-coupled differential pairs, based on routing density, total board-thickness constraints, and ability of the fab vendor to maintain tight laminate thickness control. Performance-wise, they can be equivalent.

25. For any board-level differential pairs, there will be significant return current in the planes, so avoid all discontinuities in the return path. If there is a discontinuity, do exactly the same thing to each line in the pair.

26. Worry about terminating the common signals only if the common-mode rejection ratio of the receiver is poor. Terminating the common signals will not eliminate the common signal but will just minimize its ringing.

27. If losses are important, use the widest signal trace possible and never use a trace of less than 5 mils.

28. If losses are important, consider using loosely coupled differential pairs as this will enable a wider signal line for the same dielectric thickness.

29. If losses are important, keep traces as short as possible.

30. If losses are important, do everything possible to minimize all capacitive discontinuities.

31. If losses are important, engineer the signal vias to have a 50-Ohm impedance, which usually means doing everything possible to decrease the barrel size, decrease the capture-pad size, and increase the antipad-clearance holes.

32. If losses are important, use the lowest-dissipation-factor laminate you can afford.

33. If losses are important, consider using pre-emphasis and equalization.

A.2 Minimize Cross Talk

Strategy: Minimize mutual capacitance and mutual inductance between signal and return paths.

Tactics:

34. For microstrip or stripline transmission lines, keep the spacing between adjacent signal paths at least twice their line width. While it is really the dielectric thickness that determines the extent of the fringe field lines, a 50-Ohm line defines a ratio of line width to dielectric thickness, so a spacing based on line width is spacing based on dielectric thickness.

35. Minimize any discontinuities in the return path the signals might cross over.

36. If you have to cross a gap in the return path, only use differential pairs. Never cross a gap with single-ended signals routed close together.

37. For surface traces, keep the coupled lengths as short as possible and use as much solder mask as practical to minimize far-end cross talk.

38. If far-end cross talk is a problem, add a laminate layer to the top of the surface traces to make them embedded microstrip.

39. For long, coupled lengths where far-end cross talk may be a problem, route the traces in stripline.

40. If you can't keep the coupling length less than the saturation length, changing the coupling length will have no impact on the near-end cross talk, so don't worry about decreasing coupling length.

41. Use the lowest-dielectric-constant laminate you can afford so the dielectric spacing to the return planes can be kept to a minimum for the same target characteristic impedance.

42. In a tightly coupled microstrip bus, the deterministic jitter can be reduced by keeping the spacing at least as wide as twice the line width or by routing timing-sensitive lines in stripline.

43. For isolations in excess of –60 dB, use stripline with guard traces.

44. Always use a 2D field solver to evaluate whether you need to use a guard trace.

45. If you use a guard trace, make it as wide as will fit and use vias to short the ends to the return path. Add additional shorting vias along the length if it is free and easy to do so. They are not as critical as the two on the ends. Guard traces in microstrip do not help much.

46. Minimize ground bounce by making the return paths in any packages or connectors as short and as wide as possible.

47. Use chip-scale packages rather than larger packages.

48. Minimize ground bounce in the power return path by bringing the power plane as close to the return plane as possible.

49. Minimize ground bounce in the signal return paths by bringing the signal path as close to the return path as acceptable, consistent with matching the impedance of the system.

50. Avoid using shared return paths in connectors and packages.

51. When assigning leads in a package or connector, reserve the shortest leads for the ground paths and space the power and ground leads uniformly among the signal paths, or closest to those signal paths that will carry a lot of switching current.

52. All no-connect leads or pins should be assigned as ground-return connections.

53. Avoid using resistor single inline packages (SIPs) unless there is a separate return path for each resistor.

54. Check the film to verify that antipads in via fields do not overlap and there is a well-defined web between clearance holes in the power and ground planes.

55. If a signal changes return planes, the return planes should be as closely spaced as you can afford. If you use a decoupling capacitor to minimize the impedance of the return path, its capacitance value is immaterial. Select it and design it in for lowest loop inductance.

56. If many signal lines switch return planes, space the signal path vias as far apart as possible rather than clustering them all in the same location.

57. If a signal switches return layers, and the planes are the same voltage level, place a via between the return planes, as close to the signal via as possible.

A.3 Minimize Rail Collapse

Strategy: Minimize the impedance of the power-distribution network.

Tactics:

58. Minimize the loop inductance between the power and ground paths.

59. Allocate power and ground planes on adjacent layers with the thinnest dielectric you can afford.

60. Get the lowest impedance between the planes by using the highest dielectric constant between the planes that you can afford.

61. Use as many power- and ground-plane pairs in parallel as you can afford.

62 Route the same currents far apart and route opposite currents close together.

63. Place each power via as close as practical to a ground via. If you can't get them at least within a pitch equal to their length, there will be no coupling and no value in proximity.

64. Route the power and ground planes as close as possible to the surface where the decoupling capacitors are mounted.

65. Use multiple vias to the same power or ground pad but keep the vias as far apart as possible.

66. Use vias as large in diameter as practical when routing to power or ground planes. However, it is better to use multiple small diameter vias spread apart than to use one large via.

67. Use double-bonding on power and ground pads to minimize the loop inductance of the wire bonds.

68. Use as many power and ground connections from the chip as you can afford.

69. Use as many power and ground connections from the package as you can afford.

70. Use chip-interconnect methods that are as short as possible, such as flip-chip rather than wire-bond.

71. Use package leads as short as possible, such as chip-scale packages rather than QFP packages.

72. Keep all surface traces that run between the pads of the decoupling capacitors and their vias as short and wide as possible.

73. Use a total amount of bulk-decoupling capacitance to take over from the regulator at low frequency.

74. Use a total number of decoupling capacitors to reduce the equivalent inductance at high frequency.

75. Use as small a body size for a decoupling capacitor as you can afford and minimize the length of all connections from the capacitor pads to the power and ground planes.

76. Place as much decoupling capacitance as you can afford on the chip itself.

77. Place as many low-inductance decoupling capacitors as you can afford on the package.

78. Use the FDTIM to select capacitor values to provide a flat impedance profile to damp out the on-die capacitance and package lead inductance parallel peak impedance on the Vdd planes.

79. Use differential pairs for I/Os to reduce the switching dI/dt currents.

A.4 Minimize EMI

Strategy: Reduce the voltage that drives common currents, increase the impedance of the common current paths, and shield and filter as a quick fix.

Tactics:

80. Reduce ground bounce.

81. Keep all traces at least five line widths from the edge of the board.

82. Route traces in stripline when possible.

83. Place the highest-speed/highest-current components as far from the I/O connections as possible.

84. Place the decoupling capacitors in proximity to the chips to minimize the spread of high-frequency-current components in the planes.

85. Keep power and ground planes on adjacent layers and as close together as possible.

86. Use as many power- and ground-plane pairs as you can afford.

87. When using multiple power- and ground-plane pairs, recess the power planes and then stitch shorting vias between the ground planes along the edges.

88. Use ground planes as surface layers, where possible.

89. Know the resonant frequencies of all packages and change the package geometry if there is an overlap with a clock harmonic.

90. Avoid signals switching different voltage return planes in a package. This will drive package resonances.

91. Add ferrite filter sheets to the top of a package if it might have a resonance.

92. Minimize any asymmetries between the lines in each differential pair.

93. Use a common-signal-choke filter on all external differential pair connections.

94. Use a common-signal-choke filter around the outside of all peripheral shielded cables.

95. Filter all external I/O lines to use the longest signal rise time that is tolerable for the timing budget.

96. Use spread-spectrum clock generator to spread the first harmonic over a wider frequency range and decrease the radiated energy within the bandwidth of the FCC test.

97. When connecting shielded cables, try to keep the shield as an extension of the enclosure.

98. Minimize the inductance of the shielded cable connections to the enclosure. Use a coaxial connection right from the end of the cable and to the enclosure.

99. Equipment bays should not penetrate the integrity of the enclosure.

100. Only interconnects need to break the enclosure integrity.

101. Keep aperture diameters small—significantly smaller than a wavelength of the highest-frequency radiation that might leak. More and smaller holes are better than fewer and larger holes.

102. The most expensive rule is the one that delays the product ship date.

100 Collected Rules of Thumb to Help Estimate Signal-Integrity Effects

A rule of thumb should be used only when an okay answer *right now* is more important than a good answer later.

A rule of thumb is only meant to provide a very rough approximation. It is designed to help feed our intuition and to help find a quick answer with very little effort. It should always be the starting place for any estimate. It can help us distinguish between a 5 or a 50. It can help us see the big picture early in the design phase. In the balance between quick and accurate, a rule of thumb is quick; it is not meant to be accurate.

Of course, you can't use a rule of thumb blindly. It must be coupled with an understanding of the principles and good engineering judgment.

When accuracy is important, such as in a design sign-off, where being off by a few percent may have a $1 million impact, why use anything other than a verified numerical simulation tool?

This appendix provides a collection of the most useful rules of thumb, separated by the chapters in which they are introduced.

B.1 Chapter 2

1. The rise time of a signal is ~10% the clock period and ~1/10 × 1/F_{clock}. For example, a 100-MHz clock has a rise time of about 1 nsec.
2. The amplitude of the nth harmonic of an ideal square wave is ~2/(πn) times the magnitude of the clock voltage. For example, the first harmonic amplitude of a 1-v clock signal is about 0.6. The third harmonic is about 0.2.
3. The bandwidth, BW, and rise time, RT, of a signal are related by BW = 0.35/RT. For example, if the rise time is 1 nsec, the bandwidth is 350 MHz. If the bandwidth of an interconnect is 3 GHz, the shortest rise time it can transmit is about 0.1 nsec.
4. If you don't know the rise time, you can estimate the bandwidth of a signal as roughly five times the clock frequency. For example, if the clock frequency is 1 GHz, the bandwidth of the signal is about 5 GHz.

B.2 Chapter 3

5. The resonant frequency of an LC circuit is 5 GHz/sqrt(LC) with L in nH and C in pF. For example, a package lead and its return path might have a loop self-inductance of 7 nH. Its capacitance might be about 1 pF. The frequency at which it would ring would be about 2 GHz.

B.3 Chapter 4

6. An axial lead resistor looks like an ideal resistor up to about 400 MHz. An SMT 0603 resistor looks like an ideal resistor up to about 2 GHz.
7. The ESL of an axial lead resistor is about 8 nH. The ESL of an SMT resistor is about 2 nH.
8. The resistance per length of a 1-mil-diameter gold wire bond is about 1 Ohm/inch. For example, a 50-mil-long wire bond has a resistance of about 50 milliOhms.
9. A length of 24 AWG wire has a diameter of about 20 mils and a resistance per length of about 25 milliOhms/foot.
10. The sheet resistance of 1-ounce copper is about 0.5 milliOhm/square. For example, a trace 5 mils wide and 1 inch long has 200 squares and would have a series resistance of 200 × 0.5 = 100 milliOhms = 0.1 Ohm.
11. The skin-depth effects for 1-ounce copper begin at about 10 MHz.

B.4 Chapter 5

12. The capacitance of a sphere 1 inch in diameter is about 2 pF. For example, a pigtail cable hanging off a board a few inches long will have a capacitance to the floor of about 2 pF.

13. The capacitance of a pair of plates the size of a penny, with air between the faces, is about 1 pF.

14. When the plates of a capacitor are as far apart as the plates are wide, the fringe fields contribute just as much capacitance as the parallel-plate fields. For example, if we estimate the parallel-plate capacitance for a microstrip with line width of 10 mils and dielectric thickness of 10 mils to be 1 pF/inch, the actual capacitance will be about twice this value, or 2 pF/inch.

15. If we don't know anything else about a material except that it is an organic laminate, a good estimate for its dielectric constant is 4.

16. For a 1-watt chip, the amount of time, in sec, a decoupling capacitance, in F, will provide the charge with less than 5% voltage droop is $C/2$. For example, if there is a decoupling capacitance of 10 nF, it will provide decoupling for only 5 nsec. If we require 10 microseconds of decoupling, we need 20 µF of capacitance.

17. The capacitance available in the power and ground planes of a typical circuit board when the separation is 1 mils is 1 nF/square inch, and it scales inversely with the dielectric thickness. For example, in a 10-mil separation board, the total board area available for decoupling an ASIC may be only 4 square inches. The capacitance would be $4\ \text{in}^2 \times 1\ \text{nF/in}^2/10 = 0.4$ nF, and it would provide decoupling for about 0.2 nsec.

18. The effective dielectric constant in a 50-Ohm microstrip is 3 when the bulk dielectric constant is 4.

B.5 Chapter 6

19. The partial self-inductance of a round wire, 1 mil in diameter, is about 25 nH/inch, or 1 nH/mm. For example, the partial self-inductance of a via 1.5 mm long is about 1.5 nH.

20. The loop self-inductance of a 1-inch-diameter round loop, made from 10-mil-thick wire about the size of your finger and thumb held together in a circle, is 85 nH.

21. The total inductance per length for a section taken out of a 1-inch-diameter loop is about 25 nH/inch, or 1 nH/mm. For example, if a package lead is part of a loop and is 0.5 inch long, it has a total inductance of about 12 nH.

22. When the center-to-center separation of a pair of round rods is 10% of their length, the partial mutual inductance will be about 50% of the partial self-inductance of either one. For example, if we have two wire bonds, 1 mm long on 0.1 mm centers, the partial self-inductance of either one is about 1 nH, while their partial mutual inductance will be about 0.5 nH.

23. When the center-to-center separation between two round rods is about equal to their length, the partial mutual inductance between them is less than 10% of the partial self-inductance of either one. For example, if we space 25-mil-long vias more than 25 mils center-to-center, there is virtually no inductive coupling between them.

24. The loop inductance of an SMT capacitor is roughly 2 nH, including the surface traces, vias, and capacitor body. Good engineering efforts are required to reduce this below 1 nH.

25. The loop inductance per square of a pair of planes is 33 pH per square per mil of spacing. For example, if the dielectric spacing is 2 mils, there is 66 pH/square of loop inductance between the planes.

26. The loop inductance of a pair of planes having a field of clearance holes with 50% open area will increase by about 50%.

27. The skin depth in copper is 2 microns at 1 GHz and increases with the square root of frequency. For example, at 10 MHz, the skin depth is 20 microns.

28. In a 50-Ohm transmission line of 1-ounce copper, the loop inductance per length is constant at frequencies above about 50 MHz. This means characteristic impedance is constant above 50 MHz.

B.6 Chapter 7

29. The speed of electrons in copper is about the speed of an ant, or 1 cm/sec.

30. The speed of a signal in air is about 12 inches/nsec. The speed of a signal in most polymer materials is about 6 inches/nsec.

31. The wiring delay, 1/v, in most laminates is about 170 psec/inch.

32. The spatial extent of the signal is the rise time multiplied by the speed, or RT × 6 inches/nsec. For example, if the rise time is 0.5 nsec, the spatial extent of the edge as the signal propagates down a board is 3 inches.

33. The characteristic impedance of a transmission line varies inversely with the capacitance per length. The characteristic impedance is about 160 Ohms/C_len, with C_len in pF/inch.

34. All 50-Ohm lines in FR4 have a capacitance per length of about 3.3 pF/inch. For example, if a BGA lead is designed as 50 Ohms and is 0.5 inch long, it has a capacitance of about 1.7 pF.

35. All 50-Ohm lines in FR4 have an inductance per length of about 8.3 nH/inch. For example, if a connector is designed for 50 Ohms and is 0.5 inch long, the loop inductance of the signal and return path-loop is about 4 nH.

36. A 50-Ohm microstrip in FR4 has a dielectric thickness about half the line width of the trace. For example, if the line width is 10 mils, the dielectric thickness will be 5 mils.

37. A 50-Ohm stripline in FR4 has a plane-to-plane spacing about twice the line width of the signal trace. For example, if the line width is 10 mils, the spacing between the two planes will be 20 mils.

38. The impedance looking into a transmission line will be the characteristic impedance for a time shorter than the round-trip time of flight. For example, if a 50-Ohm line is 3 inches long, all drivers with a rise time shorter than 1 nsec will see a constant 50-Ohm load during the transition time when driving the line.

39. The total capacitance in a section of transmission line with a time delay of TD is $C = TD/Z_0$. For example, if the TD of a line is 1 nsec and the characteristic impedance is 50 Ohms, there is 20 pF of capacitance between the signal and return paths.

40. The total loop inductance in a section of a transmission line with a time delay of TD is $L = TD \times Z_0$. For example, if the TD of a line is 1 nsec and the characteristic impedance is 50 Ohms, there is 50 nH of loop inductance between the signal and return paths.

41. If the width of the return path in a 50-Ohm microstrip is equal to the signal line width, the characteristic impedance is 20% higher than the characteristic impedance when the return path is infinitely wide.

42. If the width of the return path in a 50-Ohm microstrip is at least three times the signal line width, the characteristic impedance is within 1% of the characteristic impedance when the return path is infinitely wide.

43. Trace thickness will decrease the characteristic impedance of a line by about 2 Ohms per mil of thickness. For example, from 1/2-ounce copper to 1-ounce copper, the thickness increases by 0.7 mil. The impedance of the line would decrease by about 1 Ohm.

44. Solder mask on top of a microstrip will decrease its characteristic impedance by about 2 Ohms per mil of thickness. For example, a 0.5-mil-thick solder mask will decrease the characteristic impedance by about 1 Ohm.

45. For an accurate lumped-circuit approximation, you need at least 3.5 LC sections per spatial extent of the rise time. For example, if the rise time is 1 nsec, in FR4, the spatial extent is 6 inches. You would need at least 3.5 LC sections per 6 inches—or about one section every 2 inches of trace—for an accurate approximation.

46. The bandwidth of a single-section LC model is 0.1/TD. For example, if the time delay of a transmission line is 1 nsec, using a single LC section to model it would be accurate up to a bandwidth of about 100 MHz.

B.7 Chapter 8

47. If the time delay of a transmission line is less than 20% of the rise time of the signal, you may not have to terminate the line.

48. In a 50-Ohm system, an impedance change of 5 Ohms will give a reflection coefficient of 5%.

49. Keep all discontinuities shorter, in inches, than the rise time, in nsec. For example, if the rise time is 0.5 nsec, keep all impedance discontinuities, such as a neck-down to pass through a via field, less than 0.5 inch long, and it may be acceptable.

50. A capacitive load at the far end will increase the rise time of the signal. The 10–90 rise time is about $100 \times C$ psec, for C in pF. For example, if the capacitance is 2 pF, typical of the input-gate capacitance of a receiver, the RC limited rise time would be about 200 psec.

51. If the capacitance of a discontinuity is less than $0.004 \times RT$, it may not cause a problem. For example, if the rise time is 1 nsec, capacitive discontinuities should be less than 0.004 nF, or less than 4 pF.

52. The capacitance, in fF, of a corner in a 50-Ohm line is twice the line width, in mils. For example, if the line width of a 50-Ohm trace is 10 mils, a 90-degree bend would have a capacitance of about 20 fF. It might cause reflection problems for rise times of 0.02 pF/4 = 5 psec.

53. A capacitive discontinuity will add a delay time to the 50% threshold point of about $0.5 \times Z_0 \times C$. For example, if the capacitance is 1 pF in a 50-Ohm line, the delay adder will be about 25 psec.

54. If the inductance, in nH, of a discontinuity is less than 10 times the rise time, in nsec, it may not cause a problem. For example, if the rise time is 1 nsec, the maximum inductive discontinuity that may be acceptable is about 10 nH.

55. An axial lead resistor with a loop inductance of about 10 nH may contribute too much reflection noise for rise times less than 1 nsec. Switch to surface-mount resistors.

56. To compensate for a 10-nH inductance, a 4-pF capacitance in a 50-Ohm system is required.

B.8 Chapter 9

57. At 1 GHz, the resistance of a 1-ounce copper trace is about 15 times the resistance at DC.

58. At 1 GHz, the attenuation from the resistance of an 8-mil-wide trace is comparable to the attenuation from the dielectric, and the attenuation from the dielectric will get larger faster with frequency.

59. The low-loss regime, for lines 3 mils or wider, is all frequencies above about 10 MHz. In the low-loss regime, the characteristic impedance and signal speed are independent of the loss and of frequency. There is no dispersion due to loss in typical board-level interconnects.

60. A −3-dB attenuation is a drop to 50% of the initial power level and a drop to 70% of the initial amplitude.

61. A −20-dB attenuation is a drop to 1% of the initial power level and a drop to 10% of the initial amplitude.

62. When in the skin-depth regime, the series resistance per inch of a signal- and return-path line, of width w in mils, is about $8/w \times \mathrm{sqrt}(f)$, with f in GHz. For example, a line 10 mils wide has a series resistance of about 0.8 Ohm/inch, and it increases with the square root of frequency.

63. In a 50-Ohm line, the attenuation from the conductor is about $36/wZ_0$ in dB per inch. For example, if the line width is 10 mils in a 50-Ohm line, the attenuation is $36/(10 \times 50) = 0.07$ dB/inch.

64. The dissipation factor of FR4 is about 0.02.

65. In FR4, the attenuation from the dielectric is about 0.1 dB/inch at 1 GHz and increases linearly with frequency.

66. At 1 GHz, a 50-Ohm line in FR4 with a line width of 8 mils has the same conductor loss as the dielectric loss at 1 GHz.

67. The bandwidth of an FR4 interconnect Len inches long, when limited by dissipation factor, is 30 GHz/Len. For example, a 50-Ohm line 10 inches long has a bandwidth of 3 GHz.

68. The shortest rise time that can be propagated by an FR4 interconnect is 10 psec/inch × Len. For example, a signal propagating down a 10-inch length of 50-Ohm line in FR4 will have a rise time of at least 100 psec.

69. The rise-time degradation from losses will be important in FR4 laminates if the interconnect length, in inches, is greater than 50 times the rise time, in nsec. For example, if the rise time is 200 psec, worry about losses when line lengths are greater than 10 inches.

B.9 Chapter 10

70. In a pair of 50-Ohm microstrip transmission lines, with spacing equal to the line width, the coupling capacitance between the signal lines is about 5%.

71. In a pair of 50-Ohm microstrip transmission lines, with spacing equal to the line width, the coupling inductance between the signal lines is about 15%.

72. The saturation length for near-end noise in FR4 is 3 inches for a 1-nsec rise time, and it scales with the rise time. For example, if the rise time is 0.5 nsec, the saturation length is 1.5 inches.

73. The loaded capacitance of a trace is constant and independent of the proximity of other traces nearby.

74. The near-end cross talk for 50-Ohm microstrip traces with spacing equal to the line width is 5%.

75. The near-end cross talk for 50-Ohm microstrip traces with spacing equal to twice the line width is 2%.

76. The near-end cross talk for 50-Ohm microstrip traces with spacing equal to three times the line width is 1%.

77. The near-end cross talk for 50-Ohm stripline traces with spacing equal to the line width is 6%.

78. The near-end cross talk for 50-Ohm stripline traces with spacing equal to twice the line width is 2%.

79. The near-end cross talk for 50-Ohm stripline traces with spacing equal to three times the line width is 0.5%.

80. In a pair of 50-Ohm microstrip traces with spacing equal to the line width, the far-end noise is 4% × TD/RT. If the time delay of the line is 1 nsec and the rise time is 0.5 nsec, the far-end noise is 8%.

81. In a pair of 50-Ohm microstrip traces with spacing equal to twice the line width, the far-end noise is 2% × TD/RT. If the time delay of the line is 1 nsec and the rise time is 0.5 nsec, the far-end noise is 4%.

82. In a pair of 50-Ohm microstrip traces, with spacing equal to three times the line width, the far-end noise is 1.5% × TD/RT. If the time delay of the line is 1 nsec and the rise time is 0.5 nsec, the far-end noise is 3%.

83. There is no far-end noise in stripline or fully embedded microstrip.

84. In a 50-Ohm bus, stripline or microstrip, to keep the worst-case near-end noise below 5%, keep the spacing between the lines more than twice the line width.

85. In a 50-Ohm bus with a spacing equal to the line width, 75% of all the cross talk on any victim line is from the trace on either side of the victim line.

86. In a 50-Ohm bus with a spacing equal to the line width, 95% of all the cross talk on any victim line is from the nearest two traces on either side of the victim line.

87. In a 50-Ohm bus with a spacing equal to twice the line width, 100% of all the cross talk on any victim line is from the trace on either side of the victim line. You can ignore the coupling from all other traces in the bus.

88. For surface traces, separating the adjacent signal traces sufficiently to add a guard trace will often reduce the cross talk to an acceptable level, eliminating the need for a guard trace. Adding a guard trace with the ends shorted can reduce the cross talk by about 50%.

89. For stripline traces, using a guard trace can reduce the cross talk to less than 10% of the level without the guard trace.

90. To keep the switching noise below an acceptable level, keep the mutual inductance less than 2.5 nH times the rise time, in nsec. For example, if the rise time is 0.5 nsec, the mutual inductance should be less than 1.3 nH for acceptable switching-noise cross talk due to coupling between only two signal/return-path pairs.

91. For a connector or package that is limited by switching noise, the maximum usable clock frequency is 250 MHz/(n × L_m) for a mutual inductance between pairs of signal and return paths, in nH, and a number of simultaneous switching lines, n. For example, if there are four pins that share the same return path and their mutual inductance is about 1 nH between pairs, the maximum usable clock frequency for the connector will be 250 MHz/4 ~ 60 MHz.

B.10 Chapter 11

92. In LVDS signals, the common signal component is more than two times the differential signal component.

93. With no coupling, the differential impedance in a differential pair is twice the single-ended impedance of either line.

94. In a pair of 50-Ohm microstrip lines, the single-ended characteristic impedance of one line is completely independent of the proximity of the adjacent line, as long as the other line is tied low or tied high.

95. In the tightest coupled differential microstrip, a line width equal to the spacing, the differential impedance drops only about 10% from the differential impedance when the traces are far apart with no coupling.

96. In a broad-side coupled differential pair, the trace-to-trace separation must be at least larger than the line width in order to have the possibility of getting as high as a 100-Ohm differential impedance.

97. The FCC Class B requirement is a far-field strength less than about 150 microV/m at 100 MHz and 3 meters distance.

98. It takes only about 3 uA of common current on unshielded twisted-pair cables to fail an FCC Class B EMC certification test.

99. A highly coupled differential pair will have 30% less differential-signal cross talk than a weakly coupled pair, from a closely spaced single-ended aggressor, provided that the line widths and dielectric thicknesses are kept the same. This means the differential impedance decreases when the traces are more tightly coupled.

100. A highly coupled differential pair will have about 30% more common-signal cross talk than a weakly coupled pair, from a closely spaced single-ended aggressor.

Selected References

Anderson, E.M. *Electric Transmission Line Fundamentals*. Reston, VA: Reston Publishing Company, Inc., 1985.

Archambeault, B. *PCB Design for Real World EMI Control*. The Netherlands: Kluwer Academic Publishers, 2002.

Bakoglu, H.B. *Circuits, Interconnects, and Packaging for VLSI*. Reading, MA: Addison-Wesley, 1990.

Bennett, W.S. *Control and Measurement of Unintentional Electromagnetic Radiation*. Hoboken, NJ: John Wiley and Sons, 1997.

Buchanan, J.E. *Signal and Power Integrity in Digital Systems*. Columbus, OH: McGraw-Hill Book Company, 1995.

Chipman, R.A. *Transmission Lines*. Schaum's Outline Series. Columbus, OH: McGraw-Hill Book Company, 1968.

Dally, W.J., and Poulton, J.W. *Digital Systems Engineering*. Cambridge, England: Cambridge University Press, 1998.

Derickson, Dennis, Muller, Marcus. *Digital Communications Test and Measurement: High-Speed Physical Layer Characterization.* Upper Saddle River, NJ: Prentice-Hall, 2008.

Gardial, F. *Lossy Transmission Lines.* Norwood, MA: Artech House, 1987.

Grover, F.W. *Inductance Calculations.* Mineola, NY: Dover Publications, 1973.

Hall, S.H., Hall, G.W., and McCall, J.A. *High Speed Digital System Design.* Hoboken, NJ: John Wiley and Sons, 2000.

Itoh, T. *Planar Transmission Line Structures.* Piscataway, NJ: IEEE Press, 1987.

Johnson, Howard, and Graham, Martin. *High Speed Digital Design.* Upper Saddle River, NJ: Prentice-Hall, 1993.

Konsowski and Helland et al. *Electronic Packaging of High Speed Circuitry.* Columbus, OH: McGraw-Hill, 1997.

Li, Mike Peng. *Jitter, Noise, and Signal Integrity at High-Speed*, Upper Saddle River, NJ: Prentice-Hall, 2007.

Mardiguian, Michel. *Controlling Radiated Emissions by Design.* The Netherlands: Chapman and Hall, 1992.

Martens, L. *High Frequency Characterization of Electronic Packaging.* The Netherlands: Kluwer Academic Publishers, 1998.

Oh, K.S., and Yuan, X. *High-Speed Signaling: Jitter Modeling, Analysis and Budgeting.* Upper Saddle River, NJ: Prentice-Hall, 2012.

Ott, Henry. *Noise Reduction Techniques in Electronic Systems.* Hoboken, NJ: Wiley-Interscience, 1988.

Paul, Clayton. *Introduction to Electromagnetic Compatibility.* Hoboken, NJ: Wiley-Interscience, 1992.

Pandit, V.S., Ryu, W.H., and Choi, M.J. *Power Integrity for I/O Interfaces: With Signal Integrity/Power Integrity Co-Design.* Upper Saddle River, NJ: Prentice-Hall, 2010.

Poon, Ron. *Computer Circuits Electrical Design.* Upper Saddle River, NJ: Prentice-Hall, 1995.

Rosenstark, Sol. *Transmission Lines in Computer Engineering.* Columbus, OH: McGraw-Hill, 1994.

Skilling, H.H. *Electric Transmission Lines*. Melbourne, FL: Krieger Publishing Company, 1979.

Smith, D. *High Frequency Measurements and Noise in Electronic Circuits*. New York: Van Nostrand Reinhold, 1993.

Smith, Larry and Bogatin, Eric. *Principles of Power Integrity for PDN Design-Simplified*. Upper Saddle River, NJ: Prentice-Hall, 2017.

Tsaliovich, A. *Cable Shielding for Electromagnetic Compatibility*. The Netherlands: Chapman and Hall, 1995.

Wadell, Brian. *Transmission Line Design Handbook*. Norwood, MA: Artech House, 1991.

Walker, C. *Capacitance, Inductance and Crosstalk Analysis*. Norwood, MA: Artech House, 1990.

Walsh, J.B. *Electromagnetic Theory and Engineering Applications*. New York: The Ronald Press Company, 1960.

Williams, Tim. *EMC for Product Designers*. Burlington, MA: Newnes Press, 1992.

Young, B. *Digital Signal Integrity*. Upper Saddle River, NJ: Prentice-Hall, 2000.

Review Questions and Answers

This appendix contains review questions and answers to help you solidify your understanding of the material covered throughout the book. Try to answer all questions before looking at the answers that follow.

Chapter 1

1.1   Name one problem that is just a signal-integrity problem.
Reflections on a transmission line are just a signal-integrity problem. This is sometimes called "self-aggression" noise from the signal in the transmission line.

1.2   Name one problem that is just a power-integrity problem.
The rail noise on the core's Vdd supply on the die is just a power-integrity problem. This is also an example of "self-aggression" noise on the Vdd rail, as it comes from the noise on the Vdd rail from the currents in the Vdd rail.

1.3   Name one problem that is just an electromagnetic compliance problem.
The source of ALL radiated emissions is from changing currents in extended conductors which are part of the system. Of course, the source of the currents is only from either signal and return currents or from power or

ground currents, so there is always interaction between these three effects. However, many features of signals that do not affect the signals can have a dramatic impact on EMI. One good example is a slight mode conversion in a differential pair, such as a CAT 5 twisted pair. A small amount of common signal on the twisted pair will have no impact on the differential signal, but can cause a complete EMC failure.

1.4 **Name one problem that is considered both an SI and PI problem.**

The most commonly occurring problem that involves both signals and power integrity is ground bounce. This is the effect in which there is a skewed-up return path—i.e., not a wide uniform plane, and there are signals that share this common return path. When the return path is a narrow path like a lead in a package, it will have more inductance than a wide plane.

Since signal return paths are typically shared by power return paths, the skewed-up return paths could be part of the signal interconnects, or part of the power distribution network. When current flows through this shared, common return path, a voltage is generated. This is "ground bounce." It is a form of cross talk between signals, or cross talk between power and signals.

1.5 **What causes an impedance discontinuity?**

An impedance discontinuity occurs when the instantaneous impedance the signal would see changes. This happens whenever the cross-sectional geometry changes. This could be when the signal conductor changes shape, when the return path geometry changes, or both. Discontinuities usually happen at interfaces of structures—from on-die, to packages, to traces on one layer, thru vias, and through connectors.

1.6 **What happens to a propagating signal when the interconnect has frequency-dependent losses?**

Generally, losses from conductors and dielectrics increase as you go up in frequency. When a signal with a fast edge passes through an interconnect with losses increasing with frequency, the higher frequency components will be attenuated more than the lower frequency components. This effectively decreases the bandwidth of the signal.

A high bandwidth signal will have a short rise time. As its bandwidth decreases, the rise time will increase. This rise time increase is the chief problem with lossy interconnects. When the rise time is comparable to the unit interval of the data pattern, the losses will affect signal quality.

1.7 What are the two mechanisms that cause cross talk?

The fundamental mechanism for cross talk is fringe electric and fringe magnetic field coupling. If all the electric and magnetic fields between the signal and return paths were confined to the vicinity of the signal paths, there would be no cross talk.

We can approximate the electric and magnetic field coupling as circuit elements with coupling capacitance and coupling inductance, which are called mutual capacitance and mutual inductance.

1.8 For lowest cross talk, what should the return path for adjacent signal paths look like?

A wide, solid plane always confines the stray electric and magnetic fields more than any other structure. For the lowest cross talk, always use a wide return plane. Any other structure increases the relative cross talk.

1.9 A low-impedance PDN reduces power-integrity problems. List three design features for a low-impedance PDN.

The power distribution network (PDN) are all the interconnects from the pads of the VRM to the pads of the power rail on the die. One way of reducing the impedance is by reducing the loop inductance in the path. This would be by 1) bringing the power path as close as possible to the ground path, 2) keeping interconnects short, and 3) using wide conductors like planes.

To reduce the interconnect lengths between the planes and the devices, 4) the power and ground planes should be on adjacent layers 5) with a thin dielectric, and 6) positioned as close as possible to the top component layer in the stack up.

1.10 List two design features that contribute to reduced EMI

One of the biggest sources of EMI is common currents on external, twisted pairs, like CAT 5 cables. This is reduced by using a shielded cable like CAT 6, with its shield connected to the chassis. It's not that the "shield" in anyway shields the fields inside from radiating. It's that the shield acts as a return path for the common currents on the twisted pair and reduces the external fields from the common currents.

A second problem arises from the connection of the shield. If it is not a 360-degree connection and does not look like an ideal coaxial connection, there will be some total inductance in the return path. When return current flows through this total inductance, it generates a ground bounce voltage and drives new common currents on the outside shield of the cable.

1.11 When is it a good idea to use a rule of thumb? When is it not a good idea to use a rule of thumb?

The starting place for tackling any problem should be applying a rule of thumb to estimate what to expect. This initial value will help drive your engineering judgment. When you are ready to sign off on a design, when the accuracy of the value to 10% or better is important, do not use a rule of thumb—use a numerical simulation with a verified process.

1.12 What is the most important feature of a signal that influences whether it might have a signal-integrity problem?

All signal-integrity problems increase with shorter rise time. Generally, when the rise time of a signal is 10 nsec or longer, the interconnects are pretty transparent and signal-integrity problems will be small. But, as the rise time decreases, signal-integrity problems will get worse. Generally, if the rise time is 1 nsec or less, interconnects are not transparent; and if you don't worry about signal integrity, your product probably will not work the first time.

1.13 What is the most important piece of information you need to know to fix a problem?

The root cause. If you don't know the correct root cause, your chance of fixing a problem is based on pure luck. Not only must you know the root cause, but you must have confidence you have the correct one. This is usually by thinking of as many consistency tests as you can, and trying them.

1.14 Best design practices are good habits to follow. Give three examples of a best design practice for circuit board interconnects.

1. Route all signals as uniform transmission lines.
2. Avoid using branches when routing signal lines.
3. Always use a continuous return path under each signal line.
4. Consider using a termination if the line is long enough or rise time short enough.
5. Always use as short a surface trace as possible for capacitors so their mounting looks as close to via in pad as practical.
6. Avoid messing up return paths or overlapping return currents of different signal paths

1.15 What is the difference between a model and a simulation?

A model is the electrical description of a physical structure, component, or device. It is the input information to the simulator, written in the language the simulator understands. For example, for a SPICE-like circuit simulator, the language of models is circuit elements such as capacitors, resistors, inductors, and transmission lines.

A simulator is an engine that computes electric or magnetic fields, S-parameters, voltages or currents in the time or frequency domain. Sometimes an S-parameter description of an interconnect can be used as a behavioral model of the interconnect and inserted in a simulation tool.

1.16 What are the three important types of analysis tools.

The simplest type of analysis tool is a rule of thumb. This is easy to use, but not intended to be very accurate. It's designed to be the starting place of any analysis and helps establish solid engineering judgment.

The next, more accurate tool is an analytical approximation. While they can sometimes look complicated, this is not measure of their accuracy. They can be integrated into a spreadsheet and what-if scenarios evaluated. Their accuracy is not well established.

The potentially most accurate analysis tool is a numerical simulation. Although the value of this type of analysis can be very high, it also comes at a higher cost in expertise, learning curve, and actual dollar cost of the tool.

1.17 What is Bogatin's rule #9?

Before you do a measurement or simulation, always anticipate what you expect to see. If you are wrong and you see something that is different than you expected, there is a reason for it. Do not use the result until you can understand why the result is not what you expect. Maybe you did something wrong, or the tool is wrong.

If you are correct, and you see what you expect, you get a nice warm feeling that maybe you do understand what you are doing. It is an important confidence builder.

The corollary to Rule #9 is that there are so many ways of screwing up a measurement or simulation, you cannot do enough consistency tests to help test the quality of your result.

1.18 What are three important reasons to integrate measurements somewhere in your design flow?

The input information to many simulators are material properties. These can often only be obtained from measurements.

The accuracy of models of components can only be validated by measurements.

Many simulation tools are intrinsically accurate. However, the process used to set up the problem and integrate it into the simulation tool may not be done correctly. The best way to validate a simulator tool and process to use it is by comparing the results to measurements on a well-characterized test vehicle.

1.19 A clock signal is 2 GHz. What is the period? What is a reasonable estimate for its rise time?

The period of a 2 GHz clock signal is $1/F = 0.5$ nsec. If the rise time is about 10% of the period, the rise time would be 0.05 nsec, or 50 psec. This is just a rough estimate of what fraction of the period the rise time might be. It could be as long as 50% the period.

1.20 What is the difference between a SPICE and an IBIS model?

A SPICE model describes an active device as a combination of voltage and current sources and capacitors, inductors, and resistors. These are circuit element models which can often be related to specific physical features that make up the microscopic model of the device.

An IBIS model of a driver is often described as a behavioral model of the device. It consists of the behavior of the voltage-current curves of the driver under different load conditions. There is no connection between the behavior of the I-V curves and any physical design features.

1.21 What do Maxwell's Equations describe?

They relate how electric and magnetic fields interact with currents, charges, and each other. Most importantly, Maxwell introduced two important new features in his equations. The first is the idea that a changing electric field acts just like a current, but is not the motion of physical charges. He termed it displacement current. This is a critically important concept in signal integrity.

The second important innovation he introduced in his equations was the coupling between a changing electric field generating a changing magnetic field, which then generates a changing electric field. It's this mutual interaction that is the origin of propagating electromagnetic fields.

When you combine the linear differential equations between E fields and B fields, you get a linear, second order differential equation in just the E field and just the B field. The solution to these equations are propagating electric and magnetic fields—light.

1.22 If the underlying clock frequency is 2 GHz, and the data is clocked at a double data rate, what is the bit rate of the signal?

With two bits in every clock period, the bit rate would be 4 Gbps. The underlying clock frequency of a data rate, when there are two bits per cycle, is called the Nyquist frequency.

Chapter 2

2.1 What is the difference between the time domain and the frequency domain?

The time domain is the real world. The frequency domain is a mathematically constructed world.

In the real world, events happen sequentially with a time stamp and interval between events. An important property of the time domain in the real world is causality. This means, a response cannot happen before a stimulus.

In the time domain, we describe a signal as a voltage at discrete instants in time. A waveform is a voltage vs time display.

The frequency domain is a mathematical construct. In this precisely defined domain, the only type of waveforms that can exist are sine waves. The collection of all the various sine waves in a signal is a spectrum. A signal is then a voltage amplitude and phase at discrete frequency values.

2.2 What is the special feature of the frequency domain and why it is so important for signal analysis on interconnects?

In the frequency domain, the only type of waveform that can exist is a sine wave. Some physical effects have a natural response that matches the shape of sine waves.

This is the case when the physical effects are described by second order, linear differential equations. The solutions for these equations are sine waves. This means sine waves are naturally occurring in such systems.

Electrical circuits with resistors, capacitors, inductors, and transmission lines are described by linear, second order differential equations. This means that sine waves are naturally occurring in electrical circuits.

Many of the voltage vs time responses of electrical circuits will look like combinations of sine waves. Often, these waveforms can be described in a simpler way by just a handful of sine waves in the frequency domain than by voltage vs time in the time domain. When this is the case, you can often have a simpler solution in the frequency domain than in the time domain. This often translates to a shorter time to the answer by passing through the frequency domain than staying in the time domain.

2.3   What is the only reason you would ever leave the real world of the time domain to go into the frequency domain?

Since the real world is in the time domain, your first choice is always to stay in the time domain to solve problems. The only reason you would ever leave the real world to detour through the frequency domain, this mathematically constructed world, is to get to the answer faster.

Not all problems can be solved more quickly in the frequency domain. But when it works, the frequency domain can be an important shortcut. This is why you should become bilingual and learn to think and act in both the time domain and the frequency domain.

2.4   What feature is required in a signal for the even harmonics to be nearly zero?

In a time-domain waveform that is repetitive, each harmonic in the frequency domain will be a multiple of the repeat frequency. If the waveform is symmetric—that is, whatever happens in the first half of the period is exactly the same, with a negative sign in front of it—all even harmonics will have a zero amplitude. There will be no even harmonics in the frequency domain.

However, if there is any asymmetry between the first and second half of the period of the waveform, some even harmonics will appear. For example, if the signal is a square wave with a duty cycle of 50%, there will be no even harmonics. If the duty cycle is NOT 50%, there will be even harmonics.

If the waveform has some features that happen in the first half of the period of the waveform that doesn't exist in the second half—like a rise time different than a fall time—there will be even harmonics.

2.5   What is bandwidth? Why is it only an approximate term?

The bandwidth of a signal is the highest sine-wave-frequency component that is significant in the spectrum of the signal. This means that you can eliminate all the frequency components of the signal above its bandwidth, and when you re-create the waveform, not have any important impact on the signal.

It is as though you send original signal through a low-pass filter with a steep high-frequency stop at the bandwidth. When you compare the original and filtered signal, they should be the same for all important features you care about.

But the terms "significant" and "care about" are vague terms. If you define the similarity of the original and low-pass filtered waveform as no difference to 10% in voltage vs time or no difference to 1% or no difference to 1%, you will get a different value for the bandwidth.

This makes the term bandwidth an approximate term. If you care about the 1% differences, you should not use the term bandwidth but include the whole signal waveform or the whole spectrum.

2.6   In order to perform a DFT, what is the most important property the signal has to have?

In order for a discrete Fourier transform to be most useful in describing a signal in the frequency domain, you should make sure the signal is repetitive and there are a whole number of cycles within the time window you use to calculate the DFT.

If this is the case, you will see harmonics as multiples of the repeat frequency of the signal and the spectrum will be independent of the number of cycles in the time interval.

If you choose a time interval that does not contain a whole number of cycles, the spectrum that results will depend specifically on the time interval you choose and will not be intrinsic to the signal.

2.7   Why is designing interconnects for high-speed digital applications more difficult than designing interconnects for rf applications?

Signals in high-speed digital applications have a wide bandwidth. This means interconnects must be well-behaved over a wide frequency range. Signals in the rf world have a narrow bandwidth, centered over the carrier frequency. This means interconnects must be well behaved at one frequency, the carrier frequency.

I think it is easier to design an interconnect for specific impedance properties at one frequency than over a wide range of frequencies.

Of course, "easier" is a subjective term. My friends working in the rf world would probably say their designs are harder to engineer than high-speed digital designs because they require much tighter control of impedance and can't tolerate cross-talk levels that are as high as in digital designs.

2.8   What feature in a signal changes if its bandwidth is decreased?
The bandwidth is like the value of the frequency cliff in the low-pass filter. When selected as the bandwidth of the signal, it will be the lowest frequency at which the original signal and the filtered signal should look the same.

If you send a signal through a low-pass filter, the chief feature in the output signal that changes when you decrease the cliff frequency is the rise time of the signal. The lower the bandwidth, the longer the rise time.

2.9   If there is –10 dB attenuation in an interconnect, but it is flat with frequency, what will happen to the rise time of the signal as it propagates through the interconnect?
It the frequency response of the –10 dB attenuation is flat, this means every frequency component in the spectrum of the signal would see the same attention. The shape of the spectrum would stay the same; it would just be scaled down, everywhere, by –10 dB, which is a drop to about 30% of the original amplitude.

Since the rise time is related to the relative drop from the frequency components in the spectrum and the shape stays the same, the rise time would be the same. The signal amplitude would just be scaled.

Attenuation by itself does not cause rise-time degradation. It's the frequency dependence of the attenuation that does.

2.10 When describing the bandwidth as the highest significant frequency component, what does the word significant mean?
The term "significant" is a vague term. It refers to the highest sine-wave-frequency component that needs to be included in any analysis. If you set all the frequency components above the bandwidth to zero and convert the signal back into the time domain, all the important properties of the signal you care about, like the rise time, will be preserved and close enough to the original signals.

Describing the bandwidth of a signal is the same as saying if you send the signal through a low-pass filter with a very steep, cliff wall drop-off, what is the lowest wall frequency you could use and still preserve the important features of the signal?

Different applications might have a different meaning of "significant." If you want to re-create the rise time of a signal, and get the same 10–90 rise time of the signal after it passes through the cliff wall filter, you would need a cliff wall at about 0.35/rise time.

2.11 Some published rules of thumb suggest the bandwidth of a signal is actually 0.5/RT. Is it 0.35/RT or 0.5/RT?

If you are worrying about the difference between 0.35 or 0.5 in the relationship, don't use the term bandwidth. You should consider the entire spectrum in your analysis. The term bandwidth, when referring to preserving a rise time, is too vague to distinguish 0.35 or 0.5.

If the rise time of the signal is 10% the period, T, of a square wave, for example, the harmonics will be at multiples of 1/T. This means the resolution of looking at the features of the spectrum will be at 1/T.

The bandwidth, defined by 0.35/RT, is the same as saying a frequency of $0.35/(0.1 \times T) = 3.5/T$. Saying the bandwidth is 0.5/RT is the same as a frequency of 5/T. The resolution of the spectrum is only 1/T and you are trying to distinguish a value of 3.5/T or 5/T. These two values are within 1.5/T of each other, close to the resolution limit of the spectrum.

2.12 What is meant by the bandwidth of a measurement?

The bandwidth of a measurement is the highest sine-wave-frequency component that you can measure in the system of the probes or fixtures and the device. To quantify this value, you usually define the frequency at which the amplitude of the detected signal is reduced by −3 dB of what is actually there.

In many digital storage scope (DSO) applications, the shape of the frequency response curve has a very sharp drop off at the high-frequency end, due to extensive signal processing. It is customary practice to define the bandwidth of the scope instrument alone as the −2 dB frequency point.

A 1 GHz scope has a −2 dB frequency point at 1 GHz. This would be based on sending sine waves into the scope and looking at the highest frequency that gets in with a signal amplitude down by only −2 dB.

A frequency-domain instrument, like a spectrum analyzer or a network analyzer, typically has a flat response, after calibration, up to the highest measurement frequency. Beyond this, the measured response, is, of course, zero. The bandwidth of the instrument is just the highest measured frequency component.

2.13 What is meant by the bandwidth of a model?

The bandwidth of a model is the highest sine-wave frequency at which the model's predictions matches any measurement of the real components response to an "acceptable" level of accuracy.

Again, this term of "acceptable" is vague. Depending on the application, it generally means an agreement to within 10% between the measured response and the simulated response of the real component.

Generally, the agreement between the measured and simulated response drops off in accuracy much faster above the model's bandwidth than below the bandwidth.

2.14 What is meant by the bandwidth of an interconnect?

The bandwidth of an interconnect is the highest frequency that can be transmitted through the interconnect and still meet the required signal performance specification.

Obviously, depending on the application, the requirement will change. If you want to preserve the rise time of the signal and not affect it by transmission through the interconnect, this generally means the frequency at which a signal would have an attenuation of −3 dB.

This assumes the attenuation is linearly frequency dependent. If the attenuation were flat at −3 dB, the rise time would not be degraded at all. If the frequency response were a cliff wall low-pass filter, the frequency at which the output 10–90 rise time were the same as the input rise time might be closer to the −2 dB point.

However, if the application required having just −20 dB of the carrier frequency amplitude at the output, compared to the signal at the input, you could tolerate a much poorer interconnect.

This is why you often add a qualifier to the term bandwidth of the interconnect and refer to the −3 dB bandwidth, or the −10 dB bandwidth, or the −20 dB bandwidth. This is based on what an acceptable attenuation might be for the application.

2.15 **When measuring the bandwidth of an interconnect, why should the source impedance and the receiver impedance be matched to the characteristic impedance of the interconnect?**

If you define the interconnect bandwidth as the highest frequency at which you have −3 dB of the input signal at the output, there are two reasons why the source and receiver impedance matter, compared to the characteristic impedance of the interconnect.

The amount of signal launched into the transmission line depends on the impedance of the source compared to the impedance of the interconnect. Increase the source impedance and less signal is launched into the transmission line. Likewise, change the receiver impedance and the output voltage at the receiver will vary. Increase the receiver impedance, and more voltage will be measured at the receiver.

If the source and receiver impedances are not matched to the characteristic impedance of the interconnect, then the reflections between the ends of the interconnect will cause ripples in the frequency response of the interconnect. These ripples are not intrinsic to the response of the interconnect. You can change their modulation depth by changing the external conditions of the source and receiver impedances.

You will minimize the effects of the terminations at the ends if you match the terminations to the characteristic impedance of the interconnect. This reveals the intrinsic performance of the interconnect.

Of course, if the interconnect is not a controlled impedance interconnect, and its instantaneous impedance varies down its length, there will be reflections inside the interconnect, which are part of the intrinsic properties of the interconnect. The best you can do is match the source and receiver impedances to the average, or low-frequency characteristic impedance of the interconnect.

This is where you sometimes must define the bandwidth of the interconnect at a specific terminated impedance. Luckily, if you know the interconnect response at one termination value, you can mathematically calculate the response at any other terminations. There is no need to measure every combination.

2.16 If a higher bandwidth scope will distort the signal less than a lower bandwidth scope, why shouldn't you just buy a scope with a bandwidth 20 times the signal bandwidth?

There are two reasons why higher bandwidth isn't necessarily better. First is cost. The higher the bandwidth of the scope, the more expensive it is. When purchasing an instrument, you always should consider the useful future use, or "headroom" of the instrument, and your budget.

Like hard drive memory space, you should always buy the highest bandwidth you can afford.

There is a fundamental problem with higher bandwidth measurements. Every receiver has some noise associated with it, even if it is just the digitizing noise of the ADC. Generally, the noise density is flat with frequency. This means that a higher bandwidth measurement lets in more amplifier noise than a lower bandwidth measurement. If there is no signal information at the higher bandwidth, all you are doing is increasing the noise.

The signal-to-noise ratio is literally the ratio of the signal energy to the noise energy. If a higher bandwidth measurement does not increase the signal energy, but only increases the noise energy, you get a lower SNR with a higher bandwidth scope.

This is why most scopes have a feature to allow the user to select the measurement bandwidth at the front of the amplifier. If you have plenty of SNR, it may not matter. But if you are pushing the limits to SNR, you want to use just enough measurement bandwidth to let in all the signal, but not so much as to let in more noise in a frequency range where there is no signal.

If you want to be a master of measurements, always pay attention to the bandwidth of your signal and match your measurement bandwidth to be at least twice the signal bandwidth so you don't lose any signal—but not too much above this, so you don't add too much noise.

2.17 In high-speed serial links, the −10 dB interconnect bandwidth is the frequency where the first harmonic has −10 dB attenuation. How much attenuation will the third harmonic have? How small an amplitude is this?

In an interconnect that is dominated by dielectric loss, the attenuation of the interconnect, in dB, drops off linearly with frequency. If there is −10 dB at a first-harmonic frequency of 1 GHz, for example, the attenuation at the third harmonic, 3 GHz, will be 3× this or −30 dB.

The amplitude of the first harmonic of the signal at the receiver will be down to 30% of the input signal with −10 dB of attenuation. At −30 dB attenuation, the amplitude will be down to:

$$\text{amplitude} = 10^{\frac{-30}{20}} = 10^{-1.5} = 3\%$$

This is not much signal left at the third harmonic.

2.18 What is the potential danger of using a model with a bandwidth lower than the signal bandwidth?

The whole purpose of a model is to represent the electrical behavior of the actual component. Its value is in how faithfully or accurately it predicts the actual performance of the real component.

The bandwidth of the model is the highest frequency at which the model's prediction matches the real component behavior. If the signal has frequency components above the model's bandwidth, the predictions of the model will no longer be accurate. Although you will get an answer, it will be wrong.

If the signal bandwidth exceeds the model's bandwidth, the answer will be wrong, but it will be very difficult to quantify how wrong is the answer. This can often be misleading either as worse than reality, in which case you end up over-designing the product, or better than reality, in which case you end up with a product that may not work.

2.19 To measure an interconnect's model bandwidth using a VNA, what should the bandwidth of the VNA instrument be?

At a minimum, the model's bandwidth should be at least as high as the signal bandwidth, and the instrument's bandwidth at least as high as the required model's bandwidth.

If you can afford it, using a model and measurement bandwidth that are 2× the signal bandwidth adds a factor of two margin, which accounts for the vagueness of the term bandwidth. It's a question of if you can afford it.

In a VNA, the bandwidth of the instrument is very well defined. It is the highest frequency the VNA goes up to. Generally, the cost of a VNA measurement goes up with higher measurement bandwidth. It's the price of the VNA, and the cables, the connectors, the calibration system, the fixture design, and the care with which the measurements must be performed.

The bandwidth of the VNA to use should be at least as high as the required bandwidth of the model and extend up to the highest frequency you can afford. But going above 2× the model bandwidth does not add any additional value, unless the application requires knowing the limits. If there is no additional value above 2× the model bandwidth, higher frequency measurements than 2× the model's required bandwidth should be used only if it is free.

2.20 The clock frequency is 2.5 GHz. What is the period? What would you estimate the 10–90 rise time to be?

The period is 1/clock frequency. If the clock frequency is 2.5 GHz, the period is 1/(2.5 GHz) = 0.4 nsec.

Of course, just because you know the period doesn't mean you have any knowledge of the rise time. You have to make some assumptions. Generally, the rise time is about 10% the period in many clocked digital systems. Using this rule of thumb, if the period is 4 nsec, the rise time would be about 0.1×0.4 nsec = 0.04 nsec = 40 psec.

However, this assumption fails at extremes. In high-speed serial links, the rise time is often 1/2 the unit interval. The signal goes up and it goes down in one unit interval. If the unit interval is half the underlying clock frequency, called the Nyquist, the rise time is $0.25 \times$ the clock period.

In simple microcontroller circuits, like the Arduino, the clock frequency is 16 MHz. The period is 60 nsec. You would expect the rise time to be about 6 nsec using the 1/10th rule of thumb. When measured, you find the rise time of signals coming off the I/O to be 3 nsec, half the expected rise time.

But sometimes, an ok answer now! is better than a good answer late. If all you know is the period and you need a rough estimate, not knowing anything else about the system or the signal, using the 1/10th rule of thumb is a good starting place.

2.21 A repetitive signal has a period of 500 MHz. What are the frequencies of the first three harmonics?

The first harmonic is the repeat frequency, 500 MHz. The second harmonic is 2× this, or 1 GHz, and the third harmonic is 3× this, or 1.5 GHz.

Of course, you don't have enough information to say anything about the magnitude of the first three harmonics, but these are the frequency values.

2.22 An ideal 50% duty cycle square wave has a peak-to-peak value of 1 V. What is the peak-to-peak value of the first harmonic? What stands out as startling about this result?

The amplitude of the first harmonic is $2/pi = 0.637$ V. But, this is the amplitude of the sine wave represented by the first harmonic. The peak-to peak-value of this sine wave is 2× this, or 1.25 V.

What is startling is that the peak-to-peak value of the square wave is 1 V. Yet, the peak to peak of the first harmonic, contained in this signal, is 1.25 V. Its 25% larger than the original signal's waveform.

It's in combination with the other harmonics that this peak-to peak-value is brought down to the value closer to 1, as more harmonics are added.

2.23 In the spectrum of an ideal square wave, the amplitude of the first harmonic is 0.63 times the peak-to-peak value of the square wave. What harmonic has an amplitude 3 dB lower than the first harmonic?

The amplitude of each frequency component in the ideal square wave's spectrum decreases as $1/n$. This means the second harmonic amplitude is down to 50% of the first harmonic, and the third harmonic amplitude is down to 1/3 or 33% of the first harmonic amplitude.

An amplitude that is reduced by −3 dB is an amplitude that is down to 71% of the original amplitude. This is somewhere between the first harmonic and the second harmonic. The second harmonic is already down much lower than −3 dB. In fact, it is down by −6 dB from the first harmonic amplitude.

This is important because saying the bandwidth is when the harmonic amplitude is down by −3 dB of the first harmonic amplitude is just blatantly wrong.

2.24 What is the rise time of an ideal square wave? What is the amplitude of the 1001st harmonic compared to the first harmonic? If it is so small, do you really need to include it?

The definition of an ideal square wave is that the rise time is 0 psec. The bandwidth of an ideal square wave is infinite. When you calculate the amplitude of any harmonic of an ideal square wave analytically, you derive the equation:

$$A_n = \frac{2}{\pi n}$$

When the square wave is symmetrical and it has a 50% duty cycle, the even harmonic amplitudes are zero, so this relationship applies to odd values of n only.

You can use this to calculate the amplitude of any harmonic, even the n = 1001 value, as

$$A_{1001} = \frac{2}{\pi 1001} = 0.000636$$

This is a tiny value compared to the first harmonic. In fact, it is less than 0.1% of the first harmonic amplitude. Surely it is small enough you can ignore it. If you drop off the higher harmonics at n = 1001 and above, even though these are tiny amplitudes, you will end up with a square wave that does not have a 0 psec rise time. It would have a longer rise time.

If you care about a 0 psec rise time, every harmonic, however small, must be included in the spectrum. They are all significant.

2.25 **The 10−90 rise time of a signal is 1 nsec. What is its bandwidth? If the 20−80 rise time was 1 nsec, would this increase, decrease, or have no effect on the signal's bandwidth?**

The bandwidth of a 1 nsec rise-time signal is 0.35/1 nsec = 350 MHz. If instead of the 10–90 rise time, you use its 20–80 rise time, the bandwidth of the signal would not change. It is intrinsic to the signal.

However, if the 20−80 rise time were 1 nsec, the 10−90 rise time would be longer than 1 nsec. A longer rise time means a lower bandwidth. You should be careful which rise time you use when you estimate the bandwidth of the signal.

2.26 **A signal has a clock frequency of 3 GHz. Without knowing the rise time of the signal, what would be your estimate of its bandwidth? What is the underlying assumption in your estimate?**

Of course, it's the rise time of the signal that determines its bandwidth, not its clock frequency. However, if you don't know anything else about the signal other than its clock frequency, the bandwidth is about 5 × the clock frequency.

For a clock frequency of 3 GHz, the bandwidth is roughly 5 × 3 GHz = 15 GHz.

In making this connection, you are assuming that the rise time of the signal is about 7% the period. This means if the rise time were actually longer than 7% the period, the bandwidth would be less than 15 GHz. This is a conservative estimate of the bandwidth and generally gives a little higher bandwidth than if the rise time were 10% the period, for example.

2.27 **A signal rise time is 100 psec. What is the minimum bandwidth scope you should use to measure it?**

The bandwidth of the scope should be at least twice the bandwidth of the signal. If the rise time is 0.1 nsec, the bandwidth is 0.35/0.1 nsec = 3.5 GHz. The scope bandwidth, including its probes, should be at least 2× this, or 7 GHz.

If the scope bandwidth is less than 7 GHz, the rise time measured by the scope will be longer than the actual rise time of the signal.

2.28 **An interconnect's bandwidth is 5 GHz. What is the shortest rise time you would ever expect to see coming out of this interconnect?**

An interconnect bandwidth of 5 GHz means that if you send in a signal with a rise time of 1 psec, the rise time coming out will have a bandwidth of 5 GHz. This means a rise time of 0.35/5 GHz = 70 psec.

The shortest rise time signal you could see coming out of the interconnect would be about 70 psec.

2.29 **A clock signal is 2.5 GHz. What is the lowest bandwidth scope you need to use to measure it? What is the lowest bandwidth interconnect you could use to transmit it and what is the lowest bandwidth model you should use for the interconnects to simulate it?**

If you don't know the rise time of the signal, you cannot get an accurate measure of the bandwidth of the signal, or the bandwidth of the other features needed. All you can do is roughly estimate. When it costs extra for more performance than you need, or you want to avoid the risk of not providing adequate bandwidth, getting the information you need is really important.

If all you know is that the clock frequency is 2.5 GHz, then you can estimate the bandwidth as 5 × 2.5 GHz = 12.5 GHz. Using this as the starting place, you want to have a scope with a bandwidth of at least 2 × 12.5 GHz, or 25 GHz. This is a bandwidth of the instrument which is 10× the clock frequency.

Generally, at clock frequencies of 2.5 GHz, the rise time is longer than 7% the period, but in this case, you are paying extra for insurance and your lack of more accurate knowledge.

To not degrade the signal rise time to any noticeable extent, you should also use an interconnect with a transmission bandwidth of 2× the signal bandwidth of 25 GHz. Likewise, the model bandwidth should be 25 GHz.

While these are the goals, without knowing the rise time, you may be spending more than you should by using a conservative value of the signal bandwidth. This is why it may be worth it to invest some time and effort into determining the actual signal bandwidth before committing to 25 GHz scopes, interconnects, and models.

Chapter 3

3.1 What is the most important electrical property of an interconnect?
While there are many electrical properties that characterize an interconnect, the most important one is its impedance. This can come in a number of forms. One in particular is the input impedance as a function of frequency.

3.2 How would you describe the origin of reflection noise in terms of impedance?
When a signal is propagating down an interconnect and encounters a change in the instantaneous impedance, a reflection is generated, and the transmitted signal is distorted. It is the impedance environment the signal sees down the interconnect that determines the signal distortion from reflection noise.

3.3 **How would you describe the origin of cross talk in terms of impedance?**

Cross talk from an aggressor line to a victim line is ultimately due to fringe electric and fringe magnetic fields. You can approximate these fields in terms of small, lumped circuit elements using capacitor and mutual-inductor elements.

The cross talk between two adjacent signal and return paths is due to coupling capacitance or mutual capacitance, and coupling inductance, or mutual inductance. The amount of coupled signal depends on the source voltage and the impedance of these coupled elements.

3.4 **What is the difference between modeling and simulation?**

Modeling is the process by which you convert a physical interconnect composed of conductors and dielectrics into an equivalent electrical circuit model. You first must construct the circuit topology that describes the structure. Each circuit element in the circuit topology has a few parameters that define the element.

The second step is to calculate a value of the parameters in each circuit element. This can be done by using a rule of thumb, an analytical approximation, or a numerical simulation tool.

Simulation is using the model to predict actual signal waveforms either in the time or frequency domain as currents or voltages.

A simulator takes the circuit model and solves the differential equations each element is a shorthand for, and predict the output waveforms based on an input signal.

3.5 **What is impedance?**

Impedance is fundamentally the ratio of the voltage across a component to the current through it. This ratio defines the connection between how the current and voltage interact with the component. Impedance usually is frequency dependent, so it is more easily described in the frequency domain, though it is also defined in the time domain.

There are at least five different types of impedance, each with a different qualifier to distinguish them. There is instantaneous impedance, characteristic impedance, input impedance in the time domain, input impedance in the frequency domain, and the impedance matrix for components with more than two terminals.

3.6   **What is the difference between a real capacitor and an ideal capacitor?**
A real capacitor is the actual physical device you call a capacitor. These physical components are mounted to a circuit board and used in filter applications to block DC voltage, or to provide local charge storage, for example.

An ideal capacitor is the electrical model of the real capacitor. It is an idealized description written in the language the intended simulation understands. Being ideal does not mean it does not describe some of the more complicated behavior of the real capacitor.

The model can have multiple levels of sophistication to take into account both the low frequency and the high frequency properties of the real capacitors.

An important metric of an ideal capacitor is the bandwidth of the model—that is, how high a frequency the predictions of the model, such as its impedance, match the measured properties of the real capacitor.

3.7   **What is meant by the bandwidth of an ideal circuit model used to describe a real component?**
The bandwidth of an ideal circuit model is the highest frequency at which you would expect good agreement between the predictions of the model's behavior to still match the measured properties of the real capacitor's.

This is a measure of how high a frequency you can use this ideal model to approximate the real capacitors. For example, at low frequency, the ideal capacitor can be model as a single ideal C element. As you go up in frequency, a better model is a series RLC circuit. This model can be further refined to account for the frequency dependence of the real part of the impedance.

3.8   **What are the four ideal passive circuit elements used to build interconnect models?**
The four most commonly used circuit elements to describe interconnects are the resistors, R; the capacitor, C; the inductor, L; and the uniform transmission lines, with a Z0 and a TD.

3.9   **What are two differences between the behavior you might expect between an ideal inductance described by a simple L element and a real inductor?**
At low frequency, the real inductance should behave pretty much like an ideal inductance. You would be surprised if there was much difference in the measured and modeled properties, if you pick the right value of L.

However, the real inductor would show a real part to the imped-ance, due to the frequency-dependent series resistance. In addition, at very high frequencies, above about 100 MHz, the impedance of a real inductance begins to flatten out and may even decrease with higher fre-quency, due to the presence of stray capacitance between the terminal of the inductor.

If you know the actual performance of the real inductance, it's pos-sible to construct a circuit model for the ideal capacitor out of building block elements, which can closely match the actual behavior of the real inductor.

3.10 Give two examples of an interconnect structure that could be modeled as an ideal inductor.

An engineering change wire added to a board that splices a new route be-tween two pins can often be modeled as a simple inductor. A wire bond inside of an IC package connecting the die pads to the package leads can often be modeled as an ideal inductor.

3.11 What is displacement current, and where do you find it?

Displacement current is the invention of Maxwell. When he looked at the properties of magnetic fields and conduction current, he found that if he treated a dE/dt as a current, he was able to preserve continuity of current through an insulation dielectric.

This was one of the important unifying principles he introduced when he collected his four equations together. He realized there are re-ally two types of current: conduction current and displacement current. The displacement current flows along changing electric field lines and is proportional to how fast the E field lines are changing.

This is how current gets through the insulating dielectric of a capac-itor. It flows as the changing E field inside the capacitor when the voltage across the plates changes.

Displacement current will flow between two conductors whenever the voltage across them changes, which means the electric field between them changes. This is how return current flows between the signal and return path as a signal travels down a transmission line.

3.12 **What happens to the capacitance of an ideal capacitor as frequency increases?**

In the model of an ideal capacitor, the capacitance is absolutely constant with frequency. The capacitance in the C element is constant with frequency. Of course, the impedance of an ideal capacitor ill change, but the value of the capacitance will be constant.

This is slightly modified in many high-end circuit stimulators. Since the real capacitor is typically filled with a dielectric constant that changes with frequency, the capacitance will slightly change with frequency. Some more complex ideal capacitor models include this slight frequency dependence effect.

3.13 **If you attach an open to the output of an impedance analyzer in SPICE, what impedance will you simulate?**

Many SPICE simulators require that there be a DC path to ground to calculate initial conditions. If there is an open to ground, SPICE will output an error message.

After all, this is the case of an immoveable object encountering an irresistible force. The constant current source will output whatever voltage it needs to in order to keep the current from the source constant. If the impedance is open, this would require an infinite voltage to generate the constant current across the open. Hence, an error message.

This is why it is a good plan to add a shunt resistance across the ends of the constant current source in SPICE. This will always provide a finite, but high resistance to ground, and the constant current can be achieved with a very high, but finite voltage.

3.14 **What is the simplest starting model for an interconnect?**

The simplest starting model for an interconnect used to be an RLC circuit. All simulators understand R, L, and C elements. But as simulators have become more sophisticated, all simulators now incorporate transmission line elements. This is a far better starting model for an interconnect.

The transmission line model of an interconnect can match the properties of the real interconnect at low frequency, and at much higher frequency. It's the simplest model that can achieve reasonably good bandwidth.

3.15 **What is the simplest circuit topology to model a real capacitor? How could this model be improved at higher frequency?**

The simplest circuit topology of a real capacitor is an ideal C element. By choosing the correct parameter for the C value, this model can match the measured impedance of a real capacitor up to the 1 MHz bandwidth and above in some cases.

If you want to get a higher bandwidth model, a second order model would be an RLC circuit topology. The C would be the same as the first order model value, while the L value is about the mounting of the capacitor to the circuit board.

3.16 What is the simplest circuit topology to model a real resistor? How could this model be improved at higher frequency?

The simplest model of a real resistor is just a single R element. This model is usually a good match to a real R into the 10 MHz range. The impedance of a real resistance is very flat with frequency.

For a higher bandwidth model, a series ideal R and L element can be used. This model takes into account the mounting inductance of the resistor. This ideal model can often match the measured impedance of a real resistor into the GHz regime.

3.17 Up to what bandwidth might a real axial lead resistor match the behavior of a simple ideal resistor element?

The presence of the leads in an axial lead resistor generally adds as much as 10 nH of series inductance. You can calculate the difference between the ideal model of a single R element and the increase in impedance from the series L. The higher frequency you go, the bigger the impedance of the series inductance and the larger the impedance of the real resistor.

As a rough metric of good enough, you can take when the impedance of the inductor is more than 10% the impedance of the series resistance. This is the frequency where neglecting the series inductance in the ideal model gives a result that is as much as 10% off from the real resistor. This condition is when

$$\omega L > 10\% R$$

Or

$$f > \frac{0.1 \times R}{2\pi L} = \frac{0.1 \times 50}{2\pi 10nH} = 80 MHz.$$

This is the case for a 50 Ohm resistor and 10 nH of mounting inductance. Below 80 MHz, the simple ideal R element will have an impedance that matches the real resistor within 10%, providing you select the correct resistance value.

3.18 In which domain is it easiest to evaluate the bandwidth of a model?
The term bandwidth is inherently a frequency-domain term; this means it is more easily evaluated in the frequency domain. When evaluating the bandwidth of a model, for example, it is easy to compare the predicted impedance over frequency and the measured impedance over frequency. The frequency above which they do not agree very well is the bandwidth of the model.

3.19 What is the impedance of an ideal resistor with a resistance of 253 Ohms at 1 kHz and at 1 MHz?
This is easy. The impedance of an ideal resistor is constant with frequency. It is exactly equal to its resistance. The resistance of an ideal 253 Ohm resistor is the same 253 Ohms at 1 kHz and at 1 MHz.

3.20 What is the impedance of an ideal 100 nF capacitor at 1 MHz and at 1 GHz? Why is it unlikely a real capacitor will have such a low impedance at 1 GHz?
The magnitude of the impedance of an ideal capacitor is

$$|Z| = \frac{1}{2\pi fC}$$

For the case of a 100 nF capacitor, at 1 MHz and 1 GHz, the impedance is

$$|Z| = \frac{1}{2\pi \times 10^6 \times 10^{-7}} = 1.6\Omega \quad \text{and} \quad |Z| = \frac{1}{2\pi \times 10^9 \times 10^{-7}} = 0.0016\Omega$$

Wow! It looks like using even just a 100 nF capacitor results in an impedance as low as 1 mOhm at 1 GHz. This looks really low!

Unfortunately, in a real capacitor, there is lead inductance that plays a role. The impedance of the series inductance starts to dominate the real capacitor's impedance above about 10 MHz. Above 10 MHz, the impedance of the real capacitor will go up, hiding the low impedance of the ideal capacitance of the real capacitor.

3.21 **The voltage on a power rail on-die may drop by 50 mV very quickly. What will be the dI/dt driven through a 1 nH package lead?**
The relationship between the voltage drop across an inductor and the transient current is

$$\Delta V = L \frac{dI}{dt}$$

For the case of a 50 mV drop and 1 nH package lead, the transient current that will be driven through the package is

$$\frac{dI}{dt} = \frac{\Delta V}{L} = \frac{0.05V}{1nH} = 50mA/n\sec$$

3.22 **To get the largest dI/dt through the package lead, do you want a large lead inductance or a small lead inductance?**
It's clear from the previous example that the highest transient current which comes into the on-die capacitance to re-supply the lost charge, occurs when the lead inductance is smallest. This is why you want low inductance in the package lead inductance. It encourages fast current transients into the on-die capacitance, which means less voltage droop on die.

3.23 **In a series RLC circuit with R = 0.12 Ohms, C = 10 nF, and L = 2 nH, what is the minimum impedance?**
The minimum impedance will always be equal to the R in the circuit. This occurs when the reactance goes to zero. In this example, the minimum impedance is 0.12 Ohms.

3.24 **In the circuit in Question 3.23, what is the impedance at 1 Hz? At 1 GHz?**
You can calculate the impedance of this series RLC circuit in a few ways. First, you can bring it into a SPICE-like simulator and simulate the impedance at any frequency. Second, you can write an analytical equation for the magnitude of the impedance at any frequency. Using a calculator, you can easily calculate the impedance at any frequency.

Finally, you can roughly estimate the impedance based on which part of the circuit is dominating the impedance. For example, at low frequency, easily in the 1 Hz range, the circuit impedance will be dominated by the capacitor. The impedance will be

$$|Z| = \frac{1}{2\pi fC} = \frac{1}{2\pi 1 \times 10^{-8}} = 16 M\Omega$$

At 1 GHz, the impedance will be dominated by the inductor. Its impedance will be

$$Z = 2\pi fL = 2\pi \times 10^9 \times 2nH = 12\Omega$$

3.25 **If an ideal transmission line matches the behavior of a real interconnect really well, what is the impedance of an ideal transmission line at low frequency? High or low?**

If the far end of the transmission line is open at the receiver, and you are looking at the front of the transmission line, it will look like a very high impedance at low frequency. As you go to higher frequency, the impedance will drop, until it looks like a short. Even though the transmission line is open at the end, when you look at the front of the line, the transmission line will look like a short.

As you further increase frequency, the transmission line will look like an inductor, increasing its impedance until it looks like an open; then it will drop again and continue this oscillating behavior between an open and a short. A transmission line open at the far end is not a very well-behaved interconnect for high-frequency signals.

3.26 **What is the SPICE circuit for an impedance analyzer?**

Creating an impedance analyzer in SPICE is easy. All you need is a constant current AC source. This will generate a sine wave of constant current amplitude using whatever voltage amplitude it needs to generate a constant current amplitude.

The voltage needed is

$$V(f) = Z(f) \times I(f)$$

As long as you make the constant current amplitude a fixed value of 1 A, the simulated voltage on the impedance analyzer is numerically the same as the impedance of whatever is attached to the analyzer. It's important to note that this is a complex relationship. The phase of the voltage is also the phase of the impedance.

Chapter 4

4.1   What three terms influence the resistance of an interconnect?
The three most important terms influencing the resistance of an interconnect are the bulk resistivity of the conductor material, the length of the interconnect, and the cross-sectional area through which the current will flow.

4.2   While almost every resistance problem can be calculated using a 3D field solver, what is the downside of using a 3D field solver as the first step to approaching all problems?
To use a 3D field solver requires that you have the solver and know how to use it. Most of these tools will output a number. But, there are multiple ways the number can be a meaningless result. Without knowing what to expect, you have no idea if the result out of the field solver is "reasonable."

If you rely on all of your results from a 3D field solve, you miss the valuable opportunity to get a feel for the number using approximations or rules of thumb. It is much faster to estimate the resistance using a rule of thumb that you can do 50 different estimates in the time it takes to load one problem in the field solver and get a result.

That's where a simple approximation or a rule of thumb is so valuable. A rule of thumb should always be the first step in estimating the resistance of an interconnect.

4.3   What is Bogatin's rule #9, and why should this always be followed?

Rule #9 is to never do a measurement or simulation without first anticipating the result. This means you should have an idea of what the answer
should be before you start the problem. This is where a rule of thumb
that allows a quick, approximate estimate of the answer is so valuable.
It calibrates your engineering judgement and gives you a feel for the
numbers.

There are so many ways of screwing up a problem, you cannot
do too many consistency tests. The first consistency test is checking if
your result is consistent with your engineering judgement, which often is
based on rules of thumb.

**4.4   What are the units for bulk resistivity, and why do they have such strange
units?**

The units for bulk resistivity are Ohms-cm. What the heck does this
mean? It's not a resistance per volume, it's not a resistance per area, and
it's not even a resistance per length.

When you calculate the resistance of a uniform length structure, the resistance is in the form of

$$R = \rho \frac{L}{A}$$

The resistance will scale linearly with the length of the interconnect and
inversely with the cross-sectional area. The proportionality constant is
the bulk resistivity of the material. It is a measure of the resistance property of the material.

The units of bulk resistivity are to make the units of resistance
come out as Ohms. The Length/Area has units of 1/length. The bulk
resistivity has to have units of length in its numerator.

If you had a cube of material, with the length of each side as Len,
the Length/Area would be 1/Len. This says the larger the length of a side
of the cube, the lower the resistance. While the distance between end
faces increases as the side of the edge increases, which would increase
the resistance, the cross-sectional area increases with the square of the
side of an edge which would drop the resistance faster than the length
increases it.

The resistance would be R = rho / Len. The units of rho has to be Ohms-length in order for the resistance to end up in Ohms.

4.5   What is the difference between resistivity and conductivity?
These are both terms that describe the electrical resistance of a piece of the material. The resistivity is a measure of how resistive the material is. The conductivity is a measure of how conductive the material is. They both measure the same material property. One is the inverse of the other:

$$\rho = \frac{1}{\sigma}$$

Since the units of resistivity are Ohms-m, the units for conductivity are 1/(Ohms-m). You call the units of 1/Ohms Siemens. The units of conductivity are Siemens/m. Again, these units don't make much sense except to get the units to come out to Ohms when calculating the resistance of an extended object.

4.6   What is the difference between bulk resistivity and sheet resistivity?
The bulk resistivity is the intrinsic material property that relates to how much resistance there would be in the material. It doesn't matter the size or shape of the material or how much you have. All pieces of the same material have the same bulk resistivity.

Sheet resistance refers to the resistance property of a section of the material shaped in a wide, thin, uniform thickness sheet, like a foil of copper.

The sheet resistivity, or the sheet resistance, used interchangeably, refers to the resistance from edge to edge of a square piece of the material cut out from the sheet. If you were to double the length of a side of the square, the distance between the faces you measure resistance increases so the resistance would increase, but the width through which the current travels would also increase, decreasing the resistance. These two features cancel out and the resistance from edge-to-edge stays the same.

This says no matter what the length of the edge of the square, the resistance from edge to edge is the same. You call this resistance of one square, the sheet resistance or the sheet resistivity, or the resistance per square. Every square cut from the same sheet has the same resistance.

4.7   If the length of an interconnect increases, what happens to the bulk resistivity of the conductor? What happens to the sheet resistance of the conductor?

This is a trick question. The bulk resistivity of a material is an intrinsic property of the material, not the geometry. If the length of the interconnect increases, the bulk resistivity of the material stays the same. It is about the material.

Likewise, if the interconnect is composed of a sheet of conductor like a trace on a layer of a circuit board, the sheet resistance is intrinsic to the material that makes up the foil and the foil thickness. If the length of the interconnect changes, the sheet resistance of the foil does not change.

4.8   What metal has the lowest resistivity?

Of all the homogenous materials, other than a superconductor, silver has the lowest bulk resistivity, with a value of about 1.59×10^{-8} Ohm-m. Copper comes in a close second with a resistivity of 1.68×10^{-8} Ohm-m. Note that copper is only 6% more resistive than silver. This is only a small difference. Many times, this slight benefit is not worth the extra cost of using silver with its higher bulk cost and higher manufacturing cost.

It's often thought that gold is the lowest resistivity material. This is far from the case. In fact, the resistivity of gold is 2.44×10^{-8} Ohm-m. This is 45% higher than copper, which is quite a difference. Why is gold used so often as an interconnect material?

It's not often the bulk material, it is often just a coating. This is because gold does not corrode or oxidize. It is a good material to have on the surface to enable good soldering and low contact resistance.

4.9   How does the bulk resistivity of a conductor vary with frequency?

Generally, the bulk resistivity of a material is very constant with frequency. This is the case until well above 100 GHz. The resistance of an interconnect will be frequency dependent, but this is not due to the resistivity changing—it is due to the cross-sectional area through which current is traveling changing.

For all applications, assume the bulk resistivity of copper and all other conductors is constant with frequency.

4.10 Generally, will the resistance of an interconnect trace increase or decrease with frequency? What causes this?

The resistance of an interconnect trace will always increase with frequency. This is due to the effect called skin depth. As the frequency of the current increases, the path through the conductor changes in order for the current path taken to reduce the loop inductance of going down through the conductor and back on the return conductor.

Within each conductor, a lower inductance is achieved when the current flows toward the outer surface of the conductor. The higher the frequency, the more the current concentrates to the outer surface. When the cross-sectional area through which the current travels gets thinner, the resistance increases, usually proportional to the square root of frequency.

4.11 If gold has a higher resistivity than copper, why is it used in so many interconnect applications?

Gold's chief property is that it does not oxidize or corrode. This means that if it is on the outer surface of a conductor, there will be low contact resistance when another gold surface comes in contact. And, since it does not oxidize when solder wets gold, it is a good surface to solder on even after a long time exposed to the air.

Usually the gold on an interconnect is very thin. The typical spec for a connector lead is 30 micro-inches, which is a little less than 1 micron of gold. It's only enough to protect the underlying metal from oxidation and corrosion.

4.12 What is the sheet resistance of ½-ounce copper?

This is one of those numbers useful to remember. The sheet resistance of ½ ounce copper foil is 1 milli-Ohm per square. You can see where this comes from based on the calculation of sheet resistance. It is given by

$$R_{sq} = \frac{\rho}{t} = \frac{1.68e-8 Ohm-m}{17e-6m} = 0.99 m\Omega / sq$$

1 mOhm/sq is an easy number to remember.

4.13 **A 5-mil wide trace in ½-ounce copper is 10 inches long. What is its total DC resistance?**

The sheet resistance of ½-oz. copper foil is 1 mOhm per square. To calculate the resistance of the trace, you need to know how many squares are along its length, as each square is 1 mOhm of resistance.

The number of squares is 10 inches/0.005 inches = 2,000 squares. The series resistance is 2,000 squares × 1 mOhm/square or 2 Ohm. This narrow trace, spanning 10 inches, has a resistance of 2 Ohms. This many not be much in a digital circuit, but if it is in a power path, it can be a lot.

4.14 **Why does every square cut out of the same sheet of conductor have the same edge to edge resistance?**

If you have a piece of sheet cut in the same of a square and measure the edge to opposite edge resistance, you would see that if the distance between the edges doubled, the resistance would double. But if you doubled the width, the resistance would be cut in half.

If you do both, double the length and double the width, you still have a square and the resistance stays the same.

You can see this when looking at the resistance based on the geometry

$$R = \frac{\rho}{t} \times \frac{Len}{w}$$

If in your interconnect, you keep the ratio of the length of the interconnect and the width of the interconnect the same, the resistance does not change. And if the ratio is 1, so you have a square shape, the resistance from edge to edge of the square is a constant that just depends on the bulk resistivity and the foil thickness, which you call the sheet resistance.

4.15 **When you calculate the edge-to-edge resistance of a square of metal, what is the fundamental assumption you are making about the current distribution in the square?**

When you calculate the edge to edge resistance, you are assuming that the current flows uniformly down the length of the square. You launch the current into the edge and take it out of the other end so that there is the same current density everywhere in the square of conductor. If you make contact with a point on the edge, the resistance you measure would be higher.

4.16 **What is the resistance per length of a signal line 5 mils wide in ½-ounce copper?**

In half-ounce copper, the sheet resistance is 1 mOhm/sq. A line that is 5 mils wide would have a resistance per length of 1 mOhm/sq/(0.005 inch) = 0.2 Ohms per inch.

4.17 **Surface traces are often plated up to 2-ounce copper thickness. What is the resistance per length of a 5-mil wide trace on the surface compared to on a stripline layer where it is ½-ounce thick?**

When the thickness of the conductor increases, there is more cross-sectional area for the current to travel and the sheet resistance and the resistance per length of the conductor decreases.

   In going from ½ oz. copper in a stripline layer to 2 oz. copper on the surface, the thickness has increased by 4×. This means the resistance has decreased by ¼. The resistance per length of a 5-mil wide trace in ½ oz. copper is 0.2 Ohms/inch. This same trace fabricated on the surface has a resistance of ¼ this, or 0.05 Ohms/inch.

4.18 **To measure the sheet resistance of ½-ounce copper using a 4-point probe to 1%, you are resolving a resistance of 1 μOhm. If you use a current of 100 mA, what is the voltage you have to resolve to see such a small resistance?**

In order to measure a resistance of 1 μOhm, with a forcing current 0.1 A, the voltage that you would have to measure is V = I × R = 0.1 A × 1 μOhm = 0.1 μV = 100 nV. This is a very tiny voltage.

   This means that for routine sheet resistance measurements, an instrument would need to be able to routinely measure 100 nV signal. It is no wonder that these measurements are rather difficult.

4.19 **Which has higher resistance: a copper wire 10 mil in diameter and 100 inches long, or a copper wire 20 mils in diameter but only 50 inches long? What if the second wire were made of tungsten?**

You could answer this question by calculating the resistance of each wire and looking to see which is larger. Alternatively, you can use scaling to estimate the different resistance.

   The first copper wire is 10 mil in diameter and 100 inches long. The second wire is 20 mils in diameter and 50 inches long. This one is obvious. The smaller cross-section wire will have a higher resistance per length. And, it's longer. Clearly the first wire will have higher resistance.

Suppose you made the second wire out of tungsten. Which wire would be higher resistance?

Now you can apply scaling. The resistance of the wire is basically

$$R = \rho \frac{\text{Len}}{A}$$

The length of the second wire is ½ that of the first wire. The cross-sectional area is 4× that of the first wire. These two factors alone would make the resistance of the second wire ½ /4 = 1/8th that of the first wire. The resistivity of copper is 1.68×10^{-8} Ohm-m. The resistivity of tungsten is 5.6×10^{-8} Ohm-m. This is a factor of $3.3 \times$ higher for the second wire.

This makes the second tungsten wire, a factor of 3.3/8 = 0.42, as large as the first wire. Even with the higher resistivity, the shorter length and larger diameter overcompensates for the higher resistivity.

4.20 **What is a good rule of thumb for the resistance per length of a wirebond?**
A wire bond, usually made from 1-mil diameter aluminum or gold wire, has a resistance of about 1 Ohm/inch. If the wire bond is 0.1 inches long, its resistance will be 0.1 Ohms.

4.21 **Estimate the resistance of a solder ball used in a chip attach application in the shape of a cylinder, 0.15 mm in diameter and 0.15 mm long with a bulk resistivity of 15 μOhm-cm. How does this compare to a wire bond?**
The resistance of a uniform cross-section interconnect is

$$R = \rho \frac{\text{Len}}{A}$$

The resistance per length, for this case of a 0.15-mm diameter cylinder, is then just

$$\frac{R}{\text{Len}} = \rho \frac{1}{A} = 15\mu\Omega - \text{cm} \frac{1}{\pi(0.015\text{cm})^2} = 0.021\,\Omega/\text{cm} = 0.05\,\Omega/\text{inch}$$

The resistance per length of a wire bond is about 20 × the resistance per length of a solder ball. This is mostly due to the larger diameter of the solder ball. If the ball is 0.15 mm long, the resistance of the solder ball will be 0.021 Ohm/cm × 0.015 cm = 0.0003 Ohms. This resistance is pretty insignificant compared to other sources of resistance in the path.

4.22 The bulk resistivity of copper is 1.6 μOhms-cm. What is the resistance between opposite faces of a cube of copper 1 cm on a side? What if it is 10 cm on a side?

The resistance from one face of a cube, 1 cm on a side, to the other, is

$$R = \rho \frac{Len}{A} = \rho \frac{Len}{Len^2} = \rho \frac{1}{Len} = 1.6 \times 10^{-6} \Omega - cm \frac{1}{1\,cm} = 1.6\,\mu\Omega$$

If the side of a cube is 10 cm, larger by 10×, the resistance will be lower by 10×, to 0.16 μOhm.

4.23 Generally, a resistance less than 1 Ohms is not significant in the signal path. If the line width of ½-ounce copper is 5 mils, how long could a trace be before its DC resistance is > 1 Ohms?

When the line width for ½ oz. copper is 5 mils, the resistance per length is 1 mOhm/sq /0.005 inches = 0.2 Ohms/inch. This means a length of Len × 0.2 Ohms/inch > 1 Ohms, means a length of > 5 inches has a DC resistance > 1 Ohms. This does not mean the interconnect will not work if it is longer than 5 inches. It just means you should pay attention to see if the 1 Ohm DC series resistance is significant.

4.24 The drilled diameter of a via is typically 10 mils. After plating it is coated with a layer of copper equivalent to about ½-ounce copper. If the via is 64 mils long, what is the resistance of the copper cylinder inside the via?

You can approach this problem in two ways. You can just take the cross-sectional area and length and calculate the end-to-end series resistance. Alternatively, you can do a simpler analysis.

If you slit the via from top to bottom and unwrap the via to flatten it out, the width of the trace would be the circumference, or 3.14 × 10 mils = 32 mils. The length of the via is 64 mils. This means it is about 2 squares long. If the sheet resistance is 1 mOhm/sq and it is 2 squares long, the series resistance of the via will be about 2 mOhms.

4.25 Sometimes, it is recommended to fill the via with silver filled epoxy, with a bulk resistivity of 300 μOhm-cm. What is the resistance of the fillet of silver filled epoxy inside a through via? How does this compare with the copper resistance? What might be an advantage of a filled via?

You can estimate the resistance of the fillet of silver filled epoxy. The resistance will be

$$R = \rho \frac{\text{Len}}{A} = 300\mu\Omega - \text{cm} \frac{0.15\text{cm}}{\left(\pi 0.025^2 \text{cm}^2\right)} = 22 \text{ m}\Omega$$

You see that the resistance of the silver filled epoxy is more than 10× larger than the copper in the wall of the via. Filling the via with silver filled epoxy will not reduce the series resistance of the via very much.

The value of using silver filled epoxy is so that the top surface is flat and solderable.

4.26 Engineering change wires on the surface of a board sometimes use 24 AWG wire. If the wire is 4 inches long, what is the resistance of the wire?

The resistance of 24 AWG wire is about 0.08 Ohms/m or 2 mOhms/inch. With a wire that is 4 inches long, the resistance of the engineering change wire would be 2 mOhms/inch × 4 inches = 8 mOhms. This is generally a small value and not a significant contribution to the electrical properties of the engineering change wire.

Chapter 5

5.1 What is capacitance?

Capacitance is usually defined as the ratio of the charge separated between two conductors to the voltage between them. While this is correct, it doesn't say much about what really is capacitance.

The principle behind capacitance is that when you separate charges on two adjacent conductors, there is a voltage difference between them. The capacitance between the plates is really a measure of how efficient the conductors are at storing charge, at the price of voltage.

The more charge you can store for a small amount of voltage difference, the more capacitance the conductors have. If you have conductors that are not very efficient, they can't store much charge before the voltage goes way up.

The capacitance of two conductors is not related to how much charge is on the conductor, nor what the voltage is. It is about the efficiency of storing the charge at the cost of voltage.

5.2   Give one example where capacitance is an important performance metric.
The input gate capacitance of a receiver is concentrated in a very small region at the pad on the die and directly below it. When a signal is received, it is coming from some impedance—either the drivers output impedance, or the transmission line's characteristic impedance.

When the signal reaches the gate capacitance, the voltage is increased as current flows in. The time it takes for the input gate capacitor to charge up is a measure of the received signal rise time.

The larger the input gate's capacitance, the longer it takes to charge and the longer the rise time at the receiver. When this rise time becomes a significant fraction of the unit interval, timing problems result.

5.3   What are two different interpretations of what the capacitance between two conductors measures.
Interpretation 1: Capacitance is the ratio of the charge stored to the voltage across the two conductors. It is a DC effect.

Interpretation 2: Capacitance is a measure of how much current will flow through two conductors separated by an insulating dielectric driven by a dV/dt. The larger the dV/dt, the larger the current through the capacitor. For a fixed dV/dt, the larger the capacitance, the more the current through the conductors.

Interpretation 3: Capacitance is the efficiency by which two conductors store a charge difference, at the cost of the voltage between them. The more efficient they are at storing charge without increasing the voltage much, the more their capacitance.

5.4   A small piece of metal might have a capacitance of 1 pF to the nearest metal, inches away. There is no DC connection between these pieces of metal. At 1 GHz, what is the impedance between these conductors?
The impedance of a capacitor at some frequency, f, is

$$Z = \frac{1}{2\pi f C} = \frac{1}{2\pi 10^9 \times 10^{-12}} = 160\ \Omega$$

It is rather remarkable that a piece of metal just sitting in air, with no connection to another piece, can have as low as 160 Ohm impedance between it and an adjacent conductor, at 1 GHz. This is a very low impedance between two conductors that are not touching.

5.5   How does conduction current flow through the insulating dielectric of a capacitor?

Conduction current does NOT flow through the insulating dielectric of a capacitor. It's an insulating dielectric. There is a very small amount of leakage current that flows through a capacitor due to ion motion, or leakage current, but this tiny amount plays no role in signal quality—just in very low-level currents less than 1 nA.

5.6   What is the origin of displacement current, and where will it flow?

Displacement current is the term Maxwell introduced to account for continuity of current through an insulating dielectric. He said displacement current flowed along electric field lines whenever the E field changed. The amount of displacement current was precisely the dE/dt.

Between any two conductors with a voltage difference, there will be an electric field. If the voltage changes, the electric field changes and the displacement current flows along the field lines between the conductors.

In his world view, Maxwell thought even empty space was filled with the ether. Between the two plates of a capacitors, filled with vacuum, was really ether. And to Maxwell, the ether was polarizable.

When the electric field between the plates changes, the E field polarized the ether particles and pulled the + and − charges of the ether particles farther apart, displacing them. This displacement of charges in ether particles, when the E field changed, he coined as displacement current.

It referred to the motion of bound charges, as distinct from free charges that could move in a conductor, as conduction currents.

To this day, we still call the current that flows when E fields change displacement current, but attribute it to a property of space-time, rather than to the polarization of ether particles.

5.7   If you wanted to engineer a higher capacitance between the power and ground planes in a board, what three design features would you change?

To increase the capacitance between two conductors, you bring the conductors closer together and increase the overlap width and overlap length of the two planes.

5.8 What primary property about the chemistry of a dielectric most strongly influences its dielectric constant?

The dielectric constant of a material is a measure of how much it increases the capacitance of two conductors when inserted between them. The way it does this is to shield a little of the electric field between the two conductors by polarizing the bound charges inside the material.

A high dielectric constant material usually is a very polarizable material. This means it has a lot of bound charges that can be separated. The most common process is to have large dipoles in the material that can rotate in the electric field.

High Dk materials usually have large dipoles that are mobile and can rotate and align in an electric field. Water is a perfect material with a large dipole that can rotate in an electric field. This is why water has a Dk of 80.

Polymer materials with a high Dk have dipole groups, either tied off the backbone chin, or integrated in the backbone chain that can rotate and align in an external electric field.

The epoxy polymers incorporate a C-O-C group in the polymer backbone. This group has a large dipole and can rotate in an external E field.

Materials with a low Dk value do not have many dipoles that can rotate. Teflon, or PTFE, polytetrafluoroethylene, is just a carbon backbone with fluorine atoms along the chain. There is not dipole that can rotate and very little structure that can be polarized.

5.9 What happens to the capacitance between two conductors when the voltage between them increases?

Trick question. Nothing. Capacitance is the ratio of the charge stored to the voltage difference between two conductors. This ratio is independent of the amount of charge stored, or the voltage between the conductors.

When the voltage increases, the capacitance, which is just about the geometry, stays the same. It's the charge separation between the conductors that changes.

5.10 In a coax geometry, what happens to the capacitance if the outer radius is increased?

If two conductors are moved farther apart, their capacitance decreases. In a coax, if the outer conductor radius is increased and the outer conductor moves farther away from the inner conductor, the capacitance of the coax decreases.

5.11 What happens to the capacitance per length in a microstrip if the signal path is moved away from the return path?

When the signal and return paths of any conductor topology are pulled farther apart, the capacitance decreases.

5.12 Is there any geometry in which capacitance increases when the conductors are moved farther apart?

No.

5.13 Why does the effective dielectric constant increase as the thickness of the dielectric coating increases in microstrip?

As the thickness of the dielectric over a microstrip trace increases, the capacitance of the microstrip will increase. Where some fringe fields were in air before the dielectric coating, after the coating, these fringe fields see a higher Dk. This means the overall capacitance increases.

The geometry hasn't changed, just the distribution of dielectric material. You describe the increase in capacitance as an increase in effective dielectric constant.

5.14 What is the lowest dielectric constant of a solid, homogenous material? What material is that?

The lowest dielectric constant of a solid homogenous material is 2. One example of such a material is Teflon. It has no net dipoles that can move. It is only the electrons within the molecules that can displace and polarize in an electric field.

When a material has no dipoles and it's only the electrons in the molecules that can polarize in an external electric field, its dielectric constant will be the lowest possible, with a value of about 2.0.

5.15 What could you do to a material to dramatically reduce its dielectric constant?

One trick that is often used in interconnects to reduce the Dk of a dielectric is to foam it. By foaming the material, you add air to it and decrease its density. You can take Teflon or Polyethylene, foam it and reduce its dielectric constant to nearly 1, close to air. This is as low as it is possible to get. Foamed Teflon is often used in high-performance cables.

5.16 What happens to the capacitance per length of a microstrip if solder mask is added to the top surface?

When solder mask is added to the top of a microstrip, some fringe field lines that were in air now see a higher DK material. This increases the capacitance between the signal and return path of the microstrip.

5.17 What happens to the capacitance per length of a microstrip if the conductor thickness increases?

If the conductor thickness increases, there will be a few more fringe field lines from the edge of the thicker signal trace to the return plane. This increases the capacitance between the signal and return path a little bit.

5.18 What happens to the capacitance per length of a stripline if the line width increases?

If the line width increases, the area of overlap between the signal and return paths increases, and the capacitance per length increases.

5.19 What happens to the capacitance per length of a stripline if the trace thickness increases?

If the trace thickness increases, and the separation between the two planes of the stripline stays the same, then the distance between the top surface of the signal line to either plane will decrease. The spacing between the signal line and return will be closer, and this will increase the capacitance between the two conductors.

5.20 For the same line width and dielectric thickness per layer, which will have more capacitance per length: a microstrip or a stripline?

A microstrip signal line will have a capacitance to just one plane, while a stripline trace will have a similar value capacitance to each of two planes in parallel. The stripline trace would have a higher capacitance per length than the microstrip trace, if the line widths were the same and the dielectric thickness per layer were the same.

5.21 What is theoretically the lowest dielectric constant any material could have?

No material can have a dielectric constant lower than 1, which is the dielectric constant of air. Some materials made of a foam and used in cable applications can approach this value. If the material is solid and homogenous, the lowest Dk it can have is about 2. This is limited by the polarizability of electrons in the molecules that make up the material. As dipoles that can rotate are added to the material, the Dk will increase.

5.22 **Why is the capacitance per length constant in a uniform cross-section interconnect?**

The capacitance between two conductors depends on the spacing between the two conductors and their area of overlap. In a uniform transmission line, the spacing between the signal and return paths are constant down the length, and the width of the conductor is constant. This means the capacitance for a fixed length will be constant.

5.23 **On die, the dielectric thickness between the power and ground rails can be as thin as 0.1 micron. If the SIO2 dielectric constant is also 4, how does the on-die capacitance per square inch compare to the on-board capacitance per square inch if the power and ground plane separation is 10 mils?**

The capacitance of two plates with the same area and Dk value will scale inversely with the dielectric thickness. A 10-mil thick separation is really 250 microns. This is 2500 times larger than the dielectric thickness on die. This means the on-die capacitance per area between the power and ground rails will be 2500 times higher than in a typical circuit board.

5.24 **What is the capacitance between the faces of a penny if they are separated with air?**

A penny has faces that are about 0.5 square inches in area and are separated by about 0.1 inch. The capacitance of two conductors can be calculated from

$$C = 0.225\text{pF} / \text{in}\,\frac{A}{h} = 0.225\text{pF} / \text{in}\,\frac{0.5\text{in}^2}{0.1\text{in}} = 1.1\,\text{pF}$$

The capacitance between the two faces of a penny if they were isolated planes, would be about 1 pF. Of course, in a penny, the two faces are not separated by a dielectric of air—they are connected together with the rest of the copper, so they have no capacitance. But if two plates are in the shape of a penny, separated by the same distance, they would have a capacitance of about 1 pF.

Thinking about the size and shape of a penny is a good calibration for how much capacitance there is in 1 pF.

5.25 **What is the minimum capacitance between a sphere 2 cm in diameter suspended a meter above the floor? How does this capacitance change as the sphere is raised higher above the floor?**

The capacitance of a sphere when other conductors are 100 diameters away is just

$$C = 4\pi\varepsilon_0 r = 4 \times 3.14 \times 0.225 \times \frac{2}{2.54} = 2.2 \text{ pF}$$

If you hold the 2-cm diameter sphere 1 m over the surface of the floor, its capacitance will be slightly higher than this. But as you move it farther and farther away, it will roughly approach this value. You see that the capacitance, in pF, is about equivalent to the diameter of the sphere. This is a rough calibration of the capacitance in a piece of metal just suspended away from the floor. You can never get away from stray capacitances between any piece of metal and some grounded surface somewhere. Everything couples.

5.26 **What is the capacitance between the power and ground planes in a circuit board if the planes are 10 inches on a side, 10 mils separation, and filled with FR4?**

To calculate the capacitance between two planes, you can always start with the parallel plate approximation. It takes less than 1 minute to set it up, put in the values, and calculate the capacitance.

In addition, you can get to an estimate for the special case of two planes with an FR4 dielectric between them using the simple approximation that

$$C\left[\frac{nF}{in^2}\right] = \frac{1}{h[mils]}$$

With a dielectric thickness of 10 mils, the capacitance per unit area is just 0.1 nF/in^2. If the area of the plates is 10 in $\times$ 10 in = 100 in^2, the capacitance between the planes is $0.1 \text{ nF/in}^2 \times 100 \text{ in}^2 = 10 \text{ nF}$.

5.27 Derive the capacitance per length of the rod over a plane from the capacitance per length of the twin rod geometry, in air.

The capacitance between two parallel rods, with radius r and center to center separation of s, in air, is

$$C_{L-rod-rod} = \frac{\pi\varepsilon_0}{\ln\left(\dfrac{s}{r}\right)}$$

You should never use a relationship without first applying rule #9. Is it reasonable? Does it match your expectations? As you pull the rods farther apart, the capacitance should decrease. When s increases, the denominator increases, and the capacitance per length will decrease. This matches what you expect. It doesn't mean this is a good approximation, just that it is consistent with one simple test.

If you place a conducting plane in the middle between the two rods, the electric field distribution will not change. The capacitance between the two rods will not change. But, you can describe it as the series combination of the capacitance between one rod and the plane in series with the capacitance of the second rod and the plane.

These two capacitances in series add up to the capacitance of the two rods. This means the capacitance between the rod and plane in this configuration is 2 × the capacitance between the two rods:

$$C_{L-rod-plane} = 2 \times C_{L-rod-rod}$$

If you replace the separation, s with 2 × h, the capacitance of the plane to rod can be calculated as

$$C_{L-rod-plane} = 2 \times C_{L-rod-rod} = \frac{2\pi\varepsilon_0}{\ln\left(\dfrac{2h}{r}\right)}$$

This is the capacitance between a rod and plane.

Chapter 6

6.1   What is inductance?

Inductance is the efficiency with which a conductor can create rings of magnetic field lines at the cost of the current through it. If a conductor can generate a lot of rings of magnetic field lines with a little current, it has a high inductance. If it generates only a few rings of magnetic field lines per amp of current through it, it has a low inductance.

6.2   What are the units used to count magnetic field lines?

You count rings of magnetic field lines in units of Webers. A Weber of rings of field lines is some number of rings.

6.3   List three properties of magnetic field lines around currents.

The magnetic field lines around a current in a wire are in the shape of complete circles or closed rings.

These rings have a direction of circulation. The direction is based on the right-hand rule. Point the thumb of your right hand in the direction of the positive current and your fingers curl in the direction of circulation of the rings of magnetic field lines.

If you count the number of rings of field lines around a wire, you find this number, measured in Webers of field lines, is directly proportional to the current in the wire. Double the current, and the number of Webers of field line rings will also double.

The density of rings of field lines will drop off as you move away from the wire. The rings get farther apart and you see fewer and fewer rings, the farther you move from the conductor.

6.4   How many field line rings are around a conductor when there is no current through it?

This is a trick question. With no current in the conductor, there will be no rings of magnetic field lines around the conductor.

6.5   If the current in a wire increases, what happens to the number of rings of magnetic field lines?

If the current in a wire doubles, the number of rings of magnetic field lines will exactly double as well. The number of Webers of rings of field lines around a conductor is directly proportional to the current in the wire.

**6.6   If the current in a wire increases, what happens to the inductance of
the wire?**

Trick question. If the current in a wire doubles, the number of Webers of
rings of magnetic field lines will double, but the ratio of the Webers of
field lines to the current stays the same. Inductance is independent of the
current in the wire. A conductor has an inductance even with no current
in the wire. It is a measure of the efficiency of generating rings of magnetic field lines, not of the number of field lines present.

6.7   What is the difference between self-inductance and mutual inductance?

Self-inductance is a measure of the rings of magnetic field lines around
a conductor per amp of current through that conductor. It is a measure
of the efficiency of a conductor at creating rings of magnetic field lines
around itself.

Mutual inductance is a measure of the number of rings of magnetic field lines around another conductor, per amp of current in the
first conductor. It is a measure of the efficiency of creating rings of field
lines around another conductor at the cost of current through the first
conductor.

**6.8   What happens to the mutual inductance between two conductors when the
spacing between then increases? Why?**

When you pull two conductors apart, you decrease the mutual inductance between them. If one conductor has current in it, there will be
rings of magnetic field lines around it. The density of the rings will drop
off the farther away you go from the conductor.

When there is an adjacent conductor present, some of the rings
of field lines from the first conductor will also be around the second
conductor. These are the mutual field lines. The farther away the second
conductor, the fewer rings of field lines will be present to go around this
second conductor.

The farther apart the two conductors, the lower the mutual
inductance.

**6.9   What two geometrical features influence the self-inductance of a
conductor?**

The two most important design terms that influence the self-inductance
of a conductor are its length and the current distribution inside the conductor.

The longer the length, the more distance you have to count rings of field lines for the same current in the conductor and the higher the self-inductance.

The tighter you constrict the current distribution in the conductor—by using a narrower conductor, for example—the more rings of magnetic field lines you can count around the conductor, and the higher the self-inductance.

6.10 Why does self-inductance increase when the length of the conductor increases?

The self-inductance is a measure of how many Webers of field lines you count around the conductor per amp of current through the conductor. The longer the conductor, the more length you have to count field lines around.

6.11 What influences the induced voltage on a conductor?

The inducted voltage between two points on a conductor depends on how fast the total number of rings of field lines change around the conductor. If the number of Webers of field lines around the conductor between the two points is the same, there will be no voltage induced.

But, if the number of Webers of field lines changes, for whatever reason, there will be an induced voltage.

6.12 What is the difference between partial and loop inductance?

Inductance, in general, is about the efficiency of creating rings of field lines around a conductor at the cost of the current through it. When you set up a real conductor and send current through it, you can only do this with current flowing in a complete loop.

The question you have to address is what part of the loop are you counting the rings of field lines around, and are you counting the total number of rings of field lines around each section of the loop as you walk down the length of the loop?

After all, if you are at one piece of the loop, counting rings of field lines, you will see some of the rings as coming from the current in that section of the loop, but some of the field lines will also be coming from the section of the loop that has the return current flowing in the exact opposite direction. Its rings of field lines around your section of the loop will be circulating in the opposite direction. These two different sets of rings of field lines around the section of the loop you are sitting on will subtract.

When you walk along the loop from one end to the other, counting the entire number of rings of field lines, taking into account the direction of circulation of each one, the inductance you get is the loop inductance of the loop. This is unambiguous. There is only one value for the loop inductance of the loop.

But, when you only want to consider a part of the loop, and count only those specific field lines coming from the currents in that part of the loop, ignoring the other rings of field lines around your part of the loop from other currents, you call this inductance the partial inductance, and specifically, the partial self-inductance.

6.13 **Why does the mutual inductance subtract from the self-inductance to give the total inductance when the other conductor is the return path?**

A ring of magnetic field lines has a direction of circulation. The total number of rings of field lines around a section of a conductor depends on how many circulate one way and the other. They subtract from each other, and a ring in one direction cancels a ring circulating in the other direction.

When counting the total number of rings around a conductor, all the rings from its own current, the self-field lines, will be in the same direction and add together. But, the mutual field lines from the return current, since the return current is flowing in the opposite direction, will circulate in the opposite direction around the conductor.

The mutual-field lines, being in the opposite direction around the current than the self-field lines, will subtract. This means the presence of the adjacent return current will result in fewer net rings of magnetic field lines around the first conductor, per amp of current in the conductor.

6.14 **In what cases should the mutual inductance add to the self-inductance to give the total inductance?**

Whenever the two currents flow in the same direction, the mutual field lines between them will add to their self-field lines to create a larger total number of field lines.

Not all adjacent currents are return currents. If there is another adjacent conductor with current from a different source flowing in the same direction as the first current, its self-field line rings will circulate in the same direction as the first current.

Those mutual field lines from this adjacent current, in the same direction as the first current, around the first current, will also circulate in the same direction as the self-field lines of the first current.

The total number of rings of field lines around the first current will be the sum of its self-field lines and the mutual-field lines from the adjacent current.

6.15 What three design features will decrease the loop inductance of a current loop?

The total number of rings of magnetic field around a loop depends on the length of the loop, the width of the conductor through which current is flowing, and the proximity of the return half of the loop to the first half of the loop.

The longer the loop, the larger the loop inductance. The smaller diameter the wire, the larger the loop inductance. The farther away the return half of the loop is from the first half of the loop, the larger the loop inductance.

6.16 When estimating the magnitude of ground bounce, what type of inductance should be calculated?

It's the total inductance of the return path that contributes to ground bounce. After all, ground bounce is the voltage from one end to the other in the return path, due to the rate of change of the total number of rings of field lines around the return conductor.

6.17 If you want to reduce the ground bounce in a leaded package, which leads should be selected as the return leads?

There are only three knobs that affect the total inductance of the return path: the length of the return path, the width of the return path, and the proximity of the return path to the signal paths.

The length is the term that has the largest impact on the total inductance. The return pins in a package should be selected as the shortest leads. This is usually the case for the central leads in square package.

6.18 If you want to reduce the loop inductance in the power and ground paths in a connector, what are two important design features when selecting the power and ground pins?

Use pins that are short, and select the power pins adjacent to the ground pins.

6.19 **There are 24 Webers of field line rings around a conductor with 2 A of current. What happens to the number of field lines when the current increases to 6 A?**

The number of Webers of field lines around a conductor is directly proportional to the current through the conductor. If the current increases by 6 A/2 A = 3×, the number of Webers of field lines will increase by the same amount. It will change from 24 Webers to 24 × 3 = 72 Webers.

6.20 **A conductor has 0.1 A of current and generates 1 microWeber of field lines. What is the inductance of the conductor?**

The definition of inductance is that it is the ratio of the number of rings of field lines per amp of current. If there are 10^{-6} Webers of field lines for 0.1 A, the inductance is $10^{-6}/0.1 = 10^{-7}$ H, or 100 nH of inductance.

6.21 **A current in a conductor generates 100 microWebers of rings of field lines. The current turns off in 1 nsec. What is the voltage induced across the ends of the conductor?**

The induced voltage across a wire is dN/dt. If the number of Webers changes linearly from 100 μW to zero in 1 nsec, the induced voltage during this period of time is V = 100 μW/1nsec = 100 μW/0.001 μsec = 10^5 V, a huge number! This is an indication that you can sometimes generate large voltages when currents turn off quickly.

This happens in relays. When the relay has current flowing through it and it opens, the current turns off very quickly and you can generate large voltages that can arc across the gap in the open relay. This is the dominant source of degradation in relays. To avoid this problem, either engineer the relay to open when there is little current flowing through it, or add a capacitor or diode across the gap to provide an alternative current path to reduce the dI/dt.

6.22 **The return lead in a package has 5 nH of total inductance. When the 20 mA of current through the lead turns off in 1 nsec, what is the voltage noise induced across the lead?**

The voltage across an inductor is V = LdI/dt. In this case, the L is 5 nH, the dI = 20 mA, and the dt = 1 nsec. The voltage is easily calculated as V = 5 nH × 20 mA/1nsec = 100 mV.

The most important thing to watch out for are the units. In this case, the nano in nH cancels out with the nano in the nsec to turn off time. This means the units that are left are mV.

6.23 What if four signals use the lead in Question 6.22 as their return path? What is the total ground bounce noise generated?

If the four signals all have the same 20 mA of signal current and are turning off in the same 1 nsec, and their returns all flow through the same total inductance, the dI will be 4× the case of just one signal switching.

This means the ground bounce noise will be 4× the previous case, or 400 mV of ground bounce noise.

6.24 What is the skin depth of copper at 1 GHz?

As a good rule of thumb to remember, the skin depth of copper at 1 GHz is about 2 microns.

6.25 If current flows in both the top and bottom surfaces of a signal trace on a circuit board, how much does the resistance increase at 1 GHz compared to DC?

At DC, if the conductor is composed of 1 oz. copper, such as typically found on surface traces, the sheet resistance is 0.5 mOhms/square. This is when current flows through the 34 microns of conductor thickness uniformly.

If a trace, patterned from this layer, is part of a microstrip, the current distribution will have about equal currents flowing in the top surface as in the bottom surface, each to a thickness of the skin depth. At 1 GHz, this is a thickness of 2 microns in the top and bottom. The total cross section through which current flows in the microstrip at 1 GHz is 4 microns. This is compared to the cross section for current to flow at DC of about 34 microns.

The resistance per length of the conductor at 1 GHz compared to DC will scale with the ratio of the cross-section thicknesses, as all other terms are the same. The smaller the thickness, the higher the resistance.

The 1 GHz resistance will be 34/4 = 8.5 × higher than the DC resistance in 1 oz. copper.

If the copper trace thickness were ½ oz. copper, the geometrical thickness would be 17 microns and the 1 GHz resistance would be 17/4 = 4.2 × higher than the DC resistance.

6.26 Based on the simulation results in Figure 6-26, what is the percentage of decrease in inductance from DC to 1 GHz?

The loop inductance per inch in a microstrip at DC is about 10 nH/inch. At 1 GHz, it is about 8 nH/inch for this specific characteristic impedance. This is a drop in the DC inductance of 2 nH/10 nH = 20%.

Above about 100 MHz, all the current has redistributed as much as it can, and the loop inductance is the same at 100 MHz as at 1 GHz or higher.

6.27 When the spacing between two loops doubles, does the loop mutual inductance increase or decrease?
Since you are pulling the two loops farther apart, the loop mutual inductance between them has to decrease. There will be fewer mutual field lines from one loop around the other when you pull them farther apart.

6.28 What is the loop inductance of a loop composed of 10-mil diameter wire, in a circular loop 2 inches in diameter?
The loop inductance of a circular loop is approximately

$$L_{loop} = 32 \times R \times \ln\left(\frac{4R}{D}\right) nH = 32 \times 1 \times \ln\left(\frac{4 \times 1}{0.01}\right) nH = 192 \, nH$$

The radius of the loop is 1 inch.

6.29 What is the loop inductance per length of two rods 100 mils in diameter, and spaced by 1 inch? What is the total inductance per inch of each leg?
The loop inductance per inch of the pair of rods is roughly

$$L_{loop-Len} = 10 \times \ln\left(\frac{s}{r}\right) nH / inch = 10 \times \ln\left(\frac{1}{0.05}\right) nH / inch = 30 \, nH / inch$$

In this case, the radius is 50 mils or 0.05 inches. You have to be careful of the units. The s and r terms should have the same units.

This is the loop inductance of the pair of rods, assuming current goes down one and back up the other.

You can separate this into a total inductance of each leg. The sum of the total inductances of each leg in series makes up the loop inductance, so the total inductance of each leg would be about 15 nH/inch.

6.30 **A typical dielectric thickness between the power and ground planes in a 4 layer board is 40 mils. What is the sheet inductance of the power and ground planes? How does 1 square of sheet inductance compare to the typically 2 nH of mounting inductance of a decoupling capacitor?**

The sheet inductance in two planes is about 32 pH/mil × h[mils]. When the spacing between the planes is 40 mils, the sheet inductance is 32 × 40 = 1.3 nH. This is the inductance per square in the planes.

This is smaller than, but of the same order, as the mounting inductance of a capacitor, if it were 2 nH. This means that the spreading inductance in the cavity from the capacitor to the package would be significant. Doing what you can to reduce the spacing between the capacitor and the package would reduce the equivalent inductance of the capacitor. In this environment, the cavity is not transparent, and positon would matter.

If you reduce the dielectric thickness between the planes to 3 mils, the sheet inductance would be 0.1 nH, which is small compared to the mounting inductance of the capacitor, and the cavity would be transparent. Location of the decoupling capacitor would not matter.

Chapter 7

7.1 **What is a real transmission line?**

A real transmission line is composed of any two conductors with some extended length. One conductor you label as the signal path and the other conductor as the return path.

7.2 **How is an ideal transmission line model different from an ideal R, L, or C model?**

At low frequency, the ideal transmission line input impedance matches the impedance of a simple L or C element depending on if the far end is open or shorted. But, at frequencies beyond some frequency, the behaviors of the L or C model drastically departs from the ideal transmission line.

This means the L or C elements are good approximations to an ideal transmission line model at low frequency. They are terrible approximations to an ideal transmission line at higher frequency. An ideal transmission line model is a brand-new circuit element with properties that are completely different from a single L or C element, which becomes apparent at very high frequency.

7.3   What is ground? Why is it a confusing word for signal-integrity applications?

The term ground should only be used to refer to the single reference point in a circuit from which all other voltages are referenced. The ground point should not carry any current.

On a plane with current flowing, not all points on the plane labeled as ground are the same voltage. This makes it difficult to use a ground plane as a reference enplane.

The term ground is often misused in the industry and confused with return path. The return path is the conductor that carries the return current. It can be at any DC voltage, and often has a different voltage across it form one region to another.

7.4   What is the difference between chassis and earth ground?

Earth ground, often referred to as safety ground, is literally a connection to the ground. There is a low-resistance path between all points labeled as earth ground to a copper pipe sticking into the ground somewhere nearby. This pipe defines a common reference point tied to the Earth from which other points can be references. As long as all points connect to the same earth ground, they will be at the same voltage and there is less chance of someone getting a shock.

Chassis ground is the connection to the external metal housing of an instrument or device. In the case of a plastic enclosure, there is no chassis ground. For safety reasons, the UL requires that chassis ground be connected to earth ground. This reduces the risk of a user getting a shock due to widely different chassis ground voltages.

7.5   What is the difference between the voltage on a line and the signal on a line?

The voltage on a transmission line is what would be measured by a scope if the probes were touching between the signal and an adjacent point on the return path. This is a scalar voltage in the sense there is no measure of the propagating nature of this voltage. It is just the total voltage between the signal and return path.

The signal is the voltage that is propagating down the line. It has a direction associated with it. When the voltage of a signal is measured, the magnitude can be the same as the magnitude of the signal, but there is no information measured about the direction of propagation of the signal.

The real difference arises when there are multiple signals propagating on the transmission line. Two signals may be propagating in opposite directions on the transmission line. When measured at one point, the voltage measured is the sum of the two voltages. It could be less than, greater than, or the same as the two signal voltages, depending on their specific properties.

7.6   What is a uniform transmission line and why is this the preferred interconnect design?

A uniform transmission line means the cross section of the transmission line is the same up and down the length of the line. This means the instantaneous impedance of the line is the same. Engineering the instantaneous impedance the signal sees as the same up and down the line means the signal will propagate with no reflections and no distortions, improving the signal quality.

If the transmission line is not uniform, there will be impedance changes that will cause reflections and signal distortions.

7.7   How fast do electrons travel in a wire?

Surprisingly, even with as much as 1 A of current in a narrow wire, the speed of the electrons is really small, on the order of 1 cm/sec. This means that it is not the motion of the electrons which is the signal, but the propagating of the changing electric and magnetic fields.

It is like the case of a tube filled with marbles. If you push one marble in one end, another will push out the far end in the time it takes for the pressure wave to flow down between the marbles. The speed of the marble moving down the tube is slow, but the speed of the impact of the motion of the marble is very fast.

7.8   What is the difference between conduction current, polarization current, and displacement current?

Conduction current is the flow of free charges in a conductor. There can be DC current flow with a DC voltage applied.

Polarization current is the current that flows in an insulator when the polarization of bound charges changes. This is literally a motion of charges, but they are restricted to staying attached to the molecules of the dielectric. They will only flow when the polarization of the material changes, which is when the extern E field changes. These currents are transient and are proportional to dE/dt.

Displacement current is the current that flows in the air when the electric field changes. Maxwell called the displacement current the current that flows due to the changing polarization of the charges in the ether. Today, you do not have to invoke the ether, but describe displacement current, as a current that flows when the electric field changes, as a feature of electric fields built into the fabric of space time.

7.9   What is a good rule of thumb for the speed of a signal on an interconnect?
In air, the speed of a signal, or the speed of light, is 12 inches/nsec. When propagating through a dielectric with dielectric constant Dk, the speed of the changing electric field, which is light, slows down with the square root of the Dk.

The typical Dk for most interconnect materials is about 4. This means the speed of a signal on a typical laminate interconnect substrate is about 12/sqrt(4) = 6 inches/nsec.

This is a good rule of thumb for the speed of a signal on an interconnect, about 6 inches/nsec.

7.10   What is a good rule of thumb for the aspect ratio of a 50-Ohm microstrip?
With an FR4 substrate, the ratio of the line width to dielectric thickness of a 50-Ohm transmission line is 2/1.

7.11   What is a good rule of thumb for the aspect ratio of a 50-Ohm stripline?
In a stripline geometry, and FR4 substrate, the aspect ratio of the line width and dielectric spacing between the planes is ½. Since there are two planes, the dielectric thickness must be larger in stripline than in microstrip.

7.12   What is the effect called when the dielectric constant and the speed of a signal are frequency dependent?
When the dielectric constant varies with frequency, or the speed of the signal varies with frequency, you call this effect dispersion. It is a frequency-dependent speed of the signal.

7.13   What are two possible reasons the characteristic impedance of a transmission line would be frequency dependent?
The characteristic impedance of a transmission line depends on the inductance per length and capacitance per length of the line. Both of these are slightly frequency dependence.

For example, the inductance per length of the line will be higher at low frequency and drop about 10% to 20% at higher frequency. This makes the characteristic impedance higher at low frequency and lower at high frequency, by the square root of 10–20%.

The capacitance per length of a transmission line will also be slightly frequency dependent due to the frequency dependence of the dielectric constant.

At low frequency, the DK is typically a little larger than at high frequency. This makes the characteristic impedance slightly lower at low frequency, increasing at high frequency.

While these two effects move the characteristic impedance in opposite directions, their different frequency dependence, and different magnitudes, means they do not cancel out. There is still some net frequency dependence of the characteristic impedance, typically on the order of less than 10% from low to high frequency.

7.14 What is the difference between the instantaneous impedance and the characteristic impedance and the input impedance of a transmission line?
Although all three of these terms are impedances, they refer to different features. The instantaneous impedance is the impedance the signal sees each step along its path as it propagates down the transmission line. This is the impedance the propagating signal responds to at each step.

The characteristic impedance only applies to a uniform transmission line and is the one value of instantaneous impedance the propagating signal will see. If the instantaneous impedance changes, there is no one value of instantaneous impedance that characterizes the transmission line.

The input impedance of the transmission line is usually a frequency-domain term, and is the impedance you see looking into the front of the transmission line at specific frequencies. The input impedance will vary a lot with frequency.

7.15 What happens to the time delay of a transmission line if the length of the line increases by 3×?
The time delay of a transmission line is directly proportional to the length of the line. If you triple the length of the line, the time delay of the line will triple as well.

7.16 **What is the wiring delay of a transmission line in FR4?**

We sometimes refer to the inverse of the speed of a signal as the wiring delay. It is the delay per length of the line. If the speed of the signal is 6 inches per nsec, the wiring delay is 1/(6 inch/nsec) = 170 psec/inch.

This means the delay is 170 psec per inch of travel down the transmission line.

7.17 **If the line width of a transmission line increases, what happens to the instantaneous impedance?**

If the line width of a transmission line increases, the capacitance per length of the line increases and the instantaneous impedance decreases. The wider the line, the lower the instantaneous impedance. If the line is uniform, the characteristic impedance would also decrease for wider lines.

7.18 **If the length of a transmission line increases, what happens to the instantaneous impedance in the middle of the line?**

This is a trick question. If the length of the line increases, the instantaneous impedance stays exactly the same. The time delay of the line increases, but the instantaneous impedance stays the same.

7.19 **Why is the characteristic impedance of a transmission line inversely proportional to the capacitance per length?**

If the capacitance per length increases, it takes more current to charge up each small step in the transmission line. If more current is required to charge up the line to the same voltage, the impedance decreases. This means the capacitance per length is inversely related to the instantaneous impedance.

7.20 **What is the capacitance per length of a 50-Ohm transmission line in FR4? What happens to this capacitance per length if the impedance doubles?**

All 50 Ohm lines in FR4 have the same capacitance per length of about 3.3 pF/inch. If the line width increases, but the dielectric thickness increases to keep the characteristic impedance the same, the wider line transmission line will have the same capacitance per length.

If the characteristic impedance of the line increases, the capacitance goes down. They are inversely related. If the line impedance doubles, the capacitance per length is cut in half.

7.21 What is the inductance per length of a 50-Ohm transmission line in FR4? What if the impedance doubles?

All 50 Ohm transmission lines in FR4 have an inductance per length of about 8.3 nH/inch. Of course, if one dimension changes, the other dimensions need to change to keep the 50 Ohms.

And the line impedance is directly proportional to the inductance per length. If you double the impedance of the line, you will double the inductance per length of the line.

7.22 What can you say about the capacitance per length of an RG59 cable compared to an RG58 cable?

These two coax cables use the same dielectric. They have the same central conductor radius. But, they have different characteristic impedance. An RG58 cable has a characteristic impedance of about 50 Ohms, while an RG59 cable has a characteristic impedance of 75 Ohms.

The higher impedance cable will have a lower capacitance per length. This is also seen in that the outer dimeter of an RG59 cable is larger than an RG58 cable. The outer conductor, being father away, makes the capacitance per length lower in the RG59 cable.

7.23 What do we mean when we refer to "the impedance" of a transmission line?

Just referring to "the impedance" of a line is ambiguous. It could refer to the instantaneous impedance, the characteristic impedance, the input impedance in the frequency domain, or the input impedance in the time domain. Without some specific context, you don't know which one is of interest.

Unfortunately, you get lazy and often drop the qualifier. More often than not, when you refer to just the impedance of the line, you mean the characteristic impedance of the transmission line. Even though you may refer to the characteristic impedance, this does not mean that it is the same as the impedance the driver might see unless the rise time is very short and the line very long, and you only look for a very short period of time.

7.24 A TDR can measure the input impedance of a transmission line in a fraction of a nanosecond. What would it measure for a 50-Ohm transmission line, open at the end, that is 2 nsec long? What would it measure after 5 seconds?

The TDR sends a short rise-time step edge down the transmission line and measures the reflected signal. It is usually launched from a 50 Ohm source. If the transmission line it is connected to is 50 Ohms, there will be no reflection, so there will be no reflected signal when the signal enters the transmission line.

The step edge will travel down the transmission line, reach the end, and reflect from the open. The reflected edge will take about 4 nsec after being launched into the line to make it back to the front of the instrument.

Once it is into the instrument, nothing more reflects and no additional signal enters the transmission line.

The TDR will measure a 50 Ohm line initially, then 4 nsec later, an open. It will measure an open forever afterward. After 5 seconds, the TDR will measure an open.

7.25 **A driver has a 10-Ohm output resistance. If its open circuit output voltage is 1 V, what voltage is launched into a 65-Ohm transmission line?**

The driver and transmission line create a voltage divider. The transmission line looks like a 65 Ohm resistor with the 10 Ohm driver impedance in series. The voltage drop across the 65 Ohm resistor is 1 V × 65 Ohms/ (10 Ohms + 65 Ohms) = 0.87 V.

7.26 **What three design features could be engineered to reduce the impedance of the return path when a signal changes return path planes?**

When a signal transitions its return path between two planes, the return current sees the impedance between the planes in series with the impedance of the line. To minimize the discontinuity and the noise coupled into the cavity, you want to decrease the impedance of the cavity.

The most important way is to add shorting vias. This means you have to use the same voltage for the two planes.

The second feature, which applies no matter what the voltage of the two planes may be, is to use a thin dielectric between the two planes. This dramatically reduces the impedance between the planes if they are different voltages, and makes the shorting vias more effective if they are not.

Finally, if the two planes are different voltages so shorting vias can't be used, a DC blocking cap can be placed between the shorting vias. It will never have as low an impedance as shorting vias, but it is a compromise.

7.27 **Which is a better starting model to use to describe an interconnect up to 100 MHz: an ideal transmission line, or a 2-section LC network?**

An ideal transmission line is always a better model to use to describe a real interconnect than an n-section LC model. The ideal transmission line model will look like the L and C elements at low frequency, but match the properties of the real transmission line to much higher frequency than any LC model.

And the transmission line model is a much simpler model to use. All SPICE simulators understand the ideal transmission line model, and many of them have lossy transmission line models and coupled transmission line models.

7.28 An interconnect on a board is 18 inches long. What is an estimate of the time delay of this transmission line?
The speed of a signal in FR4 is about 6 inches per nsec. The delay of any transmission line is about Len[inches]/v[inch/nsec] = 18 inches/6 inches/ nsec = 3 nsec.

7.29 In a 50-Ohm microstrip, the line width is 5 mils. What is the approximate dielectric thickness?
In a 50-Ohm transmission line in FR4, the ratio of the line width to dielectric thickness is about 2 to 1. If the line width is 5 mils, the dielectric thickness should be about 2.5 mils. A common thickness laminate available is about 2.8 mils.

7.30 In a 50-Ohm stripline, the line width is 5 mils. What is the length of the transmission line?
This is a trick question. Just knowing the characteristic impedance and the line width tells you nothing about the length of the line. It could be 1 inch, or it could be 30 inches. You would need to know about the time delay of the line, or the total inductance or total capacitance. Just the line width and impedance is not enough information.

Chapter 8

8.1 What is the only thing that causes a reflection?
Reflections are only caused by the signal encountering a change in the instantaneous impedance. If there is a reflection, this means the instantaneous impedance has changed. It could be caused by a line width change, a dielectric thickness change, or some other geometry feature change.

8.2 What two features influence the magnitude of the reflection coefficient?
The reflection coefficient between two impedances is only about the ratio of the difference in impedance divided by the sum of the two impedances.

These are the only two parameters that affect the reflection coefficient, the two instantaneous impedances at the interface.

8.3   What influences the sign of the reflection coefficient?

The reflection coefficient is the second impedance minus the first impedance over their sum. If the second impedance is larger than the first impedance, the sign is positive. If the second impedance is smaller than the first impedance, the sign is negative. This is the ONLY thing that influences the sign of the refection coefficient.

8.4   What two boundary conditions must be met on either side of any interface?

When you look at the interface, you need to see continuity in the voltage between the signal and return paths. There can't be a step change in the voltage between the signal line on the left side and the right side. Otherwise, there could be an infinitely large electric field across the interface and the universe could explode.

In addition, you must see the same current circulating on the left side of the interface as on the right side of the interface. This means there is no net current circulating at the interface. If there was a net current into the interface, the interface would charge up, and if you waited long enough, the universe would explode.

You call these two conditions, boundary conditions, continuity of voltage and conservation of charge.

8.5   When viewed at the transmitter, how long does a reflection from a discontinuity last?

When a signal encounters a short discontinuity in an otherwise uniform transmission line, there will be a reflection from the front of the discontinuity and a reflection from the back of the discontinuity.

From the time of the reflection from the front interface, it will take a round-trip time down through the discontinuity, hitting the end of it, and then passing through it again for the reflection to make its way out of the discontinuity to head to the transmitter.

This means the reflected signal to the transmitter will last for a round-trip time through the discontinuity.

If the discontinuity is 1-inch long, its time delay is about 170 psec. The reflected signal will be a pulse with a width of about 170 psec $\times$ 2 = 340 psec.

8.6   **What is the difference in reflection coefficient when a signal reflection from a 50-Ohm line hits a 75-Ohm transmission line or a 75-Ohm resistor?**

Trick question. There is absolutely no difference in the refection coefficient when a signal passes from a 50-Ohm characteristic impedance transmission line into either a 75-Ohm line or a 75-Ohm resistor. The instantaneous impedance in both cases is exactly the same.

8.7   **A signal travels to the end of a 50 Ohm line and sees a 30-Ohm resistor in series with the high impedance of the receiver. What is the impact of having the 30-Ohm series resistor at the receiver?**

Trick question. Since the input impedance of the receiver is very high, adding a series resistor of 30 Ohms has no effect at all. When the signal encounters the 30-Ohm resistor, it immediately sees the open, so really the response is to the open.

8.8   **How would you terminate a bi-directional bus using source series resistance? Where would you put the source resistor?**

The simplest way of terminating a bidirectional bus is using a source series resistance at both ends of the bus. When one driver transmits, it sees a source series resistor at its output. The receiver signal upon hitting the receiver essentially sees an open, and reflects back to the transmitter to be terminated by the series resistor. Exactly the same thing happens when the other end of the line drives.

If there are no other extenuating issues, a source series resistor at both ends of the line is usually the best approach.

8.9   **What is the raw measurement actually displayed on the screen of a TDR?**

The TDR actually measures the voltage at a point right after the source-series 50-Ohm resistor, just where the signal enters a precision 50-Ohm transmission line. This voltage is composed of the incident signal, which is fixed and does not change, and the opposite traveling reflected signal.

Typically, the incident signal is 250 mV in voltage level. This is what is displayed initially. If there were no reflections for any device connected to the TDR—in other words, if you had a perfectly matched device connected—you would forever see 250 mV on the TDR screen.

Any additional measured voltage on top of the 250 mV has to have come from a reflection. The difference between what is measured and

250 mV is the reflected voltage. From the reflected voltage and incident voltage, the reflection coefficient can be measured at any location.

8.10 How do you convert this raw measurement into the instantaneous impedance the signal must have encountered?
If you know the incident voltage is 250 mV, the reflected voltage is the difference between what you see on the screen and the 250 mV incident signal.

From the reflected voltage and the incident voltage, you calculate the reflection coefficient.

Knowing the impedance of the signal from the source is 50 Ohms, and the reflection coefficient, you calculate the impedance of the second interface. After the first reflection, it's more difficult to directly measure from the front screen the impedance of each successive interface.

8.11 The unloaded voltage from a driver is 1 V. What is the output impedance if the output voltage is 0.8 V when a 50-Ohm resistor shorts the output pin?
This is a case where the output voltage drops when you load the line down. This is a DC effect. The voltage drop across the output source resistance is 0.2 V, so 0.8 V appears across the 50-Ohm resistor.

In the voltage divider circuit, the voltage across the 50-Ohm resistor is related to

$$0.8\,V = 1V\frac{50\Omega}{50\Omega + R_{source}}$$

This is straightforward to solve. The source resistance is

$$R_{source} = \frac{50(1-0.8)}{0.8} = 12.5\Omega$$

The source resistance is 12.5 Ohms.

8.12 What is the shape of the TDR response from a capacitive discontinuity? Why is it this shape?
When the step edge of the TDR signal encounters a capacitor, the very large dV/dt sees a low impedance, so the reflection is negative. As the capacitor charges from the 50 ohm source load, the voltage difference

across the capacitor decreases and the impedance increases, so the reflection is less. The resulting reflected signal is a sharp negative dip with a longer decay to no reflection.

8.13 What is the shape of the TDR response from an inductive discontinuity? Why is it this shape?

When the large dI/dt of the TDR edge encounters an inductor, it sees a large impedance and a positive voltage reflects. As the current through the inductor reaches a steady state, the impedance change is less and the reflected voltage drops off.

The reflected signal will be a sharp positive pulse with a long decay to zero.

8.14 If the output impedance of a driver is 35 Ohms, what value of source series resistor should be used in the driver when connected to a 50-Ohm line? To a 65-Ohm line?

When you source series terminate a transmission line, you engineer the series resistor so that when the reflected signal is traveling back to the source, it sees the same impedance in the transmission line as the combination of the series resistor and the source resistor. For a 50-Ohm line and 35-Ohm output resistance, the source series resistor you need to add should be 50 Ohms – 35 Ohms = 15 Ohms.

When driving a 65-Ohm line, the source series resistor should be 65 Ohms – 35 Ohms = 30 Ohms.

8.15 How can you tell the difference between a short, low-impedance transmission line and a small capacitor in the middle of a uniform transmission line using a TDR?

Sometimes this is hard to do. A short, low-impedance transmission line will look very much like a capacitive discontinuity. All you can hope for is if the length of the line is long enough, the bottom of the TDR response will start to look flat. Otherwise, it may not be possible.

8.16 By looking at the TDR response, how would you tell the difference between a really long 75-Ohm line and a short 75-Ohm line connected to a 75-Ohm resistor?

There is no way to tell the difference between a long 75-Ohm cable or a 75-Ohm cable connected to a 75-Ohm resistor. The instantaneous impedance the TDR edge sees will look the same. The only possibility to tell the difference is to run the TDR for a long enough time to possibly see the reflection from the end of the transmission line.

8.17 What are the reflection and transmission coefficients when a signal is coming from a 40-Ohm environment and encounters an 80-Ohm environment?

The reflection coefficient is the second impedance minus the first, divided by their sum. In this case, it is $(80 - 40)/(80 + 40) = 1/3$.

The transmission coefficient is twice the second impedance divided by their sum, or $2 \times 80/(40 + 80) = 1.33$.

8.18 What are the reflection and transmission coefficients when a signal is coming from an 80-Ohm environment and encounters a 40-Ohm environment?

The reflection coefficient is the same as in the previous example, just the negative. The reflection coefficient is $(40 - 80)/(80 + 40) = -1/3$.

The transmission coefficient is $2 \times 40/(40 + 80) = 2/3$.

8.19 Consider a driver with a 1-V unloaded output voltage and 10-Ohm output resistance. The transmission line it is connected to is 50 Ohms, and the line is terminated at the far end. What value resistor should be used to terminate the line? What are the high and low voltages at the receiver if the far-end resistor is tied to Vss?

The far-end terminating resistor should be 50 Ohms to prevent the reflection at the far end. In this case, if the driver output resistance is 10 Ohms, when the output signal is 0 V, the signal at the far end across the resistors will be 0 V.

When the driver is sending a 1 V signal, because of the voltage divider, the voltage across the 50-Ohm terminating resistor will be $50/(10 + 50) = 5/6$ V $= 0.83$ V.

8.20 Redo Question 19 with the far-end resistor tied to Vcc.

The same 50-Ohm termination resistor would be used independent of the voltage the resistor is attached to. When the driver outputs a 1V signal, the voltage on the 50-Ohm resistor, relative to the Vss, is 1 V.

But, when the driver is outputting a 0 V signal, the voltage on the 50-Ohm resistor is 1 V $\times$ (10 Ohms/(10 + 50)) $= 0.17$ V.
When the resistors is tied to Vcc, the low signal is 0.17 V.

8.21 Redo Question 19 with the far-end resistor tied to 1/2 × Vcc, sometimes call the termination voltage, or VTT.

When the far-end resistor is terminated to 0.5 V, the low voltage is 0.5 V × (10/(10 + 50)) = 0.08 V and the high-level signal is 0.5 V + 0.5 (50/(10 + 50) = 0.92 V.

These three examples illustrate the value in terminating to a Vtt voltage. It balances the high and low voltage levels to be symmetrical about the center voltage. This keeps the same noise margin on both the high and low sides.

8.22 If the rise time of a signal is 3 nsec, how long a line would you expect to be able to get away without having to terminate?

If the rise time is 3 nsec, it would have a spatial extent of about 3 nsec × 6 inches/nsec = 18 inches. If the transmission line length were shorter than about 1/3 the spatial extend of the rise time, the reflections will still happen, but they will be smeared out during the rising edge. A line shorter than 6 inches may not need terminations.

8.23 What happens to the size of the reflected voltage from a short transmission line discontinuity when its length gets shorter and shorter? At what length is it "transparent"?

When the round-trip time of the structure becomes shorter than the rise time of the signal, the reflections from the front and back ends of the structure begin to overlap and the magnitude of the reflection decreases.

This means that if you can keep the length of the structure low enough, smaller than ½ the rise time, the discontinuity begins to look more transparent. The shorter it is, the more transparent it will look.

8.24 Due to the capacitive loading of the nonfunctional capture pads on each layer of a via, will the via look electrically longer or shorter?

The delay of a transmission line is the square root of the product of the total inductance and total capacitance of the line. If you keep the total inductance of the path the same, but add additional stray capacitance along the via by including the nonfunctional pads, the total C will increase and the total time delay will increase. The via will look electrically longer.

8.25 **Suppose a 1-V signal from a 1-Ohm output impedance source, with a 0.1-nsec rise time and 50% duty cycle is launched on a 12-inch long 50-Ohm transmission line. What is the average power consumption with a single 50-Ohm terminating resistor to the return path?**

The signal is a square wave with 50% on-time. The source impedance of the driver is very low, so virtually all of the 1 V signal is launched into the transmission line. The end of the line is terminated with a 50-Ohm resistor to the Vss path.

You can estimate the average power consumed per clock cycle.

When the signal is a high, at 1 V across the resistor, the power dissipation is

$$P = \frac{V^2}{R} = \frac{1}{50} = 20 \text{ mWatt}$$

The power dissipation when the signal is a low, or 0 V, is 0 mW. The average power dissipation over a cycle when the duty cycle is 50% is the average, or 10 mWatt average power.

8.26 **Suppose the far-end resistor were terminated not to ground, but to a voltage 1/2 the Vcc voltage. What would be the average power consumption?**

In the same way, the voltage across the 50-Ohm resistor will be ½ Vcc, for both the low and the high signal. The polarity of the voltage flips, which is how the receiver sees a different signal.

For both the high and low voltage levels, the power dissipation is

$$P = \frac{V^2}{R} = \frac{(0.5 \text{ V})^2}{50} = 5 \text{ mWatt}$$

This is the power dissipation in each half cycle. This is also the average power dissipation, since it occurs in both cycles.

When Vss or Vcc is used as the termination, you get twice the power dissipation as if a ½ Vcc voltage were used as the termination.

8.27 **In Question 8.26, suppose a 49-Ohm source series resistor were used to terminate the line. Would the power consumption in the source resistor be larger, smaller or the same as with a far-end termination?**

By adding the large source series resistor, the current in the entire system will be reduced, and the total power consumption will be reduced.

With the 49-Ohm source resistor and 50-Ohm terminating resistor, this is about 100 Ohms. The power dissipation with the same voltage, but twice the resistance as in the case before, is half as much.

Of this half the average power, it will be split between the source resistor and terminating resistor. The average power dissipation will be ¼ the 10 mWatt = 2.5 mWatt, and this will be dissipated in both the source and terminating resistors.

Chapter 9

9.1 **What happens to a signal propagating down a transmission line with an attenuation constant with frequency of −20 dB?**

If the attenuation is a constant −20 dB at all frequencies, then the amplitude of the entire signal is reduced to −20 dB, which is 10% of it is incident value. Every frequency component is reduced to this level, but the shape is preserved.

This means the rise time will be the same coming out as going in. The signal bandwidth has not changed, just the overall signal amplitude.

9.2 **What is ISI, and what are two possible root causes?**

ISI is inter-symbol interference. It arises when information from one bit overlaps with and interferes with another bit. There are a few causes of ISI. The two most common are rise-time degradation and reflections.

If the frequency-dependent losses are so high as to increase the rise time of the signal so it is long compared to a unit interval, information from one or more bits will bleed into the information of other bits.

If there are multiple reflections in the system that cause the initial bit to be reflected back and forth, some of the reflected part of one bit will arrive at the receiver when another bit arrives. The reflection of one bit will interfere with another bit, delayed in time.

9.3 **What is the impact of losses in a channel if the resulting rise time at the receiver is still very short compared to the unit interval?**

If the losses in a channel cause only a small amount of rise-time degradation, so the rise time is still short compared to the unit interval, the increased rise time will have very little impact on the signal quality.

In this situation, losses do not play a significant role in influencing signal quality.

9.4 **What is the primary impact on a signal at the receiver due to frequency-dependent losses?**

The frequency-dependent losses always increase with higher frequency. This means the bandwidth of the transmitted signal is lower than the incident signal, and the rise time of the signal at the receiver is longer than coming from the transmitter.

When this rise time is long compared to the unit interval, the rise-time degradation can strongly affect the signal quality in the form of collapse of the eye.

9.5 **Give an example of an interconnect where you would expect to see much lower losses than in an FR4 backplane channel.**

The losses in the FR4 channel come mostly from dielectric losses and some from conductor losses. If the laminate material was replaced with a Megtron 6 laminate, or a Rogers RO1200 laminate, for example, the losses would be much less than with FR4.

9.6 **What is collapse of the eye in the horizontal direction called?**

Jitter. The horizontal axis is time and the variation in the arrival time of a 0 to 1 transition or 1 to 0 transition is called jitter.

9.7 **Why don't we include radiative losses in the attenuation term in a stripline?**

There will always be some radiative losses in any interconnect. However, the radiated losses in stripline are so small compared to the conductor and dielectric losses that you ignore them. Including the radiative losses would complicate the analysis with very little benefit returned.

9.8 **What is the most important element you need to know to efficiently fix a problem?**

The root cause. While it may be possible to fix a problem by trying some random suggestions, the most important to quickly fix a problem is to find the root cause and fix the problem at the root cause. Of course, an important step is to have confidence you have the correct root cause.

9.9   How do reflections cause ISI?

It takes two sources of reflections to change the direction of a reflected signal so it heads back to the transmitter. But if there are two large sources of reflections in a circuit, then a part of one bit will rattle around and be delayed before it reaches the receiver. If the delay time is longer than one unit interval, the piece of one bit that reflects around will interfere with a new bit appearing at the receiver. This is the very definition of inter-symbol interference.

9.10   How does a discontinuity increase the rise time of a signal?

If you think of the discontinuity as being approximated by either a C or L element, then the reflected magnitude scales with the dV/dt or the dI/dt. It is the fast edges of the incident signal which will reflect more than the low frequency components.

If you lose the higher frequency components of the signal to reflections, the bandwidth of the transmitted signal will have decreased. This means a longer rise time.

Another way to think about this is that if the signal encounters a shunt C or series L, the signal will be degraded by the RC or L/R time constant. The transmitted signal will see the LC or L discontinuity being fed by the 50 Ohms impedance of the transmission line. This will result in a rise-time degradation. For example, a 1 pF capacitive load will increase the rise time of the signal by 50×1 pF = 50 psec.

9.11   What causes the skin depth effect?

Skin depth is caused by the current redistribution in the signal and return path to reduce the impedance of the signal-return path. At frequencies above about 1MHz, the impedance is dominated by loop inductance. The current will redistribute to reduce the loop inductance This means two effects: the signal and return current wants to get as close as it can to each other, and within each conductor, the current wants to get as far away from itself as it can.

9.12   What causes dielectric loss?

Dielectric loss is caused by the rapid motion of bound dipoles in the dielectric. These rotate in the electric field of the signal and absorb energy by translating electric field energy into motion and friction energy inside the dielectric.

A material with fewer or small dipoles will absorb less energy from the electric field and have a lower loss.

9.13 Why does the leakage current from the dielectric increase with frequency?
The dielectric loss is due to the AC leakage current flowing between the signal and return path. In each half cycle of a sine-wave-frequency component, the dipoles can rotate a certain amount. This represents some charge flow.

The total charge that rotates in a cycle depends on the number of dipoles and their distribution. This is a fixed amount. The current that flows is the charge that flows in each half cycle divided by the time for half a cycle.

The higher the frequency of the signal, the shorter the time per cycle. For a fixed amount of charge that flows, if the time interval decreases, the current increases. This is why the dielectric loss is proportional to frequency, because the time interval for all the current to flow decreases linearly with frequency.

9.14 What is the difference between the dielectric constant, the dissipation factor, and the loss angle?
The dielectric constant is the material property that describes how the speed of the signal is slowed down by the presence of the dielectric material. It is also the real part of the complex dielectric constant.

The dissipation factor, often abbreviated as Df, is related to the imaginary part of the complex dielectric constant and describes the density of dipoles—how large they are and how much they can move in the presence of the external electric field. The dielectric loss is directly proportional to the dissipation factor, Df.

The loss angle is actually the angle between the complex dielectric constant vector and the real axis. This is a measure of how large the loss is compared to the real part of the dielectric constant.

The dissipation factor is also defined as the tan(loss angle), which is the ratio of the imaginary part of the complex dielectric constant divided by the real part of the dielectric constant.

9.15 In an ideal lossless capacitor, when we apply a sine-wave voltage across it, what is the phase between the voltage and the current through it? What is the power dissipation?

The phase of the voltage compared to the current is –90 degrees. They are in quadrature.

Since they are at right angles to each other, the power dissipation, which is the dot product of the voltage × the current, is zero. There is no power dissipation through an ideal capacitor.

9.16 In an ideal resistor, when we apply a sine-wave voltage across it, what is the phase between the voltage and the current through it? What is the power dissipation?

In an ideal resistor, the phase between the voltage and current is 0 degrees. They are perfectly in phase. The power dissipation, as the dot product of the voltage × the current, is just their product. This can be expressed as V^2/R or I^2R.

9.17 In a capacitor filled with FR4, with a Df = 0.02, what is the ratio of the real to the imaginary current through the capacitor? What is the ratio of the energy stored in the capacitor to the energy lost per cycle?

The dissipation factor, Df, is literally the ratio of the imaginary part of the dielectric constant to the real part.

The current flowing through the material is proportional to the dielectric constant components. The real part of the current, which is in phase with the voltage and contributes to power dissipation, is related to the imaginary part of the dielectric constant.

The imaginary part of the current that is out of phase with the voltage, and is about energy stored, not dissipated, is proportional to the real part of the dielectric constant.

The ratio of the real to the imaginary currents is just the ratio of the imaginary part of the complex dielectric constant to the real part, or Df.

The real part of the dielectric constant is a direct measure of the energy stored per cycle. The imaginary part of the dielectric constant is a direct measure of the energy lost per cycle.

The ratio of the energy stored to the energy lost per cycle, which is the Q-factor of the system, is just 1/Df.

In FR4, with a Df = 0.02, the Q-factor of the interconnect is 50. When the Df is really small, like 0.002, the Q-factor can be as high as 500!

9.18 **If you want to use an n-section lossy transmission line model for a transmission line 2-nsec long, accurate to a 10-GHz bandwidth, how many sections do you need to use?**

The number of elements you need for reasonable accuracy is 10 sections per the shortest wavelength. The time delay × the highest frequency is the number of cycles or wavelengths you have down the length of the interconnect. You want 10 × the number of cycles for adequate accuracy. This is 10 × TD × BW. In this example, the number of cycles is 10 × 2 × 10 = 200 sections. This is a huge number, and often not practical.

9.19 **What is the attenuation per inch due to just conductor loss for a 5-mil-wide line and a 10-mil-wide line at 50 Ohms? At 1 GHz and 5 GHz?**

The loss from just the conductor is about $-1/w \times sqrt(f)$, in dB/inch. This includes the smooth copper loss from the signal path—including both sides of the signal trace—a small contribution from the return path, and a factor of 2 higher resistance due to the surface texture of the copper.

For the case of a 5-mil-wide line, at 1 GHz, this is –0.2 dB/inch. For the 10-mil-wide line, it is exactly half this, or –0.1 dB/inch.

The conductor loss increases with the sqrt of frequency. At 5 GHz, the attenuation will be sqrt(5) = 2.2 higher. For the 5-mil-wideline, the attenuation at 5 GHz would be 0.44 dB/inch and for the 10-mil-wide line 0.22 dB/inch.

9.20 **What is the attenuation per inch in an FR4 channel for a 5-mil-wide trace, at the Nyquist of a 5-Gbps signal, from both the conductor and dielectric loss? Which is larger?**

The loss from conductor and dielectric loss, the transmission coefficient, is

$$S21 = -1/w \times Sqrt(f) - 2.3 \times Df \times f \times sqrt(Dk)$$

The f in this case is the sine-wave frequency at the Nyquist frequency— 2.5 GHz, in this case.

The two loss terms are:

$$S21 = -1/5 \times Sqrt(2.5) - 2.3 \times 0.02 \times 2.5 \times sqrt(4)$$
$$= -0.32 \text{ dB/in} -H \text{ } 0.23 \text{ dB/in}$$

You see by putting in the numbers that the conductor loss is a little higher than the dielectric loss at about 2.5 GHz, which is for 5 Gbps signals like USB 3.0

9.21 **What is the skin depth of copper at 1 GHz? At 10 GHz?**

As a simple rule of thumb, the skin depth of copper at 1 GHz is 2 microns, and it decreases with the square root of frequency. If you go up in frequency by 10×, the skin depth goes down by 1/sqrt(10) = 0.32, to a value of 2 u × 0.32 = 0.64 microns.

9.22 **The data rate of a signal is 5 Gbps. What is the UI? If the rise time of the signal is 25 psec, would you expect to see any ISI?**

The unit interval is 1/data rate. If the data rate is 5 Gbps, the unit interval is 1/5 Gbps = 0.2 nsec = 200 psec. If the rise time is 25 psec coming out of the channel, this is very small compared to the UI. This means there should be no ISI from rise-time degradation. No information from the previous bit leaks into the next bit from rise-time degradation.

9.23 **A signal line is 5 mils wide in ½-ounce copper. What is its resistance per length at DC and at 1 GHz? Assume that current is on both the top and bottom of the signal trace.**

In ½-oz. copper, the sheet resistance is 1 mOhm/sq. For a trace 5 mils wide, there would be 200 squares per inch, or a DC resistance of 200 mOhm/inch.

This copper foil has a thickness of 17 microns. At 1 GHz, the skin depth of copper is 2 microns. If current is flowing in both the top 2 μ and bottom 2 μ of the trace, this is a cross section of 4 μ at 1 GHz.

At DC, the current flows through 17 microns, while at 1 GHz, it flows through only 4 μ. This is a contraction of 17/4 = 4.25×. You would expect the resistance at 1 GHz to go up by 4.25 from the DC resistance.

9.24 **For the worst-case lossy interconnect with 3-mil-wide lines in FR4, above what frequency does a transmission line behave like a low loss versus lossy line?**

The condition for low loss vs lossy line is related to the relative size of the R term and the 2pifL term. The interconnect is lossy when the series resistance per length is larger than the reactive term. This is rather startling. As you go to lower frequency, the resistance stays constant with frequency, but the reactive term continues to decrease. There is a point where the resistance is larger than the reactance, and the line is loss dominated.

At low frequency, below the skin depth limit, the resistance per length is constant and roughly 330 squares/inch × 1 mOhm/sq = 330 mOhm/inch.

The inductance per length for a 50-Ohm line is about 8 nH/inch.

The condition for R > 2pifL is 0.3 Ohms/inch > 6 × 8 nH/inch × f, or f < (0.3/50) GHz = 6 MHz. This is startling. It says that this transmission line with the very narrow line is loss dominated at LOW frequency, below 6 MHz. The R term dominates performance over the L term. Above 6 MHz, it actually behaves as a low loss line.

9.25 What impact do the losses have on the characteristic impedance? In the high loss and low loss regimes?

When you think of the characteristic impedance, it is actually the real part of the characteristic impedance you deal with. The imaginary part, in the low loss, regime, is a small fraction, on the order of the dissipation factor, of no more than 2%.

The impact of an imaginary part of the characteristic impedance is to cause reflections when you use resistive components to terminate the lines. The imaginary part will reflect from a resistor, which is just a real impedance. You cannot synthesize a broadband complex impedance to match the complex characteristic impedance.

Luckily, it's only at low frequency, below 5 MHz, that the imaginary part gets much larger. But at these low frequencies, you don't have to worry about terminations. The wavelengths are so long compared to any interconnect, the interconnects will all look like lumped circuit elements at those frequencies.

9.26 What impact do the losses have on the speed of a signal? In the high loss and low loss regimes?

In addition to making the characteristic impedance complex, another impact from losses is dispersion. This means the speed will be frequency dependent. The biggest impact will be at low frequency when the line is resistance dominated. This is also in the below 5 MHz regime.

In the low loss regime, the dispersion from losses is a very small factor, and the impact from dielectric constant dispersion is usually a larger effect.

9.27 What is the most significant impact on the properties of a transmission line from the losses?

The biggest issue with losses in transmission lines is not dispersion, and it's not a complex characteristic impedance. The biggest impact is from attenuation that is frequency dependent. It's this attenuation which causes the rise time to increase and contributes to collapse of the eye.

9.28 What is the ratio of the amplitudes of two signals if they have a ratio of −20 dB? −30 dB? −40 dB?

The dB is always 10× the log of the ratio of two powers. If the value is −20 dB, the ratio of the powers is 10^{-2}. The ratio of the amplitudes is the square root of this, of 10^{-1}, which is 0.1 or 10%.

You can do this much more automatically, as:

$$fraction = 10^{\frac{dB}{-20}}$$

The case of −30 dB, the fraction is $10^{-1.5} = 0.031 = 3.1\%$.
The case of −40 dB is $10^{-2} = 1\%$.

9.29 What is the ratio of two amplitudes in dB, if their ratio is 50%, 5%, and 1%?

Going the other direction, the value in dB from a ratio of amplitudes is

$$value[dB] = 20 \times log(fraction)$$

When the fraction is 50%, the value in dB is $20 \times log(0.5) = -6$ dB
When the fraction is 5%, the value in dB is $20 \times log(0.05) = -26$ dB
When the fraction is 1%, the value in dB is $= -40$ dB.

9.30 If the line width of a transmission line stays constant, but the impedance decreases, what happens to the conductor loss? How do you decrease the impedance while keeping the line width constant? In what interconnect structures might this be the situation?

If the line width is fixed and the impedance decreases, the attenuation from conductor loss will increase. The attenuation is proportional to the

(resistance per length) /(characteristic impedance). The lower the impedance, the higher the attenuation.

One way of reducing the impedance without changing the line
width is to bring the return closer to the signal. The thinner the dielectric, the lower the impedance and the higher the attenuation due to conductor loss.

This is the case in power and ground planes. The closer the power
and ground planes can be brought together, the lower their impedance
and the higher the resistive loss. This is a powerful way of damping out
cavity resonances in power and ground planes.

**9.31 The attenuation per length per GHz from just dielectric loss depends on
material properties. This makes it a useful figure of merit for a material.
What is this figure of merit for FR4 and for Megtron 6? Select another
laminate material and calculate this figure of merit from the data sheet.**
The attenuation per inch from dielectric loss is $S21 = -2.3 \times Df \times f \times sqrt(Dk)$. The figure of merit, FoM, for the attenuation per inch per GHz is

$$FOM = 2.3 \times Df \times sqrt(Dk).$$

For FR4, the FoM = $2.3 \times 0.02 \times sqrt(4) = 0.092$.
For the case of Megtron 6, the FoM = $2.3 \times 0.004 \times sqrt(3.6) = 0.017$

**9.32 Some lossy line simulators use an ideal series resistor and an ideal shunt
resistor. What is the problem with this sort of model?**
The whole problem with lossy lines is the frequency dependence of the
loss. If the series resistance is a constant resistance and the shunt resistance is a constant resistance, the losses predicted by this model would be
constant with frequency. This means the most important feature of lossy
line are not simulated in this model. It is worthless.

**9.33 Roughly, how much attenuation at the Nyquist might be too much and
result at an eye at the receiver that is too closed? For a 5-Gbps signal,
how long an FR4 interconnect might be the maximum unequalized length?
What about for a 10-Gbps interconnect?**
When the attenuation at the Nyquist is about −10 dB, generally the eye
will be too closed and the bit error ratio too high. If you use this limit for the attenuation, and have a very lossy interconnect with about

−0.2 dB/inch/GHz, you can estimate how long you can go before you reach the −10 dB limit.

$$-10\ dB = 0.2dB/inch/GHz \times Len \times 2.5\ GHz$$

Len = 20 inches.

If the interconnect runs at 10 Gbps, the Nyquist is 5 GHz, and the length you can travel before the attenuation is −10 dB is half as long, or 10 inches.

Chapter 10

10.1 In most typical digital systems, how much cross-talk noise on a victim line is too much?
The allocation for cross talk is typically about 30% of the total noise budget. When this is 15% of the signal swing, the amount of cross talk acceptable is about 5% of the signal swing.

10.2 If the signal swing is 5 V, how many mV of cross talk is acceptable?
With 5% of the signal swing as typical allowable cross talk noise, this is $5\% \times 5V = 250\ mV$ of noise acceptable.

10.3 If the signal swing is 1.2 V, how many mV of cross talk is acceptable?
When the signal swing is as low as 1.2 V, the noise is $5\% \times 1.2\ V = 60\ mV$.

10.4 What is the fundamental root cause of cross talk?
Cross talk is due to the fringe electric and magnetic fields from one signal-return path to another. You can describe this in terms of the E and B fields, or approximate it in terms of capacitive and inductive coupling.

10.5 What is superposition, and how does this principle help with cross-talk analysis?
Superposition means that the presence of E or B fields have no effect on the E or B fields that are induced due to the presence of adjacent signals. This means, if you see fringe fields on a quiet line with no other signal present on it, you will see the same noise level when a signal is present. The total voltage will be the sum of the noise present with no other signals plus the signal.

10.6 In a mixed signal system, how much isolation might be required?

Many rf receivers have a sensitivity as low as −100 dB down from the transmitted signal. This means they are very sensitive to noise. One feature that helps to minimize the impact from cross talk is that most rf receivers are tuned to a very narrow frequency band. This means it's only the digital noise in the bandwidth of the filter the receiver is sensitive to that you need to worry about. But isolation with the transmitter has to be as low as −100 dB.

10.7 What are the two root causes of cross talk?

Cross talk is about fringe electric and magnetic field coupling. If there were no fringe electric and magnetic fields, there would be no cross talk. The reason knowing the root cause is so important is that before you can control and reduce the cross talk, you have to know the root cause.

One important way of reducing the cross talk is by engineering the geometry to minimize the extend of the fringe fields from one signal-return path to another. The two most important design knobs are to bring the return closer to the signal, and to pull the two signal lines farther apart.

10.8 What elements are in the equivalent circuit model for coupling?

While the nature of the coupling is electric and magnetic field coupling, you can approximate this behavior with circuit elements of a mutual capacitor and mutual inductor. Using these circuit elements, you can now build an equivalent circuit model that includes the signal path and the coupling and use a circuit simulator like SPICE to simulate cross-talk noise.

10.9 If the rise time of a signal is 0.5 nsec, and the coupled length is 12 inches, how many sections would be needed in an n-section LC model?

If you are using a lumped circuit model to simulate the cross talk and use an n-section transmission line model, you would need $10 \times TD \times BW = 10 \times 12/6 \times 0.35/0.5 = 14$ sections. Thus, each transmission line would need to have 14 sections, and the mutual capacitors and mutual inductors would need 14 sections as well.

10.10 What are two ways of reducing the extent of the fringe electric and magnetic fields?

Knowing that the root cause is from the fringe electric and magnetic fields, you can sculpt them to reduce their extent. The two most important ways is to pull the signal lines farther apart and bring the return path closer to the signal lines.

10.11 Why is the total current transferred between the active line and the quiet line independent of the rise time of the signal?
As long as the spatial extend of the rising edge is shorter than the coupled length, the amount of coupled current from the active line to the quiet line should be independent of the rise time.

The capacitively coupled current is related to the coupling capacitance and the dV/dt. The coupling capacitance depends on the spatial extent of the edge. It's only over the length of the edge where there is a dV/dt.

The total capacitively coupled current that flows from the active to the quiet line is proportional to $RT/v \times dV/RT = dv/v$. The shorter the rise time, the shorter the coupled region and the lower the total coupling capacitance at any one moment. But, the shorter the rise time, the larger the dV/dt. The combination of these two effects makes the coupled current insensitive to the rise time.

The same analysis applies to the inductively coupled currents.

10.12 What is the signature of near-end noise?
The near-end cross talk appears at the near end as soon as the signal enters the aggressor line. It is traveling backward in the quiet line. It will turn on with the rise time of the signal and reach a saturated or constant value when the leading part of the signal edge has traveled a time delay distance of $\frac{1}{2} \times RT$.

The near-end noise voltage will stay at this saturated value until a time equivalent to the round trip coupled time of the two lines, then will fall with the fall time.

The near-end noise will be positive if the signal edge is a rising edge. The near-end noise will be negative if the signal edge is a falling edge.

10.13 What is the signature of far-end noise in microstrip?
You will only see far-end noise in a microstrip geometry where the dielectric is asymmetrically distributed. The far-end cross talk noise will not come out of the quiet line until the signal comes out of the active line. This is one TD after the signal is launched in the active line.

The magnitude will generally be in the negative direction for a rising edge and positive for a falling edge. The width of the pulse will be about the duration of the rise time. The peak voltage will be related to the coupling length, inverse with the rise time and increase as the traces are brought closer together.

10.14 What is the signature of far-end noise in stripline?
Trick question. There should not be any cross talk in stripline if the dielectrics are homogenous. The slight difference between the Dk of the pre-preg layer and the core layer will result in a little bit of far-end cross talk, but this will typically be less than 5% the far-end cross talk you would see in microstrip.

10.15 What is the typical near-end cross-talk coefficient for a tightly coupled pair of striplines?
In a 50-Ohm stripline, with the spacing equal to the line width, the amount of near-end cross talk is about 6%. The only way to know this is to perform a 2D field solver simulation, which takes into account the shape of the fringe fields.

10.16 When the far-end of the quiet line is open, what is the noise signature on the near end of the quiet line of two coupled microstrips?
When the far end of the quiet line is open, any forward traveling cross talk, far-end cross talk, will be reflected from the open end and make its way back to the near end.

If you look at the near end of the quiet line, you will see the initial near-end noise signature. One round trip time delay later, immediately after the last bit of near-end noise has made its way back to the near end, you will see the far-end noise that reflects from the open end and make its way back to the near end. You will see this as a large negative dip in the near-end noise signature.

10.17 When a positive edge is the signal on the active line, the near-end signature is a positive signal. What is the noise signature when the active signal is a negative signal? What is the noise signature for a square pulse as the active signal?
When the signal is a falling edge, either a 1 V to a 0 V, or even a 0 V to −1 V signal, the saturated near-end noise on the quiet line will be a negative voltage.

When you send a square wave into the aggressor line, you will see a positive near-end saturated voltage signature initially turn on, then off in round trip time delay time, and then a negative saturated voltage level lasting for a round trip time delay period.

The longer the clock period, the longer time between the positive and negative signals.

10.18 Why are the off-diagonal Maxwell capacitance matrix elements negative?

The sign of the off-diagonal Maxwell capacitance matrix elements come from the definition of the matrix element. Each capacitor matrix element is always the ratio of a charge to a voltage.

In the case of the off-diagonal capacitance matrix elements, you ground all the conductors and place 1 V on the first conductor. You measure the induced charge on the second conductor. The capacitance matrix element is the ratio of the charge on the second element to voltage on the first element.

When the first conductor has a positive 1 V applied, it will repel other positive charges from all the other conductors, making them negative. The induced charge you will see on all the other conductors is a net negative charge. It's this negative charge that causes the off-diagonal matrix elements to all be negative.

10.19 If the rise time of a signal is 0.2 nsec, what is the saturation length in an FR4 interconnect for near-end cross talk?

The saturation length is a time delay down the coupled length corresponding to ½ the rise time. In this case, the rise time is 0.2 nsec, so the saturation time delay is ½ × 0.2 nsec = 0.1 nsec. This saturation length corresponds to about 6 inch/nsec × 0.1 nsec = 0.6 inches. If the coupled length is more than 0.6 inches, the near-end cross talk will have reached its maximum, saturated value.

10.20 In a stripline, which contributes more near-end cross talk: capacitively induced current or inductively induced current?

In a stripline, the relative capacitively and inductively induced currents are exactly the same. They are both equally important. This is generally why there is not far-end cross talk. The far-end cross talk is the difference between the inductively coupled current and the capacitively coupled currents. The fact that there is no far-end cross talk in stripline suggests that these two currents are equal in magnitude.

10.21 **How does the near-end noise scale with length and rise time?**

The near-end noise voltage magnitude stays the same as the coupled length increases, providing the coupled length is longer than ½ the rise time. Likewise, the rise time of the signal on the aggressor has no impact on the voltage magnitude of the near-end noise.

10.22 **How does the far-end cross talk in microstrip scale with length and rise time?**

The far-end noise will increase linearly with the coupled length. The far-end noise is effectively snowballing, growing, as the signal edge propagates down the signal line.

The far-end nose will also increase inversely with the rise time. The shorter the rise time, the shorter the width of the far-end noise and the larger the peak value.

10.23 **Why is there no far-end noise in stripline?**

You can think of the origin of far-end noise in two ways. When using the capacitively and inductively coupled model, the far-end cross talk is the difference in inductively coupled current minus the capacitively coupled current. In stripline, the inductively and capacitively coupled currents are equal, and they cancel out at the far end.

You can also think of the origin of far-end cross talk in terms of the difference in speed between the differential signal and the common signal on the two lines that make up the differential pair. If the dielectric is uniform everywhere, the speed of the differential signal will be the same as the common signal. These two signals propagate at the same speed and on the quiet line, the n line; the differential component and the common component, appearing on the n line, will exactly cancel out, and there will be no signal on the n line at the far end.

10.24 **What are three design features to adjust to reduce near-end cross talk?**

The most important design feature to adjust to reduce near-end cross talk is to pull the signal lines as far apart as possible. This will reduce the number of fringe field lines that make their way from the aggressor line to the victim line.

The second design feature to adjust is to bring the return plane as close to the signal line as possible. This will confine the fringe field lines closer to the aggressor line. This also means engineer lower characteristic impedance lines.

The last knob to adjust is to keep the coupled length shorter than the saturation length. If the lines can't be made shorter than the saturation length, there is no advantage in shorter coupling length.

10.25 What are three design features to adjust to reduce far-end cross talk?

Far-end cross talk is reduced by the same two conditions as near-end cross talk: space traces farther apart and brings the return plane closer to the signal path. In addition, far-end cross-talk scales with coupling length and inversely with rise time.

You can reduce far-end cross talk by reducing the coupled length and doing what you can to increase the rise time.

Finally, if far-end cross talk is a concern, you should route the signal paths in stripline.

10.26 Why does lower dielectric constant reduce cross talk?

It's not the lower dielectric constant by itself that reduces cross talk. In fact, if you have a stripline geometry, and just replace the dielectric with a lower dielectric, the near-end cross talk won't change.

However, if you change to a lower dielectric constant, the characteristic impedance will increase. In order to bring it back to the target value, you bring the return plane closer to the signal path. It's because you've brought the return plane closer, with the same impedance, that you get lower cross talk. By bringing the plane closer, you confine the fringe field lines closer to the aggressor signal.

10.27 If a guard trace were to be added between two signal lines, how far apart would the signal lines have to be moved? What would the near-end cross talk be if the traces were moved apart and no guard trace were added? What applications require a lower cross talk than this?

In order to fit a guard trace between the signal lines, you need to pull the signal lines apart to a spacing equal to three line widths. Just by pulling the signal lines farther apart this much, the cross talk can drop to less than 1%. This is perfectly adequate for most digital applications.

There would be no need to add a guard trace between the signal lines. Even in the case of an aggressor using a 5 V signal and a victim line running a 1 V signal, the acceptable noise level would be 50 mV on the 1 V line. This is on the order of 1% of the 5 V line.

It's primarily on sensitive analog lines that require less than 1% cross talk, such as in ADC or DAC circuits.

10.28 **If really high isolation is required, what geometry offers the highest isolation with guard traces? How should the guard trace be terminated?**

If really high isolation is required, then traces should be routed in stripline, and not pass through noisy cavities. Guard traces can dramatically decrease the coupling in stripline if the guard trace is the same length as the coupled length, and if the ends of the guard traces are shorted to the return planes. This will cause the near-end noise on the guard trace to reflect as a negative signal and propagate in the same direction as the aggressor signal. This will help to reduce the near-end cross talk in the victim line.

10.29 **What is the root cause of ground bounce?**

Ground bounce is caused by the dI/dt of the return current passing through a common return path that has total inductance. The higher the total inductance of the return path, the more the ground bounce. The more signals' return paths through this screwed-up return path, the higher the ground bounce voltage.

All those signals sharing the common return path will see this ground bounce as part of their noise.

10.30 **List three design features to reduce ground bounce.**

Since ground bounce is related to the dI/dt passing through the total inductance of the return path, this suggests that ground bounce can be decreased if you

1. Reduce the inductance of the return path using wide, short paths.
2. Reduce the number of signals using the same common return path to reduce the dI/dt
3. Add multiple return paths to reduce the common lead inductance and reduce the number of signals' return current through any one of the return paths.

Chapter 11

11.1 **What is a differential pair?**

A differential pair is composed of any two transmission lines. That's all it takes. There are some special features that will improve the signal quality for differential signals.

11.2 What is differential signaling?

Differential signaling refers to using two different signal lines, the p line and the n line, with complimentary signals to send one bit of information. When one signal line turns on, the other signal line turns off. The information is transmitted in the difference between the voltages on the two signal lines.

11.3 What are three advantages of differential signaling over single-ended signaling?

The net Vcc and Vss currents from the drivers are mostly constant in driving a differential signal. One line is turning off, and the other line is turning on. The net current through the differential driver is constant. This drastically reduces ground bounce and rail collapse noise.

The differential receivers generally are more sensitive than single-ended receivers. This means they can tolerate smaller signals, or more noise margins. This is important when you have a lot of attenuation in the channel.

Differential signals are more robust to return path discontinuities. When a differential pair crosses a gap in the return path, the net dI/dt in the return current is nearly zero, so there is no ground bounce. This is especially true when differential signals pass through vias through cavities. The net changing current is zero, and there is little cavity noise generated in differential vias.

11.4 Why does differential signaling reduce ground bounce?

Ground bounce is caused by a screwed-up return path and multiple signals' return currents sharing this screwed-up return path. It's the dI/dt of the return current, passing through the higher inductance of the screwed up return path that causes ground bounce.

When a differential signal passes through the screwed-up return path, the net dI/dt in the p and n signal lines' returns overlap and cancels out. There is no net dI/dt. This means there is no ground bounce.

11.5 List two misconceptions you had about differential pairs before learning about differential pairs.

A common misconception is that the return current of one line in a differential pair is carried by the other line. This is only the case when there is no adjacent return plane for the differential pair. This is the case for twisted pair cables or pins in a connector. However, at the board level, this is not the case.

It is also commonly believed that common currents in a differential pair cause radiated emissions. While it is true that common currents can cause radiated emissions, at the board level, common currents travel in microstrip and stripline differential pairs, just as easily and without generating any radiated emissions, as single-ended signals do.

11.6 What is the voltage swing in LVDS signals? What is the differential signal?

In an LVDS signal, the voltage on each of the signal lines swings from 1.4 v to 1.15 v. This means the p line is at 1.4 V and the n line is at 1.15 V, and then they switch. The p line is then 1.15 V and the n line is at 1.4 V. The differential signal switches from 0.25 V to − 0.25 V. This is a differential swing of 0.5 V peak to peak.

11.7 Two single-ended signals are launched into a differential pair: 0 V to 1 V and 1 V to 0 V. Is this a differential signal? What is the differential signal component, and what is the common signal component?

This is not a pure differential signal. There is a differential signal component in this, but there is also a common signal component. This signal is actually a combination. The differential signal component is the difference, while the common signal component is the average.

The difference is a +1 V swing to a −1 V swing, or a 2 V peak to peak differential signal swing. The common signal component is the average, of 0.5 V, which is relatively constant in this example.

11.8 List four key features of a robust differential pair.

While any two transmission lines make up a differential pair, there are a number of special features that improve the differential signal quality, such as:

- **Controlled impedance.** The cross section should be uniform down the line. This will prevent reflections down the line.
- **Symmetrical.** The p and n lines should be identical and of identical length. This will reduce mode conversion.
- **Tightly coupled.** While a differential pair can have any coupling and still be good quality, a tightly coupled differential pair will enable the highest interconnect density, which means the lowest cost board.

- **Routed over a solid return plane.** While a differential signal is robust to return path discontinuities, there will always be some common signal component, which is sensitive to ground bounce. Always try to design a differential pair assuming there is some common signal component propagating as well.

11.9 What does differential impedance refer to? How is this different from the characteristic impedance of each line?

The differential impedance is the impedance the difference signal sees. The difference signal is the voltage difference on the p and n signal lines. The characteristic impedance of either line is the single-ended impedance of each line.

When the two signal lines are far enough away so they are uncoupled, the differential impedance is twice the single-ended impedance of each line. As the lines are brought closer together, their coupling decreases the differential impedance.

11.10 Suppose the single-ended characteristic impedance of two uncoupled transmission lines is 45 Ohms. What would a differential signal see when propagating on this pair? What common impedance would a signal see on this pair?

When the two lines that make up the differential pair are uncoupled, the differential impedance is just twice the single-ended characteristic impedance. This would be a differential impedance of 90 Ohms.

The common impedance is the parallel combination of the two single-ended impedances. In this case the two 45 Ohm lines, in parallel, have a common impedance of 22.5 Ohms.

11.11 Is it possible to have a differential pair with the two signal lines traveling on opposites ends of a circuit board? What are three possible disadvantages of this?

The short answer is yes. A differential pair is any two transmission lines. They can be 5 inches apart and still a differential pair. But this is not such a great idea.

When they are far apart, there could easily be coupling on either the p or n lines from other signal lines or sources. This will create both differential and common signal cross talk.

The signals propagating in the two lines that make up the separated differential pair have return currents that are directly under the signal lines. If either line encounters a return path discontinuity, there will be no chance of return current cancellation, and this differential pair will be sensitive to ground bounce.

Any asymmetry between the two lines that make up the differential pair will contribute to mode conversion. When the lines are manufactured in close proximity to the board, there is a higher chance they will see the same environment and have the same features, and length. When they are routed far apart, the manufacturing variations from one edge of the board to the other might increase the manufactured asymmetry.

11.12 As the two lines that make up a differential pair are brought closer together, the differential impedance decreases. Why?

As the two lines come closer together, the fringe field lines from one line to the other increase. This means that with the higher dV/dt between the two lines, there is more capacitively coupled currents between the two lines. With more current into each line, for the same signal voltage, the impedance is going down.

In addition, the mutual inductance of one line's current loop will decrease the effective loop inductance of the other line. This lower loop inductance per length in each line means the odd-mode impedance of that line is decreasing, which makes the differential impedance less.

11.13 As the two lines that make up a differential pair are brought closer together, the common impedance increases. Why?

When a common signal is driven on both lines, each line has the same voltage on it. This means there are no fringe field lines between the two lines. As single-ended signals, there are some fringe field lines to the return path, which contributes to the capacitance per length and the characteristic impedance.

However, when the common signal is applied, the fringe field lines between the two lines decreases, and there is less capacitance per length. The even-mode impedance increases due to less capacitance per length.

In addition, with the same currents passing through each of the two lines, the effective loop inductance per length of one line is increased since the magnetic field lines from the other line are adding to the signal line's, increasing its impedance.

11.14 What is the most effective way of calculating the differential impedance of a differential pair? Justify your answer.

The differential impedance of a differential pair is due in large part to fringe electric and magnetic fields. The best way to take this into account is with a 2D field solver. This will result in the most accurate value for the differential impedance.

It's not necessary to use a 3D field solver, as long as the cross section is the same down the length of the differential pair. It will take longer, and the answer is usually in the form of S-parameters, which do not directly give the differential impedance.

11.15 Why is it a bad idea to use the terms differential mode impedance and common mode impedance?

It is not recommended to use the terms differential mode or common mode for two reasons. The modes really should be referred to as odd mode and even mode. Using these terms help to form a good engineering model for the idea of modes.

More importantly, the even and odd modes have well-defined impedances. While there is a connection between the odd-mode impedance and the differential impedance, they are not the same. If you use the term differential mode impedance, it is a little ambiguous if you are referring to the odd-mode impedance or the differential impedance.

Likewise, the common impedance and the even-mode impedance are related, but they are not the same. Using the term common mode impedance confuses the two concepts of even mode and common signal. They are different. Which one are you referring to?

It is much cleaner to refer to the differential or common impedance and the odd and even modes. Using these terms will always be unambiguous. There are enough other terms to be confused about. Why make it more confusing.

11.16 What is odd-mode impedance, and how is it different from differential impedance?

The odd-mode impedance is the impedance of one line in a differential pair of transmission lines when the pair is driven with a differential signal. The odd-mode impedance is the impedance of one of the lines. The differential impedance is the impedance of the pair of lines. It is the impedance of the two lines in series. This makes the differential impedance the series combination of the two odd-mode impedances of the two lines. The differential impedance is twice the odd-mode impedance.

11.17 **What is even-mode impedance, and how is it different from common impedance?**

The even-mode impedance is the impedance of one line in a differential pair of transmission lines when the pair is driven by a common signal. The common impedance is the impedance of the two lines in parallel. This makes the common impedance ½ the even-mode impedance of their line.

11.18 **If two lines in a differential pair are uncoupled and their single-ended impedance is 50 Ohms, what are their differential and common impedances? What happens to each impedance as the lines are brought closer together?**

When the two lines are uncoupled, the differential impedance is just twice the single-ended impedance of either line and the common impedance is ½ the single-ended impedance of either line. When the two single-ended lines are each 50 Ohms, the differential impedance is $2 \times 50 = 100$ Ohms and the common impedance is $½ \times 50$ Ohms $= 25$ Ohms.

As the lines are brought closer together, the differential impedance will decrease and the common impedance will increase.

11.19 **If two lines in a differential pair are uncoupled and their single-ended impedance is 37.5 Ohms, what are their odd-mode and even-mode impedances?**

As long as you know the two lines in a differential pair are uncoupled, you know the differential impedance is $2 \times$ the single-ended impedance of the common impedance is ½ the single-ended impedance. In this example, the differential impedance is $37.5 \times 2 = 75$ Ohms whereas the common impedance is $½ \times 37.5 = 18.75$ Ohms.

11.20 **If the differential impedance in a tightly coupled differential pair is 100 Ohms, is the common impedance higher, lower, or the same as 25 Ohms?**

If the differential impedance is 100 Ohms you know the odd-mode impedance is 50 Ohms. If you were to pull the two lines farther apart so they became uncoupled, the odd-mode impedance would increase to something larger than 50 Ohms. When uncoupled, the odd-mode and even-mode impedances are the same. This means the uncoupled even-mode impedance is higher than 50 Ohms. When brought closer together to their tightly coupled state, the common impedance would increase above 25 Ohms.

11.21 In a tightly coupled differential microstrip pair, where is most of the return current?
When the return plane in the board is on an adjacent layer, most of the return current is in the plane of the board. Only a small fraction of the return current is carried by the other line.

11.22 List three different interconnect structures in which the return current of one line of a differential pair is carried by the other line.
This is the case when there is no adjacent return plane. Three common structures where a differential pair does not have an adjacent return plane are:

- In leaded packages
- Connectors
- Twisted pairs

11.23 In stripline, how does the speed of the differential signal compare to the speed of the common signal as the two lines are brought closer together? How is this different in microstrip?
In stripline, the electric field distribution for a differential signal and for a common signal are very different. However, with a homogenous distribution of dielectric material, wherever there is electric field, there is the same dielectric constant. This means that both the differential signal and the common signal see the same effective dielectric constant and propagate at the same speed.

In microstrip, the presence of air above the traces means the differential signal, with large electric fields in the space above the traces, sees a disproportionate amount of air contributing to its effective dielectric constant. In a common signal, with most of the electric field between the signal lines and the return path, the effective dielectric constant is closer to the bulk value. This means a differential signal will travel faster than a common signal.

11.24 **Why is it difficult to engineer a perfectly symmetrical differential pair in broadside-coupled stripline?**

In broadside-coupled striplines, the dielectric between one signal line and its adjacent return plane is on a different layer in the stack up than the other signal line and its adjacent return plane.

It is difficult to ensure the two different layers have the same thickness and dielectric constant.

In addition, the two signal lines are on different signal layers. It's difficult to engineer the two lines to be the same line width when they are etched on different layers.

11.25 **What happens to the odd-mode loop inductance per length as the two lines in a microstrip differential pair are brought closer together? What about the loop inductance of the even mode? How is this different in stripline?**

The odd-mode loop inductance of one line in a pair is when the pair is driven with a differential signal. This means the current in the n line is circulating in the opposite direction than the current in the p line.

The magnetic field of one line helps to reduce the magnetic fields in the other line. This makes the odd-mode loop inductance smaller than the single-ended loop inductance. The closer the two lines in the differential pair, the more reduction in the odd-mode loop inductance.

The exact same effect happens in stripline or microstrip.

11.26 **If you only used a 100-Ohm differential resistor at the end of a 100-Ohm differential pair, what would the common signal do at the end of the line?**

Reflect. The differential signal would see the 100 Ohm termination at the far end, and there would be no reflection. However, the common signal would see an open and reflect.

11.27 **Why is on-die Vtt termination of each line separately an effective termination strategy for a differential pair? If each resistor is 50 Ohms and the differential impedance is 100 Ohms, what might be the worst-case reflection coefficient of the common signal if the two lines are tightly coupled?**

By terminating each line to Vtt with a 50-Ohm resistor, you are effectively terminating each line as a single-ended line. The advantage of doing this is that you can do the termination on die, which reduces the termination discontinuity.

This termination scheme also keeps the eye well balanced.

When the lines in the differential pair are tightly coupled and the differential impedance is 100 Ohms, the odd-mode impedance is 50 Ohms and the 50 Ohms on-die termination is a perfect termination for the differential signal.

The common impedance would be higher than 25 Ohms, and could be as large as 30 Ohms. With two 50-Ohm resistors in parallel at the far end, the common signal would see a termination resistance of about 25 Ohms. This means there would be a reflection of about (30 − 25)/(30 + 25) = 9%. This is a relatively small reflection coefficient, and it is for the common signal.

Using a Vtt termination on die is a good compromise between good signal quality with on-die termination and a far common signal termination.

Chapter 12

12.1 Fundamentally, what do S-parameters really measure?
S-parameters are ratios of the complex voltage coming out to the complex voltage going into a device under test (DUT). How you interpret them is where the real value comes in.

12.2 What is another name for an S-parameter model of an interconnect structure?
S-parameters are sometimes call behavioral models or black box models of interconnects because they contain all the information you would ever want to know about the electrical properties of an interconnect.

Some simulators can integrate S-parameter models directly into their simulation environment.

They are called black box models because the structure of the DUT is hidden inside the model. You can use the S-parameters as an electrical model, but what it says about the structure and fundamental root cause of behavior is hidden inside the numbers.

12.3 In the frequency domain, what do S-parameters describe about the properties of an interconnect?
S-parameters describe the response of the interconnect to signals and how input signals are "scattered" from the interconnect. Each S-parameter element is the response of the interconnect to an input signal, over the frequency range.

12.4 What is different about the S-parameter model of an interconnect when displayed in the time domain?

When displayed in the frequency domain or the time domain, the S-parameter data is the same. There is no difference in the behavioral model of the S-parameters between the frequency or the time domain. They are just displayed differently.

You can use any waveform you want as the stimulus in the time domain. For historical reasons, you use a step response as the source. In a few rare cases, you also use the impulse response as the source function.

The difference is in how the data looks. In the time domain, you see the time-separated response of the interconnect to a step edge incident on one port. In some cases, this information is easier to interpret in terms of a spatial mapping of properties for the DUT, either impedance, coupling, or mode conversion.

12.5 List four properties of an interconnect that could be data mined from their S-parameter model.

Contained inside the S-parameter behavioral model of an interconnect are all the properties of an interconnect structure, such as

- The instantaneous impedance profile
- The characteristic impedance of the interconnect
- The cross talk to an adjacent structure
- The attenuation
- The time delay
- The dispersion
- The input impedance in the frequency domain

12.6 What is the significance of using a 50-Ohm port impedance?

There is nothing fundamental about using a 50-Ohm port. In principle, you can use any port impedance you want. If you know the S-parameters for one port impedance, you can change them to any other port impedance.

The port impedance does not change the information contained inside the S-parameters.

However, to interpret the actual numbers inside the S-parameter file, you have to know the port impedance. If you think the port impedance is 50 Ohms, and it is actually 75 Ohms, you will interpret the S-parameters incorrectly.

This is why knowing the port impedance is so important. If you don't open up the S-parameter text file and look at the port impedance listed, you won't know what it is by looking at the S-parameters.

To avoid confusion, you should always use 50 Ohms as the port impedance unless you have a compelling reason not to.

12.7 Why is there a factor of 20 instead of 10 when converting an S-parameter, as a fraction, into a dB value?

The dB is always a unit of the log of the ratio of powers. But an S-parameter is the ratio of two voltages, which are amplitudes. When converting the amplitudes to a ratio of powers, you square the amplitudes. When you take the log of these powers, the exponent of 2 becomes a factor of 2 in front of the log.

When you convert an S-parameter into a log, there is a factor of 2 in front of it.

12.8 Which S-parameter term is the impedance of an interconnect most sensitive to?

Although the impedance of an interconnect affects all the S-parameter terms in some way, the reflected signals, each diagonal element of the S-parameter matrix, are most sensitive to the impedance profile of the interconnect.

After all, the only way you get a reflected signal is from a change in impedance down the interconnect. Contained in the return loss terms is all the information about the impedance structure looking into each port.

12.9 Which S-parameter term is the attenuation in an interconnect most sensitive to?

The attenuation of an interconnect is about the transmission of the signal through the interconnect. The impact from the attenuation is most evident in the transmission coefficient, which is the insertion loss or S21.

12.10 What are the S-parameters of an ideal transparent interconnect?

A transparent interconnect should have no reflections, and everything should transmit. The S11 should be a very large, negative dB value, and the S21 should be 0 dB.

12.11 **What is the recommended port labeling scheme for a pair of transmission lines?**

I recommend the labeling scheme of the odd port indexes on the left side and the even port indices on the right side. This way, a through path is port 1 to port 2 and port 3 to port 4.

IF you use this labeling scheme, an insertion loss of one line in a differential pair is S21—the same as for a single-ended line—and the differential response is SDD21. You are consistent in the use of the indices of "2,1" to describe a through path response.

12.12 **Which S-parameter term contains information about near-end cross talk between two transmission lines?**

If you use the recommended labeling scheme, near-end cross talk is described by the S31 term, or, on the other side, S42.

12.13 **Which S-parameter term contains information about far-end cross talk between two transmission lines?**

If you use the recommended labeling scheme, then far-end cross talk is described by the S41 term, or viewed from the other side, S32.

12.14 **How many individual numbers are contained in the S-parameter model of a 4-port interconnect, measured from 10 MHz to 40 GHz at 10-MHz steps?**

In a 4-port S-parameter file, there are 16 different combinations of going in and coming out. Not all of them are unique, but they are all contained in the S-parameter file.

Each one of these is complex, so there are two numbers associated with each matrix element. And a value for each number is provided at each of 4,000 frequency points.

This makes the total number of individual numbers in the S-parameter file, not counting the frequency values, as $16 \times 2 \times 4,000 = 128,000$ different numbers. That's a lot of information!

12.15 **Using the historically correct definition of return loss, if the return loss increases, what happens to the reflection coefficient? Is this in the direction of a more transparent or less transparent interconnect?**

The historically correct value of return loss is a positive dB value. The reflection coefficient is a negative dB value. This means that as the return loss increases and gets to be a large Db value, the return loss increases the value of the negative dB term and gets to be a smaller value.

The smaller the refection coefficient, or the larger the historically correct return loss, the more transparent the interconnect.

12.16 Using the modern, conventional definition of insertion loss, if the insertion loss increases, what does the transmission coefficient do? Is this in the direction of a more or less transparent interconnect?

Using the more conventional and common definition of the insertion loss, if the insertion loss increases, the transmission coefficient increases and more signal gets through. This means the value of S21 gets closer to 0 dB, and the interconnect looks more and more transparent.

12.17 Suppose the reflection coefficient of a cable with 50-Ohm port impedances has a peak value of −35 dB. What would you estimate its characteristic impedance to be? Suppose the port impedance were changed to 75 Ohms. Would the peak reflection coefficient increase or decrease?

A peak value of −35 dB for a reflection coefficient means the reflection from the interconnect is about 3%. As a peak value, this means there is a reflection of about 1.5% from the front of the interconnect and 1.5% from the other end.

The reflection coefficient from one end is $1.5\% = (Z2 - Z1)/(Z2 + Z1)$. And, if the source impedance is 50 Ohms, the impedance of the interconnect, Z2, would be $Z2 = 50 \times (1 - 1.5\%)/(1 + 1.5\%) = 49.2$ Ohms. This is very close to 50 Ohms.

However, if the port impedance were changed to 75 Ohms, the reflection coefficient would dramatically increase. The interconnect has not changed; but its S11 term, and others, would change just by changing the port impedance. This is why it is important to know the port impedance used to generate the S-parameters.

12.18 As two of the most important consistency tests for the S-parameter of a through interconnect, what should the return and insertion loss be at low frequency? How will this change if the impedance of the interconnect is increased?

At low frequency, the return loss, or S11, should always be a large negative dB value. The insertion loss, or S21, should be close to 0 dB. In other words, an interconnect should look like a transparent interconnect at low frequency.

If the impedance of the interconnect increases, at low frequency these two values should stay about the same. All interconnects, at sufficiently low frequency, will look transparent no matter their impedance.

12.19 What causes ripples in the reflection coefficient?

The ripples in the reflection coefficient are caused by the reflections from the front of the interconnect and the back of the interconnect returning back to the same port and interfering. When their phases are the same, they add and you see a peak. When their phases are out of phase, they subtract and the reflection coefficient has a dip.

12.20 Above what magnitude of reflection coefficient will the return loss ripples begin to appear in the transmission coefficient?

If the reflection coefficient is less than about −13 dB, the energy lost in the reflected signal is so small that it has little impact on the magnitude of the transmitted signal. But if the reflection coefficient is greater than −13 dB, you will begin to see this energy loss impact the transmitted signal by decreasing the transmission coefficient.

12.21 Why does the phase of S21 increase in the negative direction?

Phase of S21 is defined as the phase of the coming out signal minus the phase of the going in signal—NOW.

When you see a signal coming out of an interconnect, some time has passed and the phase of the signal going into the interconnect now, has advanced. This means, right now, at the instant the signal comes out of the interconnect, the phase of what is going in will be higher than the phase coming out.

Since the definition of the phase of S21 has the coming out minus the going in phase, and the going in phase is always larger than the coming out phase, the phase of S21 will always be negative; and as you increase frequency, the phase shift will increase for the same time delay, making the phase of S21 increase in the negative direction.

12.22 Why does the polar plot of the S21 spiral inward with increasing frequency?

Virtually all interconnects have attenuation that increases as you go up in frequency. This means the transmitted signal gets smaller and smaller as you go up in frequency.

When plotted on a polar plot, the magnitude of the S21 vector gets smaller as the phase advances, so the polar plot has the vector spiraling inward with higher frequency.

12.23 **As a consistency test, what insertion loss would you expect for an FR4 interconnect at 10 GHz that is 20 inches long?**

The figure of merit for the loss in an FR4 channel that includes conductor loss is about –0.2 dB/inch/GHz. For a channel 20 inches long, at 10 GHz, the loss should be about –0.2 dB/inch/GHz × 20 inches × 10 GHz = –40 dB.

12.24 **Why does S21 of one line in a microstrip differential pair shows a deep dip at some frequency? What would be an important consistency test to test your explanation?**

There are a few possibilities for why a microstrip would show a dip in the insertion loss. One possibility is if it were part of a differential pair. As the energy flows from one conductor to the other, you lose energy in the transmitted signal. A consistency test for this effect is that you should see the energy appear in the far end of the other line. If S21 decreases, you should see S41 increase.

Another alternate explanation is you have a stub resonance. If this were the case, the energy you lost in the S21 term would appear in the S11 term. This would be an important consistency test.

12.25 **What is the most common source of sharp, narrow dips in the insertion loss of through connections?**

When the dips are narrow and sharp, this is an indication of energy lost to a high Q resonance. Somewhere near the signal is coupling to a high Q resonant structure. The most common source of such a structure is a cavity composed of two planes. The signal could be passing through the cavity, and its return current couples to the cavity.

There could also be some floating conductor on the surface, which acts as a resonant structure.

12.26 **What are the two types of signal that can appear on a differential pair?**

On a differential pair, you can only have a differential signal or a common signal.

12.27 What is the only feature that causes mode conversion in a differential pair? Which two S-parameter terms are strong indicators of mode conversion?

Mode conversion is only caused by an asymmetry between the features of the two lines that make up the differential pair. However screwed up one line in a pair may be, if the other line is screwed up the same way, there will be no mode conversion.

The strongest indicators of mode conversion are the mode conversion terms in the mixed-mode S-parameters—the SCD21 and SCD11 terms. These terms are the direct measure of the mode conversion inside the interconnect.

12.28 In an uncoupled differential pair, how would the differential insertion loss compare with the single-ended insertion loss of either line?

If the two lines in a differential pair are uncoupled, the differential response of any S-parameter term would be exactly the same as the single-ended response of the S-parameter term.

After all, with no coupling, the differential response is the average response of each of the two lines. If they are symmetrical, they have the same value and their average is the same value.

12.29 When displaying the TDR response of an interconnect, which S-parameter is displayed, and how do you interpret the display?

The TDR response is really the S11 response, displayed in the time domain. You choose as the time-domain input stimulus a step edge with a rise time corresponding to the bandwidth of the S-parameters. This step edge goes on, and the reflected signal is displayed as the time-domain response.

The refection of the step edge is directly related to the impedance profile. Knowing the source impedance, you can literally rescale the reflection coefficient to display the instantaneous impedance profile of the interconnect.

12.30 What information will be in the TDT response of a through interconnect?

The two important pieces of information in the TDT response of an interconnect will be the time delay of the interconnect and the rise-time degradation from transmission. The more frequency-dependent losses, the more rise-time degradation.

Generally, if the edge is not a simple Gaussian edge, it's difficult to interpret one figure of merit that characterizes the rise time. While you can qualitatively estimate a number for the rise time, this is a case where you can get a more precise metric of the rise time distortion by looking in the frequency domain, at the frequency where the attenuation is significant.

Chapter 13

13.1 What are five elements that are part of the PDN?
In tracing the path of the PDN from the VRM to the pads on the die, you encounter the following structures:

- VRM
- Bulk decoupling capacitors
- MLCC capacitors
- The cavity composed of the power and ground planes
- The vias to the packages
- The package lead inductance
- The on-die capacitance

13.2 What are two examples of interconnect structures that are not part of the PDN?
Signal lines traveling as microstrips on the top and bottom layer of the board are not part of the PDN. Even the terminating resistors at the far end of the transmission lines are not part of the PDN.

13.3 What are three potential problems that could arise with a poorly designed PDN?
A poorly designed PDN could result in:

- Bit error noise when a receiver tries to read the voltage level transition of a received signal
- Too much jitter from a circuit part of the clock distribution network or timing circuit
- Too much cross talk when signals transition through the cavity and pick up noise from the cavity

13.4 What is the most important design principle for the PDN?
To reduce the voltage noise in the cavity, keep the impedance of the cavity as low as practical.

13.5 What performance factors influence the target impedance selection?
Select the target impedance based on the amount of voltage noise you can tolerate and the maximum current you think will flow in the PDN. This ratio is the target impedance.

13.6 What are the three most important design guidelines for the PDN?
To keep the impedance of the PDN below the target impedance, it's about

- Reducing the inductance of all interconnect elements
- Using as large an on-die capacitance as possible
- Using on-package capacitors
- Using high enough ESR in capacitors to provide damping for any high q resonances.
- Using cavities with a thin dielectric placed close to the top layers in the stack up

13.7 A 2-V power rail has a ripple spec of 5% and draws a maximum transient current of 10 A. What is the estimated target impedance?
The target impedance is the voltage noise tolerance divided by the maximum transient current. In this example, it is Ztarget = (2 V × 0.05) / 10 A = 0.1V/10 A = 0.01 Ohms.

13.8 What is one downside to implementing a PDN impedance that is well below the target impedance?
While a PDN that is everywhere below the target impedance will work with no problem, it may cost more than it could. There may be a less expensive alternative design that will still give adequate performance.

13.9 Why is the PDN easier to engineer by looking in the frequency domain?
So much of the performance of the PDN is about parallel resonances between C and L elements. These sorts of problems are so much easier to identify, understand, and engineer in the frequency domain.

13.10 What are the five frequency regions of the impedance of the PDN in the frequency domain, and what are the physical features that affect it?

- The highest frequency is the impedance from the on-die capacitance.
- The next lower frequency is the Bandini Mountain from the

parallel resonance of the on-die capacitance and package lead inductance.

- The next lower frequency is the region dominated by the MLCC capacitors.
- Next is the parallel resonance of the bulk capacitor and the VRM
- The lowest frequency region is dominated by the VRM output resistance and its effective inductance.

13.11 What feature in the PDN provides low impedance at the highest frequency?

At the highest frequency, the impedance is dominated by the on-die capacitance.

13.12 What is the only design feature you need to adjust to reduce the impedance at the highest frequency and why is this difficult to achieve?

Since the impedance at the highest frequency is due to the on-die capacitance, the only design knob to adjust to reduce this impedance is to add more on-die capacitance. This is generally expensive and not easy to achieve.

13.13 What are the three most important metrics that describe a VRM?

The VRM is typically modeled as a series resistance and inductance. These are the two most important qualities of the VRM. In the case of a switch mode power supply, the switching frequency is also an important metric.

13.14 What is the key SPICE element that enables simulation of an impedance analyzer?

Simulating an impedance analyzer uses a constant current AC current source.

13.15 Why is reducing the package lead inductance so important?

The package lead inductance is one of the two terms influencing the peak impedance of the Bandini Mountain. The lower the package lead inductance, the lower the Bandini Mountain peak impedance.

13.16 A 2-layer BGA has leads that run from one side to the other side. If adjacent power and ground leads are 0.5 inches long, and there are 20 pairs, what is the equivalent package lead inductance of the PDN?

The loop inductance per length of a pair of leads is about 20 nH/inch. If the leads are 0.5 inches long, this is a total loop inductance of about 0.5 × 20 = 10 nH. If there are 20 pairs in parallel, the equivalent inductance is 1/20 × 10 nH = 0.5 nH of loop inductance in the power/ground path.

13.17 What package style would have the lowest loop inductance? What three package features would you design to reduce the package lead inductance?

The package style with the lowest loop inductance in the power distribution is a multilayer BGA with power and ground planes on adjacent layers that are close together.

Alternatively, a chip-scale package with very short leads will also have a very small loop inductance.

13.18 Why is it sometimes noticed that if you take off all the decoupling capacitors on a board, the product will sometimes still work? Why is this not a good test?

Whether there is a failure due to a PDN problem often depends on if there is a large transient current at a frequency where there is a large peak in the impedance profile. If during the boot-up process and running the specific code at turn-on there are no large transient currents, the product may work just fine and there are no problems in the PDN. All the decoupling capacitors could be taken off and the product may work just fine.

But as soon as the right microcode is run, so there are large transient currents at a peak impedance frequency, the noise may exceed the noise limit and a failure observed. Just because the product "works" is no guarantee that it will work with all possible microcode.

13.19 What are the three most important features of an MLCC capacitor, and what physical features affect each term?

The three electrical features of an MLCC capacitor are the capacitance, the mounting inductance, and the ESR.

The capacitance is about the number of layers and their spacing.

The mounting inductance is about the loop inductance of the interconnects from the buried cavity to the leads of the capacitor.

The ESR is related to the number of plates in parallel. Generally, a capacitor with a larger capacitance will have more plates in parallel and a lower ESR.

13.20 Why is reducing inductance in the PDN elements on a board so important?

The chief problem in the PDN is high impedances at parallel resonances. One feature that influences the peak impedances is the inductance associated with the resonance. Reducing the inductance will decrease the peak impedances.

13.21 What are the three most important design guidelines for reducing the mounting inductance of an MLCC capacitor?

Reducing any loop inductance is about changing three design features:

- Use as short a perimeter as possible. This means all paths kept to the shortest length possible, like via in pad.
- Use wide conductors. Keep the surface traces as wide as possible to spread the current out.
- Bring the power and ground close together. Increase the partial mutual inductance between the power and ground paths by bringing the top of the cavity as close to the surface of the board as practical.

13.22 What three design features will reduce the spreading inductance in a power and ground cavity, and which is the most important?

The spreading inductance of the cavity from a capacitor location to the BGA is related to:

- Thin dielectric between the planes that make up the cavity. This is the first order term and most important.
- The diameter of the contact region in the cavity. This is either the size of the via into the cavity or the number of vias. Using multiple vias will increase the contact area into the cavity and reduce the spreading inductance in the contact region.
- The proximity of the capacitor and BGA. While this is a second order term and varies with the log of the distance, if it is free, closer is better.

13.23 In what combination of design features will the position of a decoupling capacitor not be important? In what case is location important?

The situation in which the location of a capacitor is important is when the spreading inductance in the cavity is a large fraction of the total mounting inductance. If the spreading inductance in the cavity is small compared to the mounting inductance, then moving the capacitor around will not change its total effective inductance.

This is the case when the cavity has a very thin dielectric between the power and ground planes, or the mounting inductance of the capacitor is high.

13.24 **A capacitor really does not behave like an ideal capacitor on a board at the higher frequencies. What is a more effective way of thinking of a capacitor?**

The simple model of a real capacitor is as a simple ideal C element. This usually works well at low frequency, but above about 1 MHz is not a very good match to a real capacitor.

At higher frequency, a better match is to an RLC series circuit. In this model, the C matches the low-frequency impedance and the L matches the high-frequency impedance of a real capacitor.

The lowest impedance of a real capacitor is matched well by the R value.

13.25 **A surface trace on a board from a capacitor to the vias to the cavity is 10 mils wide and 30 mils long. The top of the cavity is 10 mils below the surface. What is the mounting inductance of the 0603 capacitor? What would be the loop inductance of the vias if they were 10 mils in diameter?**

You can estimate the loop inductance of the surface traces, which are 10 mils wide. They are 30 mils long on each side, making them 60 mils long and a total of 6 squares. The 0603 capacitor is 2 squares long. The total number of squares for the top path is 8 squares.

The dielectric spacing between the top of the cavity and the bottom of the traces is 10 mils. If this were a piece of a cavity, the sheet inductance would be 32 pH/mil × 10 mils = 320 pH/sq. Using this rough approximation, the loop inductance of the capacitor is

$$8 \text{ squares} \times 0.32 \text{ nH/square} = 2.5 \text{ nH}$$

The pair of vias have a loop inductance of about 20 nH/inch of length. Being 10 mils long, their loop inductance contribution is about 20 nH/inch × 0.01 inches = 0.2 nH. This is less than 10% the inductance of the surface traces to the capacitor.

13.26 **What happens to the self-resonant frequency of a capacitor when 10 identical capacitors are added to a board? What are the features that change?**

The self-resonant frequency of a capacitor is related to the product of its ESL and C. When n identical capacitors are added in parallel to the board, the C increases as n and the inductance increases as 1/n. The product of L and C stays the same. The self-resonant frequency of the n capacitors is identical to the self-resonant frequency of each individual capacitor.

13.27 What new impedance feature is created when two different value capacitors are added in parallel? Why is this a very important feature for PDN design?

When two different real capacitors with different capacitance are mounted to a board, generally they will have the same mounting inductance. Their parallel impedance will show a parallel resonance, due to the interactions of the inductance of the large capacitance and the capacitance of the smaller capacitor.

This parallel resonance means a peak impedance in the PDN, higher than from either capacitor. It's this peak impedance which will often be the source of problems in the PDN.

13.28 What are three design guidelines for reducing the parallel peak impedance between two capacitors?

The first step in analyzing any problem is to find the root causes. If an equivalent circuit model can be used to describe the problem, the circuit model will often reveal the important terms.

In a parallel resonance, three terms determine the peak impedance. The L and C both contribute equally. This suggests that everything should be down to increase the capacitance and decrease the inductance of those elements.

The third term influencing the peak impedance is the loss term, described by the ESR of the components. These should be increased to damp out the peak. Going too large may increase the flat part of the impedance, so there is an optimum value, usually until the ESR is equal to about the target impedance.

13.29 What happens to the ESR of a capacitor as the value of the capacitance increases?

The capacitance of an MLCC capacitor scales with the number of parallel plates that make it up. A higher capacitance means more plates in parallel.

The ESR of a capacitor arises from the spreading resistance in the sheets of the conductor plates. Generally, the more plates in parallel, the lower the ESR. This means for higher capacitance MLCC capacitors, with more plates, the ESR will reduce.

13.30 To reduce the number of capacitors needed to achieve a target impedance, what is the most important design feature to adjust?

The number of capacitors needed is based on not increasing the loop inductance of the package lead inductance to the cavity. This means if the ESL of each capacitor is smaller, fewer capacitors in parallel will be required.

13.31 What is one advantage of engineering a flat PDN impedance profile?

A flat impedance profile means there are no peak impedances, and there are no particularly sensitive frequencies that might cause large voltage noise on the PDN.

At the board level, a flat impedance profile that looks like a resistor will help to damp out the peak of the Bandini Mountain.

13.32 What is the FDTIM process, and why is it a powerful technique?

The frequency-domain target impedance method is a process to select capacitor values so that their peaks result in a flat-looking response. Basically, you start at the lowest frequency, select the largest capacitor you have available, and keep adding smaller values—separated by 3 values per decade—and selecting the number at each value to result in an impedance profile that is relatively flat.

This way, the collection of capacitor values makes the board level PDN look like a resistor and will damp out the Bandini Mountain.